Tilapias: Biology and Exploitation

Fish and Fisheries Series

VOLUME 25

Amongst the fishes, a remarkably wide range of fascinating biological adaptations to diverse habitats has evolved. Moreover, fisheries are of considerable importance in providing human food and economic benefits. Rational exploitation and management of our global stocks of fishes must rely upon a detailed and precise insight of the interaction of fish biology with human activities.

The *Fish and Fisheries Series* aims to present authoritative and timely reviews which focus on important and specific aspects of the biology, ecology, taxonomy, physiology, behaviour, management and conservation of fish and fisheries. Each volume will cover a wide but unified field with themes in both pure and applied fish biology. Although volumes will outline and put in perspective current research frontiers, the intention is to provide a synthesis accessible and useful to both experts and non-specialists alike. Consequently, most volumes will be of interest to a broad spectrum of research workers in biology, zoology, ecology and physiology, with an additional aim of the books encompassing themes accessible to non-specialist readers, ranging from undergraduates and postgraduates to those with an interest in industrial and commercial aspects of fish and fisheries.

Applied topics will embrace synopses of fishery issues which will appeal to a wide audience of fishery scientists, aquaculturists, economists, geographers and managers in the fishing industry. The series will also contain practical guides to fishery and analysis methods and global reviews of particular types of fisheries.

Books already published are listed at the end of this book. The Publisher and Series Editor would be glad to discuss ideas for new volumes in the series.

Tilapias: Biology and Exploitation

Edited by

Malcolm C.M. Beveridge
and
Brendan J. McAndrew

Institute of Aquaculture
University of Stirling
Stirling, Scotland

KLUWER ACADEMIC PUBLISHERS
DORDRECHT / BOSTON / LONDON

Library of Congress Cataloging-in-Publication Data is available.

ISBN 0-412-80090-X (hardback)
ISBN 0-7923-6391-4 (paperback)

Published by Kluwer Academic Publishers,
PO Box 17, 3300 AA Dordrecht, The Netherlands.

Sold and distributed in North, Central and South America
by Kluwer Academic Publishers,101 Philip Drive
Norwell, MA 02018, USA

In all other countries, sold and distributed
by Kluwer Academic Publishers, Distribution Center,
PO Box 322, 3300 AH Dordrecht, The Netherlands

Printed on acid-free paper

Printed and bound in Great Britain by Antony Rowe Limited.

Contents

Series Editor's Preface

Tilapia hold a position of special importance, for very different reasons, to many people. They have occupied the attention of systematists and taxonomists for many years, and they still hold a great many significant questions to be answered in those fields. Together with the other cichlids, they are one of the most spectacular examples of adaptive radiation known among the vertebrates. They have complex social behaviour, particularly parental behaviour. They are important as food for humans, from both wild capture fisheries and various types of aquaculture. Some species of tilapia have established themselves well beyond their native ranges, as a result of accidental or deliberate human releases. In some of these cases they are viewed as serious threats to native species. Most tilapias adapt well to captive culture, and so they are widespread throughout the ornamental aquarium industry.

This volume on tilapias is extremely timely. It is the most recent link in a chain of human representations of these fishes dating back for several millennia, as evidenced by illustrations in Egyptian tombs. For me, it holds personal significance. I participated in the international symposium on the biology and culture of tilapias in 1980, and contributed to the volume resulting from that meeting. At that meeting I was able to interact at length with scientists I regarded as intellectual pioneers in the field. First and foremost among those were Ethelwynn Trewavas and Rosemary Lowe-McConnell. For that reason alone, it would have great personal significance for me as this is the first volume to appear in the Kluwer Fish and Fisheries Series since I took over the position of Series-Editor. However, the significance of this volume extends far beyond my personal conceit.

Tilapias continue to dominate both artisanal and large scale aquaculture in many tropical areas. They are increasingly recognized as the species of choice for intensive aquaculture throughout the world. Their incredible adaptability to a wide range of physical conditions and their ability to thrive and reproduce make them very attractive species for the future development of aquaculture. This requires increasing sophisticated technical knowledge of food requirements, reproductive physiology, genetics and growth.

Some tilapia species attract attention as exotic invaders far beyond their native ranges. Their impact on native species, and possible ways to manage those impacts are the concern of fishery biologists in places such as Florida and southern California. Conservation is also an increasing concern for tilapias themselves, as example of the broader issue of biodiversity. Tilapias have been subjected to large scale manipulations by humans. Introductions of species beyond their native ranges, deliberate or accidental production of hybrids, and alteration of habitats have been visited as much upon tilapia as any other fishes. Large scale alterations of natural water bodies by constructions of dams, and uncontrolled harvest of wild stocks have had major impacts on some species.

The taxonomy and systematics of tilapia continue to attract some of the best minds in the business. In particular, the great challenge posed by tilapias and other cichlids, require the latest and most powerful biochemical and analytical tools of those attempting to decipher questions of evolutionary relationships.

We are indeed fortunate to have this volume, as the state of the art and science of tilapia, to show the way forward for these eternally fascinating fishes. Like its predecessors, it will ensure that our interest and understanding of tilapia will continue to grow and develop. Like them, it will be a landmark not only for tilapia biology, but for the much broader range of science represented by these fishes.

Dr. David L. G. Noakes
Editor, Kluwer Fish and Fisheries Series
Professor of Zoology and Axelrod Institute of Ichthyology
University of Guelph

Preface

According to tilapia taxonomist Ethelwynn Trewavas the word 'tilapia' was coined by A. Smith in 1840 who, in attempting to write down the Bushman word for 'fish', which begins with a click, rendered it as '*til*'. Although a number of key taxonomic issues remain unresolved (Chapter 1), the term 'tilapia' is today generally applied to members of what is regarded by many as a Subfamily (Tilapiinae) of the cichlids indigenous to Africa and the Levant.

The earliest archaeological remains of fish associated with humans are from Africa and indicate that tilapias featured in the diets of early hunter-gatherer societies (Stewart, 1994). However, tilapias have played more than just a utilitarian role in our lives. To the ancient Egyptians, it was the association of tilapias with the annual flood cycle of the Nile upon which agriculture and society so depended that led to their role in religious ceremonies associated with death and re-birth and their depiction in tomb paintings (Chimits, 1957; Brewer and Friedman, 1989). St Peter's fish, as *Sarotherodon galileaus* is sometimes called, features in the Bible's account of Christ's feeding of the five thousand. Nevertheless, it is their characteristics as a food fish that have largely been responsible for the introduction of tilapias throughout the tropics and subtropics and the current interest among aquaculturists in the possibilities of farming a truly 'global' fish.

The tilapias are unusual among fish in having been relatively well served by science. For over 150 years, the hundred or so tilapiine fishes have attracted the interest of evolutionary biologists, ethologists, ecologists and geneticists. In recent times the emphasis has shifted to aspects of their exploitation, particularly in relation to farming. A number of key texts have dealt with cichlids in general (e.g. Baerends and Baerends-van Roon, 1950; Fryer and Iles, 1972; Keenleyside, 1991) or tilapias in particular (Balarin and Hatton, 1979; Pullin and Lowe-McConnell, 1982; Trewavas, 1983; Jauncey, 1998). The excellent series of conference proceedings – the International Symposia on Tilapias in Aquaculture (ISTA) – produced under the auspices of the International Centre for Living Aquatic Resources Management (ICLARM), did much to raise awareness among the aquaculture community of the potential of tilapias and mirror changing research interests over the past two decades. The World Aquaculture Society (WAS) has also recently responded to growing North

American interests in tilapia aquaculture through initiation of a series of volumes edited by Barry Costa-Pierce and James Rakocy (1997).

The present book evolved out of an idea to try to pull together our knowledge of this important group of species – their biology and fisheries and aquaculture – in a single volume, something that had not been done comprehensively since Roger Pullin and Lowe-McConnell's landmark volume two decades ago (Pullin and Lowe-McConnell, 1982). The first part of the book deals with tilapia biology, the latter with fisheries and aquaculture. The book begins with a consideration of the evolution, phylogenetic relationships and biogeography of tilapias and focuses on the recent contributions that molecular genetics has made to the subject. In Chapter 2, George Turner and Rosanna Robinson apply their considerable experience to a review of tilapia reproductive biology, mating systems and parental care from ethological, evolutionary and ecological perspectives. A consideration of the current state of knowledge of feeding behaviour and digestive physiology by Malcolm Beveridge and Donald Baird follows. Lindsay Ross summarizes our current understanding of ecophysiology and synthesizes disparate publications on physiological energetics. Tilapias and their role in ecosystems, especially in Africa, is reviewed by Rosemary Lowe-McConnell, who considers why they have been such a successful group. In Chapter 6, Kai Lorenzen reviews life history strategies, including the phenomenon of 'stunting', and discusses population dynamics in the context of management of both wild and farmed stocks. Chapter 7 by David Penman and Brendan McAndrew considers applied genetics and the management of cultured stocks, focusing on the genetics of sex-determination and the techniques used to produce all-male populations for aquaculture. In the subsequent chapter Dave Little and Gideon Hulata conduct a comparative analysis of global broodstock management, hatchery and nursery techniques. Tilapia nutritional requirements are summarized by Kim Jauncey in Chapter 9. Kwei Lin, Peter Edwards and Amara Yakupitiyage bring their unparalleled knowledge of Asian rural food production systems to bear on a review of semi-intensive pond culture of tilapias while in the subsequent chapter James Muir, Jaap van Rijn and John Hargreaves look at the opposite end of the scale of intensity of tilapia production, reviewing intensive and recycle systems and considering the potential offered by this technology for the future. The book concludes with a consideration of tilapia production economics and marketing by Jimmy Young and James Muir.

This volume has had a long gestation; longer, indeed, than any of us had envisaged or wished. We thank Maggie Beveridge for editorial assistance and both the contributors and the publishers for their forbearance and for their continued support and enthusiasm for the project.

REFERENCES

Baerends, G.R. and Baerends-van Roon, J.M. (1950) An introduction to the study of the ethology of cichlid fishes. *Behaviour* (Supplement 1), 1–242.

Balarin, J.D. and Hatton, J.P. (1979) *Tilapia. A Guide to their Biology and Culture in Africa*, Unit of Aquatic Pathobiology, University of Stirling.

Brewer, D.J. and Friedman, R.F. (1989) *Fish and Fishing in Ancient Egypt*, Aris and Phillips, Warminster, UK.

Chimits, P. (1957) Tilapia in ancient Egypt. *FAO Fisheries Bulletin* **10**, 211–215.

Costa-Pierce, B.A. and Rakocy, J.E. (eds) (1997) *Tilapia Aquaculture in the Americas*, Vol. 1, World Aquaculture Society, Baton Rouge, Louisiana, USA.

Fryer, G. and Iles, T.D. (1972) *The Cichlid Fishes of the Great Lakes of Africa: Their Biology and Evolution*, Oliver and Boyd, Edinburgh.

Jauncey, K. (1998) *A Guide to Tilapia Feeds and Feeding*, 2nd edition, Pisces Press, Bridge of Allan, Stirling.

Keenleyside, M.H.A. (ed.) (1991) *Cichlid Fishes. Behaviour, Ecology and Evolution*, Chapman and Hall, London.

Pullin, R.S.V. and Lowe-McConnell, R.H. (eds) (1982) *The Biology and Culture of Tilapias* (ICLARM Conference Proceedings, Vol. 7), ICLARM, Manila.

Stewart, K.M. (1994). Early hominid utilization of fish resources and implications for seasonality and behaviour. *Journal of Human Evolution* **27**, 229–245.

Trewavas, E. (1983) *Tilapiine Fishes of the Genera* Sarotherodon, Oreochromis *and* Danakilia, British Museum (Natural History), London.

Malcolm Beveridge
Brendan McAndrew
Stirling, December 1999

Contributors

Donald J. Baird
Institute of Aquaculture, University of Stirling, Stirling FK9 4LA, UK
d.j.baird@stir.ac.uk

Malcolm C. M. Beveridge
Institute of Aquaculture, University of Stirling, Stirling FK9 4LA, UK
m.c.m.beveridge@stir.ac.uk

Peter Edwards
School of Environment, Resources and Development, AIT, GPO Box 4, Klong Luang, Pathumthani 12120, Thailand
pedwards@ait.ac.th

John Hargreaves
Mississippi State University, Department of Wildlife and Fisheries, Box 9690, Mississippi State, MS 39762-9690, USA
jhargreaves@CFR.MsState.Edu

Gideon Hulata
Agricultural Research Organisation, Institute of Animal Science, Department of Aquaculture, PO Box 6, Bet Dagan 50250, Israel
vlaqua@volcani.agri.gov.il

Kim Jauncey
Institute of Aquaculture, University of Stirling, Stirling FK9 4LA, UK
kim.jauncey@stir.ac.uk

C. Kwei Lin
School of Environment, Resources and Development, AIT, GPO Box 4, Klong Luang, Pathumthani 12120, Thailand
lin@rccsun.ait.ac.th

David C. Little
Institute of Aquaculture, University of Stirling, Stirling FK9 4LA, UK
d.c.little@stir.ac.uk

Kai Lorenzen
Huxley School of Environment, Imperial College, RSM Building, Prince Consort Road, London SW7 2BP, UK
k.lorenzen@ic.ac.uk

Rosemary H. Lowe-McConnell
Streatwick, Streat Near Hassocks, Sussex BN6 8RT, UK
romcconnell@compuserve.com

Brendan J. McAndrew
Institute of Aquaculture, University of Stirling, Stirling FK9 4LA, UK
bjm1@stir.ac.uk

James F. Muir
Institute of Aquaculture, University of Stirling, Stirling FK9 4LA, UK
j.f.muir@stir.ac.uk

David J. Penman
Institute of Aquaculture, University of Stirling, Stirling FK9 4LA, UK
d.j.penman@stir.ac.uk

Rosanna L. Robinson
School of Biological Sciences, University of Southampton, Bassett Crescent East, Southampton SO16 7PX, UK
rlr@soton.ac.uk

Lindsay G. Ross
Institute of Aquaculture, University of Stirling, Stirling FK9 4LA, UK
l.g.ross@stir.ac.uk

George F. Turner
School of Biological Sciences, University of Southampton, Bassett Crescent East, Southampton SO16 7PX, UK
gft@soton.ac.uk

Jaap van Rijn
The Hebrew University of Jerusalem, Faculty of Agriculture, Food and Environmental Quality Sciences, Dept of Animal Sciences, PO Box 12 Rehovot 76100, Israel
vanrijn@agri.huji.ac.il

Amararatne Yakupitiyage
School of Environment, Resources and Development, AIT, GPO Box 4, Klong Luang, Pathumthani 12120, Thailand
amara@emailhost.ait.ac.th

James A. Young
Department of Marketing, University of Stirling, Stirling FK9 4LA, UK
jay1@stir.ac.uk

Chapter one

Evolution, phylogenetic relationships and biogeography

B. J. McAndrew

1.1 DEFINITION OF A TILAPIA

Today tilapia is used as a common name for a large number of species within the cichlid tribe Tilapiini particularly the species in the three genera *Tilapia*, *Sarotherodon* and *Oreochromis*, especially the larger species most commonly caught in wild fisheries or those used in aquaculture. Trewavas (1983), in her major monograph, loosely defined the tribe Tilapiini as an African and Levantine assemblage which included the following genera: *Tilapia*, *Sarotherodon*, *Oreochromis*, *Danakilia*, *Iranocichla*, *Tristramella*, *Pelmatochromis*, *Pterochromis*, the endemic genera of Barombi Mbo in Cameroon (*Konia*, *Myaka*, *Punga*, *Stomatepia*), and some specialized genera from the rapids (*Steatocranus*, *Gobiocichla*). However, Stiassny (1991) puts forward a more restrictive grouping based on morphology of the lower pharyngeal jaw and believes that Trewavas's inclusion of *Pelmatochromis*, *Pterochromis*, *Steatocranus* and *Gobiocichla* in this grouping is premature, but she concedes that they may be included in the future. The tilapiines are distinguished from the other great African assemblage the Haplochromini by the structure of the apophysis on the base of the skull for the articulation of the upper pharyngeal bones. In tilapiines the facets are formed from the parasphenoid alone whereas in the Haplochromini the basioccipital can form up to one half of each facet (Trewavas, 1983).

M.C.M. Beveridge and B.J. McAndrew (eds), Tilapias: Biology and Exploitation, 1–32.

1.2 EVOLUTION OF THE CICHLIDAE

The tilapiines have been grouped into a tribe within the family Cichlidae. The Cichlidae form one of the four main families (Cichlidae, Embiotocidae, Pomacentridae, Labridae) that Kaufman and Liem (1982) included within the suborder Labroidei, which includes between 5 and 10% of all known fish species. The evolutionary relationships of the families within the Labroidei are still in some doubt. Stiassny and Jensen (1987) derived a cladogram of Labroid interrelationships based on morphological characteristics. This cladogram placed the Cichlidae as the ancestral lineage. Although the cladogram is the most parsimonious for the traits analysed (Fig. 1.1), it required the loss or independent gain of many derived characters. This suggests a large degree of homoplasy in the morphological trait transformations observed and should only be seen as indicative of one possible explanation of the familial relationships within this suborder (reviewed by Stiassny, 1991).

Cichlid fish have stimulated much anatomical, behavioural and, more recently, molecular genetic research in an attempt to identify their phylogenetic relationships. Despite 100+ years of research into this group the evolutionary and taxonomic relationships are still unclear. Stiassny (1991) has reviewed much of this work; she compares a cladogram produced by Oliver (1984) (Fig. 1.2A) using classical techniques with a summary cladogram she produced of cichlid interrelationships based on morphological characters using a computer generated tree (Fig. 1.2B). The trees differed in the placement of Madagascan and Indian taxa and the position of the neotropical genus *Cichla*. In both of these

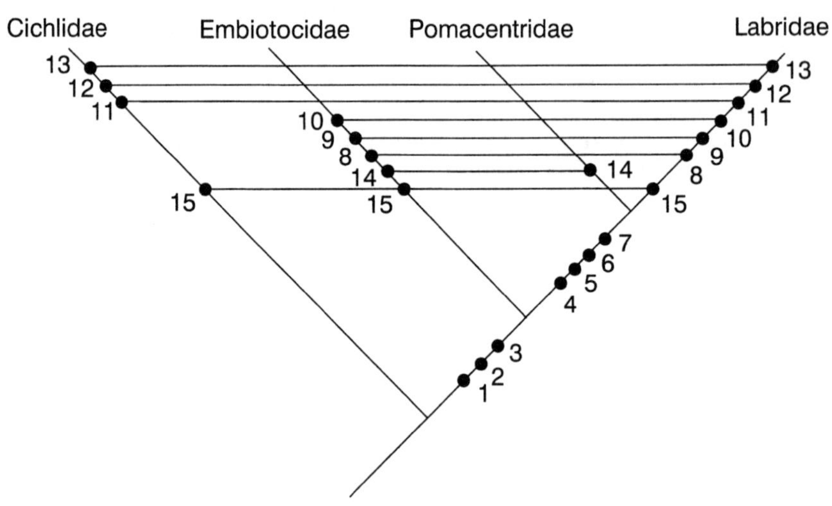

Fig. 1.1 Cladograms showing evolutionary relationship among the families within the suborder Labroidei based on morphological characters (1–15). Lines between families show homoplasy (after Stiassny and Jensen, 1987; Stiassny, 1991).

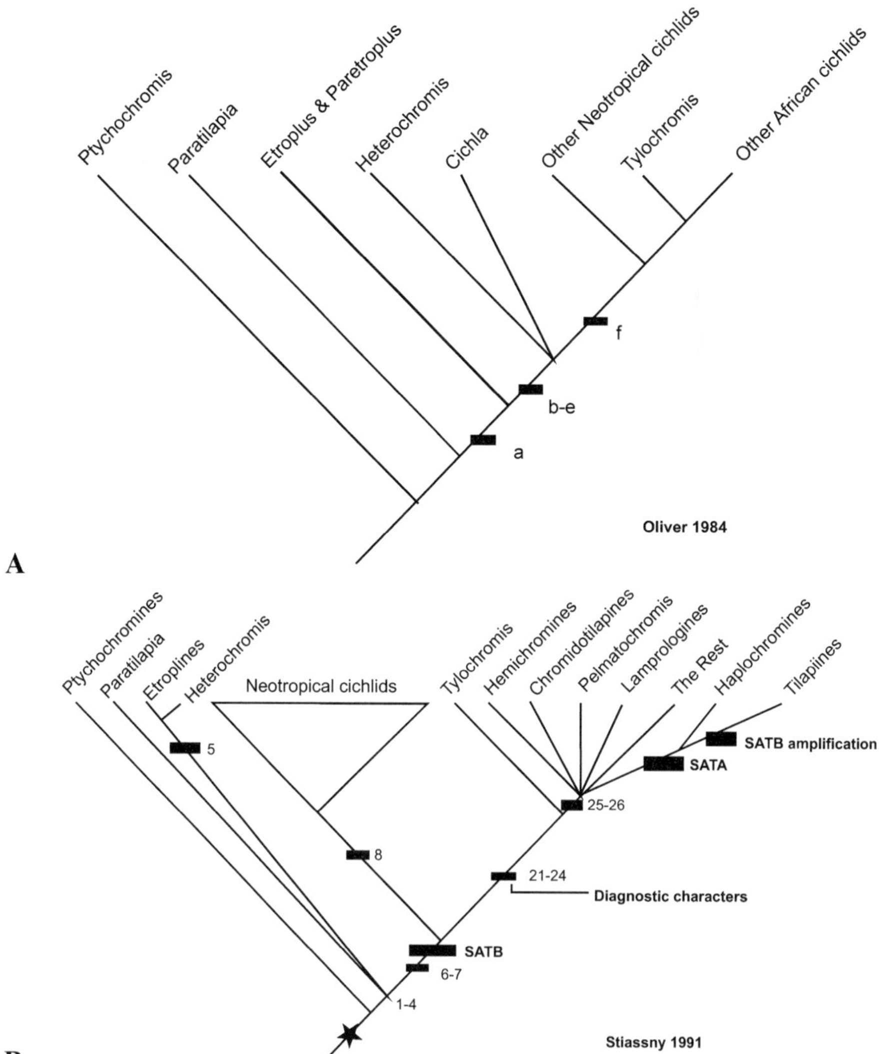

Fig. 1.2 Cladograms based on morphological characteristics showing generic interrelationships within the Cichlidae. **A**, After Oliver (1984). **B**, Modified from Stiassny (1991) to include satellite DNA (SATA and SATB) that helps to resolve haplochromine and tilapiine relationship.

trees the Asian, Madagascan taxa and the Zairian genus *Heterochromis* form a separate clade to the monophyletic lineage that includes all the remaining African and neotropical genera. The neotropical and African species form separate radiations of the family on different continents. This cladogram (Fig. 1.2B) has been modified to include molecular information that further elucidates these generic relationships (see later). It is interesting to note at this point that neotropical species did not undergo the amazing degree of speciation seen in the African lake genera.

The re-analysis of published anatomical and life-cycle characteristics using advanced computer techniques enables data sets to be infinitely manipulated and the relative importance of individual traits on inter- and intrageneric relationships to be assessed. The development of molecular techniques and the rapid developments in molecular systematics (Hillis *et al.*, 1996) now offers a means of testing congruence of relationships derived from classical and molecular data sets. This is seen as a very robust approach to solving systematic problems (Miyamoto and Fitch, 1995).

The identification of two families of satellite DNA in cichlids (SATA and SATB) has helped to confirm some of the generic relationships already described (Franck *et al.*, 1994). SATB is an ancient conserved repeat sequence that is found in both the African and neotropical cichlid genera but is absent in the Asian species *Etroplus maculatus*. This suggests that the Asian species were already isolated when SATB evolved in the ancestors of the African and neotropical lineages. The SATB satellite has also undergone change that helps to show the chronology of evolution within the African cichlids. The first of these was an amplification that is only seen in the tilapiine lineage and must have occurred after the divergence of the haplochromines. Subsequent changes have also occurred in the tilapiine tribe and will be discussed later. The SATA satellite is a more recently evolved repeat sequence and is only found in the haplochromine and tilapiine species again confirming the past monophyly of these two African tribes. The molecular markers generally confirm the evolutionary relationships suggested using classical techniques at the generic level (Fig. 1.2B).

1.3 TAXONOMIC CLASSIFICATION

The African cichlids include some of the most speciose genera (Haplochromini) with nearly 80% of all cichlid species being found in African lakes and rivers. The African cichlids display an enormous range of morphological and ecological adaptation yet despite this their underlying anatomy is very similar; there is little doubt that they form a monophyletic clade. The speciation of the African genera has been described (Stiassny, 1991) as 'evolution by concertina' rather than by major anatomical reorganization and innovation. The basic functional and behavioural blueprint of this group has enabled them to adapt and occupy many different freshwater niches through small changes in the relative growth rates of different parts of their basic anatomy or modification of their behaviour. This means that as these species evolve they retain many shared characters (synapomorphy), making it extremely difficult for taxonomists using morphological characteristics to accurately define evolutionary relationships.

The theories on tilapiine evolution and the many changes in the taxonomy show some of the problems evolutionary biologists and taxonomists have encountered even in what could be considered as one of the species-poor cichlid tribes.

The genus *Tilapia* was first described by Smith (1840) and the name is derived from the African Bushman word for fish (Trewavas, 1982b). The genus *Sarotherodon* (brush toothed) and *Oreochromis* (mountain cichlid) were first described by Rüppel (1852) and Günther (1889) respectively, with *S. melanotheron* and *O. hunteri* as the type species. The species in these two genera were lumped together in the subgenus *Sarotherodon* of the genus *Tilapia* by Regan (1920) and Trewavas (1966). Thys van den Audenaerde (1968) had recognized seven mouthbrooding subgenera (*Sarotherodon, Oreochromis, Alcolapia, Neotilapia, Nyasalapia, Loruwiala* and *Danakilia*) within the genus *Tilapia* and by 1971 he had added another subgenus, *Nilotilapia* (Thys van den Audenaerde, 1971a). Later Trewavas (1973a,b, 1980, 1982a,b) raised *Sarotherodon* to a genus which included all of Thys van den Audenaerde's (1968) mouthbrooding subgenera. Subsequently, Trewavas (1983) also believed this to be unsound and proposed the present classification with the biparental and paternal mouthbrooding *Sarotherodon* and maternal *Oreochromis* as separate genera (with *Alcolapia, Neotilapia, Nyasalapia, Loruwiala* as subgenera), with the proviso that the *Sarotherodon* and *Oreochromis* probably arose from substrate brooding *Tilapia*, possibly from different species. She also raised Thys van den Audenaerde (1968) *Danakilia* to a genus because it had a number of morphological characters that made it generically distinct. She suggests that *Danakilia* is clearly related to *Iranocichla* (Coad, 1982), which although a mouthbrooder, she thinks has a lineage independent of either *Sarotherodon* or *Oreochromis*. *Iranocichla* is the south-western Iranian tilapiine which Coad (1982) thinks is related to the Jordanian and Syrian genus *Tristramella* (Trewavas, 1942).

Trewavas (1983) also included and raised *Pelmatochromis* (Steindachner) to a genus within the Tilapiini because she felt that Thys van den Audenaerde's (1968) classification of *Pelmatochromis* as a subgenus of the *Tilapia* to be wrong. She noted that the species within this genus were the least specialized, particularly *P. nigrofasciatus* (Pelligrin), and were probably representative of the ancestral group from which the *Tilapia* diverged by an ancient dichotomy based on diet. *Pelmatochromis* retained morphological traits considered primitive in the Cichlidae (Trewavas, 1973a, 1983) and there were important differences in their diet – *Tilapia* are mainly vegetarian as adults whereas *Pelmatochromis* usually eat small invertebrates and have different structural features associated with their diets, especially dentition and intestinal morphology.

The foregoing classification presented in Table 1.1 is not complete and is not accepted by all taxonomists and other scientists working with these species. Stiassny (1991) disputes the inclusion of *Pelmatochromis, Pterochromis, Steatocranus* and *Gobiocichla* because they lack the median ridge and foramina on the ventral faces of the lower pharyngeal jaw found in all of the other genera.

Table 1.1 The present classification (Trewavas, 1983) of some genera in the tribe tilapiini. Compiled from Steinitz and Ben-Tuvia (1960), Thys van den Audenaerde (1968), Trewavas (1973a, 1982a,b, 1983) and Coad (1982). [Scientific names are given with authorities. An authority given in brackets means the generic status of that species has been changed from the original use]

	Subgenus	Type species	Other examples	Breeding habit
Pelmatochromis Steindachner	–	*P. buettikoferi* Steindachner	*P. ocellifer* Boulenger *P. nigrofasciatus* (Pellegrin)	Substrate spawners (details unknown)
Tilapia A. Smith	*Tilapia* A. Smith	*T. (T.) sparrmanii* A. Smith	*T. (T.) ruweti* Poll and Thys *T. (T.) busumana* (Günther) etc.	Substrate spawners and guarders
	Trewavas Thys	*T. (Tr.) guinasana* Trewavas	–	As *Tilapia*
	Pelmatolapia Thys	*T. (P.) mariae* Boulenger	*T. (P.) cabrae* Boulenger *T. (P.) brevimanus* Boulenger etc.	As Tilapia
	Heterotilapia Regan	*T. (H.) buttikoferi* (Hubrecht)	*T. (H.) cessiana* Thys	As *Tilapia*
	Dagetia Thys	*T. (D.) rheophila* Daget	–	Probably as *Tilapia*
	Coptodon Gervais	*T. (C.) zillii* (Gervais)	*T. (C.) rendalli* (Boulenger) *T. (C.) tholloni* (Sauvage) etc.	As *Tilapia*
Tristramella Trewavas	–	*Tr. sacra* Günther	*Tr. magdalenae* (Lortet) *Tr. simonis* (Günther)	Biparental mouthbrooders
Danakilia Thys	–	*D. franchetti* (Vinciguerra)	–	Mouthbrooders (details unknown)
Iranocichla Coad	–	*I. hormuzensis* Coad	–	Mouthbrooders (details unknown)
Sarotherodon Rüppell	–	*S. melanotheron* Rüppell	*S. galilaeus* (Linnaeus) *S. linnellii* (Lönnberg)	Paternal, biparental or maternal mouthbrooders

Table 1.1 (cont).

	Subgenus	Type species	Other examples	Breeding habit
Barombi-mbo endemics				
Konia		*K. dikume*	*K. eisentrauti*	
Myaka		*M. myaka*		
Punga		*P. maclareni*		
Stomatepia		*S. mariae*	*S. pindu, S. mongo*	
Oreochromis Günther	*Oreochromis* Günther	*O. (O.) hunteri* Günther	*O. (O.) niloticus* (Linnaeus) *O. (O.) aureus* (Steindachner) *O. (O.) spilurus* (Günther) *O. (O.) urolepis hornorum* (Trewavas) *O. (O.) mossambicus* (Peters) *O. (O.) mortimeri* (Trewavas) *O. (O.) andersonii* (Castelnau) *O. (O.) placidus* (Trewavas) *O. (O.) shiranus* (Boulenger) *O. (O.) jipe* (Lowe) etc.	Arena spawners and maternal mouthbrooders
	Alcolapia Thys	*O. (A.) alcalicus grahami* (Boulenger)	*O. (A.) alcalicus alcalicus* (Hilgendorf)	Arena spawners and maternal mouthbrooders
	Vallicola Trewavas	*O. (V.) amphimelas* (Hilgendorf)	–	Probably as *Oreochromis*
	Nyasalapia Thys (*Loruwiala*)	*O. (Ny.) squamipinnis* (Günther)	*O. (Ny.) karongae* (Trewavas) *O. (Ny.) lidole* (Trewavas) *O. (Ny.) macrochir* (Boulenger) etc.	Arena spawners and maternal mouthbrooders
	Neotilapia Regan	*O. (Ne.) tanganicae* (Günther)	–	Maternal mouthbrooders (details unknown)
Pterochromis				
Steatocranus			*S. casuarius*	Substrate
Gobiocichla		*G. wonderi* (Konazawa)	*G. ethelwynnae*	Unknown

Stiassny does, however, admit that Trewavas may in fact be proved right in time as more specimens are collected and the work is completed on more species in these genera. Information about some of these genera is still very scarce or based on small numbers of preserved specimens.

In a study which focused on the African tilapiines, Sodsuk (1993) re-analysed the 13 published dietary, anatomical, breeding behaviour and biogeographical characteristics used to define seven of the tilapiine genera (*Tilapia, Sarotherodon, Oreochromis, Danakilia, Iranocichla* and *Tristramella)* into 67 binary coded multistate characteristics using *Pelmatochromis* as the putative ancestral genus. The analyses, Wagner parsimony with bootstrap resampling, were run a number of times using all possible combinations of character states for breeding behaviour, biogeography and morphology for *Danakilia* and *Iranocichla* because there was insufficient or conflicting evidence for the correct character state in these single species genera. The effect of these changes and the impact they had on the putative phylogeny were assessed. In all dendrograms *Tilapia* was the genus closest to *Pelmatochromis,* the presumed outgroup, giving rise to a monophyletic evolution of all other genera. The *Tilapia* shared more ancestral characters (state 0) with *Pelmatochromis*. Within the monophyletic clade the closest relationships were always between *Sarotherodon* and *Oreochromis,* they form sister taxa in all dendrograms. The relationship of the remaining three genera is less clear and fluctuates with the changes in character coding. The two most frequently derived dendrograms are presented in Figs 1.3A and 1.3B. In general there appears to be a consistently closer relationship between *Danakilia* and the *Sarotherodon-Oreochromis* clade than between the others. Coad (1982) suggested a relationship between *Tristramella* and *Iranocichla* but Trewavas (1983) identified more similarities between *Iranocichla* and *Danakilia*. It is clear that more samples and studies are needed to confirm the relationships between the three minor genera and how they are related to the *Sarotherodon* and *Oreochromis* clade. The data from this reanalysis of the published data would tend to confirm the relationships suggested by Trewavas (1983).

1.4 EVOLUTION OF MOUTHBROODING TILAPIA

There have been many revisions of the classification of the species within the three tilapia genera in recent years (see earlier). The main argument has been over the classification of the mouthbrooding genera; whether they should be retained in a single broad genus *Tilapia* and only split at the subgeneric level (Peters and Berns, 1978, 1982; Thys van den Audenaerde, 1978, 1980) or whether the different mouthbrooding forms should have generic status (Trewavas, 1983). This is despite the general agreement that the ancestor of the mouthbrooding species was one or more substrate spawning and guarding *Tilapia*. The taxonomic arguments have arisen because there are two different theories on the evolution of mouthbrooding behaviour in this tribe. Trewavas

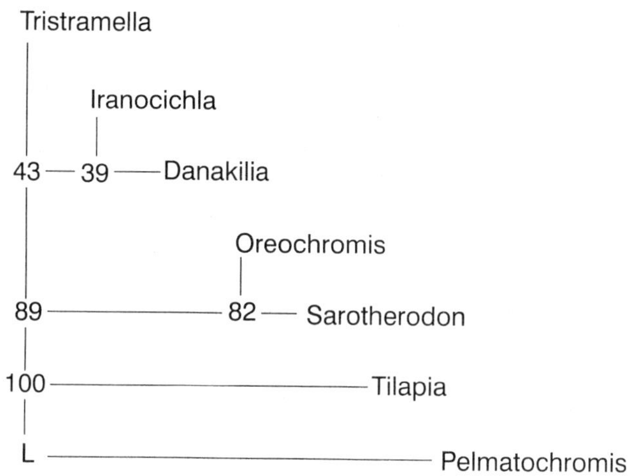

A

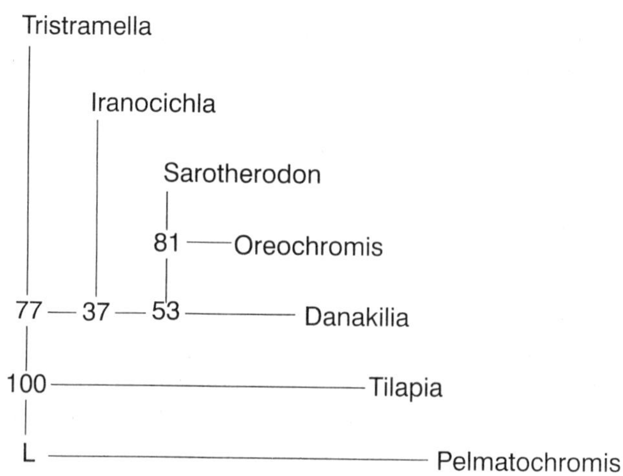

B

Fig. 1.3 Cladograms based on 67 binary coded morphological, behavioural and biogeographic character states for seven tilapiine genera. The lack of detail available for some character states for *Danakilia* and *Iranocichla* required using assumed states for breeding behaviour and biogeographical origin. The cladograms **A** and **B** are the most common produced after all character states are tested (after Sodsuk, 1993).

(1980, 1983) believes that the mouthbrooding *Sarotherodon* and *Oreochromis* genera arose from a single, or possibly, separate splits from the *Tilapia*. Trewavas also believed that *Danakilia* arose independently from either of these genera. Although she argues that the *Sarotherodon*, although containing both paternal and biparental mouthbrooders, have many characteristics that make them intermediate between the *Tilapia* and the evolutionary more progressive maternal mouthbrooding *Oreochromis*. A monophyletic origin for mouthbrooding is made more difficult to justify because biogeographic data would suggest that any overlap in the ranges of the *Sarotherodon* and *Oreochromis* today is probably secondary. There is no obvious extant *Tilapia* species that has a range that is likely to explain present day distributions. However, Trewavas (1983) believed that the breeding, brooding, feeding and anatomical differences that she found between the genera warranted separate generic status

Peters and Berns (1978, 1982) believe that there have been a number of splits from the ancestral substrate brooders; the 'older' more ancient of these having had time to evolve into maternal mouthbrooders (*Oreochromis*) compared with the more recently separated lineages which have only progressed to the less efficient paternal and biparental mouthbrooding state (*Sarotherodon*). Peters (1979) could therefore not agree with the new classification proposed by Trewavas, believing that the various species should all be *Tilapia* and at best given subgeneric status. Thys van den Audenaerde (1968, 1971a,b) also preferred this more conservative approach until there was more evidence on the evolution of the group; until this time *Tilapia* could be retained for general usage and the subgeneric status would only be used for specialist issues. The American Fisheries Society still includes all tilapia under a single genus *Tilapia* despite the almost total acceptance of Trewavas's (1982) classification by most authors who have published books and papers on these species.

Many of the theories about the classification and evolution of the tilapia are based on morphological analysis of museum specimens and behavioural studies often on a small number of individuals. This may or may not give sufficient systematic or evolutionary information to clearly reveal the evolution of this group. In fact Ethelwynn Trewavas's classification was strongly influenced by direct observation of the breeding behaviour of the different genera in fish ponds (Trewavas, personal communication). In common with many other areas of the biological sciences the development and widespread application of molecular genetic techniques has revolutionized the study of taxonomy and population genetics. The ability to be able to study, relatively easily, large numbers of different genes more or less randomly within the nuclear and mtDNA of these organisms has significantly added to our understanding of the evolution and biogeography of this group. Congruence between the findings from classical anatomical, breeding and behaviour and molecular genetics approaches would provide strong evidence for the evolutionary history of this group of fish.

1.5 TILAPIINE EVOLUTION

Allozyme studies

The analysis of isozyme and allozyme variation has provided a wealth of information on the population structure and dynamics, particularly of aquatic organisms. In tilapia, allozyme studies have been widely used to help manage farmed populations and have been successfully applied in species identification and assessing the species purity, or composition of farmed populations. The technique has also been used to assess levels of inbreeding in farmed stocks prior to the start of selective improvement programmes and in the monitoring of genetic manipulations in these species. For a detailed review on all aspects of the use of protein electrophoretic studies in tilapia see McAndrew and Sodsuk (in press) also Penman and McAndrew (Chapter 7). This particular section is restricted to how allozymes studies have helped to increase our understanding of the evolution and biogeography of this tribe.

The first study to look at a number of the African cichlid genera was a cytological and allozyme study of Levantine cichlids by Kornfield *et al.* (1979) and included species from the genera *Tristramella*, *Haplochromis*, *Tilapia* and *Sarotherodon* (the latter included both *Sarotherodon* and *Oreochromis* accepting Trewavas's 1983 classification). Their results showed evidence of a *Tilapia-Haplochromis* divergence and the levels of genetic similarity between *Tristramella* and *Sarotherodon* species pairs suggested recent divergence, but other generic level similarities indicated varied and prolonged periods of independent phyletic evolution. They also asserted that their findings were compatible with the formal taxonomy based on morphology and the raising of *Sarotherodon* to generic level by Trewavas (1973a).

As the technology developed, more detailed studies using more species and loci and more powerful analytical techniques began to be used. McAndrew and Majumdar (1984) studied nine species from three genera using 25 allozyme loci in an attempt to discriminate between the two main hypotheses, Peters and Berns vs. Trewavas, on the evolution of mouthbrooding in tilapiines. Both a phenetic and cladistic analysis of the allozyme data clearly showed a dichotomy between the substrate spawning and guarding *Tilapia* species and the mouthbrooding *Sarotherodon* and *Oreochromis*. However, this study consistently placed a single mouthbrooding species *Oreochromis jipe* outside the mouthbrooding clade in all phylogenetic analyses. This made it equivocal in terms of rejecting or accepting either hypothesis. This study was limited in the number of species analysed and was under-represented in species from the *Tilapia* and *Sarotherodon* genera.

A follow up study (Sodsuk and McAndrew, 1991), which included 15 species analysed for 44 allozyme loci, showed a clear dichotomy between the substrate spawning and guarding *Tilapia* and mouthbrooding *Sarotherodon* and *Oreochromis*. This time *O. jipe* was consistently placed with the other *Oreochromis* species possibly because of the inclusion of other more closely related species from this genus within the sample. This study tended to confirm that mouthbrooding had a monophyletic origin at least for the species used within these genera.

Sodsuk (1993) compared the phylogenetic relationship of 22 different species from three genera (*Tilapia*, *Sarotherodon* and *Oreochromis*) using morphological traits, coded as 94 binary coded multi-state characters (Fig. 1.4A), and allozyme data (Fig. 1.4B) to produce cladograms both separately and in combination to produce a consensus phylogeny (Fig. 1.4C). The main conflicts between the molecular and morphological data were resolved using the strict and majority consensus technique of Margush and McMorris (1981). Some discrepancies between the trees generated by the different character sets could not be fully resolved because some subgenera were only represented by single species. However, both approaches showed that there was a high degree of similarity between the use of these different character states. Both showed some *Tilapia* species to be congenerically grouped (*Coptodon*) and the mouthbrooding genera *Sarotherodon* and *Oreochromis* to be a monophyletic clade. The *Sarotherodon* formed a separate congeneric clade intermediate between the *Tilapia* and the maternal brooding *Oreochromis*. There was also strong evidence for subgeneric grouping within the *Oreochromis*, particularly a monophyletic origin for the subgenus *Nyasalapia*. This was confirmed in a more detailed study of Malawian

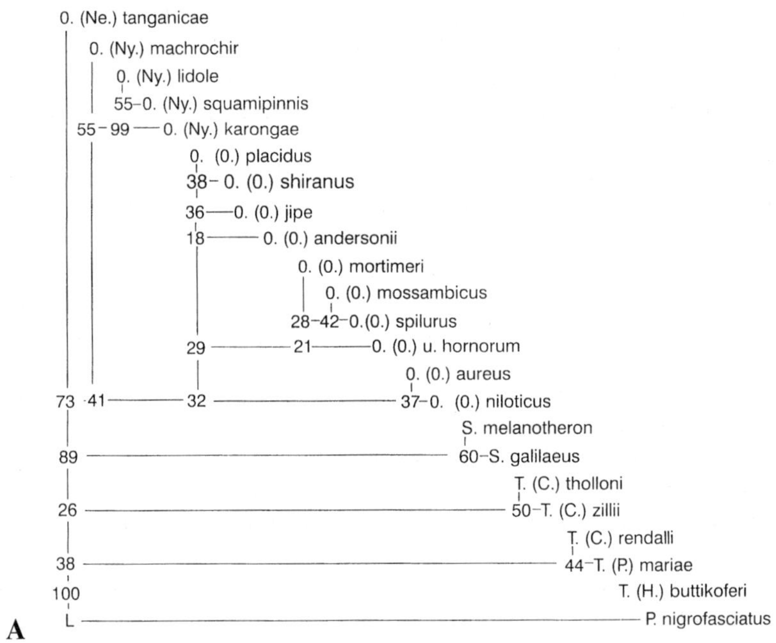

Fig. 1.4 (and opposite) **A**, Most parsimonious cladograms based 94 binary coded morphological, behavioural, and biogeographic multistate characters for 23 different tilapiine species. **B**, Cladogram constructed using Wagner parsimony method (binary coded presence or absence of alleles) for 23 tilapiine species studied at 43 allozyme loci. **C**, Consensus tree based on morphological and allozyme data following the technique of Margush and McMorris (1981). The numbers at the forks give the number of times the group to the right appeared in the 4 trees generated.

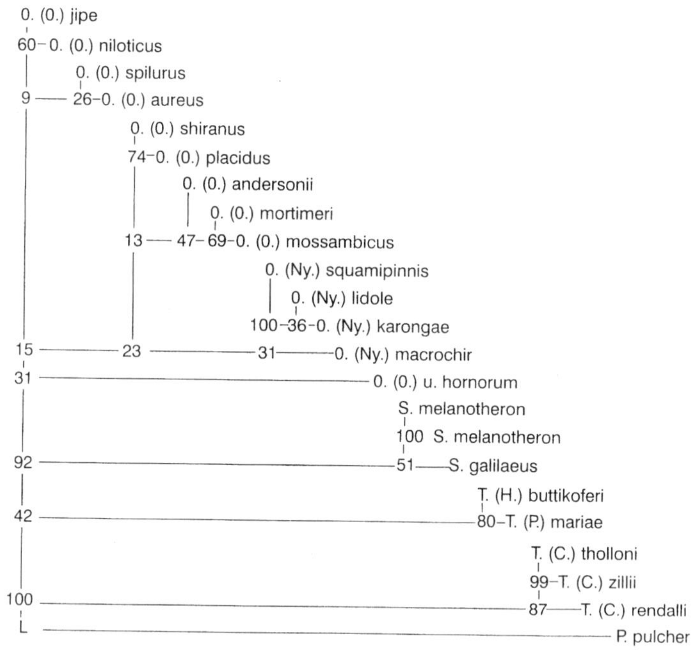
0. (0.) jipe
60 0. (0.) niloticus
0. (0.) spilurus
9 26 0. (0.) aureus
0. (0.) shiranus
74 0. (0.) placidus
0. (0.) andersonii
0. (0.) mortimeri
13 47 69 0. (0.) mossambicus
0. (Ny.) squamipinnis
0. (Ny.) lidole
100 36 0. (Ny.) karongae
15 23 31 0. (Ny.) macrochir
31 0. (0.) u. hornorum
S. melanotheron
100 S. melanotheron
92 51 S. galilaeus
T. (H.) buttikoferi
42 80 T. (P.) mariae
T. (C.) tholloni
99 T. (C.) zillii
100 87 T. (C.) rendalli
L P. pulcher

B

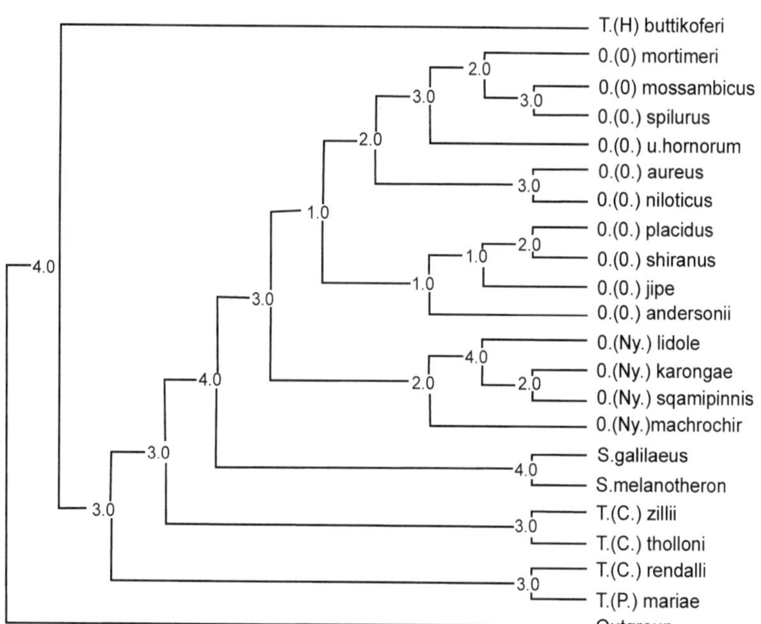
T.(H) buttikoferi
0.(0) mortimeri
0.(0) mossambicus
0.(0.) spilurus
0.(0.) u.hornorum
0.(0.) aureus
0.(0.) niloticus
0.(0.) placidus
0.(0.) shiranus
0.(0.) jipe
0.(0.) andersonii
0.(Ny.) lidole
0.(Ny.) karongae
0.(Ny.) sqamipinnis
0.(Ny.)machrochir
S.galilaeus
S.melanotheron
T.(C.) zillii
T.(C.) tholloni
T.(C.) rendalli
T.(P.) mariae
Outgroup
4.0
3.0
3.0
4.0
3.0
1.0
2.0
3.0
2.0
3.0
3.0
1.0
1.0
2.0
2.0
4.0
2.0
4.0
3.0
3.0

C

tilapia species by Sodsuk *et al.* (1995). The study by Sodsuk (1993), although the largest to date, was still biased toward *Oreochromis* and East African species and under represented *Tilapia* and *Sarotherodon* species, the predominant genera in West Africa.

Pouyaud and Agnèse (1995) included a large number of tilapiine species in an allozyme study of West African cichlids which included 12 *Tilapia*, 4 *Sarotherodon* and 5 *Oreochromis* species as well as 5 from other related genera (*Tylochromis jentinki*, *Hemichromis fasciatus*, *Chromidotilapia guntheri*, *Chilochromis duponti*, and *Pelamatochromis buettikoferi*) which acted as outgroup species. The paper presents a number of rooted trees that clearly show the species grouped by genus with the *Sarotherodon* being placed between the *Tilapia* and the *Oreochromis*, once again suggesting a monophyletic origin for the mouthbrooding genera. This paper shows that the *Tilapia* species split into two main subgeneric groupings *Coptodon* and *Heterotilapia* as was proposed by Thys van den Audenaerde (1968). Sodsuk (1993) also observed this for some of the *Tilapia* species she analysed. The outgroup species used by Pouyaud and Agnèse (1995) clustered within but were generally ancestral to the *Tilapia*. The other interesting finding was the unexpected position of the paternal mouthbrooding species *S. melanotheron*, which clustered within the *Oreochromis* rather than with the other *Sarotherodon* species. Sodsuk (1993) found that *S. melanotheron* clustered with *S. galilaeus* the only other member of that genus she analysed. It might be expected that in Sodsuk's (1993) data set, which contained many *Oreochromis* species, that any similarities *S. melanotheron* might have with species in that genus would have been exaggerated.

DNA studies

The rapid developments in molecular genetics has meant that other types of genetic variation can now be analysed and combined with existing data to increase our understanding of the evolution of the group. These include the direct analysis of mtDNA and chromosomal DNA using a variety of different techniques, which look at fragments or individual base sequences (see Hillis *et al.*, 1996; Ferraris and Palumbi, 1996). There have been a number of recent reviews on the application of these techniques in fisheries and aquaculture (Carvalho and Pitcher, 1995; O'Connell and Wright, 1997) and in the molecular systematics of fishes (Kocher and Stepien, 1997). Cichlids and particularly the highly speciose haplochromine flocks found in the African rift valley lakes (see Sültmann and Mayer, 1997) have been extensively studied.

There is a recent general review on the application of DNA techniques in tilapia research by McConnell (in press). In terms of increasing our understanding of cichlid evolution the isolation of two families of satellite DNA (SATA and SATB) has proved useful (Wright, 1989; Franck *et al.*, 1992, 1994). SATB is made up of a large number of 1900bp tandem repeats and is found in both African and neotropical cichlids but not Asian species which supports the monophyletic evolution of the former two groups. The SATB sequence has also undergone two amplifications of its copy number one just after the divergence of the haplochromine tribe and a subsequent amplification in the tilapiine tribe. The SATA is made up of 237bp consensus sequence with two smaller size variants, that have arisen by deletions, up to as many as 5000 repeats have been observed in some arrays. SATA is only found in the African haplochromine and tilapiine tribes which confirms the monophyly of this African cichlid clade (Fig. 1.2B).

Franck *et al.* (1994) analysed the SATA satellite DNA family to derive phylogenetic relationships from consensus sequence data for five species within three genera (*Oreochromis, Sarotherodon* and *Tilapia*) using the haplochromine *Protomelus similis* as an outgroup species. The most parsimonous explanation produced a tree in which the mouthbrooding *Sarotherodon* and *Oreochromis* species formed a single clade separate from the *Tilapia* species analysed.

A similar study by Sültman *et al.* (1995) identified new molecular markers, based on the Randomly Amplified Polymorphic DNA (RAPD) technique they identified polymorphic fragments of DNA for which they developed locus specific primers so they could sequence these fragments. The trees derived from one locus *DXTU1* confirmed the monophyly of the African and neotropical species and that the East African cichlid groups form a monophyletic clade in which the tilapiines form a monophyletic sister group to the haplochromines (also see Sültman and Mayer, 1997).

It now seems clear that there is strong evidence from the anatomical, protein and DNA studies that mouthbrooding arose as a single split from the ancestral substrate spawning and guarding *Tilapia*. The mouthbrooding line then splitting again resulting in the two genera *Sarotherodon* and *Oreochromis*. There is strong evidence for a number of monophyletic subgeneric groupings in both *Tilapia* and *Oreochromis*. There are still some doubts as to the final picture regarding the extent of the tribe as a number of the genera included within the tilapiines have not been studied at the molecular level, mainly because of the difficulties in obtaining material.

1.6 BIOGEOGRAPHY

General

The tilapiines are exclusively an African assemblage of species. They are essentially pan-African and are only absent from the northern Atlas Mountains and south-west Africa. Africa can be divided into approximately seven ichthyofaunal regions (see p. 133). Tilapiine species can be found in all of these regions and the distribution of the genera, species and subspecies can tell us much about how this group of fish has evolved.

Tilapia are widely distributed but no native populations have been reported from the eastern slope of the eastern rift valley and the rivers which flow into the Indian Ocean north of the Zambesi River. They are common in the rivers and lakes of Central and West Africa. *Sarotherodon* make up a relatively small genus in which all but one of the species (*S. galilaeus*) is restricted to West Africa. *Oreochromis* are abundant in the rift valley lakes and rivers and the rivers that drain into the Indian Ocean. Individual water bodies often contain endemic species. Apart from *O. niloticus* and *O. aureus* in the Nilo-Sudanian region and a small number of species belonging to the subgenus *Nyasalapia* in Angola *Oreochromis* species are rare in the western regions of Africa.

In general, *Tilapia* and *Sarotherodon* species are more common in West Africa and *Oreochromis* tends to be found in the central and eastern water bodies. There are however a small number of species which have a much larger and overlapping distribution: these are *Tilapia zillii, Sarotherodon galilaeus* and *O. niloticus* and, to a lesser extent, *O. aureus*. They have what is known as a Nilo-Sudanian distribution made possible by the interconnection of the major river systems (Nile, Chad, Niger) in the north of the continent. Much of this movement of species between watersheds would have occurred during the much wetter periods that were common up to and including the upper Pleistocene when the Sahara would not have been such a significant barrier to the spread of aquatic organisms (Fig. 1.5).

Philippart and Ruwet (1982) detail the probable sequence of colonization by these species. It appears that *O. niloticus* had colonized most of the western rift valley from its origin in the upper Nile in present day Uganda. From this source it colonized the lower Nile and associated lakes, such as Lake Turkana in the eastern rift valley, as the opportunities arose with the retreat of the ancient Tethys Sea back north towards the Mediterranean. It also expanded its range eastwards into the Ethiopian rift valley and to the south to colonize all of the major rift valley lakes as far south as Lake Tanganyika, although not the Lake Victoria watershed above the Murchinson falls barrier, where it was subsequently introduced by man in the1950s along with *T. zillii, T. rendalli* and *O. leucostictus*.

T. zillii and *S. galilaeus* would have been similarly expanding their range eastwards at about the same time from their sources of origin in West Africa. It was the connections that occurred during the wetter periods between the Chad and Nile watersheds that enabled the final pan-African spread of the three species. The presence or absence of these three species in the various rivers and lakes show the complicated hydrology and geology of Africa during this period. In the western rift valley *O. niloticus* is found in Lake Albert, Lake George, Lake Edward, Lake Kivu and Lake Tanganyika, whereas T. *zillii* and S. *galilaeus* are

---

Fig. 1.5 (opposite) Distribution of *Oreochromis niloticus* and all subspecies showing major geographical and hydrological features which may have influenced the present geographic spread and genetic differentiation (after Trewavas, 1983; Beadle, 1981; Adamson and Williams, 1980; Talbot, 1980).

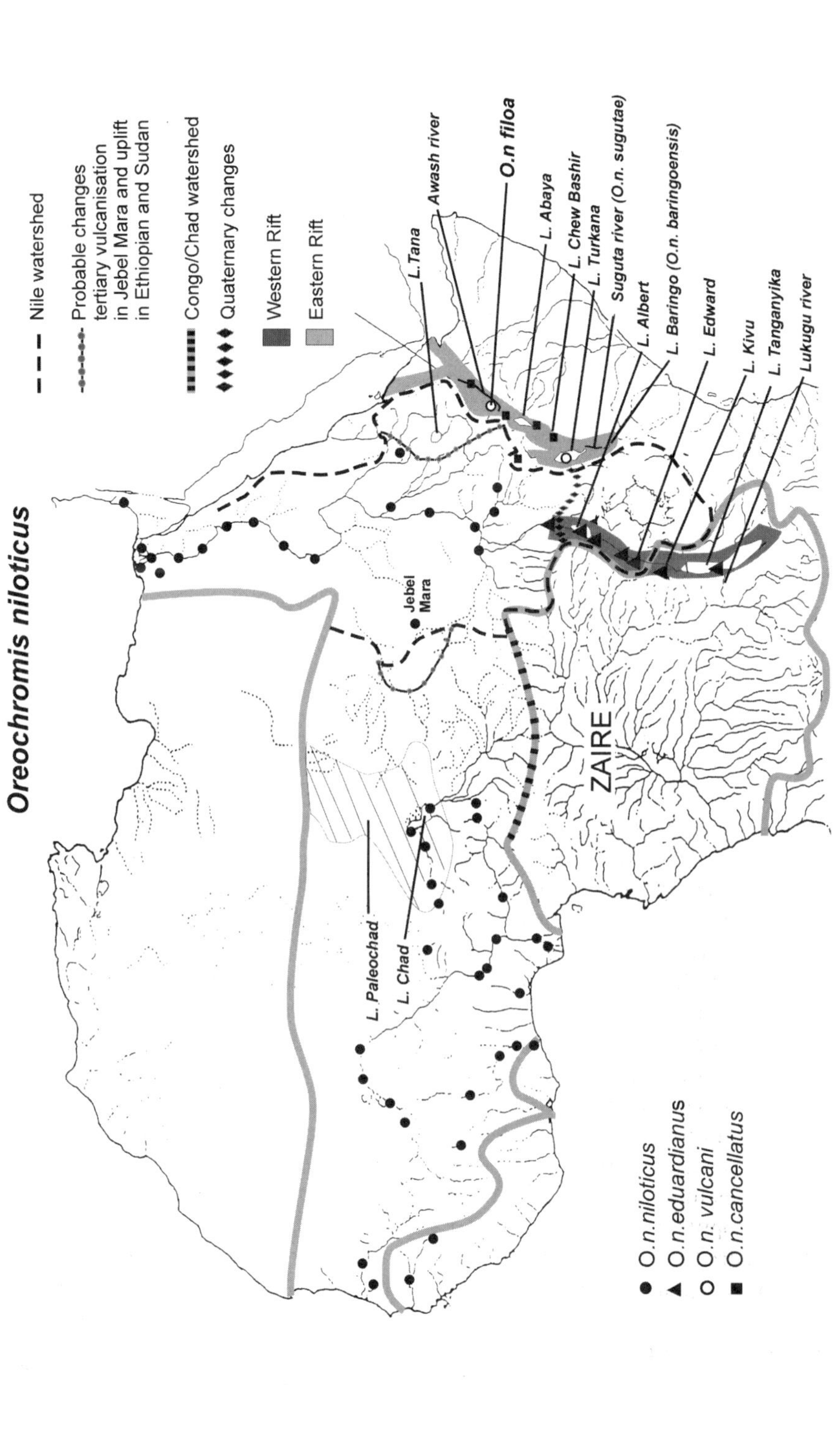
Oreochromis niloticus
Nile watershed
Probable changes tertiary vulcanisation in Jebel Mara and uplift in Ethiopian and Sudan
Congo/Chad watershed
Quaternary changes
Western Rift
Eastern Rift
L.Tana
Awash river
O.n filoa
L. Abaya
L. Chew Bashir
L. Turkana
Suguta river (O.n. sugutae)
L. Albert
L. Baringo (O.n. baringoensis)
L. Edward
L. Kivu
L. Tanganyika
Lukugu river
Jebel Mara
ZAIRE
L. Paleochad
L. Chad
O.n.niloticus
O.n.eduardianus
O.n. vulcani
O.n.cancellatus

only found as far south as Lake Albert. This suggests that the Chad-Nile connection that enabled these two species to invade the Nile must have occurred after *O. niloticus* had spread to the south and after the connection between Lake Albert and the more southerly lakes had disappeared (Fig. 1.5).

The spread of *T. zillii* and *S. galilaeus* in the Chad drainage must have been well established before any Nile connection because the capture of the former Chad Ubangi-Uele rivers by the Zaire only introduced the western species and not *O. niloticus* into that river. The connections between the Nile and Chad tributaries eventually enabled the westward migration of *O. niloticus* via the Chad/Niger/Volta/Senegal rivers into West Africa. This is a process that is still apparently taking place, as *O. niloticus* is still rare in some parts of the Upper Niger and coastal rivers of West Africa.

There is some doubt as to whether *O. niloticus* actually entered and is established in the Zaire basin naturally. Poll (1963) describes it and a number of other non-cichlid Nilo-Sudanian species in the upper Lualaba River. Poll (1963, 1976) suggested that this might have been due to this river being part of the Nile drainage before it was captured by the Zaire. This scenario has been rejected by a number of authors (see Beadle, 1981, p. 167) but the presence of the Nilo-Sudanian fish species in the Lualaba remains to be explained. A reassessment of the fauna in the Lualaba (Thys van den Audenaerde, 1968; Bannister and Bailey, 1979) suggests that the apparent similarity to a Nilo-Sudanian species was incorrect. In particular it appears that fish originally identified as *O. niloticus* were actually *O. upembae*. The distribution of *O. upembae* is the upper Zaire drainage to the west of Lake Tanganyika but a very closely related species *O. ny. malagarasi* is found in the Malagarasi River and swamp on the east side of Lake Tanganyika but neither have been found in the intervening lake. This suggests that the ancestor of both species was present in the upper tributaries of the Zaire River before the rifting which formed the lake some 9–12 million years ago (Cohen *et al.*, 1993) separated the two populations.

The other more likely possibility is that *O. niloticus* might have entered the Lualaba River and the upper tributaries of the Zaire from Lake Tanganyika via the Lukugu River outflow. About 10 000 years BP the Virunga volcanic complex dammed an affluent river of Lake Edward resulting in the formation of Lake Kivu, with a nilotic fauna, with an overflow to the south via the Ruzizi River to Lake Tanganyika. This new inflow enabled *O. n. eduardianus* to enter Lake Tanganyika where it has become established in shallower areas. It has certainly been found in the Burton Gulf and the area around the outlet to the Lukugu River (Trewavas, 1983, p. 158). The Lukugu River flows during periods of high lake levels into the Lualaba. Under these conditions *O. niloticus* might well be able to populate the upper regions of the Zaire (Fig. 1.5). Reports of *O. niloticus* in other Zaire tributaries such as the Ituri and Uele are more likely to be manmade introductions from Lake Albert (Thys van den Audenaerde, 1964). Trewavas (1983) is clear that she thinks that there are no natural *O. niloticus* populations in the Zaire basin at present.

O. niloticus and its subspecies

Of all the Nilo-Sudanian species *O. niloticus* has been the most intensively studied because of its commercial importance to aquaculture world-wide. Trewavas (1983) identified and sub-divided *O. niloticus* into seven subspecies. The most widely distributed of these is *Oreochromis niloticus niloticus* with a Nilo-Sudanian distribution from the Nile to the Senegal river basins. *O. n. eduardianus* is restricted to the upper White Nile and the rift valley lakes south of the equator (George, Edward, Albert, Kivu, Tanganyika and their associated watersheds). There is a possibility that Lake Albert might also contain *O. n. niloticus* (Thys van den Audenaerde, 1964) but Trewavas (1983) thinks this is unlikely. *O. n. cancellatus* is restricted to a few lakes and rivers in Ethiopia (Lake Tana, Lake Zwai to Gandjule and Stefanzi, Omo and Awash river systems). *O. n. filoa* is restricted to hot springs in the Awash system. *O. n. vulcani* is restricted to Lake Turkana and the crater lakes on the central island as well as a stream entering the lake at Loiengalani. *O. n. baringoensis* is restricted to Lake Baringo in Kenya. *O. n. sugutae* is restricted to the Sugata River and its tributaries which flows between Lake Baringo and Lake Turkana but is no longer connected to either. Trewavas (1983) makes the point that her classification of *O. n. filoa* and *O. n. sugutae* as separate subspecies is more for convenience and to identify them as possibly thermally modified phenotypes as they are associated with hot springs and as potential genetic resources for the future (Fig. 1.5).

Morphologically these subspecies have overlapping ranges for most meristic traits (Trewavas, 1983), the main differences being in the coloration of the mature males and the banding pattern of the caudal fin. In general, *O. n. niloticus* is very similar over the whole of its range from the Senegal River through the middle Niger and the lower Nile. The main differences in colour and morphology occur in the southern rift valley and the eastern rift populations.

The advent of aquaculture and the dominant role that *O. niloticus* is now assuming as the species of choice by most farmers has encouraged a number of studies to look at the levels of genetic variation and genetic differentiation in wild populations of this species. Most studies have utilized molecular genetics tools to assess levels of allozyme and mtDNA variation although morphological (Vreven *et al.*, 1998) and commercial traits have also been compared (Bentsen *et al.*, 1998; see Chapter 7). It would be interesting to see how the phylogenetic relationships determined from molecular data fit in with the classical taxonomy and what we know about the geological and hydrological history of this region.

The first studies to look at all the *O. niloticus* subspecies, but no populations from West Africa, were those of Seyoum and Kornfield (1992a,b). They present data from Restriction Fragment Length Polymorphism (RFLP) analysis of total mtDNA. They produced a key based on composite RFLP phenotypes that could be used to discriminate between all Nile and rift valley subspecies. The phenetic and cladistic analysis of these data generated trees that placed most of the putative *O. n. cancellatus* and *O. n. filoa* samples in a separate clade. The Lake

Tana (*O. n. cancellatus)* sample was always more closely associated with the Nile rather than the other Ethiopian samples. Trewavas (1983) justified her designation of the Lake Tana species on the associated ichthyofauna of the lake which was more Ethiopian than Nilotic in character. These results suggest that her classification may need to be revised. In an associated paper, Seyoum and Kornfield (1992a) suggest reclassifying some of the subspecies based on the mtDNA results. They suggest that *O. n. cancellatus* should be raised to a subgenus *O. cancellatus cancellatus* making *O. n. filoa* into *O. c. filoa* and that the Lake Tana population, on the basis of its distinctive haplotypes, should be made a new subspecies *O. n. tana* which shows its Nilotic affinity.

In a comparable study, Agnèse *et al.* (1997) collected samples from 17 different population of *O. niloticus*, which included samples from all of the subspecies and analysed allozyme variation and mtDNA RFLP of a 1 kb portion of the mtDNA D- loop region. They generated a number of trees from the allozyme data using genetic distance matrices from which they produced a single consensus tree. This split the populations into three main groups: 1, West African populations (Senegal, Niger, Volta and Chad) nominally all *O. n. niloticus*; 2, Ethiopian rift valley populations (Lake Koka, Lake Awash, Lake Ziway and Awash River) nominally *O. n. cancellatus* and *O. n. filoa*; 3, The remaining Nile drainage (Nile and Lake Edward) *O. n. niloticus* and *O. n. eduardianus* and Kenyan rift valley populations of Lake Turkana (*O. n. vulcani*), Lake Baringo (*O. n. baringoensis*), and Sugata River (*O. n. sugutae*).

The mtDNA RFLP data of Agnèse *et al.* (1997) identified 9 different mtDNA haplotypes. A consensus tree derived from the 18 most parsimonious networks showed that all the West African populations clustered on one side of the tree with just a single haplotype. Furthest from these were the Ethiopian rift valley populations and between were the Kenyan and Ugandan rift valley lake populations. The Nile populations showing affinities with both the West African and Lake Turkana and Lake Tana populations.

These two studies, although using different types of molecular data, show that the Ethiopian rift valley species *O. n. cancellatus* and *O. n. filoa* apart from the Lake Tana population form a distinct cluster from the other subspecies. The Lake Tana population despite its morphological similarity with the Ethiopian subspecies seems to have closer genetic affinities with the Nilotic subspecies, despite the Ethiopian association of the other fish in the lake used to justify Trewavas's subspecies classification.

The differentiation between East and West African populations of *O. n. niloticus* has also now been observed in other studies. Macaranas *et al.* (1995) used allozymes to analyse the wild populations collected for the Genetically Improved Farmed Tilapia programme (GIFT). Genetic distance data suggested a dichotomy between the Nile and Lake Turkana (East African) and the Volta and Senegal River (West African) populations of *O. n. niloticus*.

Rognon *et al.* (1996) used allozymes to analyse 27 natural and farmed populations of *O. n. niloticus* and *T. zillii* across their full east-west distribution.

A dendrogram, based on genetic distance, of the natural populations of both species shows a divergence between the West African and the East African Nile and Lake Turkana populations of *O. n. niloticus*. The mean divergence times given for the genetic distances between the different populations suggest, depending on the substitution rate used, about 0.3–1.1 million between east and western forms and 0.2–0.7 million between the different Nile populations.

The divergence between the West African and lower Nile and rift valley populations is also shown by Vreven *et al.* (1998). They studied allozyme and morphometric traits in 9 wild populations of *O. niloticus*, six West African, two lower Nile and a Lake Edward population. The morphometric analysis (25 traits), using principal component analysis consistently separated the Nile and Lake Edward populations from the West African on the shape of the caudal peduncle and toothed pharyngeal bone length and width. The allozyme (25 loci) data, analysed as genetic distance, always placed the Nile and Lake Edward populations in a separate clade from the six West African populations.

Trewavas (1983) combined the West African and lower Nile populations *of O. n. niloticus* essentially on the basis of male breeding coloration. The allozyme and mtDNA data (Macaranas *et al.*, 1995; Agnese *et al.*, 1996; Vreven *et al.*, 1998), the evidence from meristic (Vreven *et al.*, 1998) and comparative performance trials under aquaculture conditions (Bentsen *et al.*, 1998), particularly for growth, has shown there to be real differences between East and West African populations of this subspecies.

Agnèse *et al.* (1997) also analysed the levels of genetic variation in the populations they studied. This showed that the populations could be divided into two groups on the level of their allozyme polymorphism (P). The Nile and West African (*O. n. niloticus*) populations had a $P > 0.08$ whereas the East African populations (all other subspecies) were typically $P < 0.04$. These values are similar to those observed in other allozyme based studies of this species (Rognon *et al.*, 1996; also see McAndrew and Sodsuk, in press).

The allozyme data would suggest that the spread of *O. n. niloticus* across the Nilo-Sudanian region was relatively easy since the West Africans populations seem to share the same common alleles at the polymorphic loci as the Nile populations. The colonization of West Africa by *O. niloticus* obviously occurred after a connection was established between the Nile and most probably one or more Chad tributaries, most likely in the region around Jebel Marra (Williams *et al.*, 1980) (Fig. 1.5). However, it is interesting to note that although there appears to be no overall loss of allozyme polymorphism in the westward expansion of *O. n. niloticus* there was an apparent loss of mtDNA diversity. Agnèse *et al.* (1997) only observed one out of the four mtDNA haplotypes seen in the Nile fish in the West African populations. This would suggest that the lower effective population size of the mtDNA genome, 25% that of the nuclear genome, is showing evidence that some bottleneck in numbers did occur in the original Nile to Chad colonization event.

It appears that the colonization of the lakes and rivers to the south and east was also associated with a genetic bottleneck, in that these populations have similar but low heterozygosity values, which possibly suggests a single ancient colonization event of *O. niloticus* that eventually resulted in the remaining subspecies. The higher allozyme and mtDNA diversity in the Nile populations could be seen as evidence that this region was probably the centre of origin for the species.

The combined results from these studies are interesting in that they test some of the assumptions made to justify the earlier classifications of the subspecies. It appears from the studies described that the West African populations of *O. n. niloticus* are more similar to each other, on the basis of allozyme and mtDNA variation and meristic data, than they are to the same subspecies in the Nile. The rift valley and Nile populations of all subspecies cluster together but it appears the Ethiopian subspecies (*O. n. cancellatus* and *O. n. filoa)* apart from the Lake Tana population generally form a separate clade from the other Nile and rift valley populations. The distribution of the other subspecies of *O. niloticus* in the Nile and rift valleys has to be seen against a background of large tectonic, hydrological and climatic changes, many of which are still not fully understood (Street and Grove, 1976; Gasse, 1977; Williams, 1977; Gasse and Street, 1978; Gillespie, 1983). How do these major events fit with what we now know about the genetic divergence we are observing in these populations?

The Nile drainage has undergone major changes since Tertiary times. The Nile has lost 60 000 km^2 to the Chad drainage as volcanic activity changed the position of the divide in the region around the Jebel Marra volcanic massif. The Blue Nile gained 180 000 km^2 of catchment and substantial extra flow with the increased precipitation caused by the uplift of the Ethiopian uplands that moved the watershed between the Nile and Afar drainages further south. More recent additions to the Nile drainage include the outflow from Lake Victoria and Lake Albert during the Quaternary, which substantially increased the flow down the White Nile (Fig. 1.5).

Essentially, East Africa consists of a plateau dissected by the Great Rift Valley and two major tributaries of the Nile. The Ethiopian arm of the rift valley is now hydrologically completely isolated from the rest of the eastern rift valley with any drainage towards the Awash River and the Red Sea to the north-west (see Beadle, 1981, p.188). The Nile drainage extends southwards over 2600 km from the Mediterranean to Khartoum where it splits into two main tributaries: the Blue Nile which extends up to its source in Lake Tana in the Ethiopian highlands and the White Nile which would have included, at one time, all of the major western rift valley lakes as far south as the affluent streams entering Lake Edward. Some of the affluent streams of Lake Edward were subsequently lost after they were dammed by the volcanic activity that formed Lake Kivu and its new drainage pattern to the south into Lake Tanganyika. The Nile would also have included the waters from Lake Turkana and its associated eastern rift valley drainages of Lake Baringo and the Sugata River (Adamson and Williams, 1980).

Along with the physical changes in the catchment there have been very significant changes in flow rates and areas of open water associated with climatic changes. The most recent wet period reached its peak about 12 000 years BP but there is evidence of a series of wet and dry periods over the past 40 000 years (Livingstone, 1980; Gasse *et al.*, 1980).

How well does the geological and recent genetic information help to explain the present day distribution of the remaining *O. niloticus* subspecies? Of all the lakes in the rift valley the more northerly complex in the Ethiopian rift have been the ones isolated the longest from the influence of the Nile, since well before the Pleistocene (Trewavas, 1983, p. 163). Two subspecies, *O. n. cancellatus* and *O. n. filoa*, are present in this area (Trewavas, 1983), *O. n. filoa* being restricted to hot spring areas and having the characteristically low meristic counts associated with warmwater conditions. This made her unsure as to whether these characteristics were environmentally or genetically determined.

It is apparent that the Ethiopian rift valley flooded during wetter periods which unified the present day isolated northern lakes (Ziway, Abiyata, Langano, Shalla and Awasa) to form Lake Galla which drained to the north into the Awash River, the last time as recently as 5000–6000 years BP. The drier periods resulted in the retreat of the waters to present day or even lower lake levels (Gasse *et al.*, 1980). The lakes in the south of the Ethiopian rift (Abaya, Chamo, Stefanie) appear to have had a connection via the Sagan River to what would have been a much more extensive Lake Turkana, some 80 m above present levels. The lake is thought to have had an outflow to the Nile via the Sobat River which, most recently, lasted from 10 000–7000 years BP (Beadle, 1981).

Seyoum and Kornfield (1992b) collected both Ethiopian subspecies from a number of locations and showed that they could be separated by the use of mtDNA RFLP. The results showed that these subspecies were closely related and consistently clustered together in a single clade in all trees with the lowest sequence divergence estimated as $180\,000 \pm 105\,000$ years BP. However the Ethiopian subspecies had the highest divergence value from all other rift subspecies, estimated at over 1 million years.

The relationship between *O. n. cancellatus* and *O. n. filoa* is still unclear. The genetic data suggest two species fixed for different mtDNA haplotypes. Seyoum and Kornfield (1992 a,b) have shown populations with fixed mtDNA haplotypes with no evidence for any polymorphism in the populations selected with no apparent mixing between the species. However, no sample sizes or meristic data are presented for any of the populations collected. They suggest that the finding of a *O. n. cancellatus* population in Lake Beseka, part of the Awash drainage, is anomalous, since this area is primarily supposed to contain *O. n. filoa*, a statement not confirmed by Trewavas (1983). They suggested this relict population is evidence of the mixing of the subspecies caused by the expansion of Lake Galla. However, without confirming morphological evidence and a wider and larger survey of this area it difficult to say whether the haplotype data are a reliable means of identifying these two subspecies. However, both

Seyoum and Kornfield (1992a) and Agnèse *et al.* (1997) disagree with Trewavas' (1983) classification of the Lake Tana population as *O. n. cancellatus* as both studies show much greater affinity with Nile rather than Ethiopian subspecies. Both groups suggest it should be reclassified as *O. n. tana* (Seyoum and Kornfield, 1992a) or *O. n. niloticus* (Agnèse *et al.*, 1997).

The other subspecies in the southern part of the Ethiopian rift include *O. n. vulcani*, in Lake Turkana, *O. n. sugate* in the Sugata River and *O. n. baringoensis* in Lake Baringo. The faunistic associations in and around Lake Turkana are very similar to those of Lake Albert and the Nile below the Murchison and Semliki falls; 26 of the 29 species in the lake are also found in the Nile, 19 of these are also found in West Africa (Worthington and Ricardo, 1936). As already mentioned, Lake Turkana has had a relatively recent and prolonged connection to the Nile. However, it appears that Lake Turkana has been isolated from Lake Baringo and its associated drainages to the south by a volcanic eruption in the middle-upper Pleistocene (Fuchs, 1950) and before the expansion of *T. zillii* and *S. galilaeus* into the Nile as they are not present in these waters. Today, Lake Baringo is all that remains of a much larger Lake Kamasia that eventually drained away to the north (Fuchs, 1950), as faulting reduced the height of its northern rim, into the Sugata River and eventually Lake Turkana and presumably resulted in the introduction of the pre-Chad connection Nilotic species into these waters.

The results from Seyoum and Kornfield (1992a) suggest a close association between *O. n. vulcani* and *O. n. sugutae* in all dendrograms whereas *O. n. baringoensis* tended to cluster with their sample of *O. n. eduardianus* (Lake Edward). Both of these appear to be equally diverged from O. *n. niloticus*. Agnèse *et al.* (1997) allozyme results suggest the Lake Baringo population (*O. n. baringoensis*) clustered with the Lake Turkana and Lake Edward samples and the mtDNA results placed it between the Ethiopian and Lake Edward subspecies. It is possible that the original colonization of this area was by a *O. n. eduardianus* or *O. n. cancellatus* or a common ancestor which was subsequently isolated but was secondarily mixed with the Nile population during wetter periods.

The molecular data seem to fit with what we know about the geology of the region. The underlying similarity in the allozyme data suggests that there were probably only two major relatively ancient colonization events from the Nile population: one to the west which enabled the spread of *O. niloticus* into West Africa and one southwards into the rift valley. The rapid spread of the West African colonization resulted in an apparently homogenous population over a vast geographic area. The colonization of *O. niloticus* into what became a complex and geologically active rift valley system resulted in the dispersion of the species into a number of lakes and rivers. Some of these maintained intermittent contact with adjacent populations as hydrological conditions allowed whereas others were completely isolated and so became more divergent with time.

Other tilapiine species

The other tilapiine species which took advantage of the connections between the Niger/Chad/Nile drainages to extend their range were *T. zillii* and *S. galilaeus* and to a lesser extent *O. aureus*. The former two species have almost identical gross geographic distributions as *O. niloticus*, despite the fact that they colonized the same river systems from west to east. Do we see the same level of morphological and genetic differentiation in *T. zilli* and *S. galilaeus* that we now know exists in *O. niloticus*? The differences in the distribution between these species and that of *O. n. niloticus* would be informative about the hydrography in this region.

Trewavas (1983) divided *S. galilaeus* into a number of subspecies throughout its range. The most widespread is *S. galilaeus galilaeus* with a Nilo-Sudanian distribution from Senegal to Israel. It is replaced by *S. g. multifasciatus* in the freshwater reaches of the coastal rivers of the Côte d'Ivoire and Ghana and in Lake Bosumtwi. This subspecies is seen as an historical divergence between the savannah populations of *S. g. galilaeus* to the north and the Guineen forest form which has slight meristic and body form differences. *S. g. borkuanus* is a much more pigmented form found within some isolated lakes within the tectonic basin of a formerly much more extensive Lake Chad. *S. g. sanagaensis* is a subspecies restricted to the Sanaga River in Cameroon that can be distinguished from the introduced *S. g. galilaeus* by its colour. *S. g. boulengeri* is the subspecies found in the Lower Zaire River and which may have also been introduced into the lower Ogowe. It is characterized by having a scaly caudal fin. It appears that this subspecies is replaced in the upper Zaire tributaries by *S. g. galilaeus*. It is not clear why *S. galilaeus* or one of its subspecies did not manage to colonize the whole of the Zaire basin as it has the Chad, Niger and Nile. One reason might be that as the Zaire is essentially a forest river it might be an environment hostile to microphagous tilapiines.

There is relatively little molecular information on *S. galilaeus*. However, a study by Schlienen *et al.* (1994) using mtDNA sequence data, from a 340 bp fragment of the cytochome b gene and a 350bp fragment of the mtDNA control region sequence has shown some of the best evidence for sympatric speciation in any cichlid. Schlienen *et al.* (1994) showed that a local river population of *S. g. galilaeus* was the most likely ancestor of the eleven endemic tilapiine species of Lake Barombi Mbo, Cameroon. The species flock has evolved in what is a small geologically uniform bowl shaped and ecologically homogeneous volcanic crater lake. Trewavas (1983) was uncertain about the ancestry of all the tilapiine fish in Lake Barombi Mbo and postulated that *S. g. galilaeus*, or possibly a *S. melanotheron*, was the most likely ancestor for four of the species (*S. lohbergeri*, *S. steinbachi*, *S. linnellii*, *S. caroli*).

The study by Schlienen *et al.* (1994) looked at all the species in the lake and in the surrounding watersheds and included samples from the adjacent river and lake populations of *S. g. galilaeus*, the subspecies *S. g. sanagaensis* and *S.*

melanotheron. The molecular data show that all the tilapiine species in the lake are monophyletic in origin and that the most likely ancestor was *S. g. galilaeus* from the Memé River which was once connected to the lake. The sequence data identify three groups that also separate on their feeding behaviour; the first are in the predatory genus *Stomatepia*, the second are the fine particle feeders in the genera *Sarotherodon* and *Myaka* and the third are specialist feeders in the genera *Konia* and *Pungu*. The classification of these fish into four additional genera on the basis of their morphological feeding adaptations relates back to the low levels of morphological differentiation that have often accompanied speciation in the tilapiine tribe. The mtDNA variation I believe is strong evidence for the inclusion of all eleven tilapiine species in the lake to be placed within the genus *Sarotherodon* and subgenus *galilaeus*.

It is clear that more detailed studies of *S. galilaeus* and its subspecies using molecular markers would be highly informative and greatly help our understanding of the events which resulted in the differentiation in this species.

The other widely distributed Nilo-Sudanian tilapia is *T. zillii*. Rognon *et al.* (1996) is the only study to date to have analysed 9 different wild populations of *T. zillii* across its natural range. This work showed a clear divergence between the Côte d'Ivoire populations and those from the rest of the Nilo-Sudanian region; the mean genetic distance (0.121 ± 0.029) was high compared to the genetic distance (0.059 ± 0.023) between the most easterly and the most westerly populations. The latter distance was very similar to that observed between the east and west *O. niloticus* populations (0.064 ± 0.012). The Côte d'Ivoire populations came from an area known as Baoule V that contains species of both Nilo-Sudanian and Upper Guinean origins. However, it is unclear whether the differences are the result of adaptation to the different habitats, from savannah in the north to rainforest in the south, or whether they have remained remote from colonization events during pluvial periods that appear to have reduced levels of differentiation in the more northerly populations. The level of genetic differentiation in the Nilo-Sudanian populations of *T. zillii* and *O. niloticus*, and presumably *S. galilaeus*, is low considering the macrogeographic distances being compared. However, the similarity of the estimate does suggest that these species have been exposed to similar paleogeographic events albeit from different directions.

The spread of *T. zillii* appears to be very similar to that of *S. g. galilaeus*, although it appears that when it reached the Zaire basin its spread was restricted by the established *T. tholloni*. Electrophoretic analysis of *T. tholloni* by Sodsuk (1993) suggested that it was very similar to *T. zillii* (genetic identity = 0.002) but the provenance and species identification needs to be confirmed as it came via a commercial source.

In the few other species that have been studied similar patterns of differentiation were observed. Pouyaud and Agnèse (1997) studied 12 populations of *T. guineensis* and 7 populations of *S. melanotheron* from coastal lagoon sites in Senegal, Gambia and Côte d'Ivoire. The study showed high heterozygosity levels in *T. guineenis* (0.015–0.132) and dendrograms based on genetic distance show the Ivorian and Senegambian populations forming two separate groups. Nei's genetic identity ranged from 0.891–0.999 between the less disturbed natural populations. The *S. melanotheron* populations showed much lower overall heterozygosity levels (0.035–0.071) but the geographic groupings Ivorian and Senegambian, based on Nei's genetic identity (0.881–0.992), were still apparent in this species. There was also significant differentiation in the form of diagnostic allozyme loci which could be used to identify some of the subspecies (*S. melanotheron heudelotii* and *S. m. paludinosus*) proposed by Trewavas (1983). These data suggest that significant genetic differentiation has occurred in natural populations of both of these species in West Africa.

Schlienen *et al.* (1994) looked at the *Tilapia* species flock in Lake Bermin, Cameroon, and concluded that the nine endemic species in the lake were monophyletic in origin and that there is strong evidence that the Cross River population of *T. coptodon guineensis* was the sister taxon. Along with their findings on the Barombi Mbo species flock this is evidence for quite rapid and relatively radical morphological adaptation to new lake environments. No geological age is given for the lakes but they are assumed to be 'young geological speaking' (Trewavas, 1983) because of the levels of volcanic activity observed in this region. This contrasts with the only *Oreochromis* species flock of three species belonging to the *Nyasalapia* subgenus observed in the relatively stable environment of Lake Malawi (Trewavas, 1983; Sodsuk *et al.*, 1995).

The patterns of differentiation in West Africa suggest that there has been high levels of differentiation within species that have been isolated in the environmentally more stable and humid coastal lagoons and mountainous areas. This stability has enabled small monophyletic species flocks to develop in single lakes such as Barombi Mbo and Bermin. Coastal and lagoon species which have had a relatively stable environment have diverged such that there is a cline of at least 5 subspecies for *S. melanotheron* along the West African coast initially defined on their pigmentation patterns (Trewavas, 1983). The allozyme study by Pouyaud and Agnèse (1996) suggests that these colour differences between the *S. melanotheron* subspecies are also reflected at the molecular level which should prove useful in defining the genetic divergence and the ranges of the different subspecies.

A recent series of studies that looked at the soda tilapia in Lakes Natron and Magadi in East Africa have put forward evidence for the development of a species flock no more than 9000 years old. Seegers and Tichy (1999) redescribe *O. alcalicus* and describe two new species *O. ndalalani dalalani* and *O. latilabris* on the basis of head shape and dentition in Lake Natron. They suggest that these morphotypes must have evolved in the last 7000 years because the lake is young and is in an isolated basin in which this is the only tilapiine group (Tichy and Seegers, 1999). Molecular analysis of the D-loop and cytochrome-b in the mtDNA of all soda species and possible close relatives showed the soda tilapias clustered together and appeared to be distantly related to the other members of the *Oreochromis* which similarly clustered. This led them to propose that the soda tilapia subgenus status *Alcolapia* should be raised to a genus (Seegers *et al.*, 1999). The mtDNA variation does not appear to be congruent with the species radiation and therefore cannot be used alone for species discrimination. The spread of haplotypes between the lakes shows a single haplotype common to both species which could be described as ancestral to the rest. Of the remaining 17, 4 haplotypes are only observed in Lake Magadi and the remaining 13 are only present in Lake Natron. The authors suggest that the apparent speed of the morphological and mtDNA divergence in Lake Natron is being driven by the same intrinsic forces. They suggest that the radiation observed may not be adaptive or only secondarily so.

It is clear that there is still much work needed to complete our understanding of the evolution and spread of tilapia species in a geologically and climatically unstable continent. The large number of surveys and expeditions, which began in the early part of this century, resulted in the large animal collections on which much of the early taxonomic results are based. After a period of relative inactivity because of political uncertainty there has been a recent flourishing of research using molecular techniques stimulated by evolutionary, biodiversity and aquacultural interest in African fish species. It can be seen that generally the newer techniques have helped to confirm the earlier findings but that they have also opened up many new and interesting avenues for research.

REFERENCES

Adamson, D. and Williams, F. (1980) Structural geology, tectonics and control of drainage in the Nile basin, in *The Sahara and the Nile. Quaternary Environments and Prehistoric Occupation in Northern Africa* (eds M.A.J. Williams and H. Faure), A.A Balkema, Rotterdam, pp. 225–252.

Agnèse, J.F., Adépo-Gourène, B. Abban, E.K. and Fermon, Y. (1997) Genetic differentiation among natural populations of the Nile tilapia *Oreochromis niloticus* (Teleostei, Cichlidae). *Heredity* **79**, 88–96.

Bannister, K.E. and Bailey, R.G. (1979) Fishes collected by the Zaire river expedition 1974–75. *Zool. J. Linnean Soc.* **66**, 205–249.

Beadle, L.C. (1981) *The Inland Waters of Tropical Africa. An Introduction to Tropical Limnology*, 2nd edn, Longman, London.

Bentsen, H.B., Eknath, E.E., Palada-de Vera, M.S., Danting, J.C., Bolivar, H.L., Reyes, R.A., Dionisio, E.E., Longalong, F. M., Circa, A.V., Tayamen, M.M. and Gjedre B. (1998) Genetic improvement of farmed tilapias: growth performance in a complete diallel cross experiment with eight strains of *Oreochromis niloticus*. *Aquaculture* **160**, 145–173.

Carvalho G.R. and Pitcher T.J. (eds) (1994) *Molecular Genetics in Fisheries*, Chapman and Hall, New York.

Coad, B.W. (1982) A new genus and species of cichlid endemic to southern Iran. *Copeia* **1982**, 28–37.

Cohen, A.S., Sorenghan, M.J. and Scholz, C.A. (1993) Estimating the age of formation of lakes: an example from Lake Tanganyika, East African rift system. *Geology* **21**, 511–514.

Ferraris, J.D. and Palumbi S.R. (1996) (eds) *Molecular Zoology. Advances, Strategies and Protocols*, Wiley-Liss, New York.

Franck, J.P.C., Wright, J.M. and McAndrew, B.J. (1992) Genetic variability of a family of satellite DNAs from tilapia (Pisces, Cichlidae). *Genome* **35**, 719–725.

Franck, J.P.C., Kornfield, I. and Wright, J.M. (1994) The utility of SATA satellite DNA sequences for inferring phylogenetic relationships among three genera of tilapiine cichlid fishes. *Molec. Phyl. Evolution* **3**, 10–16.

Fuchs, V.E. (1950) Pleistocene events in the Baringo basin, Kenya colony. *Geol. Mag. Lond.* **87**, 149–174.

Gasse, F. (1977) Evolution of Lake Abbe (Ethiopia and T.F.A.I.) from 70,000 YBP. *Nature* **265**, 42–45.

Gasse, F. and Street, F.A. (1978) Late quaternary lake-level fluctuations and environments of the northern rift valley and Afar region (Ethiopia and Djibouti). *Palaeogeogr. Paleoclimatol. Palaeocol.* **24**, 279–325.

Gasse, F., Rognon, P. and Street, F.A. (1980) Quaternary history of the Afar and Ethiopian rift lakes, in *The Sahara and the Nile. Quaternary environments and prehistoric occupation in Northern Africa* (eds M.A.J. Williams and H. Faure), A.A Balkema, Rotterdam, pp. 361–400.

Gillespie, R., Street-Perrott, F.A and Switsur, R. (1983) Post-glacial arid episodes in Ethiopia has implications for climatic prediction. *Nature* **306**, 680–683.

Gunther, A. (1889) On some fishes from Kilimanjaro district. *Proc. Zol. Soc. Lond.* **1889**, 70–72.

Hillis, D.M., Moritz, C. and Mable, B.K. (1996). *Molecular Systematics*, 2nd edn, Sinaeur Associates Sonderland, Massachusetts.

Kaufman, L. and Liem, K.F. (1982) Fishes of the suborder Labroidei (Pisces: Perciformes): phylogeny ecology and evolutionary significance. *Breviora* **472**, 1–19.

Kocher, T.D. and Stepien, C.A. (eds) (1997) *Molecular Sytematics of Fishes*, Academic Press, San Diego.

Kornfield, I.L., Ritte, U., Richler, C. and Wahrman, J. (1979) Biochemical and cytological differentiation among cichlid fishes of Sea of Galilee. *Evolution* **33** (1), 1–14.

Livingstone, D.A. (1980) Environmental changes in the Nile headwaters, in *The Sahara and the Nile. Quaternary Environments and Prehistoric Occupation in Northern Africa* (eds M.A.J. Williams and H. Faure), A.A Balkema, Rotterdam, pp. 339–359.

Macaranas, J.M., Agustin, L.Q., Ablan, C.A., Pante, J., Eknath, A.A. and Pullin, R.S.V. (1995) Genetic improvement of farmed tilapias: biochemical characterisation of strain differences in Nile tilapia. *Aquaculture International* **3**, 43–54.

Margush, T. and McMorris F.F. (1981) Consensus n-trees. *Bulletin of Mathematical Biology* **43**, 239–244.

McAndrew, B.J. and Majumdar K.C. (1984) Evolutionary relationships within three tilapiine genera (Pisces, Cichlidae). *Zool. J. Linnean Soc.* **80**, 421–435.

McAndrew B.J. and Sodsuk P. (in press) Application of allozyme analysis in population genetics, stock management and taxonomy of tilapia (eds G.M. Mair and R.S.V. Pullin), ICLARM, Manila.

McConnell, S. (in press) (eds G.M. Mair and R.S.V. Pullin), ICLARM, Manila.

Miyamoto, M.M. and Fitch, W.M. (1995) Testing species phylogenies and phylogenetic models with congruence. *Syst. Biol.* **44**, 64–76.

O'Connell, M. and Wright, J.M. (1997) Microsatellite DNA in fishes. *Rev. Fish Biol. Fisheries* **7**, 331–363.

Oliver, M.K. (1984) *Systematics of African Cichlid Fishes: Determination of the Most Primitive Taxon and Studies of the Haplochromines of Lake Malawi (Teleostie: Cichlidae)*, PhD thesis, Yale University, New Haven, CT.

Peters, H.M. (1979) Warum ich nicht fur *Sarotherodon* bin. *Deutsche Cichliden Ges. Info.* **10**, 141–146.

Peters, H.M. and Berns, S. (1978) Über die Vorgeschiste der Maulbrutenden *Cichliden Aquarin Magazin* **5**, 211–217; **7**, 324–331.

Peters, H.M. and Berns, S. (1982) Die maulbrutpflege der Cichliden untersuchugen zur evolution eines verhaltensmusters. *Zeitschrift für Zoologische Systematik und Evolutionsforschung* **20**, 18–52.

Philippart, J.-C. and Ruwet, J.-C. (1982) Ecology and distribution of tilapia, in *The Biology and Culture of Tilapia* (eds R.S.V. Pullin and R.H. Lowe-McConnell), *ICLARM Conference Proceedings 7*, Manila, Philippines, pp. 15–60.

Poll, M. (1963) Zoogéograpie ichthyologique du cours superieur du Lualaba. Colloque sur les problèmes biogéographiques du Parc *National de l'Upemba. Publ. Univ. Elizabethville* **6**, 95–106.

Poll, M.(1976). Poissions, Explor. Du Parc National de l'Upemba. Mission G.F. Witte (1946–49), Fasc. 73, Bruxelles.

Pouyaud, L. and Agnèse, J.F. (1995) Phylogenetic relationships between 21 species of three tilapiine genera *Tilapia, Sarotherodon* and *Oreochromis* using allozyme data. *Journal of Fish Biology* **47**, 26–38.

Pouyaud, L. and Agnèse J.F. (1996) Genetic differentiation in several stocks of *Sarotherodon melanotheron* and *Tilapia guineenis* from Cote d'Ivoire, Senegal and Gambia, in *The Third International Symposium on Tilapia in Aquaculture* (eds R.S.V. Pullin, J. Lazard, M. Legrende, J.-B. Amon-Kothias and D. Pauly), *ICLARM Conference Proceedings 41*, Manila, Philippines, pp. 368–376.

Regan, C.T. (1920) The classification of the fishes of the family cichlidae. 1. The Tanganyikan genera. *Ann. Mag. Nat. His.* **7**, 230–239.

Rognon, X., Andriamanga, M., McAndrew, B.J. and Guyomard, R. (1996) Allozyme variation in natural and cultured populations in two tilapia species: *Oreochromis niloticus* and *Tilapia zillii. Heredity* **76**, 640–650.

Rüppel, E. (1852) *Verz. Mus. Senkenberg. Naturforsch. Ges. Aufges Sammlungen IV.* Fische, 40 pp, Frankfurt-am-Main.

Schlienen, U.K., Tautz, D. and Pääbo, S. (1994) Sympatric speciation suggested by monophyly of crater lake cichlids. *Nature* **368**, 629–632.

Seegers, L. and Tichy, H. (1999) The *Oreochromis alcalicus* flock (Teleostei: Cichlidae) from lakes Natron and Magadi, Tanzania and Kenya with descriptions of two new species. *Ichthyol. Explor. Freshwaters* **10**(2), 97–146.

Seegers, L., Sonnenberg, R. and Yamamoto R. (1999) Molecular analysis of the *Alcolapia* flock from Lake Natron and Magadi, Tanzania and Kenya (Teleostei: Cichlidae), and implications for their systematics ad evolution. *Ichthyol. Explor. Freshwaters* **10**(2), 175–199.

Seyoum, S. and Kornfield, I. (1992a) Taxonomic notes on the *Oreochromis niloticus* subspecies complex (Pisces: Cichlidae), with a description of a new subspecies. *Can. J. Zool.* **70**, 2161–2165.

Seyoum, S. and Kornfield, I. (1992b) Identification of the subspecies of *Oreochromis niloticus* (Pisces: Cichlidae) using restriction endonuclease analysis of mitochondrial DNA. *Aquaculture* **102**, 29–42.

Smith, A. (1840) Illustrations of the Zoology of South Africa 4. (London) Vol. 4. Pisces. pl 5. (for dates see Waterhouse, *Proc. Zol. Soc. Lond.* 1880, 89–491 and Jentink, F.A. 1893. *Notes Lieden Mus.* 15, 182).

Sodsuk, P.K. (1993). *Molecular Genetics and Systematics of Tilapiine Cichlids using Allozymes and Morphological Characters*, PhD thesis, University of Stirling, Scotland.

Sodsuk P. and McAndrew, B.J. (1991) Molecular systematics of three tilapiine genera *Tilapia, Sarotherodon and Oreochromis* using allozyme data. *J. Fish Biol.* **39** (Supplement A), 301–308.

Sodsuk, P.K., McAndrew B.J. and Turner, G.F. (1995) Evolutionary relationships of the Lake Malawi *Oreochromis* species: evidence from allozymes. *J. Fish Biol.* **47**, 321–333.

Steinitz, H. and Ben-Tuvia, A. (1960) The cichlid fishes of the genus *Tristramella* Trewavas. *Ann. Mag. Nat. Hist.* **3** (3), 161–175.

Stepien, C.A. and Kocher, T.D. (1997) Molecular and morphology in studies of fish evolution, in *Molecular Sytematics of Fishes* (eds T.D. Kocher and C.A. Stepien), Academic Press, San Diego, pp. 1–12.

Stiassny, M.L.J. (1991) Phylogenetic interrelationships of the family Cichlidae: an overview, in *Cichlid Fishes, Behaviour Ecology and Evolution* (ed. M.H.A. Keenleyside), Chapman and Hall, Fish and Fisheries Series 2.

Stiassny, M.L.J. and Jensen, J.S. (1987) Labroid intrarelationships revisited: morphological complexity, 'key note innovations', and the study of comparative diversity. *Bull. Mus. Comp. Zool.* **151**, 269–319.

Street, F.A. and Grove, A.T. (1976) Environmental and climatic implications of late Quaternary lake-level fluctuations. *Nature* **261**, 385–389.

Sültmann, H. and Mayer, W.E. (1997) Reconstruction of cichlid fish phylogeny using nuclear DNA markers, in *Molecular Systematics of Fishes* (eds T. Kocher and C.A. Stepien), Academic Press, San Deigo, CA, pp. 39–52.

Sültmann, H., Mayer, W.E., Figueroa, F., Tichy, H. and Klein, J. (1995) Phylogenetic analysis of cichlid fishes using nuclear DNA markers. *Mol. Biol. Evol.* **12** (6), 1033–1047.

Talbot, M.R. (1980) Environmental responses to climatic changes in the West African Sahel over the past 20,000 years, in *The Sahara and the Nile. Quaternary Environments and Prehistoric Occupation in Northern Africa* (eds M.A.J. Williams and H. Faure), A.A Balkema, Rotterdam, pp. 37–62.

Thys van den Audenaerde, D.F.E. (1964) Revision systématique des espèces Congolaises du genre *Tilapia* (Pisces, Cichlidae) *Ann. Mus. Roy. Afr. Cent. Ser. 8, Sci. Zool.* **124**, 155 pp.

Thys van den Audenaerde, D.F E. (1968) An annotated bibliography of *Tilapia* (Pisces: Cichlidae). *Ann. Mus. Roy. Afr. Cent. Zool.* **14**, 1–406.

Thys van den Audenaerde, D.F.E. (1971a) Some new data concerning the *Tilapia* species of the subgenus *Sarotherodon* (Pisces, Cichlidae) *Rev. Zool. Botan. Africaines* **84**, 203–216.

Thys van den Audenaerde (1971b) Some new data concerning the *Tilapia* species of the subgenus *Coptodon* (Pisces, Cichlidae) *Rev. Zool. Botan. Africaines* **84**, 160–168.

Thys van den Audenaerde, D.F.E. (1978) *Tilapia mossambicum* or *Sarotherodon mossambicum? Cichlidae* (British Cichlid Association) **126**, 1 June 1978.

Thys van den Audenaerde, D.F.E. (1980) Good advice to hobbyists on questions of nomenclature and taxonomy. *Buntbarsche Bull.* **81**, 7.

Tichy, H. and Seegers, L. (1999) The *Oreochromis alcalicus* flock (Teleostei: Cichlidae) from lakes Natron and Magadi, Tanzania and Kenya: a model for the evolution of 'new' species flocks in historical times? *Ichthyol. Explor. Freshwaters* **10**(2), 147–174.

Trewavas, E. (1942) The cichlid fishes of Syria and Palestine. *Ann. Mag. Nat. Hist.* **11**(9), 526–536.

Trewavas, E. (1966) Fishes of the genus *Tilapia* with four spines in Malawi, Rhodesia, Mozambique and Southern Tanzania. *Rev. Zool. Botan. Africaines* **74**, 50–62.

Trewavas, E. (1973a) On the cichlid fishes of the genus *Pelmatochromis* with proposals of a new genus for *P. congicus*: on the relationship between *Pelmatochromis* and *Tilapia* and the recognition of *Sarotherodon* as a distinct genus. *Bull. Br. Mus. Nat. Hist. (Zool).* **25**, 1–26.

Trewavas, E. (1973b) A new species of cichlid fishes of rivers Quanza and Bengo, Angola with a list of the known cichlidae in these rivers and a note on *Psuedocrenilabrus natalensis* Fowler. *Bull. Br. Mus. Nat His. (Zool).* **25**, 27–37.

Trewavas, E. (1980) *Tilapia* and *Sarotherodon. Buntbarsche Bulletin* **81**, 1–6.

Trewavas, E. (1982a) Genetics groupings of Tilapiine used in aquaculture. *Aquaculture* **27**, 79–81.

Trewavas, E. (1982b) Tilapias: taxonomy and speciation, in *The Biology and Culture of Tilapia* (eds R.S.V. Pullin and R.H. Lowe-McConnell), *ICLARM Conference Proceedings 7*, Manila, Philippines, pp. 3–14.

Trewavas, E. (1983) *Tilapiine Fishes of the Genera Sarotherodon, Oreochromis and Danakilia.* British Museum (Natural History), London. Publication Number 878.

Vreven, E.J., Adépo-Gourène, B., Agnèse. J.F. and Teugels, G.G. (1998) Morphometric and allozyme variation in natural populations and cultured strains of the Nile tilapia *Oreochromis niloticus* (Teleostei, Cichlidae). *Belg. J. Zool.* **128**(1), 23–34.

Williams, M.A.J. (1977) Late Quaternary lake levels in southern Afar and the adjacent Ethiopian rift. *Nature* **267**, 690–693.

Williams, M.A.J., Adamson, D.A., Willams, F.M., Morton, W.H. and Parry, D.E. (1980) Jebel Marra volcano: a link between the Nile valley, the Sahara and Central Africa, in *The Sahara and the Nile. Quaternary Environments and Prehistoric Occupation in Northern Africa* (eds M.A.J. Williams and H. Faure), A.A Balkema, Rotterdam, pp. 305–337.

Worthington, E.B. and Ricardo, C. (1936) Scientific results of the Cambridge Expedition to the East African lakes 1930–1, No. 15, The fish of Lake Turkana and Lake Baringo. *Zool. J. Linn. Soc. Lond.* **39**, 353–389.

Wright, J.M. (1989) Nucleotide sequence, genomic organisation and evolution of a major repetitative DNA family in tilapia *Oreochromis mossambicus /hornorum. Nucleic Acids Res.* **17**, 5071–5079.

Chapter two

Reproductive biology, mating systems and parental care

G. F. Turner and R. L. Robinson

2.1 INTRODUCTION

The purpose of this chapter is to provide a general outline of the reproductive strategies of the tilapiine cichlids. We will concentrate on the ethological, evolutionary and ecological aspects, rather than physiology, which will be dealt with elsewhere. As anyone remotely interested in tilapiine cichlids must be familiar with Trewavas's magnificent book on the mouthbrooding tilapias, we will concentrate on information published since that work appeared (1983) and give rather more coverage to the substrate spawning species than is warranted by their low diversity and minor economic importance.

Reproductive behaviour in tilapias is profoundly influenced by the breeding system (Turner, 1986) of the species. All *Oreochromis* species appear to be maternal mouthbrooders – indeed this forms part of the definition of the genus (Trewavas, 1983). Males give all the indications of being strongly sexually selected – they are generally larger, more brightly coloured, and more aggressive than females, build spawning pits and develop secondary sexual structures such as long extensions to the soft dorsal and anal fins, exaggerated jaws (in *Oreochromis mossambicus*) and elaborate genital papillae (in species of the subgenus *Nyasalapia*). *Tilapia* are biparental-caring substrate spawners. Sexual dimorphism is mainly apparent in size, with males being slightly larger than females in general. There may be little or no opportunity for mate choice but where this occurs, sexual selection may be stronger on males than females. Both sexes are highly aggressive during the breeding period. *Sarotherodon* are all mouthbrooders, but include species where males only, or both sexes, incubate

M.C.M. Beveridge and B.J. McAndrew (eds), Tilapias: Biology and Exploitation, 33–58.

the young. Studies of *Sarotherodon* behaviour are largely restricted to *Sarotherodon galilaeus* and *Sarotherodon melanotheron* which show little aggression or dimorphism in comparison to *Oreochromis* and *Tilapia*. The degree of sexual selection exhibited by each sex apparently depends on the sex of the carer.

2.2 MATING SYSTEMS AND MATE CHOICE

In *Oreochromis*, there is generally marked sexual dimorphism and females appear to be the more choosy sex, whereas in *Tilapia* and *Sarotherodon* dimorphism is less marked and in the paternal mouthbrooder *S. melanotheron*, males generally exert greater preference when choosing a partner. Choices may be based upon characteristics of the partner (such as size, status, colour) and features of the spawning territory (size, shape and location), but often there will be a number of factors interacting.

Oreochromis

Male *Oreochromis* are highly polygynous and may spawn with several females in quick succession. In the aquarium, females will certainly circle with several males in succession, particularly after an intruder has interrupted courtship, and there are reports that females may lay batches of eggs with different males during a single spawning (Ruwet, 1963). Genetic tests have confirmed multiple paternity within a single clutch for tilapiines (I. Kornfield, personal communication) and colonial maternal mouthbrooding haplochromines in Lake Malawi (Kellogg *et al.*, 1995, personal observation). Haplochromine females spawn with only a small proportion of the males with which they perform circling courtship displays (McKaye, 1991, personal observation) and there can be little doubt that they are exercising mate choice.

In contrast, males appear to discriminate very little between females. In *Oreochromis variabilis*, two female morphs of strikingly different appearance mate successfully with the same type of male (Lowe, 1956).

In laboratory studies, larger male *O. mossambicus* initially occupied preferred territory sites and monopolized most of the courtships. Over a period of several months, these males were overtaken in size by the males which had initially been smaller, presumably because the non-territorial males expended less energy in fighting, digging and courting. Once larger, formerly submissive males were then not able to gain access to better territories, though the increase in size enabled them to receive more courtship than previously. The distribution of time involved in courtship became less skewed, suggesting that both male size and territory quality were important in female preference (Turner, 1986). These studies have been extended by Nelson (1995) who controlled for the effect of male size and pit size in an ingenious experiment. He found that when males were isolated from each other, females preferred to mate with males which had larger spawning pits, irrespective of male size. However, when males were able to interact, females tended to prefer larger males.

Mate choice is thus limited to some extent by male-male competition. Another strategy employed by fish is 'sneaking' where a male rushes into a spawning pit while a pair is spawning and attempts to fertilize some of the eggs. Again, we know that neighbouring territorial males and occasionally non-territorial males may dash into a pit during spawning, at least in the laboratory (personal observation), but it has not been confirmed in field conditions, and there is no positive proof that eggs are fertilized in this manner rather than simply consumed by the intruder.

Male *Oreochromis* are often said to form 'leks' or breeding arenas where males crowd together to display to females (Fryer and Iles, 1972; Loiselle and Barlow, 1979; De Silva and Sirisena, 1988). Strictly speaking, a lek occurs where males actively aggregate for display purposes and leave other areas of apparently suitable habitat empty. It is not certain that male aggregation in *Oreochromis* reflects anything other than a lack of alternative sites which are suitable for pit-digging. In tanks and ponds, male territories are often packed close together, and dense aggregations are certainly observed in natural circumstances, for example in *Oreochromis macrochir* (Ruwet, 1963), *O. mossambicus* (Bruton and Boltt, 1975; De Silva and Sirisena, 1988), *Oreochromis grahami* (Coe, 1966) and *O. variabilis* (Lowe, 1956). However, this may simply be a reflection of high population density rather than active aggregation (Coe, 1966). In Lake Malawi, there is low primary production, heavy fishing and numerous algal-feeding haplochromine cichlid species. Tilapiines exist at fairly low densities. Here, male *Oreochromis* tend to nest well apart from conspecifics and interactions between neighbouring males are extremely rare – we have recorded over 24 hours of underwater video images of *Oreochromis karongae* males on their pits and not once did we see a conspecific male intrude into the territory (Turner and Robinson, unpublished). By contrast a number of haplochromine cichlids in the same area have spawning pits tightly clustered together and male-male interactions are frequent and spectacular, particularly in the large predatory *Dimidiochromis kiwinge*. Irrespective of whether males aggregate through choice or lack of space, close proximity of numerous males provides females with ample opportunity to choose between males.

Tilapia

Many biparental cichlids mate assortatively according to size with females generally being slightly smaller than their mates (McKaye, 1986). This has not been experimentally tested with *Tilapia*, but in the field there is a strong positive correlation between the sizes of males and females in pairs, with the females averaging 60% (range 45–95%) of the male's length (Schwanck, 1987a).

Access to mates is strongly influenced by dominance relationships (Turner *et al.*, 1989; Hulscher-Emeis, 1991) although it is not clear how much this is specific to laboratory conditions. There may not actually be much opportunity for mate choice and perhaps this is reflected in the lack of sexual dimorphism in the colour of most species. *Tilapia mariae* males may occasionally practise simultaneous polygyny and may also desert their mates during the parental period. This has been observed in large aquaria (Turner, personal observation) and in Nigerian rivers (Schwanck, 1987a), so there may be greater sexual selection on males than females. The trend for larger size in males may be the result of sexual selection, but it could also be explained by the greater energetic expenditure incurred by females in the manufacture of eggs when compared to sperm.

Tilapia guinasana has six colour forms which co-exist sympatrically in a single sink-hole 70 × 40 m in surface area (Ribbink *et al.*, 1991). Colour forms are not sex-linked. There appears to be a small amount of assortative mating, with fish of the dark blue form mating with each other more than expected, but there were plenty of pairings between all forms observed breeding at the time. Colour polymorphism is also found in three of the nine endemic species in Lake Bermin, Cameroon. The dark red form of *Tilapia snyderae* appeared to show weak assortative mating but none was apparent in *Tilapia gutturosa* or *Tilapia bemini* (Stiassny *et al.*, 1992).

Sarotherodon

Both male and female *S. galilaeus* prefer to mate with larger partners, although females show a stronger preference (Balshine-Earn, 1995a).

In the paternal mouthbrooder *S. melanotheron*, pairing of males and females is more likely to be successful if the females are larger than their mates (Barlow and Green, 1970). Balshine-Earn and McAndrew (1995) showed that males prefer to mate with larger females, while females do not choose mates on the basis of size. This is as would be expected, as the long duration of paternal care means that females can potentially spawn more often than males and thus access to males potentially limits female reproductive success, so males can afford to be choosy. This is precisely the reverse of the roles shown by maternal mouthbrooders, such as *Oreochromis*. In most fish species where males care for young, they only care for eggs and larvae which are stationary. Thus parental care does not interfere with the male's potential for polygamy. Perrone and Zaret (1979) cited reports that male *S. melanotheron* may brood the eggs of more than one female, leading them to suggest that they likewise retained the opportunity for multiple matings. We are sceptical of this because males of this species which we have bred frequently, seem to become quiescent immediately after spawning and do not seem to participate in further courtship activity. Balshine-Earn (personal communication), who also has extensive experience with this species, likewise knows of no instances of polygyny and it is noteworthy that neither of the references cited by Perrone and Zaret (Aronson, 1949; Lowe-McConnell, 1959) make any reference to males brooding multiple clutches.

2.3 COMPETITION AND TERRITORIALITY

In general, aggressive behaviour in tilapiine cichlids is confined to parental care and the establishment of territories for the purposes of reproduction. Although the formation of dominance hierarchies is commonplace in aquarium conditions, there is no indication that this occurs in the wild.

In natural conditions most, if not all, tilapiine cichlids are non-territorial outside the breeding season. Although some species in equatorial regions may have breeding seasons which extend over much of the year, individual fish are probably reproductively active (and therefore territorial) for shorter periods. In the wild, immature or non-breeding fishes are generally found in loose foraging groups which appear to range widely over their habitat. We have occasionally observed juvenile *O. (Nyasalapia) karongae* holding temporary feeding territories over rocky areas in Lake Malawi. These territorial juveniles adopted a dark body colour similar to territorial male conspecifics, although the latter are much more intensely coloured.

In aquarium conditions, many species are highly aggressive even as juveniles, particularly if food is delivered in predictable places. Aggressive behaviour in captive juveniles may represent the adoption of an opportunistic territorial strategy rarely required in the wild.

Oreochromis

For maternal mouthbrooding species in the wild, courtship territories are established by males and may be defended for several weeks during which time very little, if any, feeding is performed (Lowe, 1956, personal observation). In some species, males remain on the territory throughout the day and night (Lowe, 1956), though others are only present during the day. In the field, areas of substrate with particular characteristics are chosen by each species – territories are established next to solid objects in *O. grahami* (Coe, 1966), adjacent to rocks at low densities in Lake Malawi (personal observation), and both *O. variabilis* and *Oreochromis karomo* defend sites on swampy or reedy areas (Lowe, 1956). Fights between males may be rare in the field (Lowe, 1956, personal observation), though Bruton and Boltt (1975) report that aggressive displays and fights leading to injury occur in *O. mossambicus* before territory establishment.

Behaviour in the aquarium may differ in several ways. Males compete for preferred territory sites adjacent to weeds, rocks and the sides of the tank (Turner, 1986), but captive fish may also defend territories near the water surface. In addition, captive non-territorial males indulge in circling fights generally of short duration and low intensity. This probably serves to enable males to assess one another's relative strength and prowess prior to making the expensive commitment to establishing a territory. Neighbouring territorial males establish the position of their boundaries through head-to-head threat displays performed in courtship colours, but occasionally try to displace one another from presumably favoured sites through circling fights (Turner, 1986).

Although males begin to develop indications of breeding colour prior to the establishment of the territory (Baerends and Baerends-van Roon, 1950; Neil, 1964), full courtship colours are generally adopted at the time of territory establishment (Neil, 1964; Lanzing and Bower, 1974). In *O. mossambicus,* males adopt a specific 'fighting pattern' of colour during prolonged circling contests. The normal courtship colour is uniformly black, apart from red edges to the unpaired fins and a white chin and operculum. In the fighting pattern, the flanks are paler with irregular dark markings and the belly pale grey. This colour change may be effected in a few seconds.

Circling fights involve a number of characteristic behaviour patterns, best documented in *O. mossambicus* (Fig. 2.1).

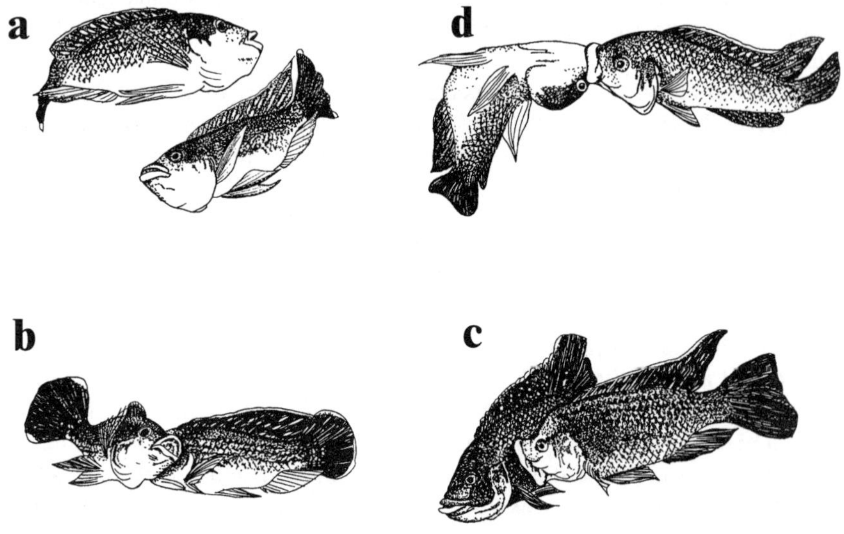

Fig. 2.1 Fighting behaviour in *Oreochromis mossambicus*: a. circling, b. tailbeating, c. ramming, d. head to head ramming.

Most of the time is spent in nose-to-tail circling. Sporadically one fish will break off to tailbeat, either in parallel to the opponent or almost at right angles in front of its head. The fins are erected during this manoeuvre. Charges with mouth open are frequent, and occasionally lead to one individual ramming the opponent's flank, probably the most dangerous activity shown in the contest. Normally physical contact consists of head-to-head ramming in which both individuals charge with the mouth open, impacting briefly, while one individual twists and pushes its rival down onto the substrate or even clear out of the water. Jawlocking, where both opponents grip their rivals' jaws and pull or push, is relatively rare in *Oreochromis* (Turner, 1987). Limited observations of *O. macrochir* (Apfelbach and Leong, 1969; Balthazart, 1973) report qualitatively similar behaviour, although mouthfighting (*combat de bouche*) is described as beginning when 'les deux poisson appliquent lentement leurs bouches ouvertes' (Balthazart, *op. cit.*), which sounds like a considerably more gentle approach than that adopted by *O. mossambicus*!

The relative frequency of performance of different behaviour during fights is related to the probability of winning the contest (Turner and Huntingford, 1986; Turner, 1994a). Different activities are predictive of winning when the contest is decided on the basis of size alone (Turner and Huntingford, 1986) or when it is decided through a combination of size and prior residency in the territory (Turner, 1994a).

Defeated male *O. mossambicus* rapidly lose their aggressive colours and perform appeasement displays. Where space permits, they rapidly flee, but in the confines of an aquarium may be severely damaged or killed by repeated rams from the victor. Once defeated, a male rarely, if ever, attempts to fight back. Contrary to reports, in earlier literature, of males being killed by a single ramming blow (e.g. Neil, 1964), we have never known male *Oreochromis* to be seriously injured during a contest, despite observing hundreds of contests.

In captive *O. mossambicus*, the rank order of mixed sex dominance hierarchies was found to be unstable from week to week, though dominant individuals were less likely to experience rank reversals than subordinate individuals (Oliveira and Almad, 1996).

In isolation, male *O. mossambicus* show consistently different levels of aggressiveness to a standard stimulus, such as a mirror image (Seow and Turner, ms). The function, if any, of these differences is not known, as aggressiveness in isolation seems not to influence the outcome of staged contests (Seow and Turner, ms), dominance in a group aquarium nor access to food (Wyatt, Seow and Turner, ms). Evolutionary biologists have recently proposed that variation in sexually selected traits, including aggressiveness, may result from genetic constraints built up during a previous period of strong directional selection and thus may have no present functional significance (Pomiankowski and Møller, 1995; Turner, 1995). It is not known if aggressiveness has a heritable component or is entirely due to individual experience.

While guarding fry, female *Oreochromis* are also aggressively territorial. Female *Oreochromis* develop a characteristic fry-guarding pattern of dark stripes across the forehead, dark opercular region and chin and a dark, almost black, eye. This pattern, which occurs in *O. mossambicus*, *Oreochromis niloticus* and the Lake Malawi *O. (Nyasalapia)* species (personal observation) is adopted prior to the release of the free-swimming young. This is helpful to those rearing the fish in aquaria, as it indicates that the female may be removed and forced to eject her fry which can now be reared independently. *Oreochromis karomo* appears to be unusual in that females also develop bright breeding colours prior to spawning (Lowe, 1956).

Tilapia

Males and females co-operate in territorial defence in substrate-spawning species. Most observations have been made with *Tilapia (Coptodon) zillii* and *Tilapia (Pelmatotilapia) mariae* in captivity. Prior to pairing, both males and females fight with conspecifics and dominance hierarchies in both sexes play at least some part in determining access to mates in *T. mariae* (Turner *et al.*, 1989; Hulscher-Emeis, 1991). In a large (10 m) flow tank, *T. zillii* males formed loose dominance hierarchies with size only becoming important after the introduction of females (Neat, 1996). Female *T. zillii* in aquaria often begin territory establishment before males (Hulscher-Emeis, 1991), but this may be related to their maturing at smaller sizes and slightly younger ages than their male siblings. Sometimes pairs may jointly establish territories (Hulscher-Emeis, 1991). As with *O. mossambicus*, non-territorial fish engaging in prolonged contests adopt a paler colour pattern than those defending territories against intruders.

Fighting behaviour in both sexes of *T. mariae* and *T. zillii* is a much more violent affair than combat between male *Oreochromis*. Damaging bites are frequent. Instead of circling, most of the time during escalated contests is spent jawlocking, which probably acts as a direct trial of strength (Apfelbach and Leong, 1970; Turner, 1987) or body size (Neat, 1996; Neat *et al.*, 1998). Particularly when kept in small groups, substrate spawning *Tilapia* are far more likely to injure or kill conspecifics in aquaria than are *Oreochromis*.

Studies of *T. zillii* by Neat (1996) showed losers of contests suffered more scale loss than winners, especially if they were larger than their opponents. Fighting was shown to result in a significant depletion of sugar reserves from muscle and liver, and losers were found to have accumulated more lactate than winners, suggesting a higher cost incurred.

According to Schwanck (1980) dominance in dyadic contests between males of aquarium populations of *T. mariae* is influenced more by the length of the genital papilla, which may be an indication of readiness to reproduce, than by overall length or weight.

In *T. zillii*, Neat *et al.* (1998) found that relative total weight was less important in deciding the outcome of fights than male gonad weight. There was a trade-off between gonad weight and body fat and a weak relationship between gonad weight and mouth size. In addition, 11-ketotestosterone levels were found to be negatively related to gonadosomatic index but there was no significant difference between winners and losers. When smaller individuals won fights, they had escalated their level of aggression higher than their larger opponents. Winners and losers differed in mouth-locking behaviour, suggesting that this is related to intention to persist, but there was no statistical evidence that this persistence is related to gonad weight.

Sarotherodon

Sarotherodon melanotheron is the only known exclusively-paternal mouthbrooding tilapia. Since males are unavailable for mating during brood care, females should have a greater potential reproductive rate and thus compete for males. Aquarium experiments by Balshine-Earn and McAndrew (1995), showed that females were indeed more aggressive, biting, chasing and initiating mouth fights more often than males. Dominant females were more likely to gain access to mates than were dominant males. Denight and Ward (1982) found that the size of a male's black chin spot was related to his dominance status in aquaria.

Barlow and Green (1970) report that *S. melanotheron* is unusually peaceful for a cichlid fish. Aggressive behaviour between males and females mainly involved tailbeating, charging with an open mouth, chasing and occasionally ramming. Apfelbach and Leong (1970) observed that contests in *S. galilaeus* consisted mostly of displays, and found it difficult to determine a clear end to the fight.

2.4 COURTSHIP AND SPAWNING

All male *Oreochromis* construct spawning pits, the characteristics of which may vary with substrate and depth as well as species, and courtship will only begin once the territory is established. In contrast, *Tilapia* courtship usually begins before either sex has established a territory. Spawning may be in a pit, a series of tunnels or on the substrate. In *Sarotherodon*, pair-bonding also occurs before territory establishment and both sexes co-operate in digging and defending a spawning pit.

Oreochromis

The central focus of the territory of a male *Oreochromis* is the spawning pit. So far as is known all male *Oreochromis* construct spawning pits. In the crowded conditions of an aquaculture pond, male *O. mossambicus* pits may be so closely packed together that there is no space between them, and they assume a hexagonal shape, which permits the closest possible packing of any two dimensional geometric shape (May and Harvey, 1988). We have observed hexagonal close packing of pits by males of the non-tilapiine cichlid *Enantiopus melanogenys* in Lake Tanganyika, despite there being ample unoccupied habitat, but this has never been observed for any tilapiine in natural conditions.

Males of *O. mossambicus* and *O. niloticus*, among others, construct simple craters in the sand or mud. Males of most *Oreochromis (Nyasalapia)* species which have been studied construct a raised central platform in the middle of their crater nest, though *O. variabilis* and *O. karomo* may simply clear a flat area (Lowe, 1956). Within a species, nest size often increases with the size of male and with water depth, and nest forms may also differ with substrate. In *O. karomo*, variation in nest size at one depth is not related to the size of the male and Lowe (1956) concludes that it results from differences in excavation behaviour between males. Spawning sites are prepared using three different methods (singly or in combination). Pushing sediment to the side using the nose creates a small nest, fanning the area with pectoral fins clears a larger area and some fish pick up and eject sediment using the mouth. Variation in nest form may perhaps play a role in the establishment and maintenance of reproductive isolation between species (Lowe, 1956), as has been suggested for nest-building haplochromine cichlids (Stauffer *et al.*, 1993). Nest form may differ between closely related species (Fig. 2.2) and along with differences in male colour and shape, was one of the major characters used by Schwanck (1994) to justify his elevation to the specific level of *Oreochromis mweruensis* Trewavas, which had formerly been regarded as a subspecies of *O. macrochir* (Boulenger). *Oreochromis macrochir* constructs a cone-shaped central platform, while that of *O. mweruensis* is star-shaped when viewed from above.

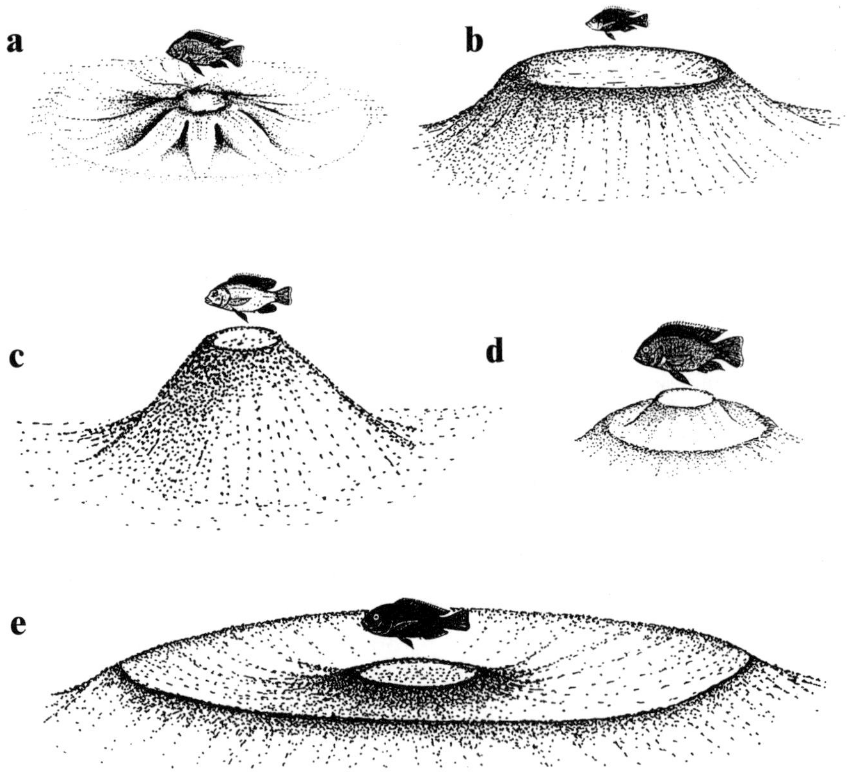

Fig. 2.2 Nest forms in *Oreochromis* species: a. *O. mweruensis*, b. *O. mossambicus*, c. *O. karomo*, d. *O. macrochir*, e. *O. lidole*. All drawings are to scale.

Lowe (1956) reports that the diameter of the spawning platform in *O. karomo* is much smaller than the body length of the male. Trewavas (1983) reports that even large *O. niloticus* (one of the largest species) build relatively small nests – with the diameter only about twice as great as the fish's length. The largest cichlid nest known was previously considered to be a crater 1.8 m in diameter dug by *O. mossambicus* in Lake Sibaya (Bruton, 1979). However, we have reported far larger pits occupied by *Oreochromis (Nyasalapia) lidole*, a species endemic to Lake Malawi. The largest was 3.1 m in diameter and 1.1 m deep with a flat central platform 75 cm across (Turner *et al.*, 1991). The sympatric sibling species *O. (Ny.) karongae* and *Oreochromis (Nyasalapia) squamipinnis* build smaller pits, but of similar shape.

Courtship behaviour has been most closely studied in *O. mossambicus*. Males in full black, white and red breeding dress swim up out of their spawning pits towards approaching females. Males adopt a head-down posture and close the unpaired and pelvic fins, while performing a high frequency low amplitude tail-quivering display. With the approach of the female to the nest, males often attempt to 'lead' the female to the nest, breaking away to swim towards the pit with exaggerated swimming movements. Returning to the female, the male generally performs a high amplitude tailbeating display, with its body in a horizontal orientation. If the female follows the male into the spawning pit, the male will circle round, initially in the head down posture, often leaning away from the female, but later, in a horizontal orientation skimming his ovipositor over the substrate. The ovipositor is white and very conspicuously contrasts with the male's black body colour. Receptive females generally circle with the male and perform similar skimming movements before spawning. The female lays a batch of eggs and immediately begins to take them into her mouth, while the male is fertilizing them.

Species differ in the details of their courtship displays: field studies indicate that *O. karomo* females are unusual in that they develop bright courtship colours and are apparently recognized by males immediately (Lowe, 1956), unlike *O. mossambicus* in captivity. Lowe records that both *O. karomo* and *O. variabilis* males lead the female straight to the nest, without any invitation displays. The females of these species do not perform any feeding or digging movements at the spawning site, and males perform 'nosewagging' displays instead of the tail-quivering of *O. mossambicus*. Spawning is rapid, lasting only a few minutes. Lowe (1956) suggests that these differences could be explained by a closer relationship between *O. mossambicus* and biparental nest-builders, than either of the other two species. *Oreochromis niloticus* consistently erects the spinous dorsal fin to a greater degree than does *O. mossambicus* (Baerends and Bolkzijl, 1963). Males of the *O. (Nyasalapia)* group have elaborate genital papillae up to 15 cm in length (Fig. 2.3). These are believed to serve as egg-mimics (Fryer and Iles, 1972). This process is taken a step further by *O. macrochir*, in which males produce long sticky sperm threads which are taken up into the female's mouth immediately after spawning (Fryer and Iles, 1972). These structures probably

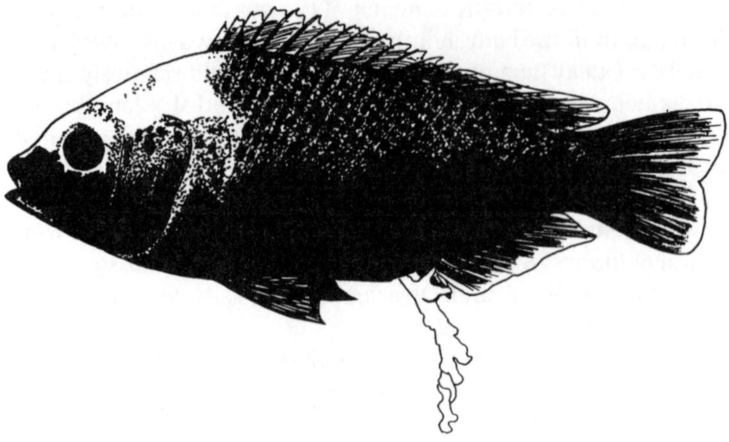

Fig. 2.3 Drawing of male *Oreochromis (Nyasalapia) squamipinnis* showing genital tassel.

evolved to increase fertilization rates of eggs, as has been proposed for the egg-spots on the anal fin of haplochromine cichlids. However, Hert (1989) found no difference between the fertilization rates of male haplochromine cichlids and those in which the egg-spots had been removed, but showed that females preferred to mate with males with egg-spots. Similar experiments have not been performed with *Oreochromis*, but it is possible that elaborate male genital papillae may also function in mate choice.

Tilapia

Males and females of substrate spawners tend to establish pair bonds gradually, unlike male maternal mouthbrooders which perform flamboyant and energetic courtship displays. Some early ethologists, particularly of the Lorenz school, thought that females had to perform a specific appeasement display towards males (Oehlert, 1958; Baldaccini, 1973), but others maintained that this was largely a laboratory artefact (Baerends and Baerends-van Roon, 1950). It is certainly not necessary for pairing in *T. mariae* (Turner *et al.*, 1989) and indeed Hulscher-Emeis (1991) has observed that courtship in *T. zillii* is generally initiated by the female permitting the male to remain in her territory. In both *T. mariae* and *T. zillii* courtship is often initiated while neither fish is in possession of a territory and mainly consists of the pair performing a 'greeting display' where, as it returns to its partner, a fish slows down and passes by at close range performing a slight undulation of the body (Turner *et al.*, 1989; Hulscher-Emeis,

1991). Pairing before territory establishment is common in biparental cichlid fishes – including *T. zillii* and *T. mariae* in the field (Schwank, 1987a; Hulscher-Emeis, 1991) – and it seems likely that a pair would have a much better chance of securing a space in a crowded breeding area than a single fish of either sex (McKaye, 1984). There are many reports of males of biparental cichlids establishing territories before attracting females (e.g. in *T. zillii*, Bruton and Gophen, 1992).

The types of territory defended by substrate spawners are variable both within and between species. Substrate spawning *Tilapia* may lay their eggs on hard surfaces such as rocks, tree roots or submerged vegetation, in a depression in a muddy or sandy substrate, or in a series of tunnels excavated by the parents. Eggs may be concealed or out in the open. There may be substantial variation within species, even in the same water body. In rocky areas with heavy wave action, *T. zillii* spawn directly onto rock surfaces in the open. In shallow sandy or muddy areas with heavy wave action, nests are simple saucer shaped pits, but in sheltered pools or lagoons, each pair excavates a complex of tunnels (Bruton and Gophen, 1992). Such behavioural flexibility is shared by *Tilapia rendalli* in Lake Malawi which may dig a simple saucer-shaped nest (Lowe, 1959), a complex of tunnels (Ribbink *et al.*, 1981) or spawn on a rock surface (Robinson, personal observation).

Sarotherodon

In the paternal mouthbrooder, *S. melanotheron*, females court more than males (Aronson, 1949; Heinrich, 1967). In contrast to the maternal mouthbrooders, courtship is protracted and may last hours or even days. Both sexes co-operate in digging and defending a spawning pit (Barlow and Green, 1970) and perform nodding, quivering, nipping (the substrate), and skimming behaviour.

Male and female *S. galilaeus* also co-operate in preparing the spawning site, which is a simple depression, either by clearing away vegetation or excavating a pit, but courtship in this species is brief (Fryer and Iles, 1972; Balshine-Earn, 1995). The female is said to be more active in courtship and nest building (Balshine-Earn, 1995). Territorial defence is weak (Balshine-Earn, 1995a). During the early stages of pair formation males have a darker dorsal fin.

2.5 PARENTAL CARE

All *Oreochromis* are female mouthbrooders and males play no part in parental care. During the time when the eggs are developing, females remain relatively inactive. Most *Oreochromis* females aggressively guard their free-swimming fry after first release and parental care may continue for several weeks. The eggs and young of substrate spawning cichlids, including *Tilapia*, are usually guarded by both parents, though the division of labour between the sexes varies.

Although nominally biparental, uniparental care often results from mate desertion and communal care has also been reported. In *Sarotherodon*, the sexes do not remain together during the brooding period and in contrast to *Oreochromis*, there does not appear to be any post-release care of fry. In *S. galilaeus* either or both parents may brood the eggs, but *S. melanotheron* is a paternal mouthbrooder.

Oreochromis

After spawning, female *Oreochromis* depart from the area of the male territories, carrying the eggs in their mouths. Perhaps this is simply to avoid harassment by the males, as, both in the aquarium and in the field, males of maternal mouthbrooding cichlids are particularly attentive in their courtship displays toward brooding females (personal observation). Brooding females generally form into small groups and are relatively inactive. As the young absorb their yolk sacs and require exogenous food, females move singly into shallow areas, to release and stand guard over clouds of fry. In Lake Malawi, fry are generally released in areas with dense weed cover. This seems to be typical for *Oreochromis* (Lowe-McConnell, 1959), but *O. variabilis* in Lake Victoria prefers shallow sloping bedrock (Fryer and Iles, 1972). Females become territorial and develop a characteristic fry-guarding pattern. When threatened, the fry swarm around the female's head and are admitted into her mouth. During this time, females often adopt a head-down posture which has been interpreted as a 'calling signal' (Fryer and Iles, 1972). However, fry seem to actively seek to enter the female's mouth, often before she adopts this posture, and in haplochromine cichlids fry may swarm all over the female's head for some time before she opens her mouth to admit them. As the fry-guarding period progresses, the female is less likely to respond to this behaviour (Robinson, 1995). In the laboratory, tilapiine fry will respond to stationary models and previous exposure to models of different patterns affects the preference of the fry, indicating that there is a learned component to recognition of the female's appearance (Russock, 1986).

Parental care, including guarding of free-swimming fry, generally lasts about 3 weeks (Trewavas, 1983). *Oreochromis niloticus* continues to guard fry until they attain around 14 mm total length (Trewavas, 1983), which seems to be a fairly typical size for most species. The Malawian endemics brood the young to much larger sizes. Trewavas (1983) cites 24 mm for *Oreochromis saka* (=*O. karongae*), and up to 58 mm for *O. lidole*. Parental care clearly lasts longer in these species, up to 6 weeks for *O. squamipinnis* (Trewavas, 1983). *Oreochromis grahami*, a dwarf species of soda lakes, seems to be unique among the genus in that fry are not permitted to return to the female's mouth after first release (Coe, 1966).

Tilapia

For *T. zillii* in areas with strong wave action (whether on rocky or sandy/muddy sites), parental care of free-swimming fry lasts no more than 2–3 days. However, in sheltered pools or lagoons, where pairs excavate a complex of tunnels and hollows, parental care is extended (Bruton and Gophen, 1992). Adults are fiercely territorial and may plug the entrance to the brood chambers with their bodies. Pairs nest in dense colonies, often in water so shallow that adults have to swim on their sides or jump out of the water in order to pass from the brood chamber to the open water. In captivity, parental care may last for 2 weeks or even longer (Beveridge, personal communication).

The eggs of substrate-spawning cichlids, including *Tilapia*, are usually guarded by both parents. Females typically perform most of the intimate care, fanning and cleaning eggs, while males generally remain at the periphery of the territory. In some pairs of *T. mariae*, however, sex roles are more uniform, either with both sexes staying close to the eggs, or with sexes regularly alternating position (Schwanck, 1987a). In aquaria, males tend to perform the majority of attacks on intruders during the egg and wriggler phases, but once the fry are free-swimming, the female's aggression increases and generally surpasses that of the male. In the field, however, it appears that females perform more brood defence than males at all stages (Schwanck, 1987a). Attacks are mostly directed towards conspecifics, particularly non-territorial individuals, and also to large predators, such as *Hemichromis fasciatus*. Smaller, less predatory species, such as *Alestes longipinnis* are generally ignored. Males generally direct a larger proportion of their attacks towards large fish.

Tilapia (Coptodon) discolor is unusual for a biparental substrate spawner. After finding males with eggs and young in the buccal cavity, Thys van den Audenarde (1970) concluded that this species was a paternal mouthbrooder. However, more recent observations both in captivity and in natural conditions (Whyte, 1975; Owusu-Frimpong, 1987) reveal that both parents guard the eggs on the substrate, but after a few days, they pick the eggs up in their mouths and transfer them to a series of narrow tunnels only a few centimetres wide and deep. The transfer often involves several trips by both parents. Although other *Tilapia* may transfer juveniles in their buccal cavity, *T. discolor* is thought to be alone in also transferring eggs. This species may represent a preliminary transitionary phase towards biparental mouthbrooding.

Although nominally biparental, uniparental care by substrate-spawning *Tilapia* often results from mate desertion. In Nigerian rivers, Schwanck (1987a) observed that 30% of broods of eggs were tended by single females, but only 10% of broods of fry. The reduction may be due to mortality of young, as 80% of all broods (of single- or two-parent families) disappeared per week. Not surprisingly more uniparental broods were lost than biparental. However, Schwanck also found that males sometimes returned to their broods after a period of absence ranging from 1 to 15 days. It is tempting to suggest that males were seeking opportunities for further matings, and perhaps, if unsuccessful, returning to invest in the original brood.

Communal care of young by two pairs of *T. rendalli* has been observed in Lake Malawi (Ribbink *et al.*, 1981). On one occasion two pairs of *T. mariae* in an aquarium were seen to initially defend separate territories, but fry regularly exchanged between the pairs, eventually leading to the formation of a single huge shoal which was defended by all four parents and also a single, apparently unmated male (Turner, personal observation). Schwanck (1987) observed a single male *T. mariae* participating in brood care with two females occupying adjacent territories.

Cannibalism of eggs occurs commonly in pairs kept in aquaria, although it is difficult to determine in the field, where the disappearance of a brood is generally regarded as resulting from predation by fish other than the parent. Schwanck (1986) found that cannibalism generally took place during the egg stage, and was almost always by males. Small spawns, less than 100 eggs, were invariably eaten, and males also tended to eat the eggs of females very much smaller than themselves. Pairings between fish of approximately equal size also tended to break down, resulting in egg eating.

Sarotherodon

While very little is known of the majority of species of *Sarotherodon* which are confined to remote areas of West Africa, two widely distributed species have been investigated in some detail. In contrast to *Oreochromis*, neither of these species practise post-release care of fry, although Brestowsky (1968) has shown that fry of *Tilapia heudeloti macrocephala (=S. melanotheron)* would respond to a parent if removed from the male's mouth and rendered able to swim by puncturing the yolk sac on the 9–10th day.

Sarotherodon melanotheron is the only cichlid species known in which males alone perform parental care, although this is the norm in most other care-providing fish taxa (Perrone and Zaret, 1979; Baylis, 1981; Gross and Sargent, 1985). Males seem to be able to feed to some extent during the brooding period (Oppenheimer and Barlow, 1968).

In *S. galilaeus* either or both sexes brood the eggs (Balshine-Earn, 1995b). The entire clutch is generally laid on the substrate before either partner takes the eggs into its mouth and eggs are weakly adhesive, in contrast to those of maternal brooders. The sexes do not remain together during the brooding period. We are not aware of any other animal species in which male and female separately care for part of a brood. Balshine-Earn (1995b) has shown that parental care is costly to both sexes. In males and females, growth was depressed and the interval between spawnings was increased. Females suffered the additional cost of lost fecundity. Fry are brooded until around 11 mm long, which takes 10–15 days depending on temperature (Fryer and Iles, 1972).

Although model parents are very effective in attracting fry of maternal mouthbrooders, *Sarotherodon* juveniles are only weakly attracted to them, probably as they are not generally permitted to return to their own parent's mouth once they have been released. Parents of *Tilapia multifasciata* (=*Sarotherodon galilaeus multifasciatus*, a subspecies confined to Ghana and the Ivory Coast according to Trewavas, 1983) are reported to remain together during the brooding period and jointly defend free-swimming fry (Apfelbach, cited in Fryer and Iles, 1972).

Other genera

The little-known genus *Tristramella* is confined to Israel and Syria, and reportedly includes both biparental and maternal mouthbrooders (Iles and Holden, 1969). In Lake Kinneret, *Tristramella simonis simonis* males have brilliant blue heads during the breeding season while females stay drab yellow with a few faint stripes on the flanks and a white lower lip (Balshine-Earn, pers. comm.). This degree of sexual dimorphism suggests the species is a maternal mouthbrooder. Ben-Tuvia (1960) suggests that previous reports of mouthbrooding males may be the result of misidentification of specimens of *Tristramella simonis intermedia*, which occasionally enters the lake. *Tristramella sacra* is also apparently a biparental mouthbrooder (Ben-Tuvia, 1960).

The monotypic genus *Danakilia* is confined to the hot springs and saline lakes of the Danakil depression in Ethiopia. It is thought to be a mouthbrooder, as is the equally obscure *Iranocichla hormuzensis* (Trewavas, 1983).

2.6 HABITAT USE DURING THE LIFE CYCLE

Substrate spawning *Tilapia* generally remain in shallow weedy areas throughout their lives, but maternal mouthbrooding *Oreochromis* demonstrate markedly different habitat preferences according to the stage of their life cycles (Lowe-McConnell, 1959). This is clearly demonstrated in Lake Malawi.

Females of all Malawian *Oreochromis* species release their young in shallow vegetated areas, such as stream mouths, reedy lagoons or pools. *Oreochromis (Oreochromis) shiranus* generally remains in shallow vegetated areas throughout its life, but the endemic *Oreochromis* (*N.*) species move more extensively. At around 7–10 cm total length, the young fish are more frequently seen on sandy and rocky shores, and particularly feeding on epiphytes from the surface of macrophytes such as *Potamogeton* and *Vallisneria*. Trawl surveys (Turner and Mwanyama, 1992) indicate that the modal size of fish progressively increases with increasing water depth – 9 cm at 15–20 m, 18 cm at 20–40 m, and 22 cm at 40–50 m. However, the majority of mature fish (25–35 cm) were found at 20–30 m, probably indicating an inshore movement associated with reproduction. Apart from size-related depth preferences, immatures do not seem to have particularly strong habitat preferences and can be found on all types of substrate, rocky, muddy or sandy. During the breeding season, adult male *O. (Ny.) karongae* excavate pits at depths of less than 1 m to more than 28 m (Turner *et al.*, 1991). Nesting grounds are known to occur in a variety of habitats, ranging from steeply shelving rocky shores, such as those at Domwe Island, to sheltered muddy areas, such as Lake Malombe. Lowe-McConnell (1959) had previously recorded *O. squamipinnis* from sandy areas at depths of 16–20 m, but we have also found this species on rocky (Turner *et al.*, 1991) and muddy (Turner and Mwanyama, 1992) habitats, and at depths of less than 4 m. Territories of *O. lidole* have never been recorded from shallower than 17 m depth (Turner *et al.*, 1991) and the maximum breeding depth of this species may lie below the depth of safe SCUBA diving. Its main breeding areas all lie to the north of Boadzulu Island, where the bottom rapidly drops to more than 50 m deep close to the shoreline. Remarkably, spent and brooding females of this species are numerous in Lake Malombe, where ripe fish are unknown. It thus seems probable that after spawning some females carry their broods at least 25 km before releasing their fry (Turner and Mwanyama, 1992).

Most other *Oreochromis* show similar, if less dramatic, movements between habitats. In both Lakes Malawi and Sibaya, *Oreochromis* tend to move into deeper water during the cool season and return to shallow waters during the warmer period (Bruton and Boltt, 1975, personal observation). In Lake Sibaya, subadult *O. mossambicus* dominated the shelf areas at 0–15 m depth, but were virtually absent from 15–40 m (Bruton, 1979). Male territories were mainly confined to depths of 0.5–3 m, maximum 8.5 m. *Oreochromis niloticus* prefers to nest at depths of 0.6–2 m including peripheral lagoons (Trewavas, 1983). Both species tend to release their young in marginal pools and vegetated areas.

2.7 BREEDING SEASONALITY

In many parts of the tropics, such as Lake Victoria, tilapiine cichlids appear to breed all year round, although there is often a peak in activity during the rainy season (Lowe, 1959; Trewavas, 1983). Lowe (1956) reports that individual *O. karomo* females can spawn several times per season. In Lake Sibaya in South Africa, breeding peaks in the warm rainy season, but extends over 7 months (Bruton and Boltt, 1975; Bruton, 1979). Rising temperature also seems to be the trigger for breeding in Lakes Kinneret and Chad, where tilapias begin spawning around April or May (Trewavas, 1983; Bruton and Gophen, 1992). In some species at least, females may spawn several times within the season (Lowe, 1956). Further information on seasonality is provided by Philippart and Ruwet (1982).

In many water bodies, all species – irrespective of parental care strategy – tend to spawn at the same time of year, indicating a common response to environmental triggers and perhaps common selective forces. In Lake Malawi, there are two principal breeding seasons for most cichlid fishes (Marsh *et al.*,

1986; Robinson, 1995), the end of the windy season (July–September) when upwelling nutrients have greatly increased primary productivity, and a smaller secondary peak associated with nutrient enrichment from run-off during the rainy season (February–March). Breeding in *Oreochromis lidole* starts at the end of the windy season, but peak activity is in the hot season (October–December) before the rains. *Oreochromis squamipinnis* peaks during the rains (January–March). *Oreochromis karongae* has a protracted breeding season, peaking during both windy and rainy seasons and is really only quiescent during April and May (Turner and Mwanyama, 1992). Such protracted breeding seasons are unusual and the explanation is unclear.

Schwank (1987b) found lunar cycles in the spawning periodicity of *T. mariae* in the Ethiop River in Nigeria, most clutches being laid shortly before the full moon. This may serve to reduce predation on broods by (i) synchronizing spawning and (ii) increasing the effectiveness of defence against nocturnal predators such as catfishes by the largely visual parental cichlid fishes.

2.8 EVOLUTION OF REPRODUCTIVE STRATEGIES

The breeding biology of the tilapias simultaneously exhibits both plasticity and conservatism. Plasticity is exemplified by the variation within and between species in parental roles, mating systems, choice of spawning site, reproductive seasonality and size at maturity. Males of substrate spawners can invest heavily in parental care, remain with the fry but leave most of the work to the female, share brood guarding duties with two different females, desert or even desert and subsequently return. The biparental *S. galilaeus* shows almost as much plasticity in parental roles. Plasticity is often evolutionarily advantageous, as it allows a flexible response to variations in environmental and ecological conditions. By maturing precociously in small ponds and lagoons which may dry out, tilapias maximize their fitness since thousands of tiny fry have a much better chance of surviving in a small puddle than a few large adults. They trade-off somatic growth for early maturation. In large water bodies, maturation at large size allows for increased growth rates and rapid attainment of high fecundity or competitive ability. Monogamous pairing ensures higher brood survival from biparental care, but mate desertion allows males to take advantage of excess local females, or to stop wasting energy in defending small broods. So flexibility in breeding strategies, as in diet and feeding structures, is a feature of tilapias.

However, as Barlow (1991) has noted, cichlid breeding systems have not been substantially altered by ecological conditions, to the extent that substrate spawners in Lake Tanganyika occupy similar niches to mouthbrooders both in the same lake and also in Lakes Malawi and Victoria. Why should this be so? First, it is worth considering the possibility that it is actually only the polygynous maternal mouthbrooding strategy that is conservative once evolved. Molecular phylogenetic techniques (see McAndrew, this volume) are now providing phylogenies which are obtained independently from morphological and behavioural traits, which may be strongly influenced by natural selection to adapt to local environmental conditions. Some consistent patterns are emerging. The maternal mouthbrooding *Oreochromis* seem to be a monophyletic group – in other words, maternal mouthbrooding may have only evolved once (or perhaps twice if some *Sarotherodon* are true maternal brooders) within the tilapias. Similarly, the haplochromines also appear to be monophyletic and all of them, so far as is known are polygynous maternal mouthbrooders. One or two other maternal mouthbrooding lineages may be represented in Lake Tanganyika (Kuwamara *et al.*, 1989; Sturmbauer and Meyer, 1993; Meyer, 1993). Overall there is as yet no convincing evidence that any polygynous maternal mouthbrooding lineage has ever given rise to a species which adopts any other parental strategy. Yet there are indications that polygamous maternal mouthbrooding has frequently evolved from substrate spawning, perhaps through the intermediate states of cave-brooding, biparental mouthbrooding or largely monogamous maternal brooding.

Many biologists think that the correct way to approach evolutionary problems is to compare the mean fitness of the population if all individuals adopt one strategy with that of another where they all adopt another strategy. This kind of rationale underlies questions like 'does mouthbrooding pre-adapt cichlids for living in lakes?' Often it is a good method for considering the organism's struggles against the environment or competing species, but there are some problems with this approach. Evolution is the replacement of one allele with another in a population, so strictly the correct way to think about evolutionary problems, particularly in social behaviour, is to consider the phase where one strategy is in the process of invading a population where all the other individuals adopt a different one. This way of thinking underpins the branch of theoretical biology known as evolutionary game theory (Maynard Smith, 1982) which would pose a question like 'once the whole population has adopted maternal mouthbrooding, is it possible for any set of circumstances to favour its replacement by any other strategy?' Adopting this approach, it becomes apparent that the progression from female- to male-mouthbrooding would be difficult to achieve. It would require females to desert their broods and simultaneously for males to switch to parental care. A male which is willing to provide parental care will have no advantage until brood-deserting females have evolved and vice versa. In contrast, a male of a biparental species may sometimes benefit from deserting his mate, particularly if mouthbrooding increases the survival of the young under single-parent care. The benefits of desertion are likely to be greatest for the males which are most attractive to females – they will have least trouble in finding another mate. Already this establishes the conditions for sexual selection where females become increasingly choosy about the appearance and competitive ability of males, while males become more exaggerated in their size and strength, and in courtship colours, structures and displays.

Although it has been suggested that biparental substrate-spawning is 'less well-adapted' ecologically than maternal mouthbrooding, this argument is unnecessary. Many species with each strategy exist and prosper in a variety of environments. It may simply be that biparental care easily evolves into maternal care, but not vice versa.

Another striking evolutionary generalization is that maternal mouthbrooding lineages seem to be particularly prone to speciation. This is obvious when considering the huge diversity of the haplochromine cichlid flocks in the African Great Lakes, but the pattern emerges even within the tilapias (Lowe-McConnell, 1959). It seems likely that sexual selection may lead to speciation if there are arbitrary changes in female mate preference based on male colour or nest form. This mechanism is well supported by theoretical studies (Turner and Burrows, 1995). It is notable that in haplochromine cichlids, females are often polymorphic for colour within a population, while males are not (Turner, 1994b). This suggests that females are choosy about male appearance, but males are less discriminating. Polymorphism in tilapias is rare in maternal mouthbrooders, but generally limited to females in *O. variabilis* (Lowe-McConnell, 1959). In substrate spawners, colour polymorphism in both sexes occurs, but as we have seen, seems to have little role in mate choice. This suggests that mate choice for other traits may override any tendencies for arbitrary preferences (e.g. based on colour) to develop. In substrate spawners, since both sexes are involved in parental care, males and females may be under strong selection pressure to choose large mates which may be better at establishing territories and defending fry. It is theoretically possible that size-assortative mating could lead to speciation.

2.9 IMPLICATIONS FOR EXPLOITATION

Knowledge of the reproductive behaviour of tilapias has several implications for the design of exploitation and conservation strategies. There has been concern that fishing techniques which involve the dragging of gears along the bottom, such as trawling and seining, may have detrimental effects on populations of cichlids by destroying their nests and killing breeding fish. Similarly, it has been suggested that the large, rapid fluctuations in water level that occur in reservoirs managed for hydro-electric power generation cause tilapia nests to be stranded and consequently, only small populations of tilapia can be supported. Substrate spawning species would indeed be adversely affected by the destruction of nests, since they are the focus of parental care, where both parents and offspring are vulnerable. The situation is likely to be very different in maternal mouthbrooders. In these polygamous, arena breeding forms, the vast majority of fish on breeding grounds are males, and it is probable that most never breed at all. A few successful males could easily fertilize the eggs of the vast majority of females. Thus, mass mortality of males, or destruction of their nests is unlikely to have a significant influence on population fecundity.

The use of nursery areas by tilapias is also a significant factor for consideration in exploitation and conservation. Elimination of lagoons and reedbeds is likely to increase mortality of juveniles. Fry are also vulnerable to exploitation with small-meshed beach seines in shallow water – this was almost certainly the cause of the catastrophic collapse of the *Oreochromis* fishery in Lake Malombe (FAO, 1993). Fluctuations in water level may either open up or eliminate shallow water nursery areas, with concomitant effects on tilapia recruitment (Kolding, 1993).

In aquaculture, total fry production of substrate spawners is limited by the availability of space for territorial defence, but parents will defend their young until they reach a size where predation by conspecifics is no longer a risk. In contrast, even with limited space for male territories, all female maternal mouthbrooders will have a chance to breed (Philippart and Ruwet, 1982).

Precocious maturation of tilapias is perhaps their single most undesirable feature with regard to aquaculture. Surprisingly, there has been very little research into the ecological basis of this trait (discussed by Lowe, 1959; Fryer and Iles, 1969; Welcomme, 1970). Perhaps a better appreciation of its evolutionary role in tilapia life histories may lead to an understanding of how it may be inhibited in pond conditions. It appears that seasonal fluctuations in temperature tend to limit tilapia reproduction to a short period of the year, and this allows for a longer growing period which leads to a larger size at maturity (Philippart and Ruwet, 1982).

2.10 CONCLUSIONS

The major breeding strategies of tilapias are biparental substrate spawning (*Tilapia*) and maternal mouthbrooding (*Oreochromis*), but paternal and biparental mouthbrooding are known (*Sarotherodon*). These broad divisions influence much of the rest of the reproductive biology of the tilapias – nest building, habitat use, monogamy or polygamy, sexual dimorphism, and speciation rates. These in turn have implications for their exploitation and conservation. Despite the renowned plasticity of tilapias, maternal mouthbrooding appears to be a conservative strategy, and it is suggested that this is simply due to the ease of evolving into the strategy and the difficulty in evolving out of it, and not due to the benefits of any ecological superiority.

ACKNOWLEDGEMENTS

Thanks to Sigal Balshine-Earn and Francis Neat for sending unpublished information.

REFERENCES

Apfelbach, R. and Leong, D. (1970) Zum Kampfverhalten in der Gattung *Tilapia* (Pisces, Cichlidae). *Z. Tierpsychol.* **27**, 98–107.

Aronson, L.A. (1949) An analysis of the reproductive behaviour in the mouthbreeding cichlid fish *Tilapia macrocephala* (Bleeker). *Zoologica* **34**, 133–158.

Baerends, G.P. and Baerends-van Roon, J.M. (1950) An introduction to the ethology of the cichlid fishes. *Behaviour* **1**(Supplement), 1–242.

Baerends, G.P. and Blokzijl, G.J. (1963) Gedanken über das Entstehen von Formdivergenzen zwischen homologen Signalhandlungen verwandter Arten. *Z. Tierpsychol.* **20**, 517–528.

Baldaccini, E.O. (1973) An ethological study of the reproductive behaviour including colour patterns of the cichlid fish *Tilapia mariae. Monit. Zool. Ital.* **7**, 247–290.

Balshine-Earn, S. (1995a) *The Evolution of Parental Care in Cichlid Fishes*, PhD thesis, University of Cambridge.

Balshine-Earn, S. (1995b) The costs of parental care in Galilee St. Peter's fish, *Sarotherodon galilaeus. Anim. Behav.* **50**, 1–7.

Balshine-Earn, S. and McAndrew, B.J. (1995) Sex-role reversal in the black-chinned tilapia, *Sarotherodon melanotheron* (Ruppel) (Cichlidae). *Behaviour* **132**, 861–874.

Balthazart, J. (1973) Analyse factorielle du comportement agonistique chez *Tilapia macrochir* (Boulenger, 1912). *Behaviour* **46**, 37–72.

Barlow, G.W. (1991) Mating systems among cichlid fishes, in *Cichlid Fishes* (ed. M.H.A. Keenleyside), Chapman and Hall, London, pp. 173–190.

Barlow, G.W. and Green, R.F. (1970) The problems of appeasement and of sexual roles in the courtship of behaviour of the blackchin mouthbreeder, *Tilapia melanotheron. Behaviour* **36**, 84–115.

Baylis, J.R. (1981) The evolution of parental care in fishes, with reference to Darwin's rule of male sexual selection. *Env. Biol. Fish.* **6**, 223–251.

Ben-Tuvia, A. (1960) The biology of the cichlid fishes of Lakes Tiberias and Huleh. *Bull. Sea Fish Res. Stn. Israel* **27**, 153–188.

Brestowsky, M. (1968) Vergleichende Untersuchungen zur Elternbindung von *Tilapia*-Jungfischen (Cichlidae, Pisces). *Z. Tierpsychol.* **25**, 761–828.

Bruton, M.N. (1979) The fishes of Lake Sibaya, in *Lake Sibaya* (ed. B.K. Allinson), Monographiae Biologicae 36, pp. 162–245.

Bruton, M.N. and Boltt, R.E. (1975) Aspects of the biology of *Tilapia mossambica* Peters (Pisces: Cichlidae) in a natural freshwater lake (Lake Sibaya, South Africa). *J. Fish Biol.* **7**, 423–445.

Bruton, M.N. and Gophen, M. (1992) The effect of environmental factors on the nesting and courtship behaviour of *Tilapia zillii* in Lake Kinneret (Israel). *Hydrobiologica* **239**, 171–178.

Coe, M.J. (1966) The biology of *Tilapia grahami* Boulenger in Lake Magadi, Kenya. *Acta Tropica* **23**, 146–177.

Denight, M.L. and Ward, J.A. (1982) Relationship of chin spot size to dominance in the black-chinned mouthbrooding cichlid fish (*Sarotherodon melanotheron*). *Anim. Behav.* **30**, 1099–1104.

De Silva, S.S. and Sirisena, H.K.G. (1988) Observations on the nesting habits of *Oreochromis mossambicus* (Peters) (Pisces: Cichlidae) in Sri Lankan reservoirs. *J. Fish Biol.* **33**, 689–696.

FAO (1993). *Fisheries Management in South-east Lake Malawi, the Upper Shire River and Lake Malombe.* CIFA Technical Paper 17, 106 pp, FAO, Rome.

Fryer, G. and Iles, T.D. (1969) Alternative routes to evolutionary success as exhibited by African cichlids fishes of the genus *Tilapia* and the species flocks of the great lakes. *Evolution* **23**, 359–369.

Fryer, G. and Iles, T.D. (1972) *The Cichlid Fishes of the Great Lakes of Africa*, Oliver and Boyd, Edinbugh.

Gross, M.R. and Sargent, R.C. (1985) The evolution of male and female parental care in fishes. *Amer. Zool.* **25**, 807–822.

Heinrich, W. (1967) Untersuchungen zum Sexualverhalten in der Gattung *Tilapia* (Cichlidae, Teleostei) und bei Artbastarden. *Z. Tierpsychol.* **24**, 684–754.

Hert, E. (1989) The function of egg-spots in an African mouth-brooding cichlid fish. *Anim. Behav.* **37**, 726–732.

Hulscher-Emeis, T.M. (1991) *A Causal Ethological Analysis of the Display of Variable Colour Patterns in* Tilapia zillii (*Cichlidae*), PhD thesis, University of Groningen.

Iles, T.D. and Holden, M.J. (1969) Bi-parental mouthbrooding in *Tilapia galilaea* (Pisces, Cichlidae). *J. Zool. Lond.* **158**, 327–333.

Kellogg, K.A., Markert, J.A., Stauffer, J.R. and Kocher, J.D. (1995) Microsatellite variation demonstrates multiple paternity in lekking cichlid fishes from lake Malawi, Africa. *Proc. Roy Soc. Lond. B.* **260**, 79–84.

Kolding, J. (1993) Population dynamics and life history styles of Nile tilapia (*Oreochromis niloticus*) in Ferguson's Gulf, Lake Turkana, Kenya. *Env. Biol. Fish* **37**, 25–46.

Kuwamara, T. Nagishi, M. and Sato, T. (1989) Female to male shift of mouthbrooding in a cichlid fish *Tanganicodus irsacae*, with notes on breeding habits of two related species in Lake Tanganyika. *Env. Biol. Fish* **24**, 187–198.

Lanzing, W.J.R. and Bower, C.C. (1974) Development of colour pattern in relation to behaviour in *Tilapia mossambica* (Peters). *J. Fish Biol.* **6**, 29–42.

Loiselle, P.V. and Barlow, G.W. (1979) Do fishes lek like birds? in *Contrasts in Behaviour* (eds E. Reese and F.J. Lighter), Wiley, New York, pp. 31–75.

Lowe, R.H. (1956) The breeding behaviour of *Tilapia* species (Pisces; Cichlidae) in natural waters: Observations on *T. karomo* Poll and *T. variabilis* Boulenger. *Behaviour* **IX**, 2–3.

Lowe, R.H. (1959) Breeding behaviour patterns and ecological differences between *Tilapia* species and their significance for evolution within the genus *Tilapia* (Pisces: Cichlidae). *Proc. Zool. Soc. Lond.* **132**, 1–30.

Marsh, B.A., Marsh, A.C. and Ribbink, A.J. (1986). Reproductive seasonality in a group of rock-frequenting cichlid fishes in Lake Malawi. *J. Zool. London (A)* **209**, 9–20

May, R.M. and Harvey, P.H. (1988) Tampering with territories. *Nature* **335**, 668–669.

Maynard Smith, J. (1982) *Evolution and the Theory of Games*, Cambridge University Press, Cambridge, UK.

McKaye, K.R. (1984) Behavioural aspects of cichlid reproductive strategies: patterns of territory defence in Central American substrate spawners and African mouthbrooders, in *Fish Reproduction: Strategies and Tactics* (eds G.W. Potts and R.J. Wootton), Academic Press, London, pp. 245–273.

McKaye, K.R. (1986) Mate choice and size assortative pairing by the cichlid fishes of Lake Jiloa, Nicaragua. *J. Fish Biol.* **29**(Supplement A), 135–150.

McKaye, K.R. (1991) Sexual selection and the evolution of the cichlid fishes of Lake Malawi, Africa, in *Cichlid Fishes* (ed. M.H.A. Keenleyside), Chapman and Hall, London, pp. 241–257.

Meyer, A. (1993) Phylogenetic relationships and evolutionary processes in East African cichlid fishes. *Trends Ecol. Evol.* **8**, 279–284.

Neat, F.C. (1996) *Behavioural and Physiological Studies of Fighting in Male* Tilapia zillii (*Cichlidae*), PhD thesis, University of Glasgow.

Neat, F.C., Huntingford, F.A. and Beveridge, M.C.M. (1998) Fighting and assessment in male cichlid fish: the effects of asymmetries in gonadal state and body size. *Anim. Behav.* **55**, 883–891.

Neil, E.H. (1964) An analysis of color changes and social behavior in *Tilapia mossambica*. *Univ. Calif. Publ. Zool.* **75**, 1–58.

Nelson, C.M. (1995) Male size, spawning pit size and female mate choice in a lekking cichlid fish. *Anim. Behav.* **50**, 1587–1599.

Oehlert, B. (1958) Kampf und Paarbildung eineger Cichliden. *Z. Tierpsychol.* **15**, 141–174.

Oliveira, R.F. and Almada, V.C. (1996) On the (in)stability of dominance hierarchies in the cichlid fish *Oreochromis mossambicus*. *Agg. Beh.* **22**(1), 37–45.

Oppenheimer, J.R. and Barlow, G.W. (1968) Dynamics of parental behavior in the black-chinned mouthbreeder, *Tilapia melanotheron* (Pisces, Cichlidae). *Z. Tierpsychol.* **25**, 889–914.

Owusu-Frimpong, M. (1987) Breeding behavioural patterns of the lake fish *Tilapia (Coptodon) discolor* Günther (Teleostei, Cichlidae). *J. Fish Biol.* **30**, 1–5.

Perrone, M. and Zaret, T.M. (1979) Parental care patterns of fishes. *Amer. Natur.* **113**, 351–361.

Philippart, J.-Cl. and Ruwet, J.-Cl. (1982) Ecology and distribution of tilapias, in *The Biology and Culture of Tilapias* (eds R.S.V. Pullin and R.H. Lowe-McConnell), ICLARM, Manila, pp. 15–59.

Pomiankowski, A. and Møller, A.P. (1995). A resolution of the lek paradox. *Proc. Roy. Soc. Lond. B.* **260**, 21–29.

Ribbink, A.J., Marsh, A.C. and Marsh, B.A. (1981) Nest-building and communal care of young by *Tilapia rendalli* Dumeril (Pisces: Cichlidae) in Lake Malawi. *Env. Biol. Fish* **6**, 219–222.

Ribbink, A.J., Greenwood, P.H., Ribbink, A.C. Twentyman-Jones, V. and van Zyl, B.J. (1991). Unique polychromatism of *Tilapia guinasana*, an African cichlid fish. *S. Afr. J. Sci.* **87**, 608–611.

Robinson, R.L. (1995) *The Dynamics of Space Use in Some Lake Malawi Fishes*, PhD thesis, Rhodes University, South Africa.

Russock, H.I. (1986) Preferential behaviour of *Sarotherodon (Oreochromis) mossambicus* (Pisces: Cichlidae) fry to maternal models and its relevance to the concept of imprinting. *Behaviour* **96**, 304–321.

Ruwet, J.-C. (1963) Observations sur le comportement de *Tilapia macrochir* Blgr. (Pisces: Cichlidae) au lac de retenue de la Lufira (Katanga). *Behaviour* **20**, 242–250.

Schwanck, E. (1980) The effect of size and hormonal state on the establishment of dominance in young males of *Tilapia mariae* (Pisces; Cichlidae). *Behavioural Processes* **5**, 45–53.

Schwanck, E. (1986) Filial cannibalism in *Tilapia mariae*. *J. Appl. Ichthyol.* **2**, 65–74.

Schwanck, E. (1987a) *Reproductive Behaviour of a Monogamous Cichlid Fish* Tilapia mariae, PhD thesis, University of Stockholm.

Schwanck, E. (1987b) Lunar periodicity in the spawning of *Tilapia mariae* in the Ethiop River, Nigeria. *J. Fish Biol.* **30**, 533–537.

Schwanck, E. (1994) Behaviour and colour differences between *Oreochromis macrochir* and *O. mweruensis* (Teleostei: Cichlidae). *Ichthyol. Explor. Freshwaters* **5**, 267–280.

Shaw, E.S. and Aronson, L.R. (1958) Oral incubation in *Tilapia macrocephala*. *Bull. Amer. Mus. Nat. Hist.* **103**, 375–416.

Stauffer, J.R., LoVullo, T.J. and McKaye, K.R. (1993) Three new sand-dwelling cichlids from Lake Malawi, Africa, with a discussion of the status of the genus *Copadichromis* (Teleostei: Cichlidae). *Copeia* **1993**, 1017–1027.

Stiassny, M.L.J., Schliewen, U.K. and Dominey, W.J. (1992). A new species flock of cichlid fishes from Lake Bermin, Cameroon with a description of eight new species of *Tilapia* (Labroidei: Cichlidae). *Ichthyol. Explor. Freshwaters* **3**, 311–346.

Sturmbauer, C. and Meyer, A. (1993) Mitochondrial phylogeny of the endemic mouthbrooding lineages of cichlid fishes from Lake Tanganyika in eastern Africa. *Mol. Biol. Evol.* **10**, 751–768.

Thys Van den Audenaerde, D.F.E. (1970) The paternal mouthbrooding habit of *Tilapia (Coptodon) discolor* and its special significance. *Rev. Zool. Bot. Afr.* **LXXXII**, 3–4.

Trewavas, E. (1983) *Tilapiine fishes of the genera* Sarotherodon, Oreochromis *and* Danakilia. Brit. Mus. Nat. Hist. Publ., London.

Turner, G.F. (1986a). Teleost mating systems and strategies, in *The Behaviour of Teleost Fishes* (ed. T.J. Pitcher), Croom Helm, Beckenham, pp. 253–274.

Turner, G.F. (1986b) Territory dynamics and the cost of reproduction in a captive population of the colonial nesting mouthbrooder *Oreochromis mossambicus* (Peters). *J. Fish Biol.* **29**, *573–587.*

Turner, G.F. (1987) *Display and Deceit in Cichlid Fishes*, PhD thesis, University of Wales, Bangor.

Turner, G.F. (1994a) The fighting tactics of male mouthbrooding cichlids: the effects of size and residency. *Anim. Behav.* **47**, 655–662.

Turner, G.F. (1994b) Speciation mechanisms in Lake Malawi cichlids: a critical review. *Advances in Limnology (Arch. Hydrobiol. Beih. Ergebn. Limnol.)* **44**, 139–160.

Turner, G.F. (1995). The lek paradox resolved? *Trends Ecol. Evol.* **10**, 473–474.

Turner, G.F. and Burrows, M.T. (1995). A model of sympatric speciation by sexual selection. *Proceedings of the Royal Society B* **260**, 287–292.

Turner, G.F. and Huntingford, F.A. (1986) A problem for game theory: assessment and intention in male mouthbrooder contests. *Anim. Behav.* **34**, 961–970.

Turner, G.F. and Mwanyama, N.C. (1992). *Distribution and Biology of Chambo* (Oreochromis spp.) *in Lakes Malawi and Malombe*. FI:DP/MLW/86/013. Field Document 21, 23 pp.

Turner, G.F., Phiri, L.M. and Cawthraw, S. (1989) Partner attack inhibition in the sexually monomorphic biparental cichlid *Tilapia mariae*. *Behaviour* **109**, 1–19.

Turner, G.F., Witimani, J., Robinson, R.L. Grimm, A.S. and Pitcher, T.J. (1991) Reproductive isolation and nest sites of Lake Malawi chambo, *Oreochromis (Nyasalapia)* spp. *J. Fish Biol.* **39**, 775–782.

Welcomme, R.L. (1970) Studies on the effects of abnormally high water levels on the ecology of fish in certain shallow regions of Lake Victoria. *J. Zool. Lond.* **160**, 405–436.

Whyte, S.A. (1975) Distribution, trophic relationships and breeding habits of the fish populations in a tropical lake basin (Lake Bosomtwe, Ghana). *J. Zool. Lond.* **177**, 25–26.

Chapter three

Diet, feeding and digestive physiology

M. C. M. Beveridge and D. J. Baird

3.1 INTRODUCTION

This chapter considers feeding mechanisms, diet and trophic efficiencies in tilapias. It is intended to complement Chapter 5, which examines the role of tilapias in ecosystems, and Chapter 9, which is concerned with nutritional requirements and feeds.

The chapter begins with a brief review of the dietary preferences of tilapias; following this, the behavioural and morphological adaptations associated with feeding and quantification of ingestion rates and assimilation efficiencies are considered. The chapter concludes with a discussion of how information from feeding and dietary preference studies might be used to increase tilapia production in aquaculture.

3.2 DIET AND FEEDING ECOLOGY

The young of teleosts, including herbivores, are typically carnivorous (Lazzaro, 1987; Gerking, 1994), and tilapias are no exception. While juvenile stages (i.e. 25–40 mm standard length (SL)) of microphagous *O. aureus*, *O. mossambicus* and *O. niloticus* are demonstrably capable of ingesting planktonic algae (Trewavas, 1983; McDonald, 1985a,b; de Moor *et al.*, 1986; Northcott *et al.*, 1991), young fish show a preference for zooplankton which are captured by active visual feeding (Le Roux, 1956; Bowen, 1982; Yowell and Vinyard, 1993). Laboratory studies indicate that a distinct ontogenic shift from visual, particle capture to

M.C.M. Beveridge and B.J. McAndrew (eds), Tilapias: Biology and Exploitation, 59–87.

filter (suction) feeding occurs when fish are 6-7 cm SL. This is consistent with field studies of dietary preferences in a range of tilapias (Moriarty *et al.*, 1973; Spataru and Zorn, 1978; Gophen, 1980; Bowen, 1982; Gophen *et al.*, 1983; Trewavas, 1983).

Ontogenic changes in feeding resulting in a shift from zooplanktivory towards the adoption of a largely plant (phytoplankton)-based diet are common in planktivorous fishes (Lazzaro, 1987). Ontogenic dietary shifts, from plankton to aquatic plants, have also been reported in the macrophyte-feeding *T. rendalli* (Bassay *et al.*, 1997). Explanations for the change in diet with body size have emphasized developmental changes in the abilities of the fish to capture and process phytoplankton – i.e. buccopharyngeal mucus cell numbers and activity and digestive tract morphology (Bowen, 1982; Drenner *et al.*, 1984; Northcott and Beveridge, 1988). However, prey type is important in determining net energy return associated with different feeding strategies. The model of Yowell and Vinyard (1993) predicts that *O. aureus* should switch from particulate feeding to filter feeding when between 4 and 7 cm (SL), depending upon plankton density and size, supporting field and laboratory observations (see also Lazzaro, 1991).

Diet of adult tilapias

The characteristic diet of adult tilapias is plant material and/or detritus of largely plant origin. Phytoplankton, benthic algae, macrophytes and periphyton are utilized to a greater or lesser extent by all species, although *Oreochromis* species are principally microphagous while *Tilapia* species are characteristically macrophyte feeders (Caulton, 1977; Lowe-McConnell, 1982). Although *Tilapia* spp. appear to be able to filter-feed on phytoplankton as efficiently as *Oreochromis* spp., at least when small (Ndonna, 1996), with the exceptions of *T. kottae* and *T. mariae* in hypereutrophic Lake Barombi Kotto (Trewavas, 1983), adults rarely subsist on phytoplankton. Zooplankton are also important dietary components in some phytoplankton-feeding species, such as *O. lidole*, which occurs in Lake Malawi (Hartnell, 1996).

Such generalizations apart, however, it must be recognized that tilapias are highly opportunistic feeders (see also Lowe-McConnell, this volume). The realized feeding niche varies with time and location and even with sex, female mouthbrooders ceasing feeding while incubating eggs and young (Hodgkiss and Man, 1977). Adult *T. rendalli* from African and Brazilian lakes and reservoirs have been shown to ingest a wide range of prey items: macrophytes, phytoplankton, zooplankton, insect larvae, fish eggs, and embryos and detritus (Caulton, 1976, 1977a,b; Denny *et al.*, 1978; Lazzaro, 1991). Studies of *O. mossambicus* in Sri Lankan reservoirs have demonstrated that there is a great deal of intraspecific variation in diet between environments and over time (De Silva *et al.*, 1984; Maitipe and De Silva, 1985). Thus, detritus was absent in the stomach contents of populations where phytoplankton was abundant while zooplankton was the preferred food in reservoirs where phytoplankton was less abundant. Marked seasonal variations in diet were also observed, fish capitalizing on the changing abundance of food resources. Hence, detrivory predominated during the rainy season, phytophagy during the dry season. While several other studies confirm seasonality in diets (Spataru, 1976; Spataru and Zorn, 1978; de Moor *et al.*, 1986; Arthrington and Bluhdorn, 1994), others show little variation (Hodgkiss and Man, 1977; Opiyo and Dadzie, 1994; Redding, 1997).

The tilapias have evolved to exploit well-defined seasonal cycles. Ribbink (1991) has suggested that, due to seasonal water level fluctuations, African riverine fish populations may be subjected to greater levels of inter- and intraspecific competition for food than lake-dwelling populations, promoting opportunistic feeding behaviour. Ultimately, availability and competition determine food choice. The role of competition has been studied by Turner *et al.* (1991) in Lake Malawi, who noted that although there was a great deal of dietary overlap among tilapia species, it was clear that different species exploited different microhabitats for food.

3.3 FOOD CAPTURE AND PROCESSING

Liem (1991) distinguishes three methods of food capture among cichlids:

(i) *suction or pump feeding*, in which prey is sucked into the buccopharyngeal cavity by the stationary animal;
(ii) *ram feeding*, in which the fish opens its mouth and overtakes its prey;
(iii) *biting*, in which the teeth of the upper and lower jaws are used to capture food.

Suction and ram feeding are associated with planktivory. However, Lazzaro (1987, 1991) and Sanderson and Wassersug (1993), among others, consider there to be another method of feeding on plankton: visual feeding. This differs from ram feeding in that it involves location and selection of individual prey items. Biting may be sub-divided into grazing and browsing behaviours. The former involves the combing and ingestion by sucking of unicellular and filamentous algae from rocks, the ingestion of fine sand and silt often occurring incidentally. The latter term is used to describe the biting off and ingestion of macrophytes and benthic algae, usually without the ingestion of any incidental material (Lazzaro, 1987; Yamaoka, 1991).

Although suction feeding, visual feeding and biting and grazing behaviours occur in tilapias, associated morphological specializations are limited in comparison to the haplochromine cichlids (Fryer and Iles, 1972; Bowen, 1982; Liem, 1991; Goldschmidt, 1996). There is little interspecific variation in head profiles or in the shape of the buccal cavities, at least among *Oreochromis* spp. In terms of functional design, the mouths of *Oreochromis* spp. are intermediate in size between cichlids that specialize in visual feeding (large) and those that specialize in suction feeding (small). Liem (1991) and others have used the shape of the cichlid buccal cavity at rest to indicate feeding mode. The buccal cavity of tilapias most resembles that of the combined ram (cylindrical; low suction pressure) and suction (attenuated, truncated cone; capable of generating very high negative pressures) feeder (Fig. 3.1b) although the suction component perhaps dominates the design (K. Liem, personal communication).

Suction (filter) feeding

Both suction and visual feeding are utilized in the ingestion of larger, crustacean zooplankton while suction feeding alone is employed in the ingestion of free-living bacteria and phytoplankton and small zooplankton (rotifers, etc.). Observations show that fish tend to remain almost stationary during suction feeding, but that the volumes of water passed over the gills, as estimated from opercular beat rate measurements, greatly increase (Northcott *et al.*, 1991). Sanderson *et al.* (1996) describe the process of suction feeding as consisting of a repetitive series of suctions, during which water enters the mouth and exits via the opercular slits. A typical bout of suspension feeding consists of 30–40 seconds of pumping, followed by 5 seconds of buccopharyngeal movements that are probably related to prey processing. Assuming an entrapment efficiency of 100%, Hartnell (1996) estimated filtration rates during filter feeding of around 100–500 ml g body weight^{-1} h^{-1} for Lake Malawi *Oreochromis* species (*O. karongae, O. lidole*) feeding on *Anabaena* and *Ankistrodesmus*. Estimates by Drenner *et al.* (1987) and Lazzaro (1991) for *T. rendalli* filter feeding on zooplankton suggest a figure of around 130 ml g body weight^{-1} h^{-1}. However, given that the efficiency of food entrapment is likely to be considerably less than 100%, these values must be seen as the lower limits of water exchange.

While gill rakers are more numerous among microphagous tilapia species, they are small and the inter-gill raker spaces (around 40 μm in fish of 14 mm SL; 110 μm in fish of 70 mm SL) are too large to entrap bacteria or small algal cells (Fig. 3.2; see also Fig. 4.2). In a definitive experiment, Drenner *et al.* (1987) surgically removed the gill rakers from tilapias to discover that it made no difference to phytoplankton ingestion rates (Fig. 3.3). Microbranchiospines are small (100–200 μm long) bony, rib-like structures bearing fine lateral spines (20–40 μm) that occur in a single row on both faces of the second, third and fourth gill arches (Gosse, 1956; Beveridge *et al.*, 1988a) (Fig. 3.2). They are peculiar to the tilapias, occurring in all species except members of the genus *Danikilia* and the sub-genera *Alcolapia* and *Vallicola* (Trewavas, 1983). In view of the gap (0–25 μm) between adjacent spines, it had been suggested that they may be important in filter feeding (Gosse, 1956; Whitehead, 1959; Fryer and Iles, 1972; Campbell, 1981; Drenner *et al.*, 1984). However, recent studies have shown that these play little part in food entrapment (Drenner *et al.*, 1987; Beveridge *et al.*, 1988b).

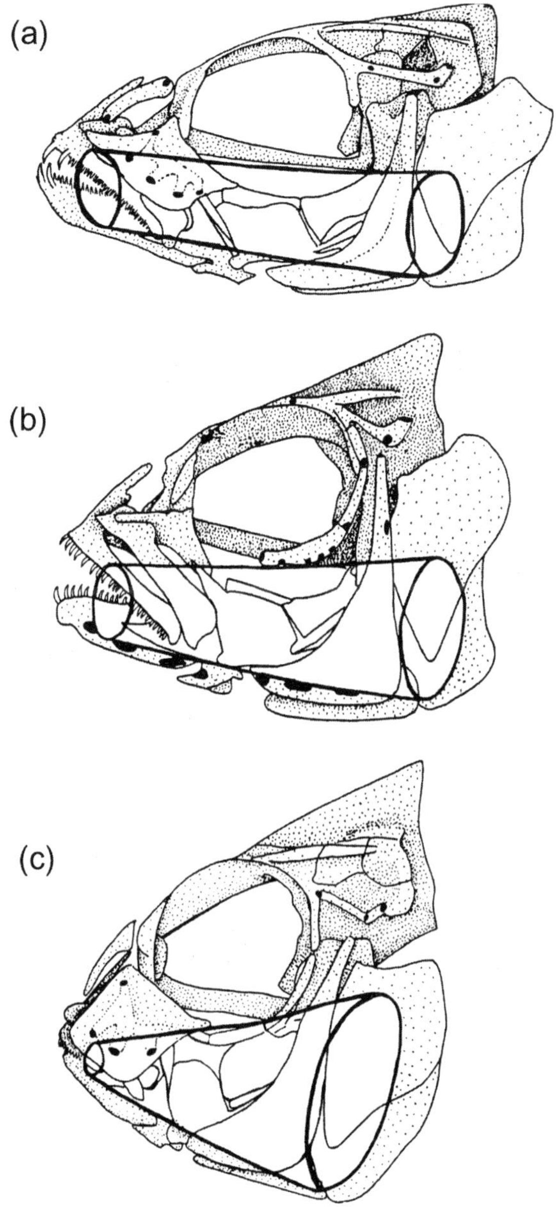

Fig. 3.1 Models of buccal cavity designs in cichlids, superimposed in heavy lines on drawings of the skull. (a) Model of a typical ram feeder, e.g. *Lamprologus elongatus*, with a cylindrically shaped buccal cavity. (b) Model of a combined ram and suction feeder, e.g. by *Grammatoria lemairei*. (c) Model of a typical suction feeder, e.g. *Tylochromis microlepis*. The buccal cavity design of tilapias most resembles that of (b). From Liem (1991) (fig. 6.1), reproduced with kind permission from Kluwer Academic Publishers.

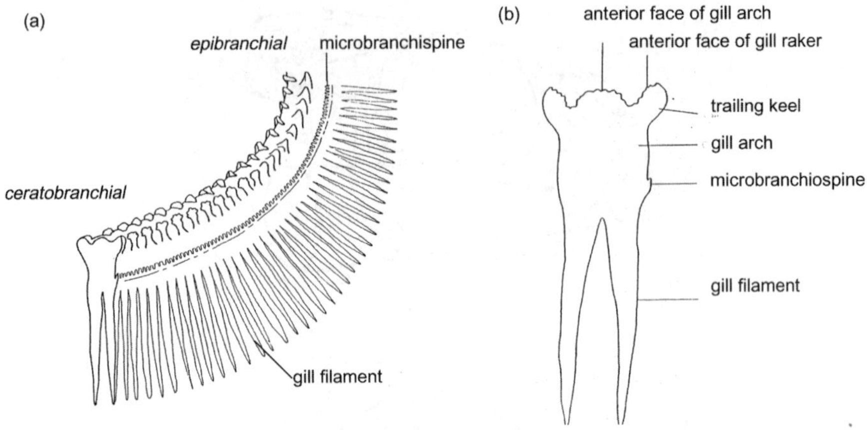

Fig. 3.2 (a) Diagram showing the general structure of a gill arch from the tilapia *O. niloticus*. (b) Diagram of transverse section through a gill arch. Modified from Northcott and Beveridge (1988).

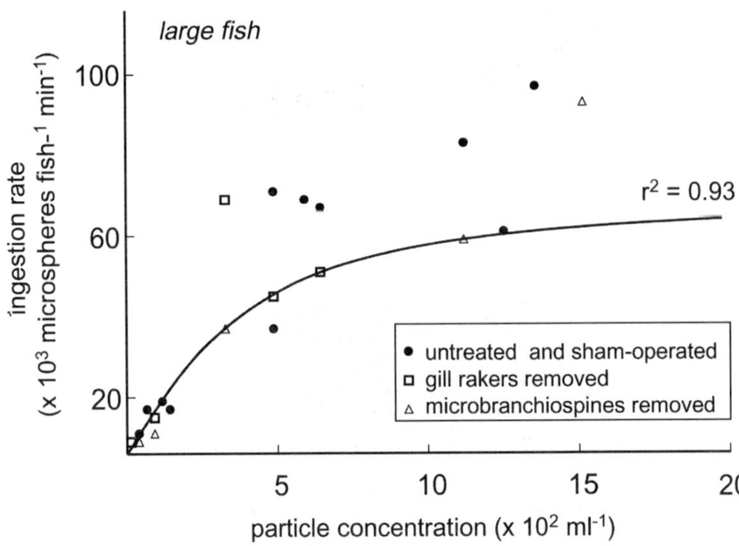

Fig. 3.3 Ingestion of microspheres by 12.6–14.3 cm *T. galilea* as a function of microsphere concentration. The curve shows ingestion rate for untreated and sham-operated animals, as predicted by the Michaelis-Menten model: $I = I_{max}C/K_{0.5} + C$, where I_{max} is the maximum feeding rate, $K_{0.5}$ is the half-saturation constant and C is the microsphere concentration, calculated as $C = C_f - C_i/\log_e(C_f/C_i)$, where C_f and C_i are final and initial microsphere concentrations respectively. The I_{max} value (84.746×10^3) was determined from linear regression of the inverse of microsphere ingestion rates versus the inverse values of microsphere concentrations. Modified from Drenner *et al.* (1987).

There is a growing consensus that suspension feeding in tilapias is achieved by mucus entrapment rather than by mechanical sieving. Two distinct types of mucous cell are associated with the gill arch; small, superficial, goblet-shaped cells and large, clavate cells. The former produce mucus high in neutral/acid glycoproteins, the latter, which seem to be absent in fish less than 14 mm SL, secrete predominantly acidic muco-substances (Northcott and Beveridge, 1988). Examination of fish held in high concentrations of suspended particles show small particles (< 1.0 mm) entrapped among the gill rakers and filaments, apparently caught in strands and aggregates of mucus, (Greenwood, 1953; Northcott and Beveridge, 1988; Sanderson *et al.*, 1996). Sanderson and colleagues used a miniature fibre-optic endoscope to study the process in detail. As water is pumped through the buccopharyngeal cavity mucus-bound particles gradually move back towards the ceratobranchials of arches I–IV. Entrapped material is transferred to the buccopharyngeal cavity every few seconds by a violent reversal of water flow over the gills.

The mechanisms involved in mucus entrapment are still not entirely resolved. Sanderson *et al.* (1996) believe that aerosol filtration, in which particles are intercepted by direct encounter or by passing extremely close to a mucus strand, is the single most important mechanism. Greenwood (1953) and Northcott and Beveridge (1988) have proposed that electrostatic interaction between the acidic – and thus negatively charged – mucus and particle surface chemistry could also be important. However, Sanderson and colleagues believe the flow velocities in the buccopharyngeal cavity are too great for electrostatic interactions to play a significant role. Some have stated that the mucus-entrapment hypothesis is inconsistent with the particle-size selectivity observed in tilapias (see below) (Fryer and Iles, 1972; Drenner *et al.*, 1987). However, the aerosol feeding model proposed by Sanderson *et al.* (1996) fully addresses these concerns: small particles are likely to encounter the filtration surface less frequently than larger particles at the same concentration, resulting in disproportionately fewer small particles being ingested.

The relative advantages of aerosol filtration compared with mechanical sieving are summarized in Table 3.1. Sanderson *et al.* (1996) predict that aerosol filtration by mucus entrapment may be important in many other cichlids and among filter feeding cyprinids. However, it remains to be explained why tilapias seem to be far superior to other phytoplanktivorous fishes in terms of their abilities to remove free-living bacteria from the water column (Beveridge *et al.*, 1989; Rahmatullah and Beveridge 1993).

Visual feeding

Visual feeding is particularly important in juvenile fish (Lazzaro, 1991; Yowell and Vinyard, 1993). The mechanisms involved are discussed above.

Table 3.1 Relative advantages of aerosol filtration *versus* sieving by suspension feeding fishes (adapted from Sanderson *et al.*, 1996)

Aerosol filtration by suspension feeding fish	Sieving by suspension feeding fish
Extraction of small particles (~5–50 μm) involves less drag	Extraction of small particles involves more drag
Wide range of particle sizes retained	Few particles below threshold size retained
Filters less prone to clogging	Filters more prone to clogging
Particles already bound in mucus for transport to oesophagus	Separate process required to prepare particles for transport to oesophagus

Biting and grazing

Tilapias possess between one and five rows of jaw teeth. The structure of the teeth is typical of generalist feeders, being uniformly small, distally flattened, unicuspid, bicuspid or tricuspid enameloid structures (Lanzing and Higginbotham, 1976; Sasagawa, 1997). The outer row of teeth in *O. niloticus* is capped with a brown material that may be iron oxide and may afford some protection during grazing or browsing (Northcott and Beveridge, 1988). The jaw teeth are employed by *Tilapia* species to bite and tear plant material (Fryer and Iles, 1972; Caulton, 1976; Bowen, 1982) and, among *Oreochromis* species, have been observed being used to tug at and detach periphyton from substrates (Dempster *et al.*, 1993; Zimudzi, 1995). Periods of periphyton grazing are interspersed with periods of buccopharyngeal processing, swallowing and foraging.

Buccopharyngeal food processing

Once in the buccal cavity, material is sorted and either ejected or swallowed. Phytoplankton contaminated with large amounts of suspended inorganic material or toxic phytoplankton is often rejected in mucus-bound boluses, the numerous taste buds associated with the gill rakers perhaps playing some role (Northcott and Beveridge, 1988; Beveridge *et al.*, 1993). Material accepted as food is masticated then raked towards the oesophagus and swallowed (Aerts *et al.*, 1986).

Liem (1991) uses the term 'hydrodynamic tongue' to describe the transport and positioning of prey items within the buccal cavity by water currents created by selective local distortions of areas of the buccal cavity. Lifting the anterior part of the buccal cavity floor, for example, results in a posteriorly directed current carrying food with it. A description of oral food processing in *O. niloticus*, based on X-ray cineradiography, is given by Aerts *et al.* (1986). As food is forced through the buccopharyngeal cavity by the action of the hydrodynamic tongue, the upper and lower jaws retract, forcing the food backwards and upwards towards the pharyngeal bones. Cichlid pharyngeal bones are highly specialized (see Liem, 1991): the upper pharyngeal bones articulate with the pharyngeal apophysis of the skull by means of a synovial joint while the lower jaw is suspended by a muscular sling. The degree of interspecific variations in the size and shape of the pharyngeal bones and apophysis and in the size, shape and architecture of the muscle sling is well-known (Trewavas, 1983; see Fig. 3.4). The pharyngeal bones prepare food prior to swallowing, breaking filamentous algae and phytoplankton into smaller fragments and disrupting cell walls, thereby increasing surface area and facilitating digestion, and rake the food towards the oesophagous (Greenwood, 1953; Liem and Osse, 1975; Caulton, 1976; Pauly, 1976; Bowen, 1982). The pharyngeal bones show extraordinary plasticity in response to diet (Kornfield, 1991). Specializations can also be seen in the fine structure of the pharyngeal teeth. Coarse, robust structures are apparent among macrophyte-feeding *Tilapia* species while thin, fine, often hooked teeth are apparent among species which tend towards microphagy, such as *O. niloticus* and *O. macrochir* (Caulton, 1976; Bowen, 1982; Huysseune, 1983; Trewavas, 1983; Northcott and Beveridge, 1988).

Food trapped between the upper and lower pharyngeal bones is masticated by the bones, which move rapidly in opposing directions. Mucus, produced in the pharyngeal teeth sockets, may play a role in lubricating the raking action of the pharyngeal bones (Northcott and Beveridge, 1988). The degree of mastication that occurs and its role in facilitating digestion is unclear. Caulton (1976, 1977a) found that disruption of macrophyte material during mastication was of variable efficiency, while Northcott (1988) found little evidence for disruption of phytoplankton cells (see also below). Swallowing is achieved by an extreme retraction of the upper pharyngeal bones that forces food through the oesophageal sphincter and into the alimentary tract (Greenwood, 1953; Aerts *et al.*, 1986).

3.4 GUT MORPHOLOGY AND DIGESTION

The arrangement of the digestive tract of cichlids, with the stomach's extendable blind pouch, the left-hand exit to the intestine and the first intestinal loop on the left-hand side, is unique (Zihler, 1982). Tilapias have a very short oesophagus that leads to a sac-like stomach that is separated from a long, coiled intestine by a sphincter. The oesophageal lining in tilapias consists of a stratified, irregularly folded epithelium, the most distinctive characteristics of which are the absence of taste buds and the abundance of numerous mucus secreting goblet cells (Cataldi *et al.*, 1988; Gargiulo *et al.*, 1996a). The oesophageal mucus is unusual in its histochemistry, affording protection of the mucosa not only against injury from ingested food but also perhaps from viruses and bacteria (Scocco *et al.*, 1998).

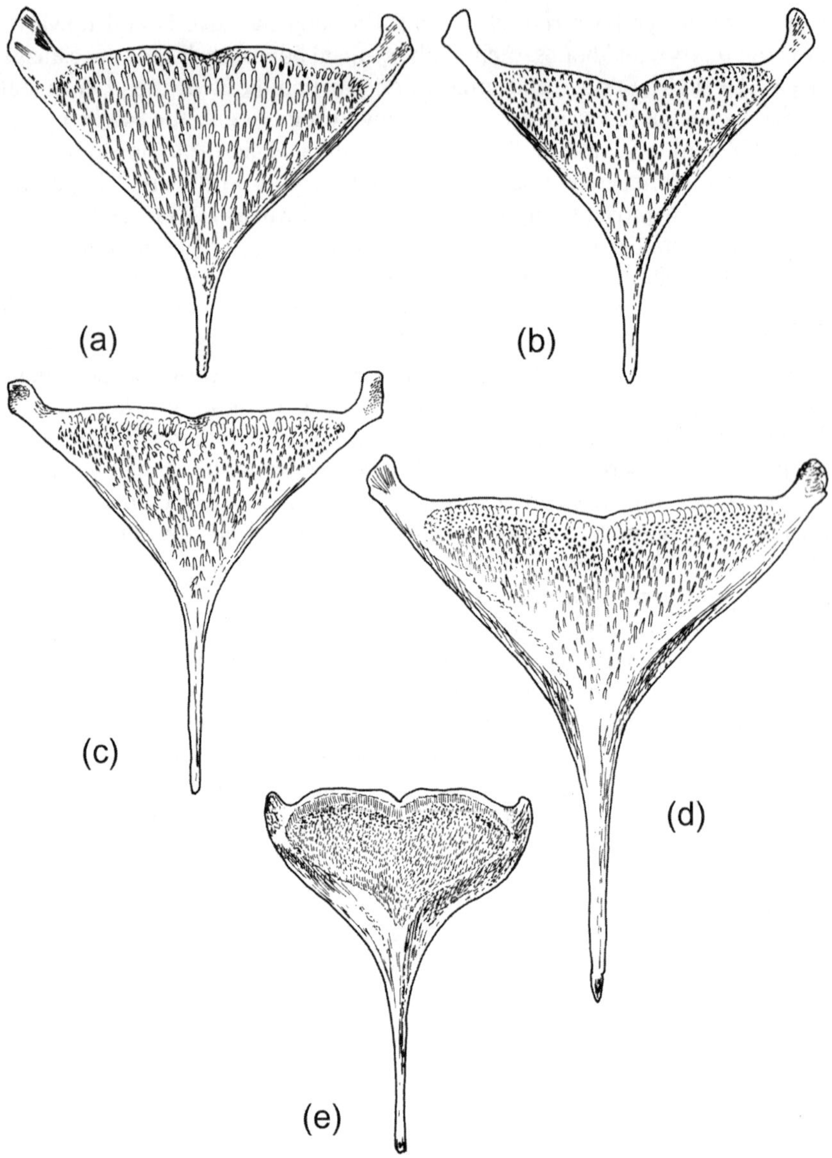

Fig. 3.4 Lower pharyngeal bones of: (a) *T. rendalli*; (b) *T. cabrae*; (c) *O. niloticus*; (d) *O. mossambicus*; (e) *S. gallilaeus*. From Trewavas (1983).

The tilapia stomach has three distinct regions, approximately corresponding to the cardiac (or initial), fundic (middle) and pyloric (terminal) portions of the mammalian stomach (Caceci *et al.*, 1998). Like the rest of the intestinal tract, the stomach is composed of four layers: a serous layer, a muscular layer and the mucosal and submucosal layers (Gargiulo *et al.*, 1996b; Caceci *et al.*, 1998). Skeletal muscle extends from the oesophagus into the proximal region of the stomach, a possible adaptation to the trituration of plant material. The mucosa of the middle region of the stomach is thrown into longitudinal folds that project into the lumen. The lumen is stellate in cross-section when empty, facilitating the accommodation of large variations in meal size and permitting considerable distension when full (Osman and Caceci, 1991; Chang and Lin, 1992; Garguilo *et al.*, 1996a). The central region of the stomach is also believed to be the site of acid secretion, the presence of food in the stomach stimulating secretion by the gastric glands (Akintunde, 1982; Getachew, 1989; Caceci *et al.*, 1998). The pH of the contents is thus a function of stomach fullness (Northcott, 1988; Getachew, 1989). Protection of the stomach epithelium from the extremely acid environment has been attributed to the presence of both mucopolysaccharides and to long chain carbohydrates, the latter forming a viscoelastic barrier (Scocco *et al.*, 1996, 1997; Caceci *et al.*, 1998).

Stomach evacuation rates range from 4 to 18 h and, as in other teleosts, there are inverse relationships between gastric emptying time, intestinal evacuation and total evacuation and temperature and body size (Ross and Jauncey, 1981; De Silva and Owoyemi, 1983; Hartnell, 1996). However, De Silva and Owoyemi (1983) observed that the pattern of intestinal filling and emptying in *O. mossambicus* also changed with diet, gastric emptying time being negatively correlated with diet specific gravity. Caceci *et al.* (1998) have drawn parallels between the anatomy of the tilapia stomach, with its large, blind-ended middle section, and that of ruminants. They propose that particularly refractory food items can be retained within the stomach for long periods in order to facilitate digestion and suggest that the thickening of the muscular layer in the terminal region of the stomach gives the animal a degree of voluntary control over the passage of food into the intestine. The observations of Moriarty and Moriarty (1973) and De Silva and Owoyemi (1983) that algal diets tend to be moved into the intestine in finite amounts while other foods are released from the stomach on a more continuous basis support this.

Tilapias have a long, coiled gut, typical of a herbivore. Intestinal length in adults is 7–13 times total body length, much longer than that of other cichlids (Fryer and Iles, 1972; Balarin and Hatton, 1979; Bowen, 1982; Zihler, 1982; Trewavas, 1983). Data collected by Frierson and Foltz (1992) for the microphagous *O. aureus* and macrophagous *T. zillii* show that the relationship between intestine length (IL) and total body length (TL) is the same for the two species:

$$IL = 89.04 \times 10^{0.0059TL}; r^2 = 0.86.$$

Electron microscopy shows the tilapia intestine to be a relatively undifferentiated muscular tube lined with a simple columnar epithelium interspersed with goblet cells (Gargiulo *et al.*, 1998). However, specializations in intestinal morphology and fine structure are apparent. There is a great deal of intestinal folding, especially in the proximal region, and the *lamina propria* is connective tissue, typical of herbivores and detrivores (Kapoor *et al.*, 1975). The microvilli of the proximal mucosa are unusually tall and exhibit the typical ultrastructural features associated with pinocytosis, namely extensive invaginations of the luminal plasma membrane and massive accumulation of vesicles in the apical cytoplasm. The microvilli of the distal intestinal mucosa are somewhat shorter and more sparsely distributed (Frierson and Foltz, 1992; Gargiulo *et al.*, 1998). The microvilli represent more than 90% of the digestive area in both *Tilapia* and *Oreochromis* species (Frierson and Foltz, 1992). Differences have been observed between the guts of microphagous and macrophagous species, the former having a more than 20% greater absorptive surface area (Frierson and Foltz, 1992). However, claims that gut length and fine structure change with diet composition (Stroband, 1977; Buddington, 1987; Kihara *et al.*, 1998) are disputed by Frierson and Foltz (1992).

Digestion and assimilation

Digestion in tilapias has been studied by Fish (1951, 1955, 1960), Moriarty and Moriarty (1973), Bowen (1976), Caulton (1976), Bitterlich (1985) and de Moor and Scott (1985) and reviewed by Bowen (1982). Digestion appears to be a two-step process, largely achieved by intestinal enzymes following acid-induced degradation of food in the stomach. Caulton (1976) incubated shredded and whole pieces of plant material in an acid (pH 1.5) environment for periods of time prior to measuring chlorophyll content, and as a result concluded that mastication of plant material prior to ingestion was important in ensuring effective digestion.

Stomach pH values can be <1, much less than that of other teleosts. The contention that cyanobacteria and algae are of poor nutritional value and pass through the gut largely undigested (Fryer and Iles, 1972; Man and Hodgkiss, 1977; Getachew, 1993; Opiyo and Dadzie, 1994) is a moot point. It has been shown that the gastric acids break down both eukaryote and prokaryote cell walls (Moriarty and Moriarty, 1973; Bowen, 1979, 1981, 1982; de Moor and Scott, 1985). Questions concerning the nutritional value of plant material may have arisen from examination of stomach and intestinal contents of fish that had only just begun to feed. Acid secretion increases exponentially with stomach fullness (Northcott, 1988; Getachew, 1989) resulting in extremely low acid pH values. Stomach pH values rise again to around pH 7 when feeding stops.

Although Bowen (1981) estimates that as much as 20% of dietary carbohydrates are digested in the stomach, most digestion appears to take place in the intestine. Bile, entering from the bile duct which opens into the intestine immediately behind the pyloric sphincter, maintains intestinal pH at 6.8–8.8 and trypsin, chymotrypsin, amylase and esterase activity have been identified from intestinal fluid (Fish, 1960; Moriarty and Moriarty, 1973; Bowen, 1982). Bowen (1980, 1981) has described the process of digestion in *O. mossambicus*. Digestion of protein is complete within the proximal quarter of the intestine although amino acids are not digested until near the distal end of the intestine. Although the intestinal tracts of tilapia species have sometimes been found to contain large numbers of obligate anaerobes, a reflection of gut length (Sakata *et al.*, 1984; Sugita *et al.*, 1985), it was widely believed that microbial fermentation was unimportant in the decomposition of plant-based material and fish nutrition. However, recent studies have recovered amylase and short-chain fatty acids from the intestine of fish fed starch- or plant-based diets, which are consistent with the occurrence of anaerobic fermentation (Kihara *et al.*, 1997; Sugita *et al.*, 1997; Henderson and Beveridge, unpublished).

Assimilation efficiencies of different food types are summarized in Table 3.2. In general, assimilation efficiencies are high for cyanobacteria and diatoms and low for green algae and macrophytes. Assimilation efficiency values for detritus are highly variable, being dependent upon source and composition. Protein and lipid are much more efficiently utilized than carbohydrate (Bowen, 1982; Bowen *et al.*, 1995).

3.5 FEEDING RATES AND TROPHIC EFFICIENCIES

Assessments of food consumption by tilapias have followed two approaches: gut contents analysis of fish captured in the field (Moriarty and Moriarty, 1973; Hofer and Schiemer, 1983; De Silva *et al.*, 1984; Maitipe and De Silva, 1985; de Moor *et al.*, 1986; Getachew 1987, 1989; Getachew and Fernando, 1989; Turner *et al.*, 1991) and direct quantification of algal ingestion by fish in the laboratory (Caulton, 1982; McDonald, 1985a,b, 1987; Robinson *et al.*, 1990, 1995; Golder, 1991; Northcott *et al.*, 1991; Dempster *et al.*, 1993; Hartnell, 1996). Quantitative, experimental studies of feeding have been largely concerned with filter feeding on phytoplankton, although there has been some recent work on zooplankton grazing.

Field observations of stomach contents have shown that while diel feeding rhythms in tilapias are influenced by environmental conditions and food supply, intense feeding activity is almost entirely restricted to daytime (Moriarty and Moriarty, 1973; Caulton, 1982; de Moor *et al.*, 1986; Northcott, 1988; Getachew, 1989; Opiyo and Dadzie, 1994) (see also Chapter 5).

Zooplankton ingestion

As in other planktivorous fishes, Yowell and Vinyard (1993) suspect that both particulate (i.e. visual feeding; see above) and filter feeding (suction feeding)

Table 3.2 Assimilation efficiencies of tilapias feeding on different food sources. Various methods for assimilation efficiency (AE) determination were used, including mass balance and the use of reference materials (ash, cellulose). Modified and updated from Bowen (1982)

Species	Food source	AE %
T. rendalli	*Ceratophyllum demersum*	53–60[1]
T. zillii	*Najas guadalupensis*	29[2]
S. melanotheron	detritus (1200 cal g^{-1})	50[3,4]
O. aureus	*Chlamydomonas* sp.	38[5,6]
	Ankistrodemus falcatus	53[5,6]
	Anabaena flos-aquae	83[5,6]
O. niloticus	*Microcystis* sp.	15[7]
		70[8]
	Anabaena	75[8]
	Nitzschia	79[8]
	Chlorella	49[8]
	Botryococcus brauni	28[7]
	Lake George suspended material >100 μm	43[8]
O. mossambicus	*Microcystis aeruginosa*	51[9]
	Spirogyra maxima	50–78[10]
	benthic detritus	38[11]
	periphytic detritus	63[11]
	detritus	15[12]

Source: [1]Caulton (1978); [2]Buddington (1979); [3]Pauly (1976); [4]Pauly *et al.* (1988); [5]McDonald (1985a); [6]McDonald (1985b); [7]Getachew (1987); [8]Moriarty and Moriarty (1973); [9]de Moor and Scott (1985); [10]Mathavan *et al.* (1976); [11]Bowen (1979); [12]de Moor and Scott (1986)

methods of feeding are used interchangeably by adult *O. aureus*, filter feeding being the usual mode, visual feeding on less abundant, larger prey occurring whenever the opportunity arises. Yowell and Vinyard (1993) quantified ingestion rates and net energy returns for both methods by video-recording *O. aureus* feeding on *Daphnia magna*. For particulate feeding, a range of prey sizes was used and attack distance, velocity and acceleration determined. Energy gains per unit time were estimated from data on capture efficiency (50–100%) and prey energy content and assimilation efficiencies, while energy expenditure was estimated from data on fast start swimming in fishes. Net energy gains associated with filter feeding were estimated from data on filter feeding efficiencies, energy value of prey and assimilation efficiencies while energy expenditure was derived from simultaneous measurements of oxygen consumption in filter feeding fish.

Yowell and Vinyard's study shows that net energy gain from particulate feeding on zooplankton decreases with increasing fish size as well as with distance travelled during attacks. Their model shows that the profitability of particulate feeding on zooplankton among adults is highly dependent upon prey size and prey density and predicts that under most circumstances it is more profitable for adult *O. aureus* to feed on phytoplankton (Fig. 3.5). It is worth pointing out that *D. magna* is among the largest species of Cladocera and does not usually occur in habitats with planktivorous fish. Yowell and Vinyard (1993) will therefore have overestimated the energy value of the typical zooplankton prey, perhaps by as much as an order of magnitude. It is a moot point whether their conclusions would have been the same had a smaller zooplankton species been used.

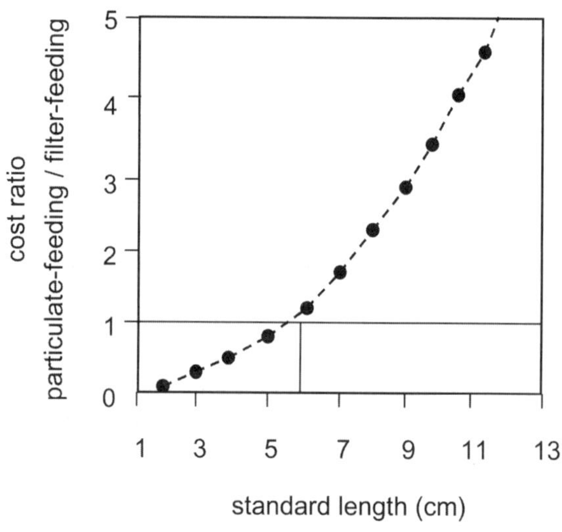

Fig. 3.5 Results from simulation of filter-feeding and particulate-feeding *O. aureus*, showing relative cost ratio of each feeding mode while acquiring equivalent net calorific intake values. The horizontal line at y = 1 indicates the level at which net profitability of each feeding mode are equivalent. For ratios <1 particulate feeding is most profitable. At ratios >1 filter feeding is most efficient (see text). Modified from Yowell and Vinyard (1996).

Phytoplankton

There have been numerous quantitative, laboratory-based studies of phytoplankton feeding, establishing many of the general principles involved in determining ingestion rate. Northcott *et al.* (1991) studied feeding in 40 mm (2.3 g) and 85 mm (22.3 g) SL *O. niloticus* exposed to different concentrations of small (*M. aeruginosa*) and large (*Anabaena cylindrica*) cyanobacteria. Particle ingestion by filter feeding was shown to increase with phytoplankton density, following a first-order saturation curve, in this instance reaching a maximum at concentrations of around 1.5×10^8 μm^3 algae ml^{-1} (equivalent to 2.4×10^6 colonies ml^{-1}). The cell diameters of the small, two-cell colonies of *M. aeruginosa* were in the range 3.5–4.5 μm and mean cell volume was 61.6 μm^3 (n = 30) while mean colony length of *A. cylindrica* ranged from 126.5 to 192.6 μm (n = 150). Algal ingestion per unit time was much higher for the large, colonial *A. cylindrica* than for *M. aeruginosa* and, apparent for *A. cylindrica* at least, ingestion rates per unit body weight increased with fish body mass. The so-called 'incipient limiting food concentration' at which filtering activity ceases to increase also falls as particle size increases, as suggested by Lehman (1976) (Northcott *et al.*, 1991).

Results of feeding rates from different studies are summarized in Table 3.3. To facilitate comparison, ingestion rates are expressed throughout in terms of cell numbers, biovolume and dry weight ingested g^{-1} fish body weight h^{-1} using conversion factors detailed in Dempster *et al.* (1995). Irrespective of the measure of filter feeding, it is readily apparent from Table 3.3 that there are enormous inter- and intra-specific differences in ingestion, in part because of differences in fish size, algal species, temperature and the presence of anti-nutritional factors. Studies differed too, however, in the degree to which fish were starved prior to experimentation.

Species and strains of many of the cyanobacteria are known to be toxic. It was widely assumed that tilapias ingested toxic cells and that the microcystins were prevented from inflicting damage by being denatured in the highly acidic environment of the stomach. However, *O. niloticus* exposed to toxic cyanobacteria (*Microcystis aeruginosa*) ingest very few cells (Beveridge *et al.*, 1993). When fish are exposed to mixtures of toxic and non-toxic *M. aeruginosa* cells, there is an inverse linear relationship between the proportion of toxic cells in the population and ingestion rate (Keshavanath *et al.*, 1994) (Fig. 3.6). While tilapias do not seem feed on toxic cyanobacteria it remains unknown how they detect the presence of toxic cells and stop ingesting them. Entrapment of algal cells is primarily by mucus and it seems unlikely that mucus synthesis can be rapidly switched off. Rejection of mucus-bound algal boluses was observed in both of the above studies.

Table 3.3 Ingestion rates by filter-feeding tilapia species determined under a range of experimental conditions. Numbers (column 1) refer to Fig. 3.7 (modified from Dempster *et al.*, 1995)

No	Fish	Phytoplankton	Fish size g (cm)	T°C	Ingestion rate: cells × 10^6 g^{-1} h^{-1}	Ingestion rate: μm^3 × 10^8 g^{-1} h^{-1}	Ingestion rate: mg g^{-1} h^{-1}	Source
1	*O. lidole*	*A. falcatus*	8.9 (6.3)*	24	40.8	21.1	0.69	1
2	*O. niloticus*	*A. falcatus*	9.4 (6.4)*	24	12.8	6.6	0.22	1
3	*O. mossambicus*	*A. falcatus*	11.4 (6.8)*	24	18.8	9.7	0.32	1
4	*O. niloticus*	*M. aeruginosa*	2.3 (4.0)	26	6.5	4.0	0.13	2
5	*O. niloticus*	*M. aeruginosa*	22.3 (8.5)	26	16.2	10.0	0.33	2
6	*O. niloticus*	*A. cylindrica*	2.3 (4.0)	26	1.9	25.2	0.97	2
7	*O. niloticus*	*A. cylindrica*	22.3 (8.5)	26	2.9	38.5	1.48	2
8	*O. niloticus*	*M. aeruginosa*	2.78 (4.2)	26	17.1	10.5	0.35	3
9	*O. niloticus*	*M. aeruginosa*	3.43 (4.6)	26	17.0	10.5	0.35	3
10	*O. niloticus*	*M. aeruginosa*	4.70 (4.7)	26	6.4	3.9	0.13	4
11	*O. aureus*	*A. falcatus*	15.0 (7.4)*	20	13.3	6.9	0.23	5
12	*O. aureus*	*Chlamydomonas* sp.	13.7 (7.2)*	20	2.1	2.1	0.07	5
13	*O. aureus*	*A. flos-aquae*	13.0 (7.1)*	20	6.6	2.8	0.09	5

1, Robinson *et al.* (1990); 2, Northcott *et al.* (1991); Golder (unpublished); 4, Dempster *et al.* (1993); McDonald (1985a)
*estimated from a regression for *O. niloticus* (n = 30), where total length (cm) = (wet weight/0.36)$^{0.33}$; (Dempster, unpublished)

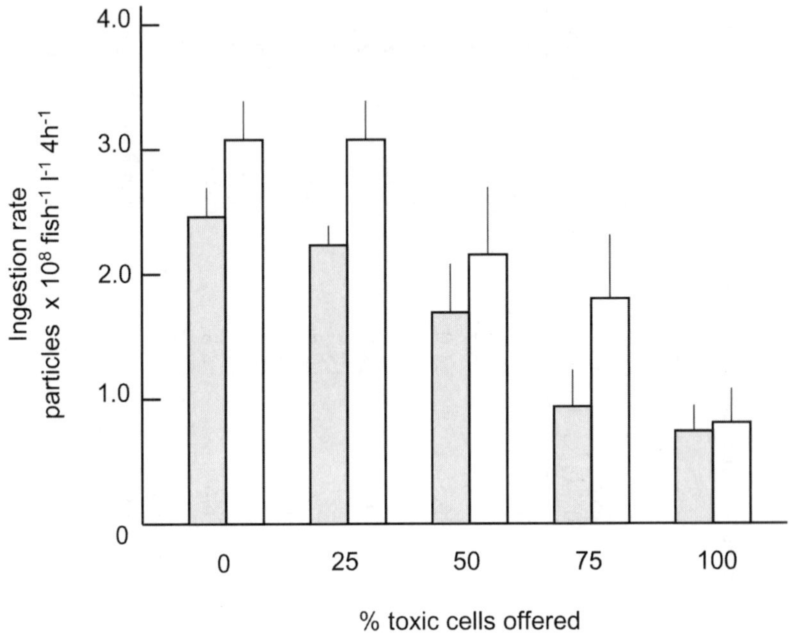

Fig. 3.6 Ingestion rates (mean ± S.D.) of *O. niloticus* exposed to two concentrations (dotted bars = 5×10^5 particles ml^{-1}; empty bars = 1×10^6 particles ml^{-1}) of pure populations and mixtures of toxic and non-toxic *M. aeruginosa*. Modified from Keshavanath *et al.* (1994).

Macrophytes

Caulton (1977b, 1982) estimated macrophyte consumption by *T. rendalli* from both laboratory and field studies. Field estimates of consumption of the macrophyte *Panicum repens* from Lake Kariba were around 3.3 g dry mass per day for 200 g fish, equivalent to 17 mg dry weight g body weigh^{-1} d^{-1}. This approximates with the estimate of 15 mg dry weight g body weight^{-1} d^{-1} determined from field data by Moriarty and Moriarty (1973) for *O. niloticus* feeding on phytoplankton. By maintaining *T. rendalli* on a diet of the macrophyte *Ceratophyllum demersum* Caulton (1982) was able to derive a series of relationships relating food intake to body size and to temperature. He found that the food intake of a 100 g fish at 28°C was around 10 mg dry weight g body weight^{-1} d^{-1}, and concluded that laboratory-derived satiation values are half that of fish in the field (see below). He also found that food intake increased between 18 and 30°C, but thereafter declined, ceasing at about 37–38°C.

Periphyton

Dempster *et al.* (1993) carried out a laboratory-based determination of periphyton ingestion rates in tilapias. Juvenile (47 mm SL) *O. niloticus* presented with excess quantities of a periphytic mat of cyanobacteria (*Oscillatoria* sp.) were found to ingest 4.8 mg dry weight g body weight^{-1} h^{-1}, a figure some twenty-five times greater occurred when presented with excess planktonic cyanobacteria.

Detritus and micro-organisms

Both free-living and particle-bound bacteria occur and it is generally presumed that much of the bacteria consumed by fishes such as carps and tilapias is bound to detritus particles (Schroeder, 1978). However, bacteria form only a small proportion of the organic mass of detritus (generally $<5\%$), the bulk of the material being composed of dead and decaying plant material (see Moriarty and Pullin, 1987; Bowen *et al.*, 1995, for reviews). Ingestion rates of starved *O. mossambicus* feeding on periphytic algal detritus were determined during short-term experiments in the laboratory by Bowen (1981). Consumption was around 0.28 g dry weight h^{-1} (no fish size indicated).

Laboratory-based experiments, however, have shown that *O. niloticus* can detect and ingest unattached bacteria and that there is a linear relationship between bacterial numbers and concentration in the water (Beveridge *et al.*, 1989; Matêna *et al.*, 1995). While it is well established that *O. niloticus* can ingest single cyanobacterial colonies of a similar size, tilapias may be better at feeding on unattached bacteria than other planktivores (Rahmatullah *et al.*, 1993). Ingestion rates of around 1.5×10^6 cells h^{-1}, even under high bacterial densities, however, suggest free-living bacteria are likely to be of only minor nutritional significance.

Energetics

Dempster *et al.* (1995) estimated daily ingestion rates of dry matter and energy by filter feeding from a number of published laboratory studies by assuming that the hourly filtration rates given in column 8, Table 3.3 are maintained for 12 h per day. The estimated range of food intake (1–17 mg g fish body weight^{-1} d^{-1}) is significantly lower than that of 43 mg g fish body weight^{-1} d^{-1} derived from stomach contents analysis by Hofer and Schiemer (1983) and the values predicted by the regression equation derived by Moriarty and Moriarty (1973). However, the values broadly correspond with the estimates of Getachew (1989) (6 mg g fish body weight^{-1} d^{-1}) for *O. niloticus* feeding on colonial green algae (*Botryococcus* sp.) and cyanobacteria (*Oscillatoria* sp.) in Lake Awasa. The analysis suggests that mechanisms of algal ingestion other than filter feeding are important in the natural environment.

Using the values of food ingested, Dempster *et al.* (1995) calculated daily ingestion rates of digestible energy and compared these with daily metabolic energy expenditure (e.g. maintenance, activity, N excretion) and non-faecal nitrogen losses derived from the bioenergetic model of Ross *et al.* (1988) (see also Ross, this volume). The model was applied under the same conditions of temperature and fish weight as used in the experiments in which ingestion rates of algae were determined. Assumptions of the model are: day length = 12 h; salinity = 0 ppt; fraction of ingested energy expended on routine metabolism = 70%; day-time activity levels are 50% expended on routine activity and 5–10% (10% assumed here) on active swimming (2 body lengths s^{-1} assumed), the balance being spent in inactivity.

The results of the comparison of daily ingestion of energy and the daily metabolic costs for the data from each laboratory experiment are shown in Fig. 3.7. Values were standardized by expressing data on a per gram fish body weight per day basis. Estimated metabolic costs were consistently higher than energy acquisition through ingestion of algae by filter feeding, the exception being the laboratory-derived data of Northcott *et al.* (1991) for 23.2 g *O. niloticus* presented with *A. cylindrica* as food. It was concluded that the ingestion rates determined from the majority of laboratory studies of filter feeding would result in fish losing weight over time.

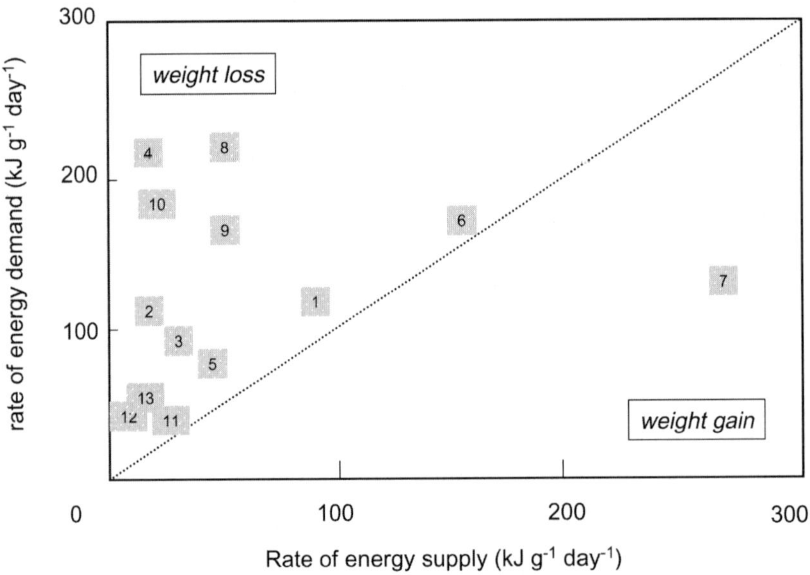

Fig. 3.7 Net energy balance of tilapias filter-feeding on phytoplankton. Each data point has been calculated using the model of Ross (1988) using data on ingestion rates given in Table 3.3 (numbers as in Table 3.3). Numbers above the line indicate a net energy loss while that below the line (i.e. *O. niloticus* feeding on *A. cylindrica*) indicates a net energy loss. Modified from Dempster *et al.* 1995.

3.6 DISCUSSION

There is a high degree of opportunism in feeding behaviour among the cichlids (McKaye and Marsh, 1983; Lowe-McConnell, 1987; Liem; 1991; Yamaoka, 1991; Arthington and Bluhdorn, 1994; Dempster *et al.*, 1995). This is as true for the tilapias as for the other cichlids, and has undoubtedly been a key to their enormous success in colonizing new habitats (Lever, 1996; Lowe-McConnell, this volume). Nevertheless, the tilapias are probably best regarded as opportunistic omnivores with a strong tendency towards herbivory, the many studies of feeding ecology showing that algae and algal-derived detritus comprise the bulk of gut contents in adult *Oreochromis*, *Sarotherodon* and *Tilapia* species under most conditions.

There has been a great deal of study of tilapia feeding ecology based on gut contents analysis and it has been assumed that because the gut contents are dominated by algae and that tilapias filter feed, ingestion of algal material is primarily by filter feeding on phytoplankton. However, as has been pointed out by Bowen (1982) and others, the material ingested by deposit feeding fish (algae, detritus, bacteria) is essentially indistinguishable from periphyton or from plankton feeding animals. Recent controversies about filter feeding, food ingestion and growth have largely arisen because of confusion about what constitutes filter feeding and the plethora of somewhat contradictory definitions of feeding methods in the literature. Adult tilapias are continuous suction feeders (*sensu* Liem, 1991; Sanderson and Wassersug, 1993; Sanderson *et al.*, 1996), entrapping food and other suspended material by aerosol filtration as water is routinely passed over the gills. This probably accounts for a significant proportion of energy intake in most species under most circumstances. While the volumes of water pumped across the gills increase with food density and ingestion increases with density, laboratory-based quantitative studies of filter feeding suggest that animals cannot ingest sufficient food per unit time to thrive and grow by filter feeding alone. Other methods of food ingestion or intermittent ingestion of more profitable food types are essential to ensure a positive energy balance and good growth. Other sources of algae include:

- sedimented algae (see Bowen, 1979, 1982);
- benthic algae;
- periphytic algae (see Dempster *et al.*, 1993);
- algal-based detritus (see Moriarty and Pullin, 1987);
- cyanobacterial surface scums.

These universally available dietary sources are more concentrated and ingestion per unit time, by suction feeding, grazing and browsing, is more efficient (see Dempster *et al.*, 1993), although as Bowen *et al.* (1995) have pointed out, issues of food quality are also important here. In feeding on one food source, other material may be ingested inadvertently: zooplankton with phytoplankton, periphyton with macrophytes (and *vice versa*), sand and silt with benthic algae. Adult tilapias undoubtedly also choose to exploit zooplankton and other foods as opportunities arise. However, until more direct observations of feeding behaviour are made, many issues related to feeding ecology are unlikely to be resolved.

Tilapias appear to be good at digesting plant-based food materials, achieving assimilation efficiencies in the order of 50–80% (see Table 3.2). Horn (1992) has argued that because herbivorous teleosts have a conventional set of digestive enzymes (see below) they must possess one or more of four digestive mechanisms in order to deal with a plant-based diet:

- pharyngeal bones ('jaws') able to grind plant-based food, thereby increasing surface for enzymatic attack and disrupting plant cell walls;
- a muscular stomach able to triturate ingested material;
- a highly acidic stomach which can lyse plant cells;
- a microbial fauna in the hindgut.

Unusually, all four mechanisms have been shown to exist in the tilapias, although Bitterlich's contention (1985) that tilapias may be the only truly herbivorous fishes must be regarded as something of an over-statement. The relative importance of each mechanism remains to be fully elucidated. Even though they possess pharyngeal jaws the trituration capabilities of tilapias, especially with regard to phytoplankton, are probably limited. The main adaptive features relevant to a herbivorous diet appear to be the possession of a stomach with a sufficiently acid environment to disrupt plant and cyanobacterial cell walls, coupled to a long intestine. The importance of microbial fermentation in the digestion and uptake of plant-based nutrients remains to be properly determined.

The understanding of tilapia feeding ecology, feeding and digestive physiology is not only of academic interest but may also help improve yields from aquaculture, especially in resource-poor countries. Acadjas are a traditionally used fishing system that originated in coastal lagoons in the Benin Republic, West Africa. They are constructed from dense masses of branches that act as artificial reefs or 'fish attractant devices' (FADs), offering not only shelter but also boosting food (periphyton) supplies (Welcomme, 1972; Konan-Brou and Guiral, 1994). Similar systems, 'katha', have also evolved in Bangladesh and elsewhere (Beveridge *et al.*, 1998). Trials in which acadja-like structures composed of bamboo or other substrates have been constructed in ponds, tanks and cages stocked with tilapias (Konan-Brou and Guiral, 1994; Shrestha and Knud-Hansen, 1994), have recently been undertaken with the aim of increasing net foraging return through increased plant food production and density. While the economic viability of periphyton-based aquaculture systems remains to be proven, they represent a novel departure from other forms of semi-intensive aquaculture which focus on the promotion of plankton or seston production for fish food.

ACKNOWLEDGEMENTS

The authors acknowledge the contributions made by students and colleagues over the years, especially Lindsay Ross, Nick Frerichs, Stuart Millar, Geoff Codd, Jim Henderson, Linda Lawton, Perar Keshavanth, Matt Briggs, Mark Northcott, Shah Rahmatullah, Md. Abdul Wahab, Patrick Dempster, Momtaz Begum, Profulla Sikdar, Israil Golder, Farayi Zimudzi, Rafael Meseguer Elizondo and Charles Ndonna. Thanks are due also to the late Humphrey Greenwood and to S. Laurie Sanderson for their invaluable correspondence over the years and to the reviewers whose comments helped improve this chapter.

We thank the Overseas Development Administration (now the Department For International Development, DFID) of the UK government and DG XII of the European Commission for supporting our research.

REFERENCES

Aerts, P.F., De Vree, F. and Vandewalle, P. (1986) Movement analysis of the pharyngeal jaw apparatus in *Oreochromis niloticus*. *Ann. Mus. Roy. Afr. Centr. Sc. Zool.* **251**, 157–160.

Akintunde, E.A. (1982) Feeding rhythm in relation to changing pattern of pH in the gut of *Sarotherodon galilaeus* (Artedi) of Lake Kainji, Nigeria. *Hydrobiologia* **97**, 179–184.

Arthrington, A.H. and Bluhdorn, D.R. (1994) Distribution, genetics, ecology and status of the introduced cichlid *Oreochromis mossambicus* in Australia. *Mitt Internat. Verein. Limnol.* **24**, 53–62.

Balarin, J.D. and Hatton, J.P. (1979) *Tilapia. A Guide to Their Biology and Culture in Africa*, Unit of Aquatic Pathobiology, University of Stirling.

Bassay, B.M., Makumyaviri, A.M. and Otikeke, L.S. (1997) Diet evolution in *Tilapia rendalli* kept in heterogesous breeding system in Lubumbashi, Congo. *Rev. Méd. Vét.* **148**, 805–807.

Beveridge, M.C.M., Briggs, M.R.P., Northcott, M.E. and Ross, L.G. (1988a) The occurrence, structure and development of microbranchiospines among the tilapias. *Can. J. Zool.* **66**, 2564–2572.

Beveridge, M.C.M., Briggs, M.R.P., Mowat, A., Northcott, M.E. and Ross, L.G. (1988b) The function of microbranchiospines in tilapias, in *Proceedings of the Second International Symposium on Tilapias in Aquaculture* (*ICLARM Conf. Proc.*, Vol. 15) (eds R.S.V. Pullin, T. Bhukaswan, K. Tonguthai and J.L Maclean), ICLARM, Manila, pp. 311–317.

Beveridge, M.C.M., Begum, M., Frerichs, G.N. and Millar, S. (1989) The ingestion of bacteria in suspension by the tilapia *Oreochromis niloticus*. *Aquaculture* **81**, 373–378.

Beveridge, M.C.M., Baird, D.J., Rahmatullah, S.M., Beattie, K.A. and Codd, G.A. (1993) Grazing rates on toxic and non-toxic cyanobacteria by *Hypophthalmichthys molitrix* and *Oreochromis niloticus*. *J. Fish Biol.* **43**, 901–907.

Beveridge, M.C.M., Verdegem, M., Wahab, Md. A., Keshavanath, P. and Baird, D.J. (1998) Periphyton-based aquaculture and the EC-funded PAISA Project. *Naga, the ICLARM Q.* **21**(4), 49–50.

Bitterlich, G. (1985) The nutrition of a stomachless phytoplanktivorous fish in comparison with *Tilapia*. *Hydrobiologia* **121**, 173–179.

Bowen, S.H. (1976) Mechanism for digestion of detrital bacteria by the cichlid fish *Sarotherodon mossambicus* (Peters). *Nature* (London) **260**, 137–138.

Bowen, S.H. (1979) A nutritional constraint in detritivory by fishes. The stunted population of *Sarotherodon mossambicus* in Lake Sibala South Africa. *Ecological Monographs* **49**, 17–31.

Bowen, S.H. (1980) Detrital non-protein amino acids are the key to rapid growth of tilapia in Lake Valencia, Venezuela. *Science* **207**, 1216–1218.

Bowen, S.H. (1981) Digestion and assimilation of periphytic detrital aggregate by *Tilapia mossambica*. *Trans. Am. Fish. Soc.* **110**, 241–247.

Bowen, S.H. (1982) Feeding, digestion and growth – qualitative considerations, in *The Biology and Culture of Tilapias* (*ICLARM Conference Proceedings, Vol. 7*) (eds R.S.V. Pullin and R.H. Lowe-McConnell), ICLARM, Manila, pp. 141–156.

Bowen, S.H., Lutz, E.V., Ahlgren, M.O. (1995) Dietary protein and energy as determinants of food quality: trophic strategies compared. *Ecology* **76**, 899–907.

Buddington, R.K. (1979) Digestion of an aquatic macrophyte by *Tilapia zillii*. *J. Fish Biol.* **15**, 449–456.

Buddington, R.K. (1987) Does the natural diet influence the intestine's ability to regulate glucose absorbtion? *Comp. Biochem. Physiol.* **157**B, 677–688.

Caceci, T., El-Habback, H.A., Smith, S.A. and Smith, B.J. (1998) The stomach of *Oreochromis niloticus* has three regions. *J. Fish Biol.* **50**, 939–952.

Campbell, K.L.I. (1981) Fishes of the genus *Sarotherodon* (Pices, Cichlidae) of springs along the northern Haso Ngiro, Kenya. *J. East Afr. Nat. Hist. Soc. Nat. Mus.* **173**, 1–12.

Cataldi, E., Crosetti, D., Conte, G., D'Ovido, D. and Cataudella, S. (1988) Morphological changes in the oesophageal epitheleum during adaptations to salinities in *Oreochromis mossambicus*, *O. niloticus* and their hybrid. *J. Fish Biol.* **32**, 191–196.

Caulton, M.S. (1976) The importance of pre-digestive food preparation to *Tilapia rendalli* Boulanger when feeding on aquatic macrophytes. *Trans. Rhod. Sci. Assoc.* **57**, 22–28.

Caulton, M.S. (1977) A quantitative assessment of the daily ingestion of *Panicum repens* L. by *Tilapia rendalli* Boulenger (Cichlidae) in Lake Kariba. *Trans. Rhod. Sci. Assoc.* **58**(6), 38–42.

Caulton, M.S. (1978) Tissue depletion and energy utilization during routine metabolism by sub-adult *Tilapia rendalli* Boulenger. *J. Fish. Biol.* **13**, 1–6.

Caulton, M.S. (1982) Feeding, metabolism and growth of tilapias: some quantitative considerations, in *The Biology and Culture of Tilapias* (*ICLARM Conference Proceedings, Vol. 7*) (eds R.S.V. Pullin and R.H. Lowe-McConnell), ICLARM, Manila, pp. 157–180.

Chang, M.H. and Lin, J.F. (1992) Gastric histological structure and localization of carbonic anhydrase of tilapia. *J. Ch. Soc. Vet. Sci.* **18**, 243–253.

Dempster, P.W., Beveridge, M.C.M. and Baird, D.J. (1993) Herbivory in the tilapia *Oreochromis niloticus* (L.): a comparison of feeding rates on periphyton and phytoplankton. *J. Fish Biol.* **43**, 385–392.

Dempster, P., Baird, D.J. and Beveridge, M.C.M. (1995) Can fish survive by filter-feeding on microparticles? Energy balance in tilapia grazing on algal suspensions. *J. Fish Biol.* **47**, 7–17.

Denny, P., Bowker, D.W. and Bailey, R.G. (1978) The importance of littoral epiphyton as food for commercial fish in the recent African man-made lake Nyumba ya Mungu reservoir, Tanzania. *Biol. J. Linn. Soc.* **10**, 139–150.

De Silva, S.S. and Owoyemi, A.A. (1983) Effect of dietary quality on the gastric evacuation and intestinal passage in *Sarotherodon mossambicus* (Peters) fry. *J. Fish Biol.* **23**, 347–355.

De Silva, S.S. Perera, M.K. and Maitipe, P. (1984) The composition, nutritional status and digestibility of the diets of *Sarotherodon mossambicus* from nine man-made lakes in Sri Lanka. *Env. Biol. Fishes* **11**, 205–219.

Drenner, R.W., Taylor, S., Lazzaro, X. and Kettle, D. (1984) Particle grazing and plankton community impact of an omnivorous cichlid. *Trans. Am. Fish. Soc.* **113**, 397–402.

Drenner, R.W., Vinyard, G.L., Hambright, K.D. and Gophen, M. (1987) Particle ingestion by *Tilapia galilaea* is not affected by removal of gill rakers and microbranchiospines. *Trans. Am. Fish. Soc.* **116**, 272–276.

Fish, G.R. (1951) Digestion in *Tilapia esculenta. Nature* (London) **167**, 900–901.

Fish, G.R. (1955) The food of *Tilapia* in East Africa. *Uganda J.* **19**, 85–89.

Fish, G.R. (1960) The comparative activity of some digestive enzymes in the alimentary canal of tilapia and perch. *Hydrobiologia* **15**, 151–178.

Frierson, E.W. and Foltz, J.W. (1992) Comparison and estimation of absorbtive intestinal surface areas in two species of cichlid fish. *Trans. Am. Fish. Soc.* **121**, 517–523.

Fryer, G. and Iles, T.D. (1972) *The Cichlid Fishes of the Great Lakes of Africa: Their Biology and Evolution*, Oliver and Boyd, Edinburgh.

Gargiulo, A.M., Dall'Aglio, C., Tsoku, Z., Ceccarelli, P. and Pedini, V. (1996a) Morphology and histology of the oesophagus in a warmwater tilapiine fish (Teleostei). *J. Appl. Ichthyol.* **12**, 121–124.

Gargiulo, A.M., Ceccarelli, P., Dall'Aglio, C. and Pedini, V. (1996b) Ultrastructural study on the stomach of *Tilapia* spp (Teleostei). *Anat. Histol. Embryol.* **26**, 331–336.

Gargiulo, A.M., Ceccarelli, P., Dall'Aglio, C. and Pedini, V. (1998) Histology and ultrastructure of the gut of the tilapia (*Tilapia* spp). *Anat. Histol. Embyol.* **27**, 89–94.

Gerking, S.D. (1994) *Feeding Ecology of Fish*, Academic Press, London.

Getachew, T. (1987) A study on an herbivorous fish, *Oreochromis niloticus* L., diet and its quality in two Ethiopian Rift Valley lakes, Awasa and Zwai. *J. Fish Biol.* **30**, 439–449.

Getachew, T. (1989) Stomach pH, feeding and ingestion rate in *Oreochromis niloticus* L. (Pisces: Cichlidae) in Lake Awasa, Ethiopia. *Hydrobiologia* **174**, 43–48.

Getachew, T. (1993) The composition and nutritional status of the diet of *Oreochromis niloticus* in Lake Chamo, Ethiopia. *J. Fish Biol.* **42**, 865–874.

Getachew, T. and Fernando, C.H. (1989) The food of a herbivorous fish (*Oreochromis niloticus* Linn.) in Lake Awasa, Ethiopia. *Hydrobiologia* **174**, 195–200.

Golder, M.I. (1991) *Influence of Cadmium on the Ingestion Rate of the Tilapia* Oreochromis niloticus *(L.) Feeding on the Blue-Green Alga* Microcystis aeruginosa, Unpublished MSc thesis, University of Stirling, Scotland.

Goldschmidt, T. (1996) *Darwin's Dreampond. Drama in Lake Victoria*, MIT Press, Cambridge, MA.

Gophen, M. (1980) Food resources, feeding behaviour and growth of *Sarotherodon galilaeus* (Linnaeus) fingerlings. *Aquaculture* **20**, 101–115.

Gophen, M., Drenner, R.W. and Vinyard, G.L. (1983) Cichlid stocking and the decline of the Galilee Saint Peter's fish (*Sarotherodon galilaeus*) in Lake Kinneret, Israel. *Can. J. Fish Aquat. Sci.* **40**, 983–986.

Gosse, J.P. (1956) Dispositions speciales de l'appareil branchial des *Tilapia* et *Citharinus*. *Ann. Soc. Roy. Zool. Belge.* **86**, 303–308.

Greenwood, P.H. (1953) Feeding mechanism of the cichlid fish *Tilapia esculenta* Graham. *Nature* (London) **172**, 207–208.

Hartnell, R.M. (1996) *Laboratory and Field Investigations into the Feeding Performance of the Lake Malawi* Oreochromis, Unpublished PhD thesis, University of Wales.

Hodgkiss, I.J. and Man, H.S.H. (1977) Age composition, growth and body condition of the introduced *Sarotherodon mossambicus* (Cichlidae) in Plover Cove Reservoir, Hong Kong. *Env. Biol. Fishes* **2**, 35–44.

Hofer, R. and Schiemer, F. (1983) Feeding ecology, assimilation efficiencies and energetics of two herbivorous fish: *Sarotherodon (Tilapia) mossambicus* (Peters) and *Puntius filamentosus* (Cuv. et Val.), in *Limnology of Parakrama Samudra – Sri Lanka.* (ed. F. Schiemer), Dr W. Junk, The Hague, pp.155–164.

Horn, M.H. (1992) Herbivorous fishes: feeding and digestive mechanisms, in *Plant-Animal Interactions in the Marine Benthos* (*Systematics Association* Specl. Vol. 46) (eds D.M. John, S.J. Hawkins and J.H. Price), Clarendon Press, Oxford, pp. 339–362.

Huysseune, A. (1983) Observations on tooth development and implantation in the upper pharyngeal jaws in *Astatotilapia elegans* (Teleostei: Cichlidae). *J. Morph.* **175**, 217–234.

Jauncey, K. (1998) *A Guide to Tilapia Feeds and Feeding*, 2nd edn, Pisces Press, Bridge of Allan, Stirling.

Kapoor, B.G., Smit, H. and Verighina, I.A. (1975) The alimentary canal and digestion in teleosts. *Adv. Mar. Biol.* **13**, 109–239.

Keshavanath, P., Beveridge, M.C.M., Baird, D.J., Nimmo, A. and Codd, G. (1994) The functional grazing response of a phytoplanktivorous fish *Oreochromis niloticus* to mixtures of toxic and non-toxic strains of the cyanobacterium *Microcystis aeruginosa. J. Fish Biol.* **45**, 123–129.

Kihara, M. and Sakata, T. (1997) Fermentation of dietary carbohydrates to short-chain fatty acids by gut microbes and its influence on intestinal morphology of a detrivorous teleost *Tilapia. Comp. Biochem. Physiol.* **118A**, 1201–1207.

Konan-Brou, A.A. and Guiral, D. (1994) Available algal biomass in tropical brackish water artificial habitats. *Aquaculture* **119**, 175–190.

Kornfield, I. (1991) Genetics, in *Cichlid Fishes, Behaviour, Ecology and Evolution* (ed. M.H.A. Keenleyside), Chapman and Hall, London, pp. 103–128.

Lanzing, W.J.R. and Higginbotham, D.R. (1976) Scanning microscopy of pharyngeal and oral teeth of the teleost *Tilapia mossambica. Hydrobiologia* **48**, 137–139.

Lazzaro, X. (1987) A review of planktivorous fishes; their evolution, feeding behaviours, selectivities and impacts. *Hydrobiologia* **146**, 97–167.

Lazzaro, X. (1991) Feeding convergence in South American and African zooplanktivorous cichlids *Geophagus brasiliensis* and *Tilapia rendalli. Env. Biol. Fishes* **31**, 283–293.

Lehman, J.T. (1976) The filter-feeder as an optimal forager and the predicted shapes of feeding curves. *Limnol. Ocenaogr.* **21**, 501–516.

Le Roux, P.J. (1956) Feeding habits of the young of four species of tilapia. *South Afr. J. Sci.* **53**(2), 33–37.

Lever, C. (1996) *Naturalised Fishes of the World*, Academic Press, London.

Liem, K.F. (1991) Functional morphology, in *Cichlid fishes. Behaviour, Ecology and Evolution* (ed. M.H.A. Keenleyside), Chapman and Hall, London, pp. 129–150.

Liem, K.F. and Osse, J.W.M. (1975) Biological versatility, evolution and food resource exploitation in African cichlid fishes. *Am. Zool.* **15**, 427–454.

Lowe-McConnell, R.H. (1982) Tilapias in fish communities, in *The Biology and Culture of Tilapias (ICLARM Conference Proceedings, Vol. 7)* (eds R.S.V. Pullin and R.H. Lowe-McConnell), ICLARM, Manila, pp. 83–114.

Lowe-McConnell, R.H. (1987) *Ecological studies of Tropical Fish Communities*, Cambridge University Press, Cambridge, UK.

Maitipe, P. and De Silva S.S. (1985) Switches between zoophagy, phytophagy and detritivory of *Sarotherodon mossambicus* (Peters) populations in twelve man-made Sri Lankan lakes. *J. Fish Biol.* **26**, 49–61.

Man, H.S.H. and Hodgkiss, I.J. (1977) Studies on the ichthyofauna in Plover Cove Reservoir, Hong Kong: feeding and food relations. *J. Fish Biol.* **11**, 1–13.

Matêna, J., Šimek, K. and Fernando, C.H. (1995) Ingestion of suspended bacteria by fish: a modified approach. *J. Fish Biol.* **47**, 334–336.

Mathavan, S., Vivekandan, E. and Pandian, T.J. (1976) Food utilization in the fish *Tilapia mossambica* fed on plant and animal foods. *Helg. wiss. Meeresunters.* **28**, 66–70.

McDonald, M.E. (1985a) Carbon budgets for a phytoplanktivorous fish fed three different unialgal populations. *Oecologia* **66**, 246–249.

McDonald, M.E. (1985b) Growth of a grazing phytoplanktivorous fish and growth enhancement of the grazed alga. *Oecologia* **67**, 132–136.

McDonald, M.E. (1987) Interactions between a phytoplanktivorous fish, *Oreochromis niloticus*, and two unialgal forage populations. *Env. Biol. Fishes* **18**, 229–234.

de Moor, F.C. and Scott, W.E. (1985) Digestion of *Microcystis aeruginosa* by *Oreochromis niloticus*. *J. Limnol. Assoc. Sth. Afr.* **11**, 14–19.

de Moor, F., Wilkinson, R.C. and Herbst H.M. (1986) Food and feeding habits of *Oreochromis mossambicus* (Peters) in hypertrophic Hartbeespoort Dam, South Africa. *South Afr. J. Zool.* **21**, 170–176.

Moriarty, C.M. and Moriarty, D.J.W. (1973) Quantitative estimation of the daily ingestion of phytoplankton by *Tilapia nilotica* and *Haplochromis nigripinnis* in Lake George, Uganda. *J. Zool.* (London) **171**, 209–255.

Moriarty, D.J.W. and Pullin, R.S.V. (eds) (1987) *Detritus and Microbial Ecology in Aquaculture (ICLARM Conference Proceedings*, 14), ICLARM, Manila, Philippines.

Moriarty, D.J.W., Darlington, J.P.E.C., Dunn, I.G., Moriarty, C.M. and Tevlin, M.P. (1973) Feeding and grazing in Lake George, Uganda. *Proc. Roy. Soc. Lond.* **184**, 299–319.

Ndonna, C. (1996) *Comparisons of Ingestion Rates of Three Tilapia Species,* Tilapia rendalli, Oreochromis niloticus *and* O. mossambicus *Feeding on* Chlorella vulgaris, unpublished MSc thesis, University of Stirling, U.K.

Northcott, M.E. (1988) *An Investigation of Filter-Feeding in the Tilapia* Oreochromis niloticus. *L.*, unpublished PhD thesis, University of Stirling, UK.

Northcott, M.E. and Beveridge, M.C.M. (1988) The development and structure of the pharyngeal apparatus associated with filter feeding in tilapias (*Oreochromis niloticus* L.). *J. Zool.* (London) **215**, 133–149.

Northcott, M.E., Beveridge, M.C.M. and Ross, L.G. (1991) A laboratory investigation of the filtration and ingestion rates of the tilapia *Oreochromis niloticus* feeding on two species of blue-green algae. *Env. Biol. Fishes* **31**, 75–85.

Opiyo, S. V. and Dadzie, S. (1994) Diet and food utilization in *Oreochromis esculentus* (Graham) in Lake Kanyaboli, Kenya. *Fish. Manage. Ecol.* **1**, 79–90.

Osman, A.H.K. and Caceci, T. (1991) Histology of the stomach of *Tilapia nilotica* (Linnaeus, 1758) from the River Nile. *J. Fish Biol.* **38**, 211–223.

Pauly, D. (1976) The biology, fishery and potential for aquaculture of *Tilapia melanotheron* in a small West African lagoon. *Aquaculture* **7**, 33–49.

Rahmatullah, S.M. and Beveridge, M.C.M. (1993) The ingestion of bacteria in suspension by fry of planktivorous Indian major carps (*Catla catla, Labeo rohita*) and Chinese carps (*Hypophthalmichthys molitrix, Aristichthys nobilis*). *Hydrobiologia* **264**, 79–84.

Redding, T.A. (1997) *The Biology and Ecology of* Oreochromis mossambicus *(Peters, 1852) (Pisces: Cichlidae) in the Sepik River, Papua New Guinea – With Reference to the Management of the Fishery*, unpublished PhD Thesis, University of Hull.

Ribbink, A. (1991) Distribution and ecology of the cichlids of the African Great Lakes, in *Cichlid fishes. Behaviour, Ecology and Evolution* (ed M.H.A. Keenleyside), Chapman and Hall, London, pp. 36–85.

Robinson, R.L., Turner, G.F., Grimm, A.S. and Pitcher, T.J. (1990) A comparison of the ingestion rates of three tilapia species fed on a small planktonic alga. *J. Fish Biol.* **36**, 269–270.

Robinson, R.L., Turner, G.F., Grimm, A.S. and Pitcher, T.J. (1995) An experimental study of phytoplankton feeding in three tilapiine cichlids. *J. Fish Biol.* **46**, 449–459.

Ross, B. and Jauncey, K. (1981) A radiographic estimation of the effect of temperature on gastric emptying time in *Sarotherodon niloticus* (L) × *S. aureus* (Steindachner) hybrids. *J. Fish Biol.* **19**, 333–344.

Ross, L.G., McKinney, R.W. and Ross, B. (1988) Energy budgets for cultured tilapias, in *Proceedings of the Second International Symposium on Tilapias in Aquaculture (ICLARM Conf. Proc.*, Vol. 15) (eds R.S.V. Pullin, T. Bhukaswan, K. Tonguthai and J.L. Maclean), ICLARM, Manila, pp. 83–89.

Sakata, T., Uno, K. and Kakimoto, D. (1984) Dominant bacteria of the aerobic microflora in *Tilapia* intestine. *Bull. Jap. Soc. Sci. Fish.* **50**, 489–493.

Sanderson, S.L. and Wassersug, R. (1993) Convergent and alternative designs for vertebrate suspension feeding, in *The Skull, Vol. 3, Functional and Evolutionary Mechanisms* (eds J. Hanken and B.K. Hall), University of Chicago Press, Chicago, pp. 37–111.

Sanderson, S.L., Stebar, M.C., Ackerman, K.L., Jones, S.H., Batjakas, I.E. and Kauffman, L. (1996) Mucus entrapment of particles by a suspension-feeding tilapia (Pisces: Cichlidae). *J. Exp. Biol.* **199**, 1743–1756.

Sasagawa, I. (1997) Fine structure of the cap enameloid of the dental epithelial cells during enameloid mineralization and early maturation stages in the tilapia, a teleost. *J. Anat.* **190**, 589–600.

Schroeder, G.L. (1978) Autotrophic and heterotrophic production of microorganisms in intensely-manured fish ponds and related fish yield. *Aquaculture* **14**, 303–325.

Scocco, P., Ceccarelli, P. and Menghi, G. (1996) Glycohistochemistry of the *Tilapia* spp. stomach. *J. Fish Biol.* **49**, 584–593.

Scocco, P., Menghi, G. and Ceccarelli, P. (1997). Histochemical differentiation of glycoconjugates occurring in the tilapine intestine. *J. Fish Biol.* **51**, 848–857.

Scocco, P., Accili, D., Menghi, G. and Ceccarelli, P. (1998). Unusual glycoconjugates in the oesophagus of a tilapine polyhybrid. *J. Fish Biol.* **53**, 39–48.

Shrestha, M.V. and Knud-Hansen, C.F. (1994) Increasing attached microorgaism biomass as a management strategy for Nile tilapia (*Oreochromis niloticus*) production. *Aquacult. Eng.* **13**, 101–108.

Spataru, P. (1976) The feeding habits of *Tilapia galilaea* (Artedi) in Lake Kinneret (Israel). *Aquaculture* **9**, 47–59.

Spataru, P. and Zorn, M. (1978) Food and feeding habits of *Tilapia aurea* Steindachner in Lake Kinneret (Israel). *Aquaculture* **13**, 67–69.

Stroband, H.W.J. (1977) Growth and diet dependent structural adaptations of the digestive tract in juvenile grass carp (*Ctenopharyngodon idella* Val.). *J. Fish Biol.* **11**, 167–174.

Sugita, H., Tokuyama, K. and Deguchi, Y. (1985) The intestinal microflora of carp *Cyprinus carpio*, grass carp *Ctenopharyngodon idella* and tilapia *Sarotherodon niloticus*. *Bull. Jap. Soc. Sci. Fish.* **50**, 1325–1329.

Sugita, Kawasaki, J. and Deguchi, Y. (1997) Production of amylase by the intestinal flora in cultured freshwater fish. *Lett. Appl. Microbiol.* **24**, 105–108.

Trewavas, E. (1983) *Tilapiine Fishes of the Genera* Sarotherodon, Oreochromis *and* Danakilia, British Museum (Natural History), London.

Turner, G.F., Grimm, A.S., Mhone, O.K., Robinson, R.L. and Pitcher, T.J. (1991) The diet of *Oreochromis lidole* (Trewavas) and other chambo species in Lakes Malawi and Malombe. *J. Fish Biol.* **39**, 15–24.

Welcomme, R. (1972) An evaluation of the acadja methods of fishing as practiced in the coastal lagoons of Dahomey (West Africa). *J. Fish Biol.* **4**, 39–45.

Whitehead, P.J.P. (1959) The feeding mechanism of *Tilapia nigra*. *Nature* (London) **184**, 1509–1510.

Whitfield, A.K. and Blaber, S.J.M. (1978) Resource segregation among iliophagous fish in St Lucia, Zululand. *Env. Biol. Fishes* **3**, 293–296.

Yamaoka, K. (1991) Feeding behaviour, in *Cichlid fishes. Behaviour, Ecology and Evolution* (ed M.H.A. Keenleyside), Chapman and Hall, London, pp. 151–172.

Yowell, D.W. and Vinyard, G.L. (1993) An energy-based analysis of particulate-feeding and filter-feeding by blue tilapia. *Env. Biol. Fishes* **36**, 65–72.

Zihler, F. (1982) Gross morphology and configuration of digestive tracts of Cichlidae (Teleostei, Perciformes): phylogenetic and functional significance. *Neth. J. Zool.* **32**, 544–571.

Chapter four

Environmental physiology and energetics

L. G. Ross

4.1 INTRODUCTION

The tilapias are extremely successful fishes and this success is largely due to their robustness, tolerance, flexibility and overall plasticity. This plasticity of growth, reproductive and developmental processes is evident from their well-known diversification and radiation into available niches, and is characterized by a remarkable physiological hardiness, adaptability and general levels of tolerance to most potentially limiting environmental variables. Although essentially freshwater species, many tilapias are euryhaline and so can be cultured in fresh, brackish or salt water. While they are not cold tolerant, they are eurythermal over a wide range, and this only limits their distribution to tropical, sub-tropical and warm temperate climates. They also have a strong reputation for tolerance of low dissolved oxygen (DO) and are quite resistant to reasonable physical handling, more so than most other fishes.

Perhaps the earliest wide-ranging review, which attempted to cover every aspect of tilapia biology and culture, was that of Balarin and Hatton (1979). Other reviews of aspects of tilapias, particularly dealing with environmental requirements and tolerances, are those by Caulton (1982), Chervinski (1982), Philippart and Ruwet (1982) and Coche and Edwards (1989). Philippart and Ruwet (1982) reviewed the natural distribution of tilapias and pointed out the wide range of colonized habitats, particularly in terms of temperature range, current velocities, salinity and alkalinity.

In some cases, a good deal of information exists on aspects of tilapia physiology. In other cases this is not so, especially on close examination of the

M.C.M. Beveridge and B.J. McAndrew (eds), Tilapias: Biology and Exploitation, 89–128.

data. For example, up to the late 1970s some data existed on respiration (Job, 1969; Farmer and Beamish, 1969; Kutty, 1972; Caulton, 1977, 1978) but it was rather fragmentary and certainly did not address itself to the problems of culture.

This chapter focuses on the physiology of tilapia at the level of the whole animal. Reproductive and nutritional physiology are covered in other chapters and these topics are not considered here, except where they impinge upon consideration of energy budgets. Many features of tilapia physiology are not significantly different from other fishes and so in some cases discussion is limited to consideration of comparative tolerances within the group. There are, however, some aspects of tilapia physiology, structure and function which are unique, and these will be discussed in more detail. The underlying theme of much of this book is the *exploitation* of the species and there are some features of their physiology, e.g. respiratory requirements and capabilities, which clearly require more emphasis than others.

4.2 BODY DESIGN AND GENERAL ANATOMICAL FEATURES

The overall body design is of a fairly conventional, laterally compressed, deep-bodied cichlid. Although the *Cichlidae* are among the most advanced types of fish, Trewavas (1983) noted that the tilapias had retained some features considered primitive, such as low vertebral count and only four lateral line openings. The scales are relatively large and are not easily dislodged, an important feature in cultured animals. The median fin rays have many hard, strong spines as well as soft ones. The large paired fins are anterior in an advanced configuration, which confers superb control over swimming, turning and manoeuvring. The fins are used a lot in locomotion of this group and consequently they have much red muscle, designed for relatively low-speed but continuous action. The spectacular locomotor control with which cichlids are endowed due to their body shape, musculature and fin layout and control is far greater than in any other fish group.

The species are generally vertically barred, but in relatively subdued colours that blend extremely well with their background. This is assisted by a modest ability to change colour by controlling the chromatophores in the skin; thus fish can modify their overall appearance to become relatively pale (blanched) or dark and this is frequently seen in response to stressors.

The sense organs are well developed and there are prominent nares and a notable lateral line. The eyes are relatively large and the species have excellent visual capabilities. The visual system is probably very important in feeding and in many aspects of courtship and reproductive behaviour. This is clear from the success of the arena-based breeding systems described by Balarin and Hatton (1979) which were developed from careful observations of lek behaviour in a number of tilapia species (see also Turner and Robinson, this volume). The swimbladder is almost certainly involved in enhancing sound detection and perhaps sound production, although the role of sounds in the life of tilapias is unknown.

The species have a series of fine mandibular and maxillary teeth, but the most notable feature of the mouth is the pharyngeal bones and teeth, which vary in design according to feeding habit (Fryer and Iles, 1972). The gut is of variable length but is usually several times the body length, typical of a herbivorious animal. The stomach is present but is correspondingly thin-walled and has a weak musculature. There are no pyloric caecae (see Beveridge and Baird, this volume).

The swimbladder

The swimbladder is a well-developed physoclist type, with a notable gas gland and a moveable diaphragm, similar to designs found in many perciforms. Both gas secretion and resorption are under vascular control. A well-developed swimbladder may confer one or more of the following capabilities: buoyancy, respiratory support, sound production, sound detection and pressure detection.

Initial swimbladder inflation

The swimbladder in tilapias seems to develop from a morula of cells adjacent to the dorsal surface of the oesophagus (Doroshev *et al.*, 1981; Cornacchia, 1982). This hollows out and is simultaneously invaded by a developing pneumatic duct arising from the oesophageal surface. The pneumatic duct eventually atrophies, leaving a physoclist arrangement. First filling is thought to be achieved totally internally by secretion of gas from a glandular epithelium which develops inside the cell mass (McEwen, 1940; Doroshev *et al.*, 1981). The result of this arrangement is that larval tilapias do not require access to the water surface to fill their swimbladders. Al-Abdul-Elah (1990) considered that this was a feature of many demersal species which, as a consequence, do not suffer the high initial mortalities in culture which have been related to air-gulping and surface access in many other species (Chatain and Dewavrin, 1989; Chatain and Ounais-Guschemann, 1990).

Buoyancy

Whatever the principal function of a swimbladder in a given species, the controllable physoclist swimbladder inevitably confers buoyancy, which should, in turn, enable exploitation of a wider range of depths. Philippart and Ruwet (1982) reviewed known depth ranges for a number of species, noting that most remain in relatively shallow water, while the maximum depth visited was of the order of 40 m, depending on species. It is likely that the maximum depth of tilapias is limited to some extent by their thermal preferences, deeper waters being in general cooler. Caulton (1982) suggested that tilapias exploited shallower, warmer waters during the daytime for feeding but that they retired to deeper, cooler waters at night to carry out digestion. He calculated that this migratory process might have an energetic advantage when the daily energy budget was considered.

The buoyancy conferred by the swimbladder is put to great effect in minimizing the energy needed to maintain station within its vertical range and this allows tilapias, in common with other cichlids, to develop a superb ability to manoeuvre, change direction, and stop dead in the water. Grazing habits and other station-maintaining requirements are all facilitated by the high state of development of the buoyancy function of this organ.

Sound production

Rodman (1966) recorded sounds between 50 and 600 Hz from *O. mossambicus*. Sounds in this frequency range are typically produced in fish by a range of mechanisms involving the swimbladder and are used in a number of forms of communication between individuals. Later attempts to record sounds from tilapias have only produced results from higher frequency ranges. Lanzing (1974) recorded sounds limited to the higher frequencies, from 1 to 16 kHz, with notable pulses of 7 kHz noise. These sounds are characteristic of stridulatory noises and have been attributed to sounds produced by the jaw teeth when rasping on pebbles and to the rasping effect of the pharyngeal teeth on each other. Noises from these pharyngeal structures have also been recorded in the author's laboratory on many occasions using hydrophones, but no other lower-frequency sounds have ever been heard. Lanzing (1974) described the musculature of the dorsal pharyngeal pad as 'sonic' musculature, although this probably belies the primary function of the muscles in the grinding of food.

Sound detection

This leaves the possibility that sound detection may be enhanced. A close connection between anteriorly developed diverticulae of the swimbladder and the auditory apparatus in the genus *Etroplus* was described by Dehadrai (1959). Although not so well developed, a similar function is highly likely in tilapias because of the close connection between the swimbladder and an apophyseal peg which descends from the third vertebra (Fig. 4.1). This peg varies in structure and has been used to assist classification in the past (Trewavas, 1983). Pressure waves in the audio range will certainly be intercepted by the swimbladder because of its phase difference and the direct and close attachment of the swimbladder to this peg means that sounds will be directly transmitted to the bony structures just posterior to the auditory apparatus. Although this arrangement is not as highly developed as the Weber's ossicles in cyprinids, it is nevertheless a significant structural enhancement of hearing, which is even

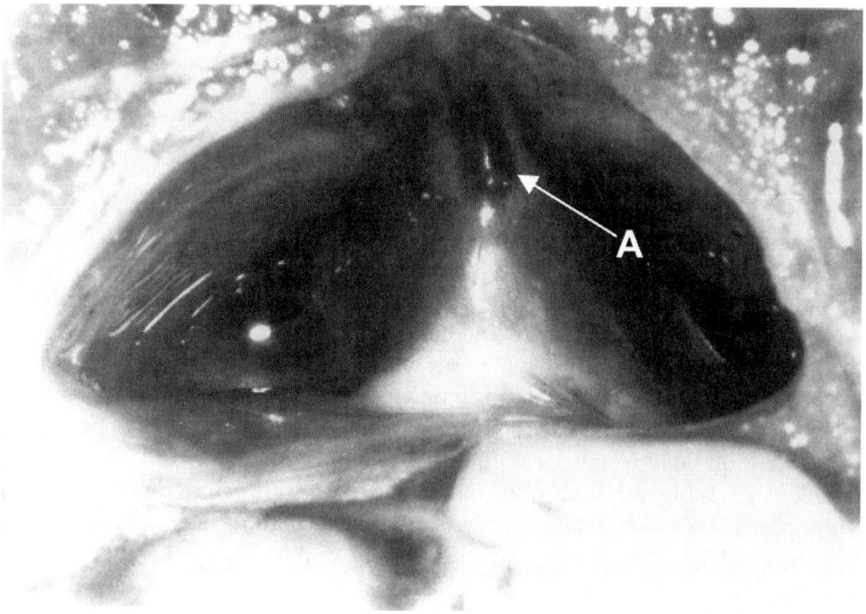

Fig. 4.1. Attachment of swimbladder to the apophyseal peg (A) in the swimbladder of *Oreochromis niloticus*, seen from the rear.

more highly developed elsewhere in the phylum. The benefit of this anatomical arrangement was confirmed by Huang *et al.* (1996) who measured microphonic potentials in the sacculus and cerebellar–medullary junction in tilapia. They showed that the fish responded to sounds between 20 and 900 Hz, with a peak response between 63 and 130 Hz and that ablation of the swimbladder reduced sensitivity.

Accessory respiratory function

The swimbladder of all physoclists is inflated by secretion of oxygen gas from the gas gland, often following initial inflation in the larval stage by air gulping at the water surface. In fish with a stable vertical location, this oxygen is usually slowly replaced by nitrogen. However, species which migrate vertically need to secrete oxygen regularly and the oxygen content of the swimbladder thus remains high (often over 80%). The use of this gas as a major respiratory source is unlikely in any long-term sense, as there simply is not enough of it. However, there is little doubt that the high oxygen content, which can occur in the gas mixture, could be used as a short-term reservoir. Resorption of the gas into the well-vascularized absorptive part of the swimbladder requires only adjustment of the tissue windows and the dilation of arterioles and capillaries, features which have been well documented in other species (Ross, 1979a,b). This may well contribute to the remarkable tolerance to short-term hypoxia of the tilapia group as a whole

4.3 HANDLING AND ANAESTHESIA

The sturdy *perciform* body design, large scales and temperature tolerance of tilapias makes them very robust and relatively easy to handle, although handlers need to beware the strong fin spines. Given the usual precautions against changes in temperature and desiccation (Ross and Ross, 1999), tilapias can be handled for all normal purposes without trouble. Where needed, tilapias respond well to a number of common anaesthetic agents. Ross and Geddes (1979) found benzocaine to be very effective and, in general, dose level is dependent on body weight, less anaesthetic being required with larger animals (Table 4.1). Table 4.2 shows the times for sedation, handling and recovery at these dose levels. Ross and Geddes (1979) noted that even yolk-sacced fry recovered after immersion for 5 minutes in 1:10 000 benzocaine and that repeated anaesthesia did not appear to cause growth checks. Extended anaesthesia at lower dose rates of approximately 1:40 000 also works well in these species (Ross and Ross, 1999).

Table 4.1 Dose rates for benzocaine anaesthesia for routine handling of tilapia at 25°C (from Ross and Geddes, 1979)

Body weight (g)	Dose rate
0–0.5	1:15 000
0.5–1.0	1:15 000
1.0–5.0	1:10 000
5–10	1:10 000
10–50	1:10 000
50–100	1:5 000

4.4 OXYGEN CONSUMPTION AND UTILIZATION

Oxygen is the first limiting variable in the aquatic environment, other than water itself. In general, it is considered to be a scarce resource and is available in much lower quantities than are at the disposal of terrestrial animals; a few milligrams per litre of environment rather than 20% of the total, as in air. Aquatic biologists and aquaculturists need to manage the environment of their animals and so need to know the oxygen requirements of animals with which they are experimenting or working. Once armed with this information, controlled husbandry becomes possible both in the laboratory and on the farm.

Table 4.2 Time for sedation, anaesthesia and first recovery of 25 g *Tilapia mariae* at different temperatures and benzocaine concentrations (from Ross and Geddes, 1979)

		Temperature (°C)		
Dose level	Stage	20	25	30
1:10 000	Sedation I-2	40	35	30
	Anaesthesia II-1	99	60	44
	Recovery	20–150	40–120	–
1:5 000	Sedation I-2	25	22	20
	Anaesthesia II-1	32	30	25
	Recovery	60–150	20-60	60–150
1:2 500	Sedation I-2	15	19	–
	Anaesthesia II-1	25	23	17
	Recovery	60–300	60–150	45–180

The accumulated biochemical processes occurring within an animal are known as *metabolism* and this can proceed at different rates. It is known that a very wide range of biotic and abiotic factors affect the metabolic rate of fishes (Brett, 1979). The rate of metabolism is reflected by, and can be measured by, the rate of oxygen uptake, carbon dioxide release or even direct measurement of heat output from an animal. Measurement of heat output is especially difficult in aquatic animals because the specific heat of the medium requires a very large energy output from the animal in order to effect a measurable change in temperature. In consequence, in most cases oxygen consumption is used as the indicator, although in some cases carbon dioxide levels have also been measured simultaneously. Factors affecting oxygen consumption and utilization are very important in the design and management of intensive and semi-intensive fish culture systems (Meyer and Brune, 1982). They can be crucial in the tropics and sub-tropics where fish growth and survival in the aquatic environment is frequently oxygen-limited (Boyd, 1982). The major factors to be considered here are body weight, temperature, salinity, photoperiod, dietary heat increment and motor activity.

Tilapia gills

Oxygen uptake is initially via the skin in early fry but the gills take over in juveniles and adults. The gill arrangement is quite conventional and the gillrakers are relatively undifferentiated. However, tilapias have the additional feature of rows of microbranchiospines on the gill arch (Fig. 4.2) which are rather beautiful microstructures of uncertain function (Northcott *et al.*, 1991).

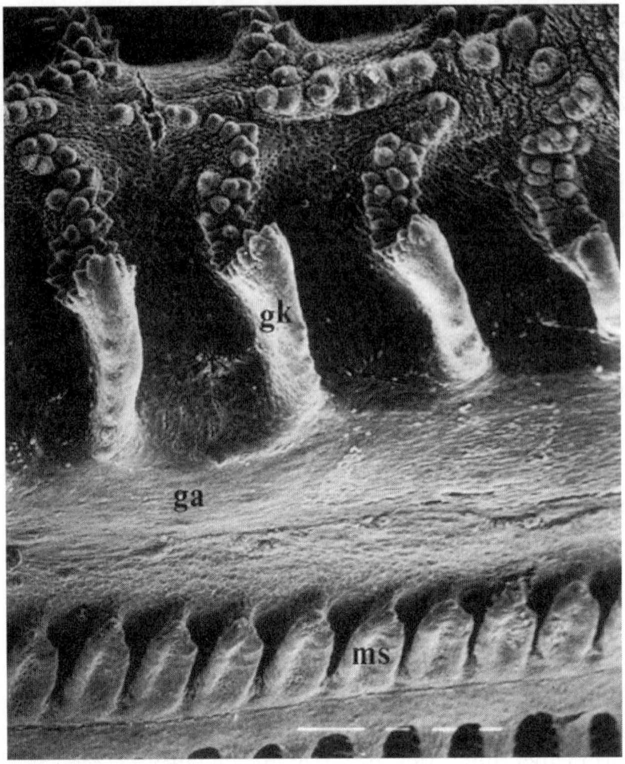

Fig. 4.2 Scanning electron micrograph of microbranchiospines (ms) and gill rakers (gk) on the gill arch (ga) of *Oreochromis niloticus* (from Northcott *et al.*, 1991).

Oxygen tolerance and anoxia

Tilapias can tolerate poor water quality and low dissolved oxygen concentrations. Survival after short-term exposure to 0.1 mg l^{-1} or less is common and widely reported (see Chervinski, 1982). This tolerance of low DO is assisted by haemoglobins which can bind O_2 at low tensions; 0.1 ppm in *S. macrochir* (Dussart, 1963) and 0.17 ppm in *S. esculentus* (Fish, 1956). They will also withstand relatively elevated CO_2 concentrations; 72.6 ppm in *S. macrochir* (Dussart, 1963) and 50 ppm in *S. esculentus* (Fish, 1956).

It is well known that tilapia will gulp air at the water surface when DO falls, for example in the hours before dawn in the tropics. Atmospheric oxygen dissolves in the buccal water and passes quickly to the gills thus enhancing *apparent* tolerance of low DO. Based principally upon evidence from respiratory quotient and muscle lactate levels, it has been suggested that tilapias may be able to respire anaerobically over short periods (Kutty, 1972) although the assertion by Karuppannan (1981) that *O. mossambicus* is a more tropical species and hence more anaerobic is somewhat questionable.

The implied message of extreme tolerance to low DO has led many to assume that tilapia will grow and thrive in such conditions. In fact there is a great deal of hard evidence to the contrary and they will not tolerate low DO in the long-term, nor will growth, feeding, digestion or reproduction be normal in these circumstances.

The effects of temperature and body weight on oxygen consumption

In common with many other fish studies, Ross and Ross (1983) showed that there is a direct inverse relationship between $\log_{10}$ body weight and $\log_{10}$ respiration rate (Fig. 4.3). In addition, there is a significant increase in the rate of oxygen consumption with increasing temperature. The regression equations and correlation coefficients for these relationships are shown in Table 4.3. Many previous studies have used closed respirometers, giving unreliable results because fish cannot be properly acclimated to the system. Ross and Ross (1983) used a hybrid open/closed system and corroborated their laboratory measurements with field estimates at 30°C, finding very close correspondence (Fig. 4.3). The Q_{10} was approximately 2.3 between 25°C and 35°C. This is similar to values of 2.2 between 20°C and 30°C and 1.6 between 20°C and 25°C obtained by Caulton (1977) for *Tilapia rendalli* Boulenger.

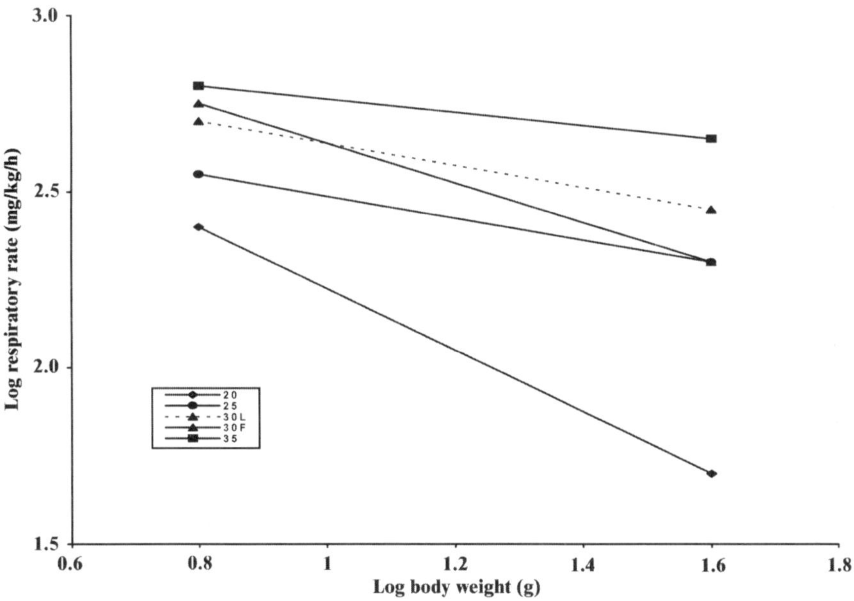

Fig. 4.3 Resting respiratory rate of *Oreochromis niloticus* at different temperatures.

Table 4.3 Regression equations for resting oxygen consumption rate *in Oreochromis niloticus* (from Ross and Ross, 1983)

Temperature (°C)	Regression equation	n	r	Significance of r
20	y = 3.00–0.777x	16	–0.979	>0.0.001
25	y = 2.80–0.350x	16	–0.822	>0.01
30 lab	y = 2.99–0.378x	24	–0.918	>0.001
30 field	y = 3.34–0.586x	14	–0.892	>0.001
35	y = 3.03–0.255x	32	-0.713	>0.001

Thermal history and acclimation is very important in respirometry. Fernandes and Rantin (1986) showed that instantaneous changes in temperature of 10°C resulted in rapid increase or decrease in metabolic rate, followed by a slow acclimation to a stable value. In the case of temperature increase, this adjustment required 4 to 5 days and in the case of temperature decrease, adjustment required 7 to 14 days (Fig. 4.4). This effect is well worth considering when moving stocks to a new location.

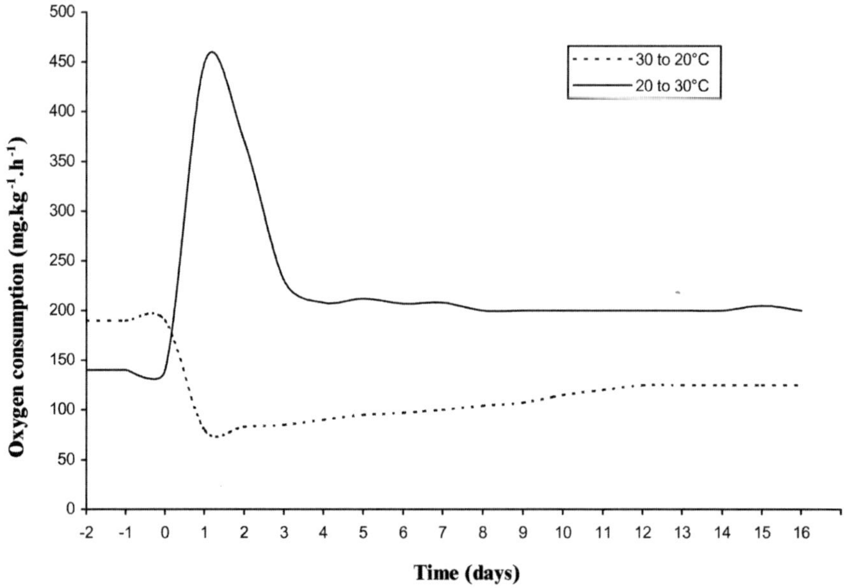

Fig. 4.4 Variation in metabolic rate of *Oreochromis niloticus* following instantaneous change in temperature (after Fernandes and Rantin, 1986).

Effects of hypoxia on oxygen consumption

O. niloticus is an efficient respiratory regulator in water containing more than 3.0 mg l^{-1} (Fig. 4.5). Ross and Ross (1983) showed that the respiratory response to rapidly-induced hypoxia was not obvious until dissolved oxygen levels were below 4 mg l^{-1} and there was no evidence of an increase in rate as the critical oxygen tension (pC) was approached. Between 2.5 and 1.0 mg l^{-1} dissolved oxygen there was an increase in spontaneous activity accompanied by a darkening of body coloration and erection of the dorsal fin. Below DO levels of about 1.0 mg l^{-1} there was a substantial decrease in oxygen consumption to approximately 5% of that in normoxia. This was accompanied by a marked vasodilation of the snout and fin sinuses. The critical point, pC, is expressed as an absolute oxygen tension and depends upon temperature. Recalculation of these data show that pC was approximately 60 mmHg and below this value the rate of oxygen consumption declined markedly. Becker and Fishelson (1986) carried out more controlled trials to determine pC in *O. niloticus* and found slightly lower values of 42 and 35 mbar (= 31 to 26 mmHg) at 25 and 36°C, respectively. Fernandes and Rantin (1994) measured oxygen uptake in tilapias in hypoxic conditions and calculated that the cost of ventilation increased from 3% of resting rate in normoxia to 18% in extreme hypoxia.

All of these observations are in close agreement with the practical findings of Coche (1977) and Melard and Philippart (1980) who, respectively, reported that DO values in culture systems of 2.3 and 3.0 mg l^{-1} caused depression in growth. The ability of *O. niloticus* to survive at very low oxygen levels with a concomitant

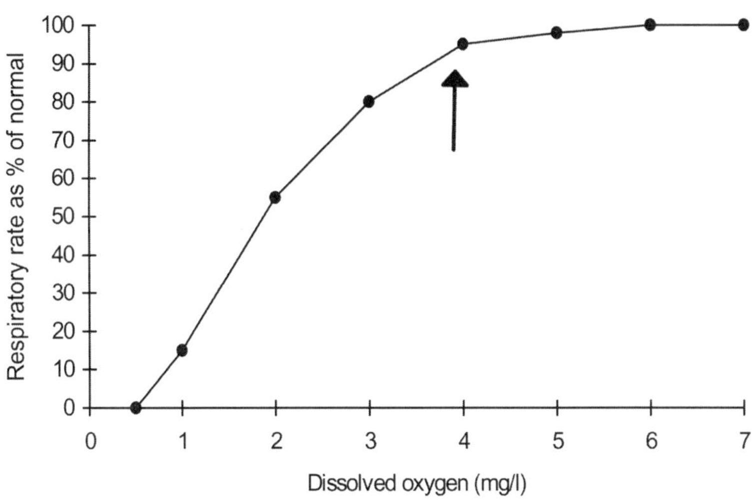

Fig. 4.5 The effect of hypoxia at 30°C on *Oreochromis niloticus*. The critical point, pC, is indicated by the arrow.

low respiratory rate suggests that they may be capable of some anaerobic respiration (Kutty, 1972). This may be the case in the short term, and the effect may be further enhanced because of oxygen available from the swimbladder on a temporary basis. However, in reality, survival times below 3.0 mg l^{-1} will be progressively limited as DO falls and it is advisable that culture conditions are maintained well above this level.

Effects of handling stress on oxygen consumption

In simulated trials, Ross and Ross (1983) held fish out of water for 2 minutes and then returned them to the respirometers. There was a pronounced post-handling increase in oxygen consumption rate ranging from +150% to +300% of the resting value (Fig. 4.6). Fish of different body weights were used and the general pattern was for the initial high respiratory rate to decrease significantly over the first hour, post-stress. Further decline in rate was, however, much slower and in many cases had not fully returned to the resting value after 3 h.

Fish that had been sedated during the handling process responded in essentially the same fashion as those that had not been sedated (Fig. 4.6). It is thus clear that the use of sedatives does not ameliorate the high oxygen demand consequent to handling, although this does not detract from the other desirable features of use of sedatives during stressful procedures.

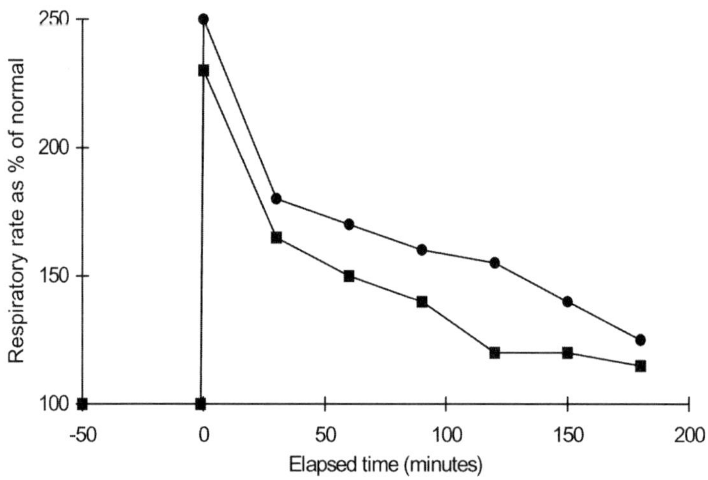

Fig. 4.6 Increase in respiratory rate of *O. niloticus* following handling. ●, without sedation, ■ with benzocaine sedation.

Respiratory photoperiodicity

Respiratory cycles in fish are well known and usually resolve into either a single daily peak or a crepuscular pattern. Using computer-operated respirometers, Ross and McKinney (1988a,b) showed a single daily peak in the daytime respiration rate of *O. niloticus* which increased by more than 30% over the night-time rate under otherwise constant conditions (Fig. 4.7). Fish of approximately 100 g body weight showed a variation from 80 mg kg h^{-1} at night to 120 mg kg h^{-1} during daylight at 25°C, with a mean value of about 95 mg kg h^{-1}. Although the fish were starved throughout this procedure in order to evaluate resting oxygen consumption rate, the respiratory cycles were maintained without apparent detriment for up to 9 days (Fig. 4.8).

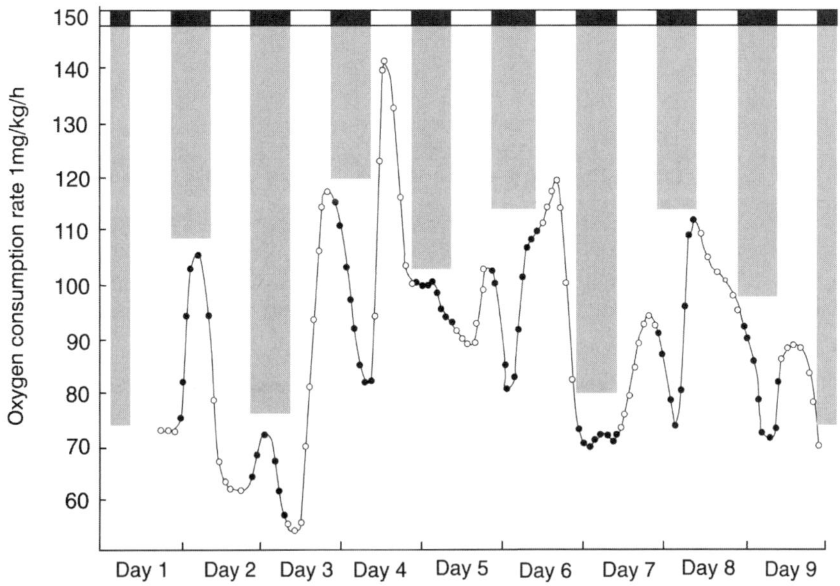

Fig. 4.7 Computer-smoothed variation in respiratory rate for an individual 96 g *Oreochromis niloticus* at 25°C and under a normal (12D:12L) photoperiod.

Exposure to constant light did not affect this respiratory rhythm over a 7-day period (Fig. 4.9). Constant darkness did not remove the respiratory rhythm either, although it was greatly reduced in magnitude and also became less clear as time progressed (Fig. 4.10).

This cyclic respiratory activity could clearly affect the calculated carrying capacity of a system depending on time of day. It is interesting to compare this performance with data for daily dissolved oxygen variation in an aquaculture pond. In Fig. 4.11, predicted respiratory cycles are compared with the predicted

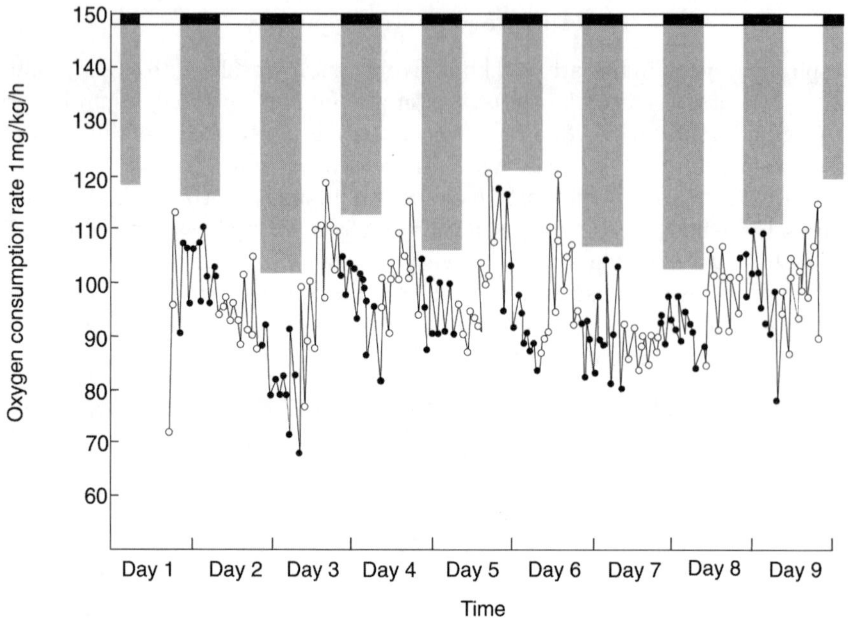

Fig. 4.8 Computer-smoothed variation in respiratory rate for *Oreochromis niloticus* at 25°C and under a normal (12D:12L) photoperiod. The data represent pooled responses from 8 fish.

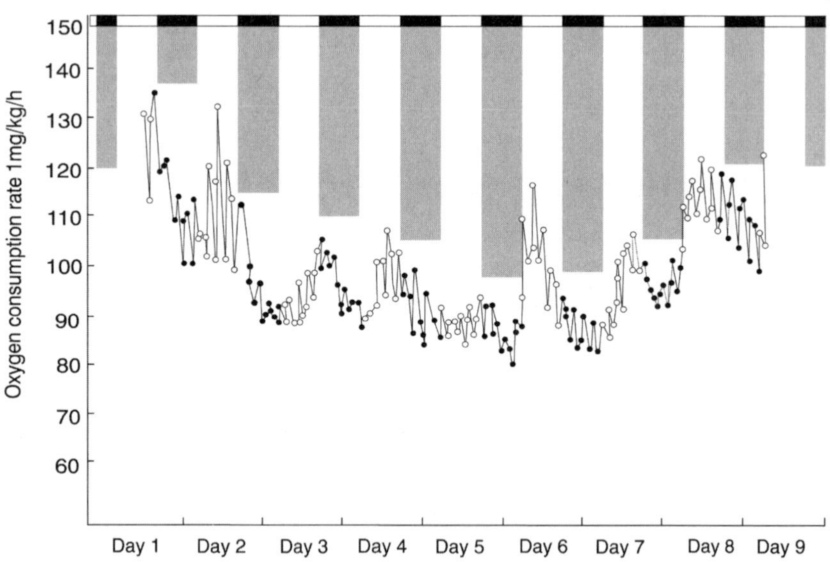

Fig. 4.9 Computer-smoothed variation in respiratory rate for *Oreochromis niloticus* at 25°C and in continous light (24D:0L). The data represent pooled responses from 12 fish.

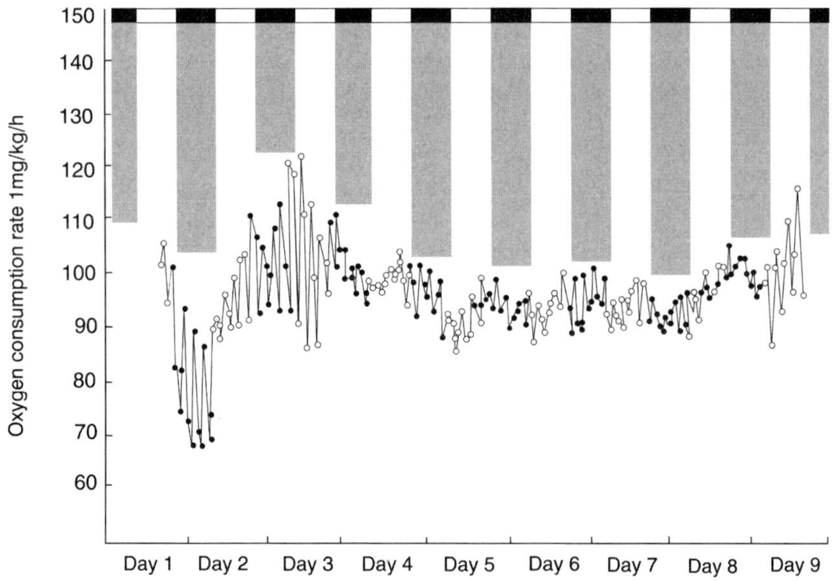

Fig. 4.10 Computer-smoothed variation in respiratory rate for *Oreochromis niloticus* at 25°C and in continuous darkness (0D:24L). The data represent pooled responses from 12 fish.

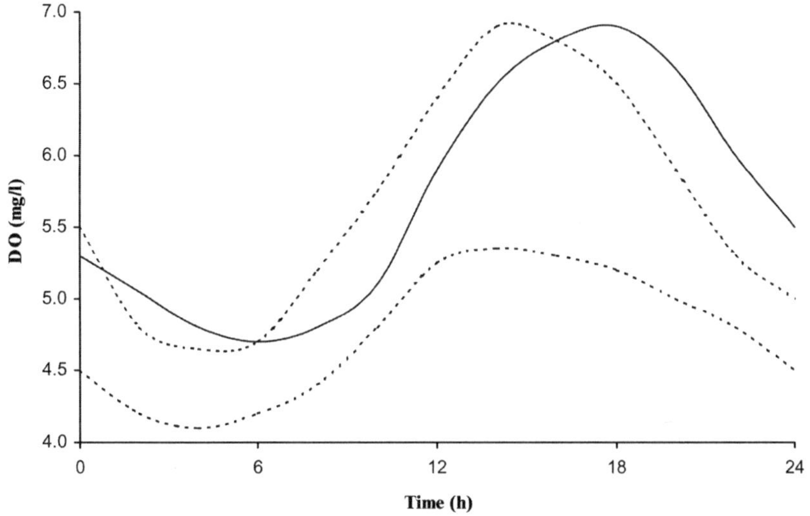

Fig. 4.11 The relationship between resting respiratory rate (solid line) and the maximum and minimum DO levels of a typical tilapia pond (pond data adapted from Meyer and Brune, 1982).

maximum and minimum DO in a subtropical pond using data derived by Meyer and Brune (1982). It is known that tilapias are oxygen conformers (see above) and it can be seen that as DO decreases in the pond, the resting respiratory rate of the fish would decrease, with a time lag of 2 to 3 h. In this example, it should be noted that the minimum DO does not fall below the critical oxygen tension (pC) which would result in a reduced, oxygen-dependent respiration. This resting respiratory behaviour has obvious adaptive significance in the natural environment but may also have implications for aquaculture in oxygen-limited systems. For example, in view of the additional oxygen requirement for feeding (SDA) it could be advisable to minimize early morning feeding with the bulk of the daily ration given later in the day so as to take full advantage of the available DO at this time.

There is some variation between studies even in the same species. For example, De Silva *et al.* (1986) have shown yolk-sac larvae of *O. niloticus* to have a crepuscular respiratory rhythm, whereas 12 mm fry have a single peak in respiration around dawn. Furthermore, Nagarajan and Gopal (1983) demonstrated crepuscular rhythm in *O. niloticus* of 5–25 g body weight. Although it is not clear why these differences should occur, the influence of developmental stage could be important, as could the respirometer techniques used.

Active metabolism

Body design and musculature

Tilapias are highly laterally compressed, have an advanced body shape and are designed for manoeuvrability rather than high speed. The majority of the body muscle is mosaic (white) muscle with only a small red muscle component, about 2% of the cross-sectional area of the body. This suggests a relatively modest routine lifestyle, with discontinuous low speed swimming, but with a high burst capacity. Indeed, this low-speed lifestyle is aided by the buoyant swimbladder and the use of the paired fins which confer remarkable control, turning and stopping ability and also a general, low-level propulsion system independent of the mass of body muscle. Not surprisingly, these fin muscles, which are in relatively constant use, are red muscle.

Swimming performance

Although tilapias have good burst swimming capabilities, they are not designed for sustained swimming at high speed. A good estimate of capacity is the critical swimming speed, Ucrit, which is the maximum velocity that can be sustained indefinitely, i.e. aerobically. This is usually estimated from flume experiments. Fig. 4.12 compares data from flume trials with tilapias and sockeye salmon and confirms that the tilapias have a generally lower Ucrit at a given size than the salmonid. Furthermore, in contrast to the salmonid, this capability does not

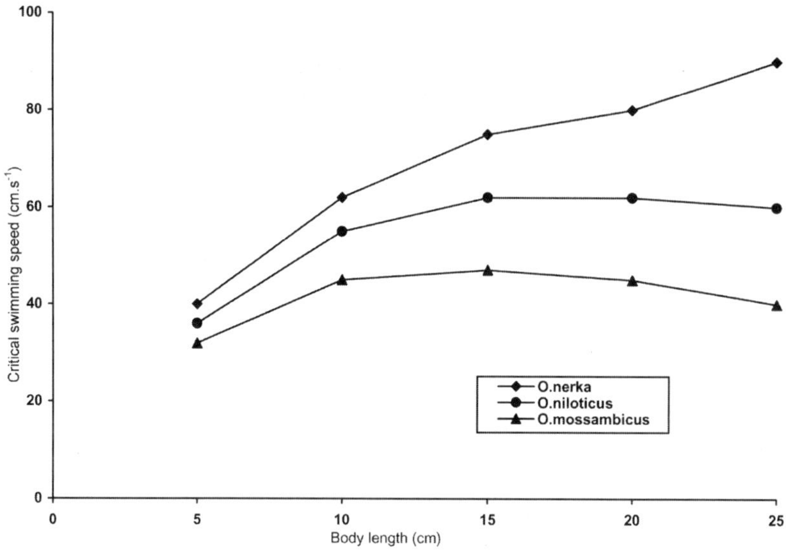

Fig. 4.12 Critical swimming speeds of tilapias (*O. niloticus** and *O. mossambicus**) and sockeye salmon (*Onchorhynchus nerka*†). *Ross, unpublished data, †adapted from Brett, 1979).

increase above a certain body length (about 15 to 20 cm), rather it appears to decrease. Such observations are not surprising in view of the habitat and lifestyle of most tilapias. Philippart and Ruwet (1982) reviewed flow characteristics of the natural environment of a number of species and only a few habitually appear in fast-running waters (Table. 4.4).

Table 4.4 Flow characteristics of the natural environment of some tilapias (from Philippart and Ruwet, 1982)

Species	Flow type
T. rheophila	fast
T. busumana	fast
S. andersonii	mid reaches
T. sparrmanii	mid-reaches

Active oxygen uptake

Farmer and Beamish (1969) measured oxygen consumption of *O. niloticus* at a range of swimming speeds and salinities using a Blazka respirometer and derived a series of regression equations relating oxygen consumption to these two parameters. Fig. 4.13 shows this relationship recalculated for a standard 80 g fish. As may be expected, oxygen consumption increased notably with swimming speed but was significantly lowered at a salinity of 11.6‰. It can be presumed that this salinity is close to the isosmotic point where the energy required for osmoregulation is least. They also showed that swimming performance improved and oxygen consumption decreased with training. Fig. 4.14. shows that, in general, oxygen consumption is lower than common carp at a given swimming speed, but similar to sockeye salmon, *O. nerka*. Much of this work was repeated over a wider mass range by Febry and Lutz (1987). They investigated how the energetic costs of maintenance, swimming and osmoregulation varied with salinity and showed that osmoregulatory costs did not limit delivery of oxygen to tissues at higher swimming speeds.

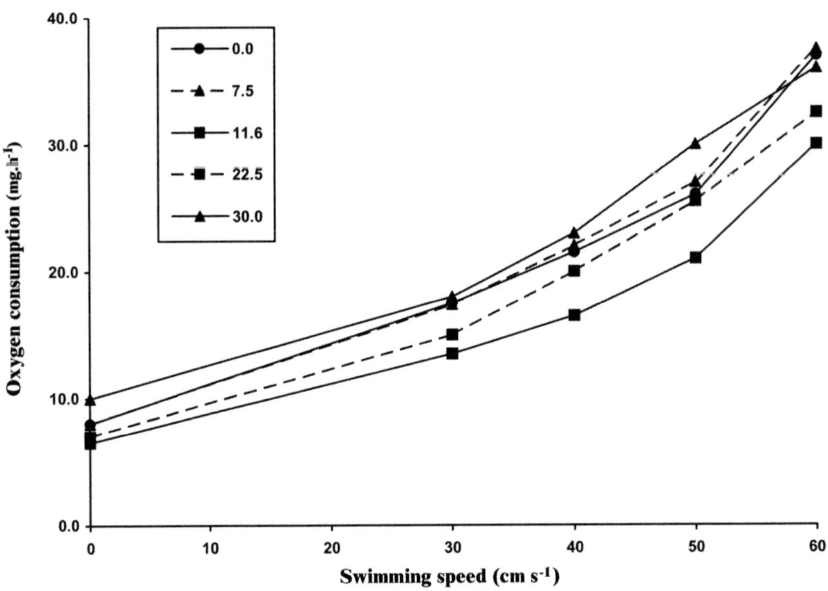

Fig. 4.13 The relationship between oxygen consumption and swimming speed at different salinities, recalculated for *Oreochromis niloticus* of 80 g body weight (after Farmer and Beamish, 1969).

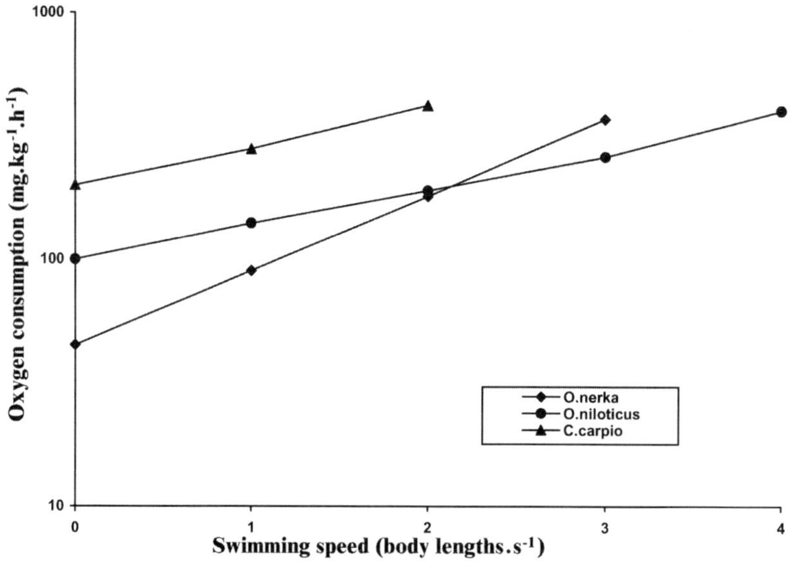

Fig. 4.14 The relationship between oxygen consumption and swimming speed for *Oreochromis niloticus*, *Cyprinus carpio* and *Onchorhynchus nerka* (adapted from Brett, 1979).

Factors affecting activity

There are two features of motor activity which have bearing on aquaculture operations; the capability of the species in the given conditions and the daily or seasonal patterns of motor behaviour of the species. From the range of abiotic and biotic factors which affect general motor activity, little systematic work has been done to quantify such effects in any fish species. Ross and co-workers (unpublished data) have used digital video monitoring to quantify motor behaviour of *O. niloticus* and to examine the effects of two major influencing factors, density and sex ratio. They used a tagged female fish in group sizes of 1, 2, 4, 8 and 16 fish m^2 which are representative of aquaculture stocking practices. Median daily swimming speeds ranging from 0.02 to 0.05 m s^{-1}, were attained, corresponding to specific swimming speeds of 0.11 to 0.29 body lengths s^{-1}. More high-speed swimming was seen at higher densities (Fig. 4.15). Motor energy loss in the tagged fish was 1295, 1547, 1840, 1691 and 2708 J day^{-1}, respectively. The effect of sex ratio was monitored in a tagged male or female fish in groups with sex ratios of 5 females:1male, 3 females:3 males or 1 female:5 males. Male fish motor energy expenditure was least at 4170 J day^{-1} in the balanced sex ratio group but was greatly elevated in unbalanced ratios. The female fish had a minimum daily motor energy expenditure of about 4400 J day^{-1}, which was much less affected by sex ratio. Overall, activity was clearly

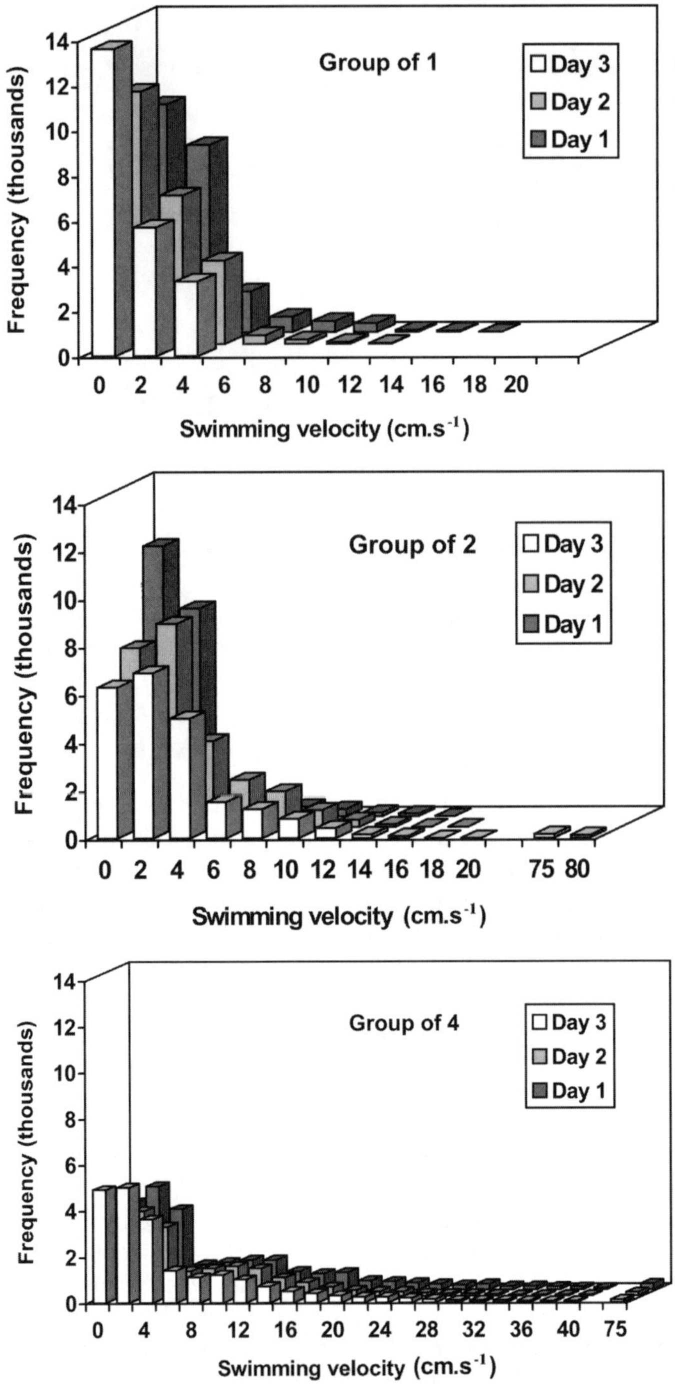

Group of 1
Day 3
Day 2
Day 1
Frequency (thousands)
0 2 4 6 8 10 12 14
0 2 4 6 8 10 12 14 16 18 20
Swimming velocity (cm.s-1)
Group of 2
Day 3
Day 2
Day 1
Frequency (thousands)
0 2 4 6 8 10 12 14
0 2 4 6 8 10 12 14 16 18 20 75 80
Swimming velocity (cm.s-1)
Group of 4
Day 3
Day 2
Day 1
Frequency (thousands)
0 2 4 6 8 10 12 14
0 4 8 12 16 20 24 28 32 36 40 75
Swimming velocity (cm.s-1)

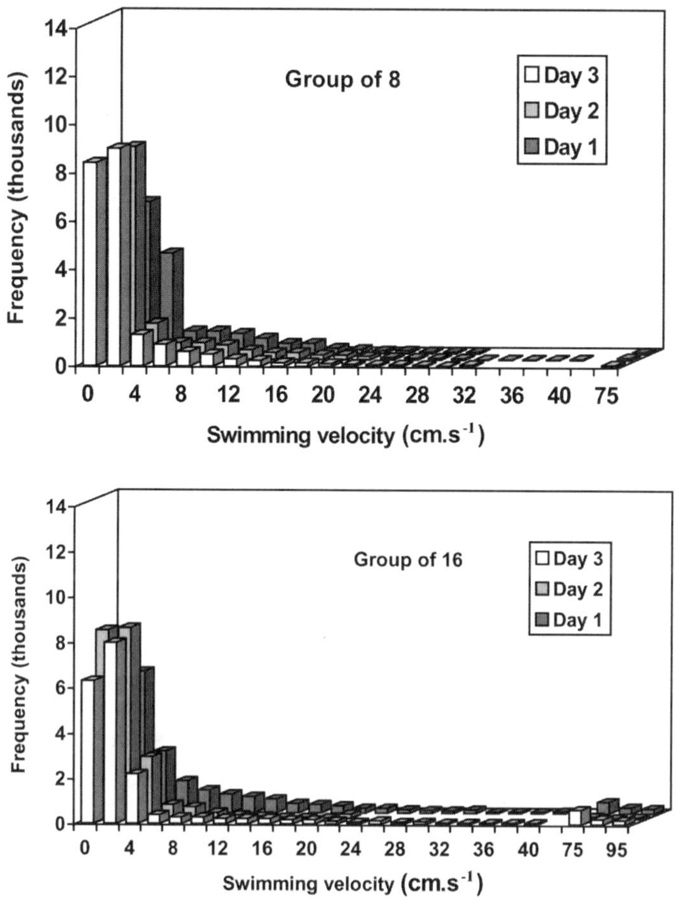

Fig. 4.15 Daily distribution of swimming speeds of a female *Oreochromis niloticus* in groups of 1, 2, 4, 8 and 16 animals. Note the high speed excursions at higher densities.

density dependent and also related to, and influenced by, individual behaviour and social interactions within the groups.

Van Ginneken *et al.* (1997) used digitized video and simultaneous direct and indirect calorimetry to investigate the link between motor activity and metabolic rate. They showed that activity was reduced, but not absent, in hypoxic conditions.

Specific dynamic action or heat increment

The biochemical processing of food is an energy-requiring step and the heat output and associated increased oxygen consumption due to this process is known as specific dynamic action (SDA) or heat increment. The main characteristics of post-prandial oxygen consumption and the SDA effect are the peak rate, the duration of the response and the overall magnitude of the response.

Ross *et al.* (1992) investigated the effects of dietary protein level, lipid level and ration level on SDA in *O. niloticus* in some detail. Oxygen consumption increases relatively rapidly after feeding, reaching a peak some hours later and eventually declines to the starting value (Fig. 4.16). The overall effect increases markedly with increasing protein level in the diet, up to 1.7 times the resting rate at the highest protein level (41%). The principal parameters of the responses are summarized in Table 4.5.

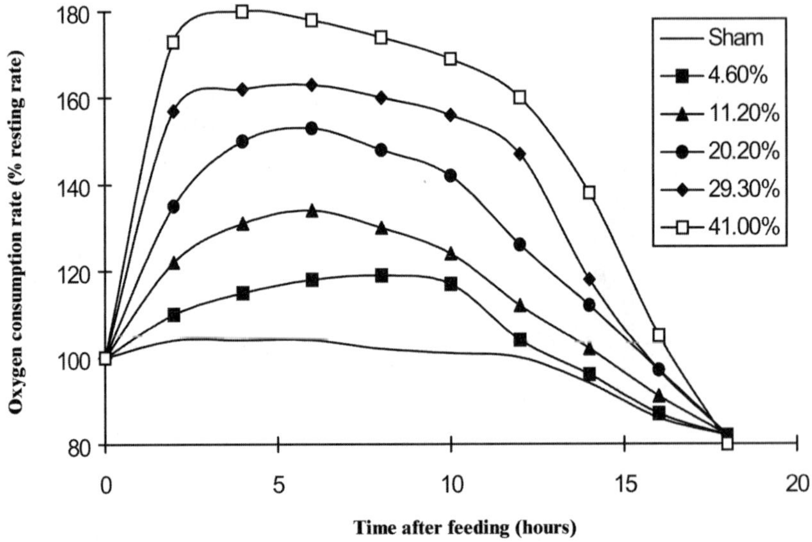

Fig. 4.16 Post-prandial increase in oxygen consumption rate in *Oreochromis niloticus* fed diets containing different levels of protein.

A good comparator of SDA is the SDA coefficient, which is the energetic value of the SDA response expressed as a proportion of ingested energy. Both the SDA coefficient and the SDA magnitude increase with dietary protein content (Fig. 4.17). The effect can be modelled for *O. niloticus* in the following equation:

$$\text{SDA coefficient} = 9.03 + 0.0502P\ 0.0541W$$

where, P = % crude protein in the diet and W = wet body weight.

Table 4.5 A summary of the effects of dietary protein level on the characteristics of the SDA effect in *Oreochromis niloticus*. Figures in parentheses are standard deviations

Protein level (%)	Time to peak (h)	Maximum (% increase)	Duration (h)	Magnitude (mg O_2/kg)	SDA coefficient
5	3.0	119	12.3	188	1.81 (0.36)
11	4.0	130	12.6	330	2.85 (0.52)
20	4.5	149	14.6	551	4.80 (0.68)
30	7.0	159	16.5	722	6.40 (1.07)
41	7.0	170	15.8	791	6.35 (1.35)

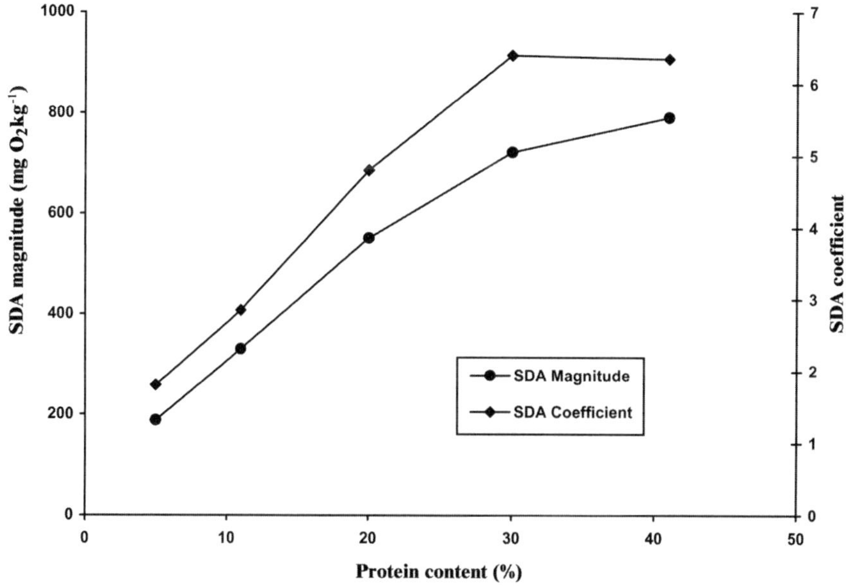

Fig. 4.17 Variation in SDA coefficient and SDA magnitude with dietary protein level in *Oreochromis niloticus*.

Although SDA magnitude and coefficient apparently decrease inversely with increase in dietary lipid, this effect has not been shown to be statistically significant in tilapia. This confirms previous observations (B. Ross, unpublished data) where no lipid sparing of dietary protein could be demonstrated in nutritional growth trials with tilapia. These results are, however, in contrast to those for rainbow trout where apparent SDA reduced with higher lipid levels.

The parameters of the SDA response also increase with increasing daily ration. Experiments designed to achieve this by repeated feeding have shown that the peak and duration of the oxygen consumption response increases in a cumulative manner and the SDA magnitude increases almost linearly with total food consumed (Figs 4.18, 4.19). The SDA coefficient was 8.96% and, as may be expected, did not vary significantly with increasing ration level.

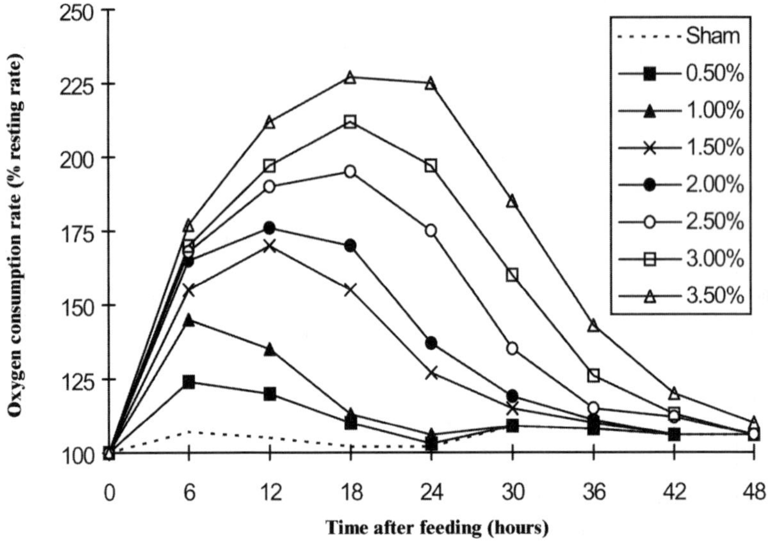

Fig. 4.18 Post-prandial change in oxygen consumption rate in *Oreochromis niloticus* fed different ration levels. The curves show the mean overall responses from a total of 46 fish measured.

These substantial post-prandial increases in metabolism and oxygen consumption have important consequences for fish culture. The maximum aerobic metabolic rate, and rate of oxygen consumption, is known as the 'active rate' and this is set for a given species by features of body design, gill size and properties of haemoglobins. The difference between the resting metabolic rate and this active rate is known as the *scope for activity* and all normal activity, including swimming, feeding and biochemical food processing, must take place within this ceiling if the fish is not to become stressed or oxygen-limited. As the scope is used up and the ceiling is approached, the potential for growth and reproduction decreases, as there is not enough aerobic scope to allow additional activities. A knowledge of these effects and the routes to management of this balance is important if fish in culture are not to become compromised and stressed.

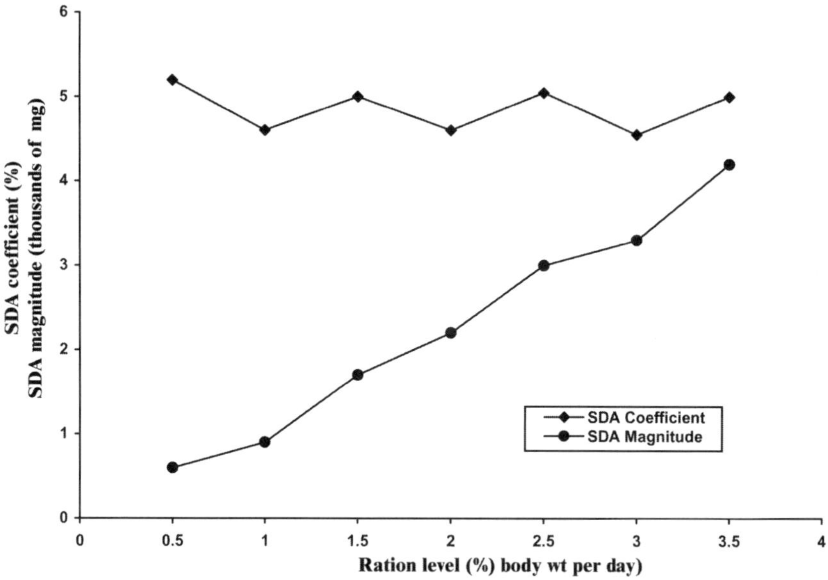

Fig. 4.19 Variation in SDA coefficient and SDA magnitude with different ration levels in *Oreochromis niloticus*.

Caulton (1982) estimated the energy costs of processing of the macrophyte *Ceratophyllum demersum* by the herbivorous *Tilapia rendalli*. He estimated that costs were low at lower temperatures and used this to support an energetic basis for offshore migration to cooler waters at night. The general applicability of this theory is unknown but *T. rendalli*, being macro-herbivores, are rather atypical.

4.5 NITROGENOUS EXCRETION

Ammonia

Freshwater fish mostly excrete nitrogenous wastes as ammonia, with only a small quantity of urea and very small quantities of other nitrogenous compounds such as uric acid, trimethylamine oxide, etc. Tilapias are no exception and have been shown to excrete 90% of their nitrogenous output from the head region (i.e. the gills) (Sayer and Davenport, 1987).

In culture systems, ammonia production needs to be managed so that it does not become toxic. In this context, it is the level of unionized ammonia which is important. Chervinski (1982) noted that, in culture, *O. aureus* is routinely able to tolerate total ammonia of 11 ppm at pH8 and 27°C, the unionized level being

0.75 ppm. Redner and Stickney (1979) determined the 48 h median lethal concentration of *O. aureus* to unionized ammonia to be 2.4 ppm. Following acclimation to low levels (0.43 to 0.53 ppm) for 35 days, the LC_{50} was slightly elevated to 3.4 mg l^{-1}. Hassan (1992) showed that 96 h LC_{50} for red *O. niloticus* was 1.64 ppm and Lin and Liu (1989) found that *O. mossambicus* × *niloticus* hybrids were stressed above 1.28 ppm.

In an energy budget study by Musisi (1984), rates of ammonia production were estimated for *O. mossambicus* at 1.72 mg N kg^{-1} h^{-1}. In the only detailed study on the physiology of ammonia production in tilapias, McKinney (1989) showed that the rate of endogenous production of ammonia in *O. niloticus* depended heavily upon body weight and temperature. The rate showed no significant variation with up to 6 days starvation. The rate varied at different times of day, but in no regular pattern and no significant difference was demonstrated between daytime and night-time rates. Regression equations resulting from this work are shown in Table 4.6.

Table 4.6 The relationship between endogenous ammonia production and body weight at different temperatures in *Oreochromis niloticus*

Temperature °C	Weight range (g)	Regression equation A = ammonia production in mg kg h^{-1} W = body weight (g)	Correlation coefficient
20	52–118	$\log_{10}A = 1.83 - (0.751 \times \log_{10}W)$	–0.885
24	26–177	$\log_{10}A = 1.49 - (0.573 \times \log_{10}W)$	–0.865
28	3–227	$\log_{10}A = 1.66 - (0.625 \times \log_{10}W)$	–0.885
32	46–184	$\log_{10}A = 1.92 - (0.545 \times \log_{10}W)$	–0.878
36	44–187	$\log_{10}A = 1.59 - (0.316 \times \log_{10}W)$	–0.660

Ammonia excretion rates increase sharply 1 to 5 h after feeding to between 190 and 360% of the endogenous rate, only returning to pre-feeding values after 5 to 12.5 h (Fig. 4.20). The time to the maximum post-feeding excretion rate, duration of the elevated rate, maximum excretion rate and magnitude of exogenous ammonia excretion all increased with increasing protein intake. This underlines the close link between ammonia excretion and protein metabolism. Brunty *et al.* (1997) have also shown that total ammonia excretion increases in proportion to dietary protein, but no significant difference in was seen in total ammonia excreted when the protein source was either fishmeal or soybean meal.

Urea

Freshwater fish excrete normally only a small quantity of urea. The tilapia *O. alcalicus grahami* is a ureotelic teleost, one of only a few which are known to

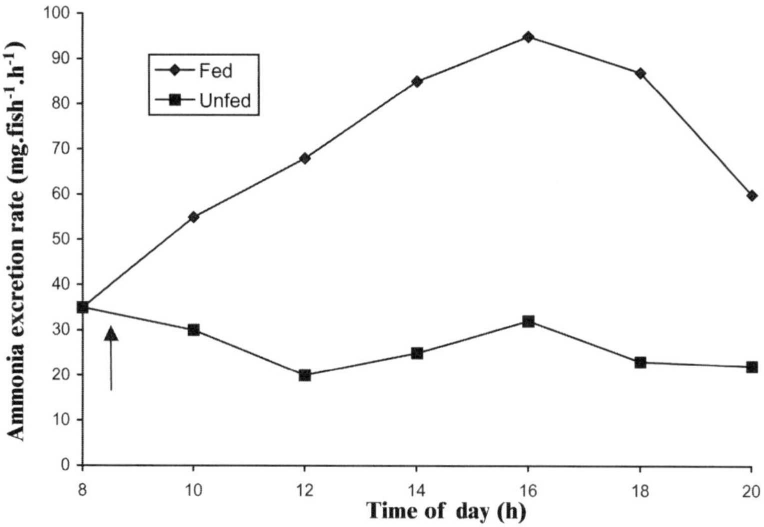

Fig. 4.20 Post-prandial elevation in ammonia excretion in *O. niloticus*. Arrow shows time of feeding. Adapted from McKinney (1989).

excrete large quantities of urea. This fish lives in the alkaline hotsprings of Lake Magadi, in the Rift Valley in Kenya at pH 9.6 to 10 (Naraha *et al.*, 1995). The species has adapted to high alkalinity waters by excreting almost all of its nitrogenous waste (more than 90%) as urea. Randall *et al.* (1989) showed that it excretes urea almost exclusively and possesses enzymes of the ornithine cycle in the liver to enable this. By contrast, Wright (1993) showed that *O. niloticus* was unable to develop enzymes of the ornithine urea cycle even when challenged with water at pH 10. Wilkie and Wood (1996), showed that the urea is excreted through the apical pits of modified seawater-type chloride cells and concluded that adjustments to nitrogenous waste metabolism and excretion patterns, as well as modifications to gill functional morphology, are necessary adaptations that permit these animals to thrive in environments considered unsuitable for other fishes.

4.6 THE MAJOR METABOLIC MODIFIERS

Temperature

Fish are thermal conformers and their body temperature closely mirrors that of the environment. Consequently, temperature is the principal factor affecting the rate of metabolic processes (Brett, 1979) and has been reviewed in tilapias by

Balarin and Haller (1982), Caulton (1982) and Chervinski (1982). Tilapias are thermophilic and so their distributions are determined by those temperatures which are not totally lethal and which also allow breeding to occur (Philippart and Ruwet, 1982).

Temperature has a powerful effect on growth, both through its influence on rates of biochemical processes and hence metabolism, and through associated factors such as feed intake. Fig. 4.21 shows the effect of temperature on feed intake for *T. rendalli*. Overall, Mires (1995) considered that their optimal growth range is 22 to 29°C, although Caulton (1979) showed that, when presented with a choice, *T. rendalli*, *O. niloticus*, *O. mossambicus* and *O. macrochir* selected temperatures of between 30 and 36°C. He noted that these temperatures are typical of those encountered in the fishes natural environment and so, a more accurate upper limit to the optimal range is probably about 35°C.

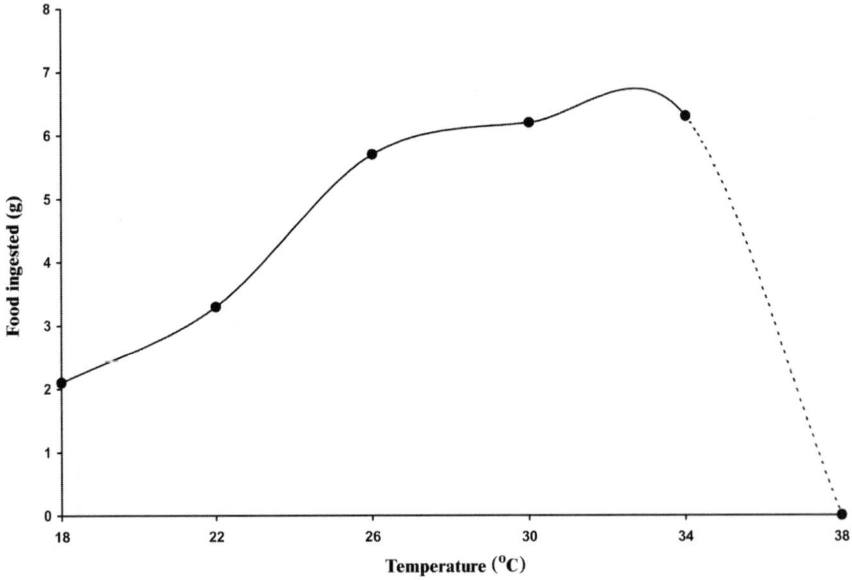

Fig. 4.21 Variation in food ingestion rate with temperature. Based on data from Caulton (1982).

In general, tilapias are quite sensitive to cold (Mires, 1995) and probably suffer at low temperatures due to osmoregulatory failure. They will not survive when exposed to temperatures below 10°C for more than a few days although some species are more tolerant than others, for example *T. sparrmanii* can tolerate temperatures as low as 7°C. They become generally inactive below about 16°C and death may occur below 12°C if exposure is maintained

(Chervinski, 1982). They grow poorly below about 20°C (Caulton, 1982) and they will reproduce only above 22°C. Low temperature tolerance depends on the thermal acclimation history (Chervinski and Lahav, 1976). It also depends on salinity and *S. mossambicus* had better low temperature tolerance at 5‰ than in freshwater (Allanson *et al.*, 1971). Table 4.7 summarizes data on upper and lower temperature limits for a range of species.

Table 4.7 Temperature limits for tilapia species. Data drawn from Chervinski (1982)

Species	Lower limit °C	Upper limit °C	Author
T. sparrmanii	7	–	Marr *et al.* (1966)
T. rendalli	11	–	Sklower, in Cherviski (1982)
S. aureus	10		Yashouv (1960)
	6.7		Lee (1979)
		41	Gleastine (1974)
S. hornorum			Lee (1979)
S. vulcani	13	–	Sarig (1969)
S. vulcani × *aureus*	8	–	Sarig (1969)
S. niloticus	11		Denzer (1968)
	7.8		Lee (1979)
S. alcalicus grahami	–	42	Coe (1966)
S. shiranus chilwae	–	42	Morgan (1972)
S. mossambicus	–	38.2	Allanson and Noble (1964)

The wide diel fluctutations in temperature often found in lake margins may account for offshore migratory behaviour found in some tilapia species as they attempt to optimize body temperature. Caulton (1982) went further, suggesting that this diel migration may be a device to optimize energy usage. In contrast to natural changes, instantaneous changes in temperature can have deleterious effects. Hocutt and Tilney (1985) showed that *O. mossambicus* could withstand sudden changes from 25°C to 30°C and 35°C with only minor stress. However, transfer to 40°C or 45°C caused death accompanied by severe gill damage including shrinkage of the secondary lamellae and collapse of the pillar cells. Transfers from 25°C to 35°C carried out at 1°C day^{-1} were achieved uneventfully. In practice, it has been found that an instantaneous temperature decrease of only 6°C can be applied to tilapia fry acclimated to 25°C, greater rates of decrease producing measurable mortalities (Okoye, 1982). Al Amoudi *et al.* (1996) found that survival of *O. mossambicus* and *O. aureus* × *niloticus* hybrids was not affected after a 10°C transfer, but that behaviour was notably compromised. By contrast, in salt water survival was affected following a 10°C transfer but not a 5°C transfer.

Salinity

Although tilapias are principally freshwater, many species are euryhaline and some even live in lakes where salinities are well in excess of those found in the sea. Myers (1938) suggested that tilapias have evolved from a marine ancestor and that they invaded freshwater secondarily.

Watanabe *et al.* (1985) found that fertilized eggs and fry of *O. niloticus* both had a 96 h median lethal salinity of 18.9‰ and survival over a week was much reduced at salinities above 10‰. The tolerance of fry to salinity progressively increased with increasing salinity of spawning, hatching or acclimatization and earlier exposure gave greater tolerance. In general, tolerance to salinity transfer does not develop until after about 45 days.

Salinity tolerances of juveniles and adult tilapia have been discussed by several authors and were reviewed for the major cultured species by Stickney (1986) and Suresh and Kwei Lin (1992). Table 4.8 summarizes data for a number of cultured species compiled by Suresh and Kwei Lin (1992) who pointed out that while many species have great potential in saline waters, tilapias become sensitive to handling and succumb to disease more readily in seawater. Optimal growth for *O. niloticus* and *O. aureus* and their hybrids is below 18‰.

Table 4.8 Salinity tolerances of cultured tilapias. Based on data from Suresh and Kwei Lin (1992). All values in ‰

Species	Tolerence limits for direct transfer	Tolerance limits for gradual transfer	Salinity range for best growth	Salinity range for reproduction
O. mossambicus	27	120	17.5	up to 49
O. niloticus	18	36	5–10	up to 32, poor hatching
O. aureus	27	40	10–15	up to 18
O. spilurus	27	36	moderate growth in S.W.	full S.W., low fecundity
Red tilapia	27–29	35	saline waters	full S.W., low fecundity

For routine husbandry, direct transfer from 0 to 10‰ is considered totally safe (Stickney, 1986) while more drastic changes require acclimation stages (Mires, 1995). Morgan *et al.* (1997) showed that the physiological changes associated with seawater acclimation in tilapia represents a significant short-term energetic cost, possibly accounting for as much as 20% of total body metabolism after 4 days in seawater.

4.7 ENERGY FLOW AND BUDGETS

There are various ways of considering energy use, either in simple terms (energy in feed, energy for swimming) or in gross terms where all the inputs to and outputs from the system are considered in energy terms. Many aspects of whole animal physiology have a bearing on energy flow and use and the energy budget is an excellent concept in which virtually all aspects of the biology and physiology of an animal are integrated into a model describing inputs and outputs from the animal. The concepts of energy partitioning in animal production systems are well-established (Petrusewicz and McFadyen, 1970). The simplest way of expressing this model is in the form: $C = P + R + F + U$, where, C = consumption, P = production, R = respiration, F = faeces and U = non-faecal losses. This budget can be expressed in terms of energy per unit time, or as a percentage of the total energy consumed, each type having its uses. In an excellent review of fish energetics, Brett and Groves (1979) summarized the findings for a range different fish species and gave an 'average' percentage daily budget of $100C = 29P + 44R + 20F + 7U$. Most recently, Jobling (1994) has summarized the status of the subject for fish.

Short-term, usually daily, budgets can be used to evaluate the immediate effects of diet, temperature, stress, etc. whereas cumulative long-term budgets, usually annual, are of value in considering strategy in the natural environment and monthly/yearly production planning and management of both farms and fisheries. Since a large number of biotic and abiotic factors affect most aspects of the physiology of animals, the production of a comprehensive energy budget based model depends on being able to quantify the effects of these factors on each of the budget variables. Much of this information can be derived from laboratory studies on respiration, nutrition, etc, while other factors must be evaluated in the field or, in the case of aquaculture, in commercial-scale production systems.

Few authors have quantified tilapia energetics in a complete form. Mironova (1976) estimated C, P and R in *O. mossambicus* and showed that assimilation efficiency increased with fish age and water temperature in fish fed to satiation. When fish were fed a restricted diet, an increase in temperature resulted in a decrease in production due to an increase in R. Caulton (1982) derived an energy budget based on his previous work for *T. rendalli* in which he suggested that the overall balance was very sensitive to environmental temperature.

Musisi (1984) constructed a series of full energy budgets over 30-day periods for *O. mossambicus*. All budget variables were measured simultaneously and fish were fed a variety of natural and artificial diets at a rate of 12% body weight per day, in three meals. Generally, values of F and U were similar to the 'average' calculated by Brett and Groves (1979) but values of R were higher and values of P were lower. Further studies have been conducted on carbon budgets by McDonald (1985) and an energy budget for a lake population of *O. mossambicus*, (Hofer and Schiemer, 1983), both studies involving algae as the sole food source. It is clear in reviewing these studies that much remains to be done and that a well-designed model of the energy budget for tilapia would be a useful tool for evaluating current knowledge and for pinpointing areas of weakness.

Based on a wide range of original experimental data, in addition to published information, Ross *et al.* (1988) compiled the major components of an energy budget for *O. niloticus* into an interactive computer model. Their model uses data on inputs and expected outputs to derive a daily production value which is then used to update body weight. From this, SGR can be calculated, and the process can be made to iterate over a selected number of days. Despite the necessary use of a number of assumptions, the model works well and is able to simulate growth over a period, building up the budget from fundamental, whole-animal, physiological relationships (Fig. 4.22). It also allows exploration of effects of any biotic or abiotic variable, including feed formulation, ration level and aspects of husbandry. For example, the effects of a single handling stress event (Ross and Ross, 1983) can be shown to add about 3% to the daily value for R. While a single such event is relatively small, cumulatively this could become significant.

On the basis of previously published energy budgets, the model appears at first to underestimate R and overestimate P. However, it is possible to approximately verify the results by comparing the SGR values calculated in energy terms with data from trials using practical diets. In this comparison, SGR falls within the expected range of values, suggesting that estimates of P, although high, are substantially correct. This may follow because the budget compiled is based on data for cultured fish and it is known that production in culture normally exceeds that in the natural environment, often markedly (Lowe-McConnell, 1982). Model output also accords well with independent estimates of daily metabolisable energy requirements by Meyer-Burgdorff *et al.* (1989). More recently, Van Dam and Pauly (1995) obtained good results with an alternative model in which feeding rate was linked to maximum oxygen uptake rate predicted from gill area and Fick's law of diffusion. Modelling within the predicted maximum uptake envelope gave useful predictions of feeding and growth.

The selection of optimal management strategies depends upon the availability of at least a minimal description of the system in an input/output form and such simple simulation studies will normally precede any attempted application of optimization techniques (Brockington, 1979). Any mathematical model de-

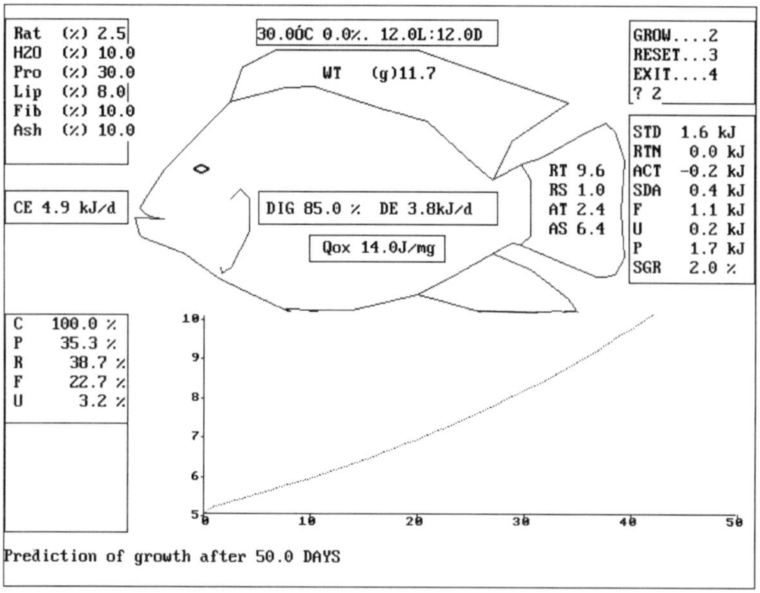

Fig. 4.22 Growth prediction modelling based on a physiological energy budget model for *O. niloticus* (Ross *et al.*, 1988).

signed for use in aquaculture or agriculture should be as simple as possible to operate and should employ the minimum number of variables necessary. The initial process of assembly is often useful in identifying missing data, thus testing logic and prompting further experiments. Once assembled, such models can provide interactive insights which may fuel further research or which may suggest alternative management strategies to achieve target output or to deal with changing environment or diet. There is little doubt that such models will be developed further, although they can never fully replace properly conducted experimental work.

4.8 EFFECTS OF WATER QUALITY PARAMETERS

Nitrite

Although nitrite occurs naturally in low concentrations in freshwaters, it can become elevated due to pollution and bacterial production. Nitrite is toxic to fish because it can pass easily into the blood through the gills, where it converts haemoglobin into the relatively stable form, methaemoglobin, which has a very poor oxygen carrying capacity (Lewis and Morris, 1986). Although the presence of methaemoglobin may not affect resting fish which will have a low metabolic rate, any additional workload or stress applied to such animals can cause death due to anoxia. This is not because of non-availability of oxygen in the environment but because of an inability to transport it to the tissues (Huey *et al.*, 1980). In general, levels of methaemoglobin over 50% of the total will lead to such problems, although the effect will be ameliorated by low activity levels.

Tilapias are less sensitive to nitrite than salmonids. Palacheck and Tomasso (1984) found the 96 h LC_{50} in *O. aureus* to be 16 mg l^{-1} $N.NO_2$ at pH 7.9 and 190 mg l^{-1} calcium hardness The 24 h LC_{50} was 3116 mg l^{-1} $N.NO_2$. They considered that differential resistance to nitrite among species may be partially due to a differential ability to prevent high concentrations of nitrite from entering the plasma.

pH

While pH of seawater is closely conserved, in freshwater it can vary more widely for a number of reasons. Tilapias of different species are found naturally in a fairly wide pH range. Bardach (1972) stated that tilapias did not grow in the acid waters of West Congo, whereas *O. alcalicus grahami* survives from pH 5 to 11 without adverse effects (see earlier). It is well known that cultured tilapia will perform badly in acidic waters, but that this can be alleviated by system management. Lin (1986) reported a four-fold increase in yield of tilapia in acid sulphate ponds, achieved by changing the water and by liming. Wangead *et al.* (1987) found that *O. niloticus* fingerlings showed behavioural manifestation of physiological stress almost immediately at pH 2 and 3 with rapid swimming and opercular movements, surfacing and gulping of air and inability to control body position. They died when exposed to pH 2 or 3, and survival at pH 4 was reduced to 58%. Murty *et al.* (1981) found the lethal limits of pH for *O. mossambicus* to be between 3.7 and 10.3. Best growth of tilapias is achieved between pH 7 to 9 and some growth reduction occurs below pH 6. In acid water, pH 4 and below, mortalities occur, associated with a fall in plasma osmolarity, sodium and calcium levels. However, after 2 weeks of exposure to such low pH, elevated plasma prolactin is thought to promote osmoregulatory homeostasis (Wendelaar Bonga *et al.*, 1984). Dijk *et al.* (1993) showed that at pH 4, *O. mossambicus* reduced oxygen consumption and maximum metabolic rate.

4.9 SUMMARY

In considering physiological data on tilapias, it is important to be critical of the conditions in which experiments are conducted. This is particularly the case where experimental systems are needed, i.e. almost always. Protocols for experimentation are frequently widely different between laboratories and this needs to be taken into account in interpreting data. Frequently, acclimation, system design or general experimental protocol differ sufficiently that comparisons can be made very difficult, even when the same species is used.

A good example is the case of oxygen consumption which, because of its importance, is frequently among the first physiological factors researched in an animal. There are many factors affecting respiratory rates, and so the conditions under which respiratory work is carried out can vary so greatly that successive workers find problems in interpreting earlier data in terms of their present situations. Although there are general rules for protocols under which such experiments should be conducted, they are rarely adhered to even by the best researchers. From the foregoing, it can be seen that it is important that careful protocols are followed if results are to be comparable between different individuals, populations, species, location, etc. At the very least, the conditions of experimentation must be fully defined and reported to enable reinterpretation of data by those wishing to use existing information. The minimum range of factors to be controlled during trials and defined in any reporting of the results are:

Biotic factors	**Abiotic factors**	**Respirometer features**
Body weight	Temperature	Dimensions
Sex	Salinity	Flow-through or closed
Pre-trial nutritional status	pH, hardness	Feeding arrangements
Feeding regime	Photoperiod, light level	Waste removal methods
Pre-trial conditioning	Diet composition	Nature of confinement
Stress level	Diet quantity	
Activity level	Pre-trial acclimation	
Genotype		

A further factor of increasing importance is the strain of animals used. Substantial variation exists within species, as will be seen elsewhere in this book, and the additional ability to engineer genetically modified stocks (GMOs) introduces another factor for consideration.

Notwithstanding these cautionary observations, there is a now a good and growing body of data on tilapias, sufficient for many purposes. The hardiness of tilapias makes them a superbly reliable workhorse in the laboratory, as well as being one of the factors which make them ideal candidates for culture. This hardiness is manifest in their general robustness to handling and in their range of tolerance to temperature, dissolved oxygen and to salinity, features that will ensure their continued place among the top cultured warmwater fish species.

REFERENCES

Al-Abdul-Elah, K. (1990) *Biotic and Abiotic Factors Influencing Initial Swimbladder Inflation of the Blue-Finned Sea Bream*, PhD thesis, University of Stirling.

Allanson, B.R. and Noble, R.G. (1964) The tolerances of *Tilapia mossambica* (Peters) to high temperature. *Trans. Am. Fish. Soc.* **93**, 323–332.

Allanson, B.R., Bok, A. and van Wyk, N.I. (1971) The influences of exposure to low temperature on *Tilapia mossambica* Peters (Cichlidae). II. Changes in serum osmolarity, sodium and chloride concentrations. *J. Fish Biol.* **3**, 181–185.

Balarin, J.D. and Hatton, J.P. (1979) *Tilapia. A Guide to their Biology and Culture in Africa*, Unit of Aquatic Pathobiology, University of Stirling, Scotland.

Bardach, J.E., Ryther, J.H. and McLarney, W.O. (1972) *Aquaculture: The Farming and Husbandry of Freshwater and Marine Organisms*, Wiley, New York.

Becker, K. and Fishelson, L. (1986) Standard and routine metabolic rate, critical oxygen tension and spontaneous scope for activity of tilapias, in *Proceedings of the First Asian Fisheries Forum* (eds J.L. Maclean, L.B. Dizon, L.V. Hosillos), Asian Fisheries Society, Manila, Philippines, pp. 623–628.

Boyd, C.E. (1982) *Water Quality Management for Pond Fish Culture*, Elsevier, Amsterdam.

Brett, J.R. (1979) Factors affecting fish growth, in *Fish Physiology Volume 8* (eds W.S. Hoar, D.J. Randall and J.R. Brett), Academic Press, New York, pp. 599–675.

Brett, J.R. and Groves, T.D.D. (1979) Physiological energetics, in *Fish Physiology Volume 8* (eds W.S. Hoar, D.J. Randall and J.R. Brett), Academic Press, New York, pp. 279–281.

Brockington, N.R. (1979) *Computer Modelling for Agriculture*, Oxford University Press, Oxford, UK.

Brunty, J.L., Bucklin, R.A., Davis, J., Baird, C.D. and Norstedt, R.A. (1997) The influence of feed protein intake on tilapia ammonia production. *Aquacult. Eng.* **16**, 161–166.

Caulton, M.S. (1977) The effect of temperature on routine metabolism in *Tilapia rendalli* Boulenger. *J. Fish Biol.* **11**, 549–553.

Caulton, M.S. (1978) The effect of temperature and mass on routine metabolism in *Sarotherodon (Tilapia) mossambicus* (Peters). *J. Fish Biol.* **13**, 195–201.

Caulton, M.S. (1982) Feeding, metabolism and growth of tilapias: some quantitative considerations, in *The Biology and Culture of Tilapias (ICLARM Conference Proceedings*, Vol. 7) (eds R.S.V. Pullin and R.H. Lowe-McConnell), ICLARM, Manila, pp. 157–180.

Chatain, B. and Dewavrin, G. (1989) The effects of abnormalities in the development of the swimbladder on the mortality of *Dicentrarchus labrax* during weaning. *Aquaculture* **78**, 55–61.

Chatain, B. and Ounais-Guschemann, N. (1990) Improved rate of initial swimbladder inflation in intensely reared *Sparus auratus*. *Aquaculture* **84**, 345–353.

Chervinski, J. (1982) Environmental physiology of tilapias, in *The Biology and Culture of Tilapias* (*ICLARM Conference Proceedings*, Vol. 7) (eds R.S.V. Pullin and R.H. Lowe-McConnell), ICLARM, Manila, pp. 119–128.

Chervinski, J. and Lahav, M. (1976) The effect of exposure to low temperature on fingerlings of local tilapia *(Tilapia aurea)* (Steindachner) and imported tilapia *(Tilapia vulcani)* (Trewavas) and *Tilapia nilotica* (Linne) in Israel. *Bamidgeh* **28**, 25–29.

Coche, A.G. (1977) Premiers résultats de l'élevage en cages de *Tilapia nilotica* (L.) dans le Lac Kossou, Côte d'Ivoire. *Aquaculture* **10**, 109–140.

Coche, A. and Edwards, D. (1989) *Selected Aspects of Warmwater Fish Culture*, GCP/INT/435/AGF. FAO, Rome, Italy.

Cornacchia, J.W. (1982) *Studies on the Development and Inflation of the Swimbladder in Two Physoclistous Fish,* Morone saxatilis *and* Sarotherodon mossambica, PhD thesis, University of California, Davis.

Dehadrai, P.V. (1959) On the swimbladder and its connection with the internal ear in cichlidae. *Proc. Nat. Inst. Sci., India* **25**(b), 254–261.

Denzer, H.W. (1968) Studies on the physiology of young *Tilapia. FAO Fish. Rep.* **44**, 356–366.

Dijk, P.L.M., Thillart, G.E.E.J.M. and Wendelaar-Bonga, S.E. (1993) The influence of gradual water acidification on the oxygen consumption pattern of fish. *Comp. Biochem. Physiol.* **105**C, 421–427.

Doroshev, S.I., Cornacchia, J.W. and Hogan, K. (1981) Initial swimbladder inflation in the larvae of physoclistous fishes and its importance for larval culture. Rapports et proces-verbaux de Reunion, *Cons. Int. l'Exp. Mer.* **178**, 478–500.

Dussart, J. (1963) Contribution a l'étude de l'adaptation des tilapias (Pisces, Cichlidae) a la vie en milieu mal oxygene. *Hydrobiologia* **21**, 328–341.

Farmer, G.J. and Beamish, F.W.H. (1969) Oxygen consumption of *Tilapia nilotica* in relation to swimming speed and salinity. *J. Fish. Res. Bd. Can.* **26**, 2807–2821.

Febry, R. and Lutz, P. (1987) Energy partitioning in fish: The activity-related cost of osmoregulation in a euryhaline cichlid. *J. Exp. Biol.* **128**, 63–85.

Fernandes, M.N. and Rantin, F.T. (1986) Thermal acclimation of the teleost *Oreochromis niloticus (Pisces, Cichlidae). Rev. Hydrobiol. Tropica.* **19**, 163–168.

Fernandes, M.N. and Rantin, F.T. (1994) Relationships between oxygen availability and metabolic cost of breathing in Nile tilapia (*Oreochromis niloticus*): Aquaculture consequences. *Aquaculture* **127**, 339–346

Fish, G.R. (1956) Some aspects of the respiration of six species of fish from Uganda. *J. Exp. Biol.* **33**, 186–195.

Fryer, G. and Iles, T.D. (1972) *The Cichlid Fishes of the Great Lakes of Africa: their Biology and Evolution,* Oliver and Boyd, Edinburgh.

Gleastine, B.W. (1974) *A Study of the Cichlid* Tilapia aurea *(Steindachner) in a Thermally Modified Texas Reservoir,* MSc thesis, Texas A&M University.

Hassan, R. (1992) Acute Ammonia Toxicity of Red Tilapia and Seabass. *Fish. Bull. Dept Fish., Kuala Lumpur, Malaysia.* No 73.

Hocutt, C.H. and Tilney, R.L. (1985) Changes in gill morphology of *Oreochromis mossambicus* subjected to heat stress. *Env. Biol. Fishes* **14**, 107–114.

Hofer, R. and Schiemer, F. (1983) Feeding ecology, assimilation efficiencies and energetics of two herbivorous fish: *Sarotherodon* (*Tilapia*) *mossambicus* (Peters) and *Puntius filamentosus* (Cuv. et Val.), in *Limnology of Parakrama Samudra – Sri Lanka* (ed. F. Schiemer), Dr W. Junk, The Hague, pp.155–164.

Huang, Y., Jia, J., He, D., Chai, M. and Deng, X. (1996) Electrophysiological observations on sacculus of *Tilapia* sp. *J. Oceanogr. Taiwan Straits* **15**, 182–190.

Huey, D.W., Simco, B.A. and Criswell, D.W. (1980) Nitrite-induced methaemoglobin formation in channel catfish. *Trans Am. Fish. Soc.* **109**, 558–562.

Job, S.V. (1969) The respiratory metabolism of *Tilapia mossambica* (Teleostei). I. The effect of size, temperature and salinity. *Mar. Biol. (Berl.)* **2**, 121–126.

Jobling, M. (1994) *Fish Bioenergetics,* Chapman and Hall, London.

Karuppannan, N.V. (1981) A note on the locomotory metabolism (swimming energetics) of the cichlid fish *Tilapia mossambica* (Peters). *Ind. Zool.* **5**, 107–109.

Kutty, M.N. (1972) Respiratory quotient and ammonia excretion in *Tilapia mossambica. Mar. Biol.* (Berlin) **16**, 126–133.

Lee, J.C. (1979) *Reproduction and Hybridization of Three Cichlid Fishes,* Tilapia aurea *(Steindachner),* T. hornorum *(Trewavas) and* T. nilotica *(Linnaeus) in Aquaria and in Plastic Pools,* PhD dissertation, Auburn University, Auburn, Alabama.

Lewis, W.M. and Morris, D.P. (1986) Toxicity of nitrite to fish: A review. *Trans Am. Fish. Soc.* **115**, 183–195.

Lin, C.C. and Liu, C.I. (1989) Test for ammomnia toxicity of cultured hybrid tilapia, in *Proceedings of the Second Asian Fisheries Forum* (eds R. Hirano and I Hanyu), Tokyo, Japan, pp. 457–460.

Lin, C.K. (1986) Acidification and reclamation of acid sulfate soil fishponds in Thailand, in *Proceedings of the First Asian Fisheries Forum* (eds J.L. Maclean, L.B. Dizon and L.V. Hosillos), ICLARM, Manila, pp. 71–74.

Lowe-McConnell, R.H. (1982) Tilapias in fish communities, in *The Biology and Culture of Tilapias* (*ICLARM Conference Proceedings*, Vol. 7) (eds R.S.V. Pullin and R.H. Lowe-McConnell), ICLARM, Manila, pp. 83–114.

Maar, A., Mortimer, M.A.E. and van der Lingen, I. (1966) *Fish Culture in Central East Africa*, FAO, Rome.

McDonald, M.E. (1985) Carbon budgets for a phytoplanktivorous fish fed three different unialgal populations. *Oecologia* **66**, 246–249.

McEwen, R.S. (1940) The early development of the swimbladder and certain adjacent parts in *Hemichromis bimaculatus*. *J. Morphol.* **67**, 1–59.

McKinney, R.W. (1990) *Factors Influencing Ammonia Excretion in Tilapia* (Oreochromis niloticus *L.*), MSc thesis, University of Stirling.

Melard, C.H. and Philippart, J.C. (1982) Pisciculture intensive de *Sarotherodon niloticus* (L.) dans les effluents thermiques d'une centrale nucleaire en Belgique. Paper presented at the EIFAC symposium on new Developments in Utilisation of Heated Effluents and of Recirculation Systems for Intensive Aquaculture. Stavanger, Norway. EIFAC/80/Symp./DOC.E/11.

Meyer, D.I. and Brune, D.E. (1982) Computer modelling of diurnal oxygen levels in a stillwater aquaculture pond. *Aquacult. Eng.* **1**, 245–261.

Meyer-Burgdorff, K.-H., Osman, M.F. and Guenther, K.D. (1989) Energy metabolism in *Oreochromis niloticus*. *Aquaculture* **79**, 283–291.

Mires, D. (1995) The tilapias, in *Production of Aquatic Animals: Fishes* (eds C.E. Nash and A.J. Novotny), Elsevier, Amsterdam, pp. 133–152.

Mironova, N.V. (1976) Changes in the energy balance of *Tilapia mossambica* in relation to temperature and ration size. *J. Ichthyol.* **16**, 120–129.

Morgan, J.D., Sakamoto, T., Grau, E.G. and Iwama, G.K. (1997) Physiological and respiratory responses of the Mozambique tilapia (*Oreochromis mossambicus*) to salinity acclimation. *Comp. Biochem. Physiol.* **117A**, 391–398.

Morgan, P.R. (1972) Causes of mortality in the endemic tilapia of lake Chilwa (Malawi). *Hydrobiologia* **40**, 101–119.

Myers, G.S. (1938) Freshwater fishes and West Indian zoogeography, in *Smithsonian Report for 1927. Smithsonian Institution Publication* 3465, Washington. D.C., pp. 339–364.

Murthy, V.K., Reddanna, P. and Govindappa, S. (1981) Hepatic carbohydrate metabolism in *Tilapia mossambica* (Peters) acclimated to low environmental pH. *Can. J. Zool.* **59**, 400–404.

Musisi, L.M. (1984) *The Nutrition, Growth and Energetics of Tilapia* Sarotherodon mossambicus, PhD thesis, University of London.

Nagarajan, K. and Gopal, V. (1983) Effect of photoperiod on oxygen consumption and food utilisation in *Tilapia mossambica* Peters. *Proc. Natl. Sci. Acad. India* **53B**, 217–225.

Narahara, A., Bergman, H.L., Laurent, P., Maina, J.N., Walsh, P.J. and Wood, C.M. (1995) Respiratory physiology of the Lake Magadi tilapia (*Oreochromis alcalicus grahami*), a fish adapted to a hot, alkaline, and frequently hypoxic environment. *Physiol. Zool.* **69**, 1114–1136

Northcott, M., Beveridge, M.C.M. and Ross, L.G. (1991) A laboratory investigation of the filtration and ingestion rates of the tilapia, *Oreochromis niloticus*, feeding on two species of blue-green algae. *Env. Biol. Fishes* **31**, 75–85.

Okoye, R.N. (1982) *Techniques for Transportation of Juvenile Tilapia*, MSc thesis, University of Stirling, Stirling, UK.

Palacheck, R.M. and Tomasso, J.R. (1984) Toxicity of nitrite to channel catfish (*Ictalurus punctatus*), tilapia (*Tilapia aurea*) and largemeouth bass (*Micropterus salmoides*): evidence for a nitrite exclusion mechanism. *Can. J. Fish. Aquat. Sci.* **41**, 1739–1744.

Petrusewicz, K. and MacFadyen, A. (1970) *Productivity of Terrestrial Animals*, IBP Handbook No 13, Blackwell, Oxford.

Philippart, J-Cl. and Ruwet, J-Cl. (1982) Ecology and distribution of the tilapias, in *The Biology and Culture of Tilapias* (*ICLARM Conference Proceedings*, Vol. 7) (eds R.S.V. Pullin and R.H. Lowe-McConnell), ICLARM, Manila, pp. 15–59.

Randall, D.J., Wood, C.M., Perry, S.F., Bergman, H., Maloiy, G.M.O., Mommsen, T.P. and Wright, P.A. (1989) Urea excretion as a strategy for survival in a fish living in very alkaline environment. *Nature (Lond.)* **337**, 165–166.

Redner, B.D. and Stickney, R.R. (1979) Acclimation to ammonia by *Tilapia aurea*. *Trans. Am. Fish. Soc.* **108**, 383–388.

Rodman, D.T. (1966) Sound productioin by the African cichlid *Tilapia mossambica*. *Ichthyologia* **38**, 279–280.

Ross, L.G. (1979a) The haemodynamics of gas resorption from the physoclist swimbladder: The structure and morphometrics of the oval in *Pollachius virens*. *J. Fish Biol.* **14**, 261–266.

Ross, L.G. (1979b) The haemodynamics of gas resorption from the physoclist swimbladder: II. The determination of blood flow-rate using radiolabelled microspheres. *J. Fish Biol.* **14**, 389–393.

Ross, L.G. and Geddes, J.A. (1979) Sedation of warm-water fish species in aquaculture research. *Aquaculture* **16**, 183–186.

Ross, L.G. and McKinney, R.W. (1988a) Photoperiod-mediated variation in respiratory rate of *Oreochromis niloticus* and its implications for tilapia culture, in *Proceedings of the Second International Symposium on Tilapias in Aquaculture* (*ICLARM Conference Proceedings*, Vol. 15) (eds R.S.V. Pullin, T. Bhukaswan, K. Tonguthai and J.L Maclean), ICLARM, Manila, pp. 421–428.

Ross, L.G. and McKinney, R.W. (1988b) Respiratory cycles in *Oreochromis niloticus* measured using a six-channel microcomputer-operated respirometer. *Comp. Biochem. Physiol.* **89**A, 637–643.

Ross, B. and Ross, L.G. (1983) The oxygen requirements of *Oreochromis niloticus* under adverse conditions, in *Proceedings of the First International Symposium on Tilapia in Aquaculture* (eds L. Fishelson and Z. Yaron), Tel Aviv University, Israel, pp. 134–143.

Ross, L.G. and Ross, B. (1999) *Anaesthetic and Sedative Techniques for Aquatic Animals*, Blackwell, Oxford.

Ross, L.G., McKinney, R.W. and Ross, B. (1988) Energy budgets for cultured tilapia, in *Proceedings of the Second International Symposium on Tilapias in Aquaculture* (*ICLARM Conference Proceedings*, Vol. 15) (eds R.S.V. Pullin, T. Bhukaswan, K. Tonguthai and J.L. Maclean), ICLARM, Manila, pp. 83–90.

Ross, L.G., Mckinney, R.W., Cardwell, S.K., Fullarton, J.G., Roberts, S.E.J. and Ross, B. (1992) The effects of dietary protein content, lipid content and ration level on oxygen consumption and specific dynamic action in *Oreochromis niloticus*. *Comp. Biochem. Physiol.* **103**A, 573–578.

Sarig, S. (1969) Winter storage of Tilapia. *FAO Fish Cult. Bull.* **2**, 8–9.

Sayer, M.D. and Davenport, J. (1987) The relative importance of the gills to ammonia and urea excretion in five seawater and one freshwater teleost species. *J. Fish Biol.* **31**, 561–570.

Stickney, R.R. (1986) Tilapia tolerance of saline waters: a review. *Prog. Fish-cult.* **48**, 161–167.

Suresh, A.V. and Kwei Lin, C. (1992) Tilapia culture in saline waters: a review. *Aquaculture* **106**, 201–226.

Trewavas, E. (1983) Tilapiine fishes of the genera *Sarotherodon*, *Oreochromis* and *Danakilia*, British Museum, London.

Van-Dam, A.A. and Pauly, D. (1995) Simulation of the effects of oxygen on food consumption and growth of Nile tilapia, *Oreochromis niloticus* (L.). *Aquacult. Res.* **26**, 427–440.

Wangead, C., Geater, A. and Tansakul, R. (1988) Effect of acid water on survival and growth rate of Nile tilapia (*Oreochromis niloticus*), in *Proceedings of the Second International Symposium on Tilapias in Aquaculture* (*ICLARM Conference Proceedings*, Vol. 15) (eds R.S.V. Pullin, T. Bhukaswan, K. Tonguthai and J.L. Maclean), ICLARM, Manila, pp. 433–437.

Watanabe, W.O., Ching-Ming Kuo and Mei-Chan Huang (1985) Salinity tolerance of Nile tilapia fry (*Oreochromis niloticus*), spawned and hatched at various salinities. *Aquaculture* **48**, 159–176.

Wendelaar Bonga, S.E., van der Meij, J.C.A., van der Krabben, W.A.W.A. and Flik, G. (1984) The effect of water acidification on prolactin cells and pars intermedia PAS-positive cells in the teleost fish *Oreochromis* (formerly *Sarotherodon*) *mossambicus* and *Carassius auratus*. *Cell Tiss. Res.* **238**, 601–609.

Wilkie, M.P. and Wood, C.M. (1996) The adaptations of fish to extremely alkaline environments. *Comp. Biochem. Physiol.* **113**b, 665–673.

Wright, P.A. (1993) Nitrogen excretion and enzyme pathways for ureagenesis in freshwater tilpaia (*Oreochromis niloticus*). *Physiol. Zool.* **66**, 881–901.

Yashouv, A. (1960) Effect of low temperatures on *Tilapia nilotica* and *Tilapia galilaea*. *Bamidgeh* **12**, 62–66.

Chapter five

The roles of tilapias in ecosystems

R. H. Lowe-McConnell

5.1 INTRODUCTION

The ecological roles of tilapias

All tilapias are basically herbivores or detritivores. As such they occupy an intermediate position between primary producers and piscivores. They have therefore two main ecological roles: the circulation of nutrient metabolites on which primary production depends (the scale of this role depending on limnological conditions such as water flow, lake depth, etc.), and another in supporting the piscivores in the ecosystem, whether fishes, reptiles, birds or mammals (including man). Tilapias are so important as food fish over most of Africa, and now through much of the tropical world, that research has concentrated mainly on this latter role.

Relatively few fishes feed directly on living plants or detritus: in Africa tilapias, cyprinid *Labeo*, characid *Citharinus* and *Distichodus* and a few smaller haplochromine cichlids. The wide distribution of tilapias and their large share of the ichthyomass (60% in Lake George, 74% in Sokoto River pools for example) makes them disproportionately represented in individual ecosystems, and their large populations are conspicuous components of ecosystem trophic structures. For the majority of freshwater fishes, which feed predominantly on invertebrates and small fish, the main problem is obtaining an adequate *quantity* of food at a minimal cost in time and energy. By contrast, the food of herbivores, such as tilapias, is generally present far in excess of their ability to consume it, but food *quality* is highly variable (especially its protein content), which affects tilapia growth and life history strategies.

M.C.M. Beveridge and B.J. McAndrew (eds), Tilapias: Biology and Exploitation, 129–162.

Tilapias have an important limnological role in the circulation of nutrients needed to keep water bodies productive. Beauchamp (1964) predicted that under tropical conditions where nutrient salts are circulated faster through the biota than by breakdown in anoxic bottom sediments, the removal of large quantities of herbivores (as by a large tilapia fishery) will decrease the overall productivity of the system. Introduced tilapia (*O. mossambicus*) increased the overall productivity in a Sri Lankan reservoir, being more efficient phytoplankton eaters than the indigenous cyprinids (see Schiemer, 1996). Their limnological role deserves serious consideration in the light of the recent declines in many tilapia fisheries.

Tilapia sensu stricto, characterized by coarse pharyngeal teeth and few gill rakers, eat mainly higher plants and detritus. *Sarotherodon* and *Oreochromis* species are basically microphagous, eating epiphytic or benthic algae and detritus, or phytoplankton in species inhabiting open waters. Most species grow to a good size, so are a valuable food source for man. Food does not appear to limit tilapia numbers but the quality of food does affect their growth rates and final size (Bowen, 1979, 1980). Bowen showed *O. mossambicus* to be a true detritivore, able to utilize the non-protein amino acids in detritus itself, not just the associated microorganisms; in Lake Sibaya the bacterial flora depended on the DOM (dissolved organic matter) secreted by the benthic diatom community, and both elements were necessary for the right quality of food. In Lake Naivasha detritus was found to tide over the tilapias (*T. zillii* and *O. leucostictus*) when other food was scarce (Muchiri *et al.*, 1995).

Most tilapias have some flexibility in diet, able to digest various forms of vegetarian food including cyanobacteria (blue-green algae) when this is abundant, as shown by Moriarty *et al.* (1973) for *O. niloticus* in Lake George, or other forms of algae and even small amounts of animal food in the form of insects and crustacea. Moreover, most juvenile tilapia feed on zooplankton until about 5 mm TL. As many cohabiting non-tilapiines also utilize the zooplankton, there could be a bottleneck of food available for juvenile tilapias, though little quantitative work has been done on these aspects. The size to which many species of tilapias grow, and at which they start to breed, varies greatly with the prevailing ecological conditions. Populations of dwarfed tilapias are found in small water bodies and under stressed conditions (species listed by Iles, 1973; Noakes and Balon, 1982).

Tilapias, especially the juvenile stages, also have an important role in supporting piscivores. As food fish they have been widely disseminated by man, both within Africa and other parts of the world where *O. mossambicus* and *O. niloticus* are now particularly widely distributed. Their success, or otherwise, in these waters demonstrates their roles in new ecosystems. In Africa tilapias have escaped from fish ponds, colonized rivers and often hybridized with species already there (as listed by Moreau *et al.*, 1988).

Which ecosystems are inhabited by tilapias?

Tilapias, endemic to Africa, live in a whole range of ecosystems: slow-moving lentic sections of rivers with their floodplain pools and swamps, lakes ranging from small shallow ones to the Great Lakes, newly impounded lakes, isolated crater lakes, soda lakes and thermal springs, and coastal brackish lagoons. They appear to have evolved as riverine fishes inhabiting marginal waters and floodplain pools, colonizing lakes as these formed, as demonstrated by how riverine tilapias have adapted to lacustrine conditions in hydroelectric impoundments. When geographically isolated for long periods, endemic species have evolved, both in river systems and in lakes. In shallow lakes they are very subject to changes in water levels, dissolved oxygen concentrations and other factors. In deep lakes they are generally restricted to shallow littoral zones. This littoral distribution may be associated with their physiological incapacity to live at great depths; but other ecological factors are involved such as water temperature, dissolved oxygen and needs for a special biotope in which to feed, spawn, guard young or as nursery areas, and cover from predators. Substrate spawners with macrophyte feeding habits are more dependent on shoreline habitats (macrophytes occupying a maximum depth of 7–8 m), than are the mouth-brooders with their microphagous and more pelagic habits, especially in water rich in plankton. As ecosystems have few types of fish depending on primary production, competition between cohabiting fishes is likely to be greatest between various sympatric tilapia species.

In most tilapias the alevins, juveniles and adults distribute themselves in depth zones from the shore towards deeper open water as they grow in size; there are also seasonal and daily changes in microhabitat. Within an ecosystem a tilapia species lives in many habitats, changing its biotope according to its life history stage. The ecological significance and survival value of such daily changes are thought to include tactics for avoiding predators, reducing competition for food between young and adults and remaining in water temperatures permitting fast growth in juveniles.

The physical and chemical conditions governing tilapia distributions, together with aspects of their behavioural ecology which govern selection of microhabitats and fish movements, were reviewed by Philippart and Ruwet (1982). Tilapias do not grow in water temperatures below 16°C, or survive below 10°C for more than a few days. They are more tolerant of high temperatures (to 42°C in *O. a. grahami*). Some euryhaline species can grow and reproduce in sea water up to 44‰ salinity. They tolerate pH values between 5 and 11. *O. mossambicus* can survive dissolved oxygen concentrations as low as 0.1 ppm for short periods.

The ecology of tilapias in fish communities was reviewed for over 30 species from numerous ecosystems by Lowe-McConnell (1982), with emphasis on conditions under which they switch from growth to reproduction. Tilapias' ability to do this at precocious sizes and ages in response to environmental conditions is an important way in which they survive adverse conditions in small water bodies, so aiding their dispersal across a landscape. This great plasticity in growth in natural waters (as well as in dams and ponds) suggested that concentrating research on environmental and behavioural factors affecting growth and maturation size was likely to be more helpful for fish culture than the search for faster-growing strains.

Pauly (1984) has since hypothesized that diminishing oxygen supply per unit weight in growing fish may be the over-riding factor inducing the transition from juvenile to adult, pointing out that anything (such as elevated temperature, overcrowding, osmotic stress) which in a given population raises the maintenance metabolism will result in reduced ultimate size. In this chapter the role of tilapias in ecosystems is explored to see how more recent ecological information accords with these ideas, starting with Nilo-Sudanian species which live under widely different conditions and have recently come to dominate and replace the endemic tilapia species in Lake Victoria. Tilapias in the Zambezian region (which include *O. mossambicus*, *O. macrochir* and *T. rendalli*, all used in fish culture) occupy comparable ecological roles to their Nilo-Sudanian counterparts.

5.2 NILO-SUDANIAN TILAPIAS IN RIVERINE ECOSYSTEMS AND IMPOUNDMENTS

The freshwaters of the Nilo-Sudanian region (Fig. 5.1) have three relatively common tilapias, *S. galilaeus*, *O. niloticus* and *T. zillii*, plus the smaller less common *O. aureus* (syn. *O. monodi*) which is more abundant in the Nile delta and Israel. These ecosystems are also inhabited by many large fishes, some of which may compete for resources with tilapias, such as *Labeo* and *Citharinus* which share the bottom detritus used as food (for example with *S. galilaeus* in Lake Chad; Lauzanne, 1972). Others are large piscivores, such as *Lates niloticus*, the Nile perch, the *Hydrocynus* spp. tiger fish and the mormyroid *Gymnarchus niloticus*, all known to include tilapias in their diets.

The riverine tilapias are generally restricted to pools, lagoons and the edges of the river and floodplain, places where aquatic vegetation provides cover. In the Upper Niger *S. galilaeus* predominated over sandy bottoms, *O. niloticus* over muddy ones, whereas *T. zillii*, an ecologically tolerant species, lived over rock, sand or mud, in still or running water (as Holden, 1963; also found in the Sokoto tributary pools). A macrophyte feeder, *T. zillii* lives near plant stands; riverine *S. galilaeus* are detritivorous using bottom algae, and *O. niloticus* takes epiphytic algae, diatoms or plankton according to the conditions.

In the Sokoto tributary in northern Nigeria Holden (1963) found that the tilapia populations became divided into at least two separate habitats, the main river and the pools isolated from the river. Algal investigations showed that the two habitats supported sources of food quite different in quantity and quality;

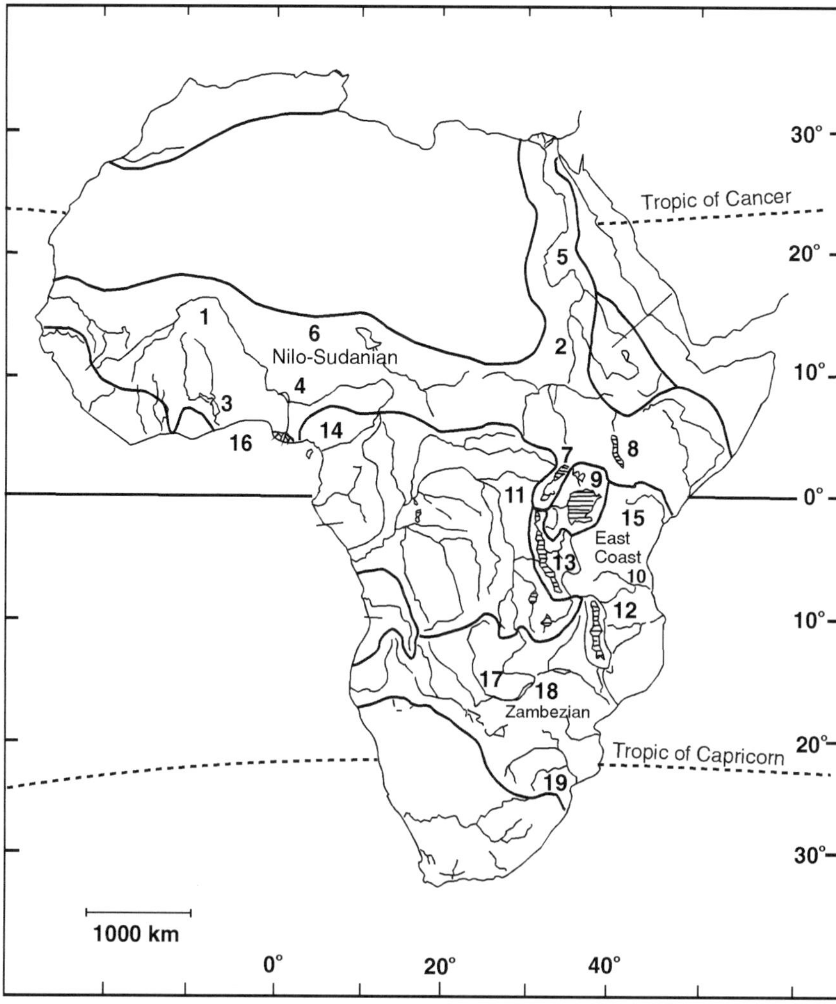

1. R Niger
2. R. Nile
3. L. Volta
4. L. Kainji
5. Nasser-Nubia
6. L. Chad
7. L Albert
8. L. Turkana
9. L. Victoria
10. Nyumba-ya-Mungu
11. George/Edward
12. L Malawi
13. L. Tanganyika
14. Cameroon Crater Lakes
15. Rift Valley Lakes
16. Coastal Lagoons
17. Zambezi System
18. L. Kariba
19. L. Sibaya

Fig. 5.1 Sites of tilapia studies in Africa discussed in text.

the concentration of algal biomass was higher in pools and backwaters than in the main stream and there was a positive relationship between the size of the fish and size of pool. These tilapias did not appear to be migratory. Male tilapia grew faster than females, scale rings indicated that they reached 15 cm in the first year, most of the growth occurring from the start of the floods in May till December.

How did the riverine tilapias respond to changed conditions in the new manmade lakes formed behind hydroelectric dams? At Kainji on the Middle Niger, in a preimpoundment survey made at low water (July–September 1965), tilapias were not common among the 66 fish species caught in experimental gillnets, though a few *S. galilaeus* were captured by cast net from the sandy shallow river and some *O. niloticus* were caught in traps at the side of the river. *O. niloticus* fry were found in mid-July in the pools cut off from the main river; gillnet catches in flooded land indicated that the start of the floods was a peak spawning period. During dam construction the 27 m deep coffer-dammed channel of the main river here yielded over 80 species of fish, but few cichlids (Ita, 1984).

The dam closed in 1968 created the 1250 km^2 Lake Kainji populated by riverine fishes. Surveys between 1966 and 1978 showed that cichlids, particularly tilapia, came to dominate the lake catch (with a mean standing crop of about 105 kg ha^{-1}), followed by characids, cyprinids, bagrid, mochokid, malapterurid, clariad and schilbeid catfishes, mormyrids, and centropomid *L. niloticus*, i.e. as part of a very rich fauna (Ita, 1984). In 1976 *S. galilaeus* made up 8% of gillnet catches, compared with only 3% *O. niloticus* and 0.5% *T. zillii*; *S. galilaeus* was still predominant in 1984 catches (Balogun and Ibeun, 1995). In this lake the build-up of tilapia populations was delayed for some years, perhaps because the large annual draw-down (10 m) affected potential spawning grounds; furthermore the large characoid *Citharinus* became very abundant when the lake first filled, flooding coinciding with, or stimulating a good *Citharinus* spawning year. Some of the numerous aquatic birds (~74 species) included *S. galilaeus*, *O. niloticus* and the catfish *Chrysichthys nigrodigitatus* as part of their diets.

The Volta River (Ghana), dammed in 1964, produced the huge (8800 km^2) Volta Lake which has much the same fish fauna as Kainji. Here too *S. galilaeus* became the dominant tilapia but populations of *S. galilaeus* and *O. niloticus* built up within the first 16 months. *O. niloticus* became fairly abundant on the eastern side of the lake and *T. zillii* in western areas. The rising water level provided optimum conditions for the growth of juveniles among the grass and bushes, and periphyton on the drowned trees proved a useful food supply in this new lake where bottom waters were at first very deoxygenated (Petr, 1967). In 1991 tilapia were still the dominant species (26%) in Volta's commercial fishery, followed by *Chrysichthys* and *Synodontis*. The influx of fishermen led to over-exploitation, while at the same time the main food source for the fish stocks, periphyton attached to submerged trees, was depleted as trees were cut down during draw-down periods to meet the increased demand for firewood (Braimah, 1995).

In the Ivory Coast the construction of hydroelectric dams has led to natural hybridization between *T. zillii* and the closely related *T. guineensis* (a brackish-water species found in the lower reaches and lagoons of three West African river systems) in a number of places where riverine conditions have been transformed into lagoons. Hybridization between these two species has also been reported in Lake Ayamé on the Bia River (Agnèse and Gourène, 1998; Gourène *et al.*, 1998).

In the River Nile's vast Sudd area in the Sudan, Bailey (1988) found *O. niloticus* to be a widely distributed and important tilapia, making up 6–9% of the experimental gillnet catches in open water and plant beds of the lakes. Adults and juveniles were also caught in the *Eichhornia* fringe. In January juveniles were relatively abundant in temporary pools in the river-flooded grasslands; ovaries indicated an extended spawning period during the flood season. *O. niloticus* spawning places were observed in the sandy substrate and along the *Eichhornia* fringe. Algae from all three major associations were identified in gut contents but periphyton and deposited algae were predominant. Fry consumed small crustaceans and rotifers. *S. galilaeus* was not as widespread or abundant and was more closely associated with submerged vegetation, as was *T. zillii*. The annual flooding of the permanent Sudd swamps inundates large areas of *Oryza* grasslands; fish migration into a portion of this seasonal floodplain was led by *Clarias gariepinus, Polypterus senegalus*; *Channa obscura* and juveniles of *O. niloticus* were prominent among late entrants. Migrants became stranded in temporary pools and subject to heavy cropping by birds (Hickley and Bailey, 1987).

On the Nile, Egypt's High Dam closed in 1964 formed the 6200 km^2 Lake Nasser/Nubia in which *O. niloticus* and *S. galilaeus* have become dominant (89%) in catches. Both species here are periphyton feeders and spawn fractionally most of the year, *O. niloticus* growing to 45 cm (estimated 5 or 6 years old), and *S. galilaeus* to 32 cm (4 years). They live as part of a very rich Nilo-Sudanian fauna with omnivorous *Labeo* and *Barbus, Synodontis* and other catfishes and mormyrids, planktivorous *Alestes*, and piscivorous *Lates, Hydrocynus, Bagrus, Clarias* and *Heterobranchus*. Rashid (1995) gives details of the fish biology and associated limnological conditions.

5.3 NILO-SUDANIAN TILAPIAS IN LAKES

Lake Chad

West Africa's largest natural lake the huge but shallow Lake Chad, has much shrunk since ORSTOM's IBP studies of trophic relationships were made here (Lauzanne, 1972; Lowe-McConnell, 1987). *S. galilaeus* was then the common tilapia species except in the north and northeast where *O. niloticus* predominated (an area which later dried up), and in the southern deltas of in-flowing rivers where *T. zillii* lived over various types of bottom. Lake Chad lies far enough north to have a seasonal drop in temperature and here Lauzanne found that the daily consumption of food by *S. galilaeus* depended on water temperature, and the conversion of detritus eaten was very low (only 3%).

Lake Turkana

The riverine lakes, Albert that drains to the Nile, and Turkana (Rudolf) now a closed basin, both have nilotic faunas, with *O. niloticus* feeding mainly on phytoplankton. Both also have populations of dwarf *O. niloticus*. These live in lagoons round Lake Albert, in crater lakes on Central Island in Lake Turkana, providing opportunities to examine conditions under which 'stunting' occurs. In Lake Albert *O. niloticus* in the main lake grew to 42 cm TL, maturing at 28 cm TL, fish in good condition. In four very alkaline lagoons *O. niloticus* in very poor condition (low weight for length, parasitized, some blinded and moribund) only grew to 23 cm TL female, 26 cm male, females maturing at 12 cm, male at 14 cm (Lowe, 1958). As in rivers, *S. galilaeus* was found off sandbanks. Lake Albert also has another inshore-dwelling *Oreochromis* (*O. leucostictus*). In Lakes Edward/George, which drain to Lake Albert, *O. niloticus* and *O. leucostictus* were abundant (the latter in the inshore zone), but *S. galilaeus* and *T. zillii* did not occur in Lakes Edward/George.

More recent studies in Lake Turkana included a 3-year survey (Hopson *et al.*, 1982) and other work collated by Kolding (1992, 1993, 1995). This 7560 km^2, 79 m deep lake, pH 9.3, has a fish fauna of only 37 species, including some endemics. The fish communities are much affected both by the seasonal inflow from the Omo River from the north, bringing in nutrients (August–October), and by long-term fluctuations in lake level. The lake has four main fish communities, the littoral one (0–4 m deep) dominated by tilapias, *O. niloticus* on sandy, *S. galilaeus* on softer and *T. zillii* on harder bottoms, with *Clarias*, *Raiamas* (syn. *Barilius*) and *Lates niloticus* as the dominant piscivores. The inshore demersal zone (4 to 10–15 m) is also inhabited by soft deposit feeders *Labeo*, *Citharinus*, *Distichodus*, with piscivorous *Bagrus docmac*. The offshore demersal (8 to 20 m according to season) has *Barbus turkanae* and *Bagrus bayad*; the epipelagic zone has small endemic zooplanktivorous species pursued by *Hydrocynus forskalii*, *Lates longispinis* and *Schilbe uranoscopus*. But at night communities are not so distinct as some species move inshore and surface-wards. The tilapias spawn in the littoral zone, most of the large otophysan species migrate north to the Omo delta to spawn in the flood season; both *Lates* species spawn offshore in the open lake.

O. niloticus (subspecies *vulcani* in this lake) is most abundant in shallow water near to the western shore, especially in the sheltered water of Ferguson's Gulf, and near the calmer eastern shore of this very windy lake (Hopson, 1982). This inshore distribution, also shown in Lake Albert, was attributed by Worthington

(1937, 1954) to the presence of large piscivorous *Lates* and *Hydrocynus* in these lakes (absent in George/Edward, and at that time from Lake Victoria). In Lake Turkana the *O. niloticus* then grew to a size unequalled (except later in the introduced population in Lake Victoria), specimens of 64 cm TL, 4.3 kg weight, being common (Worthington and Ricardo, 1936; Lowe-McConnell, 1958). These attained 39 cm TL before reaching sexual maturity, though some found later (Hopson, 1982) matured at a smaller size, and Kolding (1993) found the median (50%) size of first maturity had decreased by 13 cm over 30 years (to 26 cm TL in 1982). In the crater lakes they matured at 10–12 cm, and brooding females less than 93 mm SL were taken from hot springs near the lake.

The tilapias play an important part in the lake ecosystem and are apparently the principal prey of crocodiles, pelicans and *L. niloticus*. Ferguson's Gulf has large flocks of pelicans; consumption rates of 0.3 to 3 t fish a year per pelican have been estimated for other localities. Consumption by the plentiful *L. niloticus* was estimated as 10 000–14 000 t tilapia per year .

The abundance of *O. niloticus* followed the fluctuations in lake level. In certain years they were able to proliferate enormously in shallow sheltered areas, for example in 1976 Ferguson's Gulf yielded a staggering 16 000 t from an area about 10 km^2 (Kolding, 1993). Smaller 'booms' occurred in 1963, 1970 and 1982, but in years of low lake level Ferguson's Gulf dried out. On re-flooding, dried dung from cattle feeding on the lake shores added nutrients to those from the large bird populations, and the Gulf also produced large blooms of nitrogen-fixing cyanobacteria, not found in the open lake. Chlorophyll content was three orders of magnitude higher in the Gulf than in the main lake (where *Microcystis* was dominant).

There was no evidence of strong breeding seasonality in *O. niloticus*, but in the Gulf tilapia numbers rose to a peak during the flood season (August to November), when inundated grassy areas on the lake margin provided food and shelter for the fish fry. These grew fast and resulted in seasonal fluctuations in the fishery. Seasonal variations in the fry survival rate appeared to be the main factor affecting tilapia numbers. As neither food production nor natality appeared to limit the Gulf tilapia populations, Kolding (1993) thought food accessibility and/or predation pressure on the early stages probably did so. He pointed out that predation pressures differed in the Gulf from the main lake. In the Gulf there was intense predation by pelicans, crocodiles and man, taking mostly large specimens but few teleost predators (their access restricted by the oxygen conditions). However, during periods of decreasing water levels with progressively adverse conditions in the shallows, tilapias were forced into the deeper main lake where waiting teleost predators would crop the smallest individuals most heavily, so selection pressures would favour large sizes. Maturation sizes were smaller in the Gulf (though food was plentiful), which Kolding suggested was a direct effect of reduced or fluctuating oxygen conditions (see also Pauly, 1984). He concluded that the observed phenotypic plasticity of *O. niloticus* in Lake Turkana was closely related to the fluctuating environment and changes in mortality pressures, which resembled conditions in floodplains where this species probably evolved.

Lake George (Uganda), site of a trophic level study

Lake George, a 250 km^2, 3 m deep, equatorial aseasonal lake, was chosen as the site of a 5-year International Biological Programme study of production at all trophic levels (Greenwood and Lund, 1976). Here, as in Lake Turkana, the *O. niloticus* maturation size was found to have diminished since the initial studies, which Gwahaba (1973) attributed to the effects of intensive fishing. When commercial fishing started in the early 1950s, 50% of the females were mature at 27.5 cm TL. By 1960 this had fallen to 24.5 cm, and in 1972 was 20.5 cm.

In Lake George 95% of the biomass was of planktonic algae, mainly *Microcystis* spp. which were digested by *O. niloticus* in accordance with its diurnal physiological cycle (Moriarty *et al.*, 1973). The zoobiomass was dominated by three herbivores: *O. niloticus*, *Haplochromis nigripinnis* and the zooplankton crustacean *Thermocyclops hyalinus*. Piscivorous fishes present were *Bagrus*, *Clarias*, *Polypterus* and *Haplochromis squamipinnis* (taking juvenile fish). The two herbivorous fishes formed over 60% of the open lake ichthyomass, the piscivorous fishes about 20%. Only about 3% of the net primary production (estimated as 23 200 kJ m^{-2} yr^{-1}) was converted to secondary production (650 kJ m^{-2} yr^{-1}), 0.65% to tertiary production (150 kJ m^{-2} yr^{-1}) and only 0.21% to fish yield (50 kJ m^{-2} yr^{-1}) (Burgis and Dunn, 1978). The herbivores did not appear to be food-limited, nor was predation intense; it was suggested that tilapia numbers were limited by shortage of suitable spawning places in this soft-bottomed lake.

A trophic ecosystem ECOPATH II model of Lake George (Moreau *et al.*, 1993) confirmed the view that this ecosystem has a low ecological efficiency, with under-utilization of some sources of food (especially phytoplankton) by the fish. The gross efficiency of the fisheries (actual catch/ primary production) was computed to be 0.0057 in Lake George, between that of Lake Victoria prior to (0.0016) and after the introduction of Nile perch (0.0082). In the 1990s the Lake George tilapia were much smaller, over-fishing having led to the use of smaller-mesh gillnets; the Edward/George fishery utilized more long-lines and took more predators (*Bagrus* and *Clarias* catfishes, *Protopterus* lungfish), changes influenced by socioeconomic conditions, over-fishing and marketing (Crespi and Ardizzone, 1995).

Growth and maturation sizes in other waters

O. niloticus populations were also examined from many Ugandan lakes and dams where they had been introduced in the 1940/1950s (Lowe-McConnnell, 1958). In some of these the fish grew very large while populations were small but then multiplied rapidly and became dwarfed. In other lakes *O. niloticus* apparently never grew large, which may have been related to paucity of good quality food (cf. Iles, 1973, who took the view that 'stunting' is best summarized as 'the population stunted and the pond became overcrowded', rather than the prevailing view at the time 'the pond became overcrowded and the population stunted', as discussed later).

Data on *O. niloticus* from these lakes brought out the following points. (a) The distribution of *O. niloticus* within a water body varied with the ecological conditions. Phytoplankton, epiphytic diatoms or benthic algae were used as food according to availability. (b) The maturation sizes and size to which the tilapia grew varied greatly; fish from large lakes matured at, and attained, a larger size than those from lagoons, the largest being from Lake Turkana. (c) In the lakes male and females did not differ significantly in maturation or final size, but in smaller water bodies where the fish were dwarfed, males were larger than females, and also much more numerous than females. (d) Fish in poor condition (low weight for length) matured at a smaller size than fish in good condition; the growth rates and maturation ages were not known, but in some newly stocked waters *O. niloticus* grew to 35 g (equivalent to c. 26 cm TL) in one year, whereas in ponds they grew to 17 cm TL (98 g) and started to breed when 7 or 8 months old. (e) In equatorial waters *O. niloticus* in breeding condition might be found at any time of year, but peak spawning coincided with the two rainy seasons, a finding later confirmed by Gwahaba in Lake George; at higher latitudes this species breeds mainly in the rainy season. (f) Dwarf *O. niloticus* from shallow waters were often heavily parasitized; those from large lakes appeared relatively free of parasites.

The plasticity in size and age at which *O. niloticus* females mature has been supported by growth rate analyses in the Ivory Coast where in eight man-made water bodies (60 000 to 80 000 ha) age and size at maturity were positively correlated with reservoir area (Duponchelle and Panfili, 1998). These females all matured in their first year (5.6–10 months). The range of variation in age at maturity between populations was found to be greater than that of size at maturity. Here faster growing fish matured earlier (maturation age and, generally, size were found to be highly significantly inversely correlated with growth rate). Some intra-population variation in size at maturity in consecutive years was attributed to growth differences, probably resulting from resource availability.

Some hybridization occurred between *O. niloticus* and *O. spilurus niger* when both were introduced into Lake Bunyoni, and possibly between *O. niloticus* and *O. esculentus*, both stocked in an Ankole crater lake. Hybrids of *O. niloticus* and *O. variabilis* in Lake Victoria were described by Welcomme (1964a). *O. niloticus* and *O. macrochir* (probably escaped stocked fish) hybridized in Lake Ihema in the Kagera River system draining to Lake Victoria (Plisnier *et al.*, 1988). These last two species also hybridized when introduced to Lake Itasy in Madagascar, where later generations came to resemble *O. niloticus* (Moreau *et al.*, 1988).

Data for indigenous *O. leucostictus* in Lake George/Edward and Albert and from numerous lakes, dams and ponds where it had been introduced, also showed that maturation size varied greatly with the conditions (from 26 cm TL in Lake George to 7 cm in ponds, Lowe-McConnell, 1957; Welcomme, 1967). Nine years later its maturation size in Lake George had declined to 14 cm TL, though the reason for this was not clear as it had not been as heavily fished as *O. niloticus* (Gwahaba, 1973). In some waters male *O. leucostictus* grew to, and matured at, a larger size than the females; in ponds they bred when only 3 months old. In natural waters *O. leucostictus* occupies an inshore zone and is common in lagoons. In equatorial lakes, as in ponds, breeding occurs throughout the year without well-defined seasons.

5.4 TILAPIAS IN LAKE VICTORIA – THE REPLACEMENT OF ENDEMIC BY NILOTIC SPECIES

In recent years Lake Victoria's endemic tilapias *O. esculentus* and *O. variabilis* have become replaced in catches by introduced nilotic *O. niloticus* and *T. zillii*. *O. esculentus* has almost vanished from the main lake, though still present in some small satellite lakes (and in Nyumba ya Mungu, *vide infra*), and *O. variabilis* has become very rare in catches. How has this come about?

Lake Victoria, the world's largest tropical lake (69 000 km^2) drains to the Nile through the small Lake Kyoga and the north end of Lake Albert (Fig. 5.1). The Murchison Falls between Kyoga and Albert prevented many nilotic species from gaining access to Lakes Kyoga and Victoria, including the nilotic tilapias and the large piscivorous *Lates* Nile perch, *Hydrocynus*, *Citharinus* and others. In their absence a spectacular flock of c. 300 haplochromine cichlid species evolved within Lake Victoria, together with two endemic tilapia species, *O. variabilis* and *O. esculentus*. This burst of evolution appears to have happened in the last 12 000 years, since the lake reformed after being completely dried out for a long period (Johnson *et al.*, 1996).

As is now well known, the Lake Victoria ecosystem has undergone profound and accelerated changes in the last three decades (Kaufman, 1992; Lehman, 1998; the many papers in Pitcher and Hart, 1995). The changes, sparked by the rapidly rising human population and activities in the watershed and on the lake, included the introduction of exotic tilapias and predatory Nile perch *Lates niloticus*, over-fishing and use of destructive gear, agricultural practices which use agrochemicals and clear the land, and regional climate changes.

The most decisive impact was attributable to the explosive population increase of Nile perch *Lates niloticus* in the mid-1970s, 20 years after its introduction to the lake (Oguto-Ohwayo, 1995). This was followed by the loss of an estimated two thirds of the endemic haplochromines, as monitored by the Leiden University (HEST) team investigators (Witte *et al.*, 1995). But the introduction of nilotic tilapias *T. zillii*, *O. leucostictus*, *O. niloticus* in the 1950s, to augment the dwindling supplies of endemic tilapias that were the mainstay of the important gillnet fishery until they were over-fished, also played a part.

Lake Victoria's two indigenous tilapias were ecologically complementary species with little or no competition between them (Lowe-McConnell, 1956a; Fryer, 1961). *O. esculentus* predominated in bays where the bottom was of soft flocculant material from phytoplankton deposition, whereas *O. variabilis* was abundant in the clearer water off more exposed shores, in water-lily lagoons and in the out-flowing River Nile. Both species also occupied different biotopes according to the size and sexual condition of the fish. *O. variabilis* spawning behaviour was observed in the clear water of a water-lily swamp in Lake Kyoga (Lowe-McConnell, 1956b), *O. esculentus* appeared to spawn further offshore in less clear water, so was difficult to observe. In the equatorial Lake Victoria tilapias produced several batches of young in rapid succession (Lowe, 1955b).

The introductions and spread of the exotic tilapias were described by Welcomme (1964a, 1966, 1968). *T. zillii* first appeared in gillnet catches in 1956, *O. niloticus* and *O. leucostictus* in commercial records in 1960. Welcomme discovered suspected hybrids both between *O. variabilis* and *O. niloticus* and between *T. zillii* and *T. rendalli* (which had been introduced into, and escaped into the lake from, Kisumu fishponds). *T. zillii* spread rapidly at the north end of the lake, appearing in southern and eastern parts in 1960, and were abundant by 1964; they later became the dominant species in former *O. variabilis* areas, living in shallow marginal waters all round the lake, fry and juveniles inhabiting shelving rocky shores in sheltered places, situations favoured as *O. variabilis* nurseries. *O. niloticus* and *O. leucostictus* became much more abundant from 1964, *O. leucistictus* becoming the dominant species in lagoons round the lake, near the papyrus and in the shallows of muddy bays, *O. niloticus* on the former *O. esculentus* grounds. Introgression between *O. esculentus* and *O. niloticus* was later found to be rampant (Mwanja and Kaufman, 1995).

Welcomme investigated possible competition between both juveniles and adults for feeding and breeding places. There appeared to be little competition for food between adults, *T. zillii* taking higher plant material, *O. esculentus* planktonic diatoms, while *O. niloticus* and *O. variabilis* had more flexible diets, feeding either on the bottom or on epiphytic or planktonic diatoms depending on the habitat. *O. leucostictus* fed exclusively on bottom material. Spatial segregation also helped to reduce competition, *O. leucostictis* being confined to shallow lagoons near and often behind the papyrus fringe (in water 30 cm to 5 m deep), while *T. zillii* and *O. variabilis* spread into harder-bottomed habitats and more exposed water. The larger *O. esculentus* lived in mud-bottomed bays (c. 5–10 m deep). *O. niloticus* was found in most habitats. All the introduced species grew to a large size and were in good condition (high weight for length).

Competition for breeding grounds was more apparent and the lack of suitable places for spawning and nursery grounds appeared to be a factor limiting population expansion. This view was supported by the dramatic increase in *O. esculentus* in catches in 1964–65 following the unprecedented increase in lake level (1.4 m above the previous highest recorded level) in 1961–63; this flooded behind the papyrus fringe and became used as *O. esculentus* nurseries

(Welcomme, 1970). *O. esculentus* evidently spawned where it could find firm enough bottom in the sheltered gulfs, but the juveniles lived in lagoons at the lake's edge. Clean, firmer substrates were needed on which to spawn by the substratum-spawning *T. zillii*, and by the very small *O. variabilis*. The ousting of *O. variabilis* by *T. zillii* appeared to be due to competition between them for such places for breeding and nursery grounds (Fryer, 1961). *O. leucostictus* spawned in shallow (c. 30 cm deep) water in mud-bottomed areas at the edge of lagoons. *O. niloticus* appeared more catholic in its habits and was found in association with the other species.

Welcomme's (1964b) study of juveniles on nursery beaches, combined with laboratory experiments, differentiated between gently sloping 'gradient' and more steeply-shelving 'non-gradient' beaches. Juvenile tilapias were abundant on the former, haplochromines on the latter. The small tilapias could withstand temperatures up to 39°C, lethal to haplochromines. These juvenile tilapia moved on and off the gradient beaches in accordance with diurnal temperature fluctuations, the shallows cooling rapidly at night and the fish then moving into deeper water. A clear relationship was found between size of fish and depth of water. Inhabiting very shallow water enabled the small tilapia to feed without disturbance from the larger individuals. Also they were out of reach of the piscivorous catfishes, *Clarias gariepinus* and *Bagrus docmac*, which lived off the tilapia beaches.

The nursery beaches, though sharing high temperatures, differed in their dissolved oxygen tensions (DO), for which the various tilapias had characteristic, though to some extent overlapping, preferences. *O. leucostictus* young were confined to areas of low DO (0.6–2.0 ppm), *O. esculentus* and *O. niloticus* lived in slightly better aerated water (1.0 to 3.0 ppm and 2.0 to 4.0 ppm respectively), and *T. zillii* and *O. variabilis* were the common species in well-aerated (5–8 ppm) more exposed areas (Fig. 5.2).

The Nile perch (*Lates niloticus*), introduced into Lake Victoria c. 1954 ostensibly to crop the haplochromines, fed mainly on cichlids including tilapia. The explosion in population numbers in the mid 1970s (in Kenya) and early 1980s (Tanzanian waters) was followed by the loss of an estimated two thirds of the c. 300 haplochromine species (Witte *et al.*, 1995), of which 60% of the biomass was of detritivous/algivorous fish which probably had an important role in circulating nutrients from bottom deposits. This loss, together with pollution from various sources, and possibly slight warming of the surface waters, combined to cause radical changes to the Lake Victoria ecosystem. International interest in the great loss of biodiversity and inter-related changes has spawned over 2000 references (Crul *et al.*, 1995).

The net result is that the lake appears to be stratified throughout the year, instead of oxygen being carried down to bottom waters seasonally (by June-August wind mixing, as in the 1950s). Bottom waters are now deoxygenated, reducing the fish- inhabitated water to the surface 20–30 m, and surges of deoxyygenated water into shallow areas are causing increased fish kills. Algal

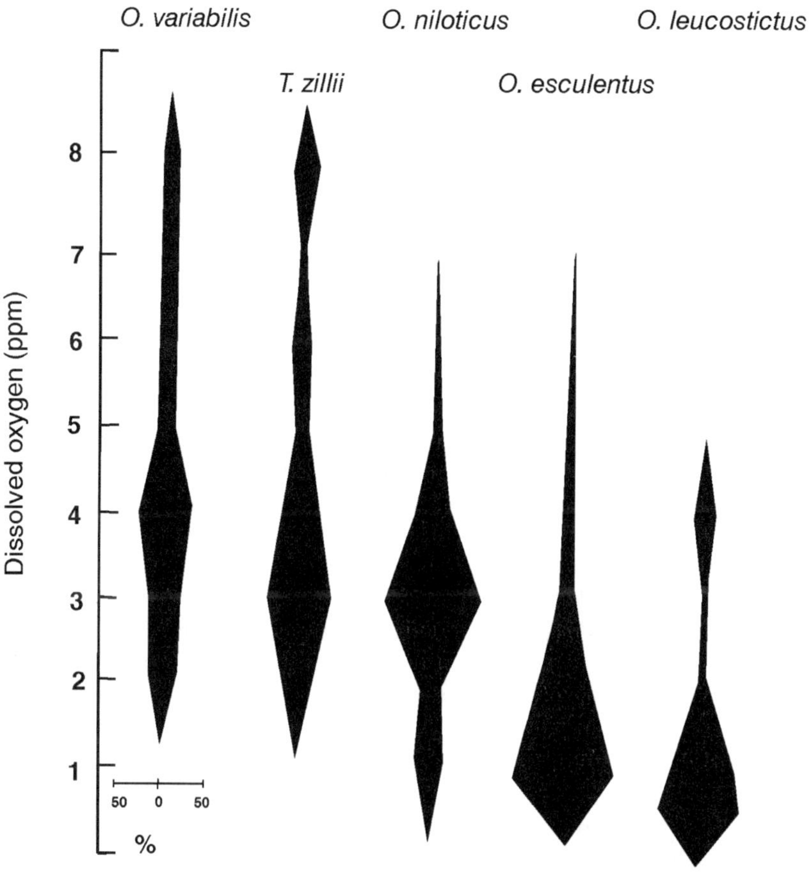

Fig. 5.2 Comparative distributions of juvenile tilapia species in relation to dissolved oxygen concentrations on Lake Victoria's beaches (redrawn from Welcomme, 1964b).

blooms have intensified and algal biomass has greatly increased, now mainly of cyanobacteria (compared with a dominance of large diatoms, particularly *Melosira* and *Stephanodiscus* in the 1950s, Hecky, 1993). Since 1989 another exotic invader, water hyacinth *Eichhornia crassipes*, has become a major problem.

The foods now eaten by the various tilapias were investigated by Gophen *et al.* (1993) together with the zooplankton and phytoplankton from Kenyan waters of the lake. They found that the major part of the food in *O. niloticus* now consisted of bottom fauna or deep water organisms (Trichoptera, *Caridina*, prawns, chironomids, ostracods, clams, *Moina*, *Chydoras*, plant debris, *Microcystis*, *Melosira*, small diatoms and small chlorophytes) with mud particles. The gut contents of *O. variabilis* were dominated by Trichoptera, mud particles, algae and plant debris probably collected from the bottom. The food composition in *T. zillii* was similar to that in *O. variabilis*, predominantly Trichoptera, *Caridina* and plant debris. Gophen *et al.* (1993) concluded that tilapias, which had been pelagic planktivores in Lake Victoria before the Nile perch invasion, are now more benthophagous feeders. Comparing their data with records from the 1960-1970s, they found shallower Secchi disc depths, algal concentrations enhanced, zooplankton densities slightly declined, phytoplankton assemblages shifted from diatoms to cyanobacteria dominance, and primary production greatly increased. They agreed with the Leiden University (HEST) team's suggestion that the decline in endemic cichlid species by Nile perch piscivory most probably enhanced the eutrophication of the lake.

O. niloticus and *T. zillii* co-evolved with Nile perch in lakes where they were indigenous, and their young stages live inshore which probably aids their survival, whereas the endemic *O. esculentus* which fed on offshore phytoplankton would have overlapped with the habitat of *Lates* in this lake, so be more at risk. *O. niloticus* also appears to be more aggressive than *O. esculentus*, and grows larger which would help it to avoid predation. Twongo (1995) who investigated the replacement of *O. esculentus* and *O. variabilis* by nilotic species in Victoria and Kyoga concluded that the vast reproductive potential of *O. niloticus* is the key to its proliferation in these lakes. According to Balirwa (1998) the Nile tilapia in Lake Victoria now mature at larger sizes, have a high body condition index and produce comparatively more eggs than the other species. Balirwa (1998) stressed that the flexibility of feeding habits in *O. niloticus*, demonstrated by the large proportion of invertebrates in its adult diet in Lake Victoria (whereas *O. esculentus* concentrated on planktonic diatoms as food) has probably contributed to this and thus its replacement of the endemic species. Commercial catch records from Lake Kyoga into which Lake Victoria flows (Fig. 5.3) show the changes in proportions of native and introduced tilapia species between 1963 and 1988 (Ogutu-Ohwayo, 1992).

5.5 NYUMBA YA MUNGU RESERVOIR, TANZANIA, NEW HOME OF *O. ESCULENTUS*

In the 1950s endemic tilapias from Lake Victoria were introduced into many small lakes, dams and fish ponds throughout East Africa (Bailey, 1966). Fry less than 20 mm TL taken from the mouths of *O. esculentus* and *O. variabilis* brooding females caught near Mwanza were introduced to a small experimental pond at Korogwe June 1950, where they bred at 16–19 cm TL 6 months later (Lowe, 1955a). From here they escaped into the Pangani River system. This already had two endemic tilapias, *O. pangani pangani* and *O. jipe*, with another subspecies *O. p. girigan* living downriver in Lake Jipe, where it was sympatric with *O. jipe* and fed on water weeds (*Najas*), the *O. jipe* feeding on finer particles.

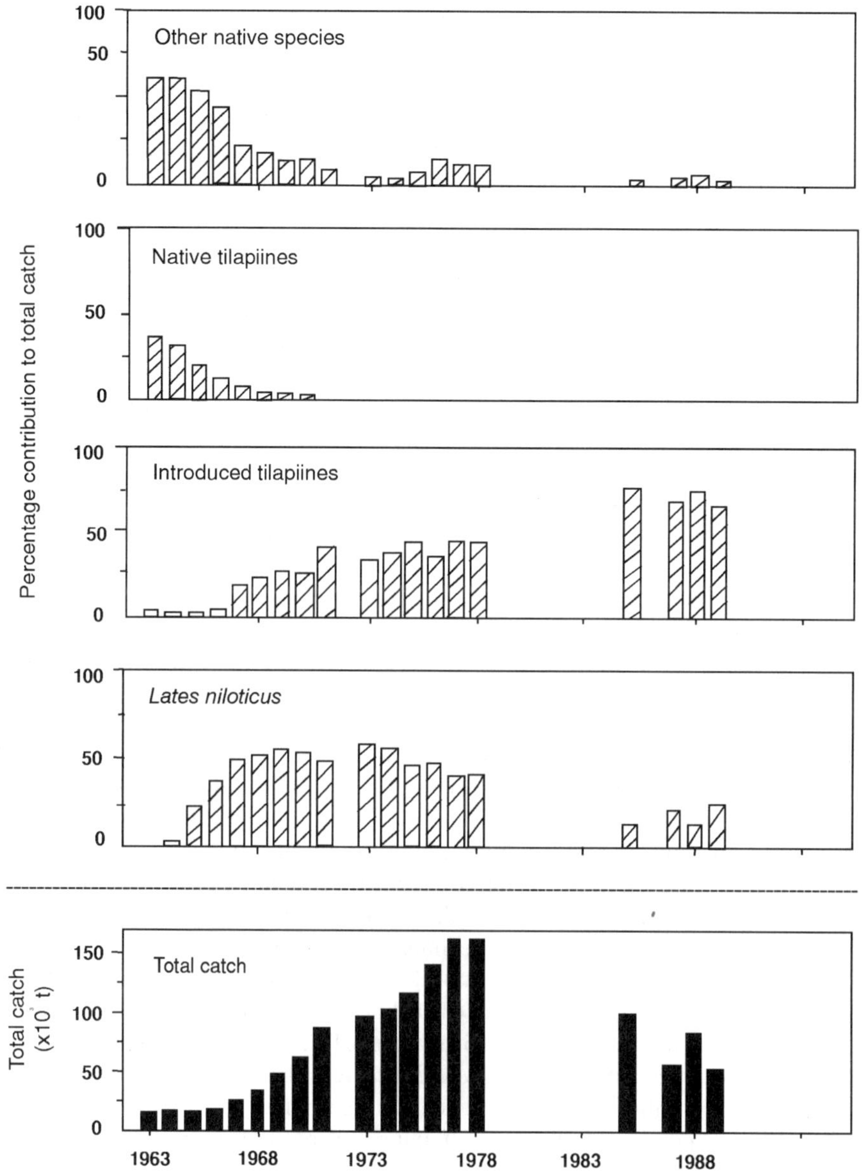

Fig 5.3 Replacement of native tilapiines by introduced nilotic tilapias (mainly *O. niloticus*) in Lake Kyoga's catches between 1963 and 1988 (redrawn from Oguto-Ohwayo, 1992).

Tanzania's largest reservoir, the 180 km^2 Nyumbu ya Mungu, 6 m deep, created in 1964 on the Pangani River system, provides a well-studied case history of tilapia populations which gained access to the new lake (Bailey *et al.*, 1978; Bailey, 1996). In 1974, 10 years after its formation, a team of biologists found three *Oreochromis* species here, the indigenous *O. pangani* and *O. jipe* and the escaped *O. esculentus*, together with *T. rendalli* (another pond escape). The indigenous tilapias, living mainly in the southern littoral where *O. jipe* predominated, were chiefly browsers of periphyton on the drowned bushes and phytobenthos from bottom deposits. They grew to a larger size (c. 50 cm TL and 1.75 kg) than recorded from Lake Jipe, males growing much larger than females (*O. jipe* males to 48 cm TL, females 27.5 cm; *O. pangani* males 47 cm, females 31.5 cm). The *O. esculentus* were living in deeper offshore waters where they consumed phytoplankton. *T. rendalli* lived inshore close to fringing reed swamp. The only natural predators in the new lake were *Clarias gariepinus*, crocodiles and huge numbers of piscivorous birds (kingfishers, cormorants, herons, pelicans, fish eagles) and man. (Fig. 5.4). When the lake was revisited in 1994 (30 years after impoundment) *O. esculentus* was found to be the common species in local markets.

In common with many new lakes, a high level of fertility influenced early developments; the tilapias spawned in sheltered backwaters, releasing swarms of young that grew in the floodplain shallows. In the 1960s a flourishing tilapia fishery attracted so many fishermen that at its 1970 peak an estimated 34 000 fishermen landed 28 500 t of fish, mainly from gillnets since the drowned scrub still prevented widespread use of beach seines. By 1974 the boom was over, the estimated yield only 2282 t. Eight of the 20 fish species identified here were common. *O. esculentus*, first noticed in 1973, was relatively abundant in 1974, inhabiting open water and feeding on plankton, a life style complementary to that of the inshore-dwelling periphyton-grazing endemic *O. pangani* and *O. jipe*. By 1994 the water level was low, the drowned trees had disappeared and seines had become a major fishing gear. The indigenous tilapias and the zooplanktivore *Rhabdalestes* were now rare in experimental gillnet catches. In 1984 *O. esculentus* had made up over 80% of the fishermen's catches; most of these were 12–15 cm TL (compared with 14–15 cm *O. jipe*). In 1994 the fish market was still dominated by huge numbers of *O. esculentus* less than 14 cm TL. Over 1000 fishermen were still active on the lake. Reduced mesh sizes were catching a substantial proportion of immature tilapias, but both *O. esculentus* and *O. jipe* were spawning at a smaller size than formerly.

Seegers (1996) records other Tanzanian waters now inhabited by *O. esculentus*, including Lake Rukwa (stocked 1969) where it now far outnumbers the endemic *O. rukwaensis*).

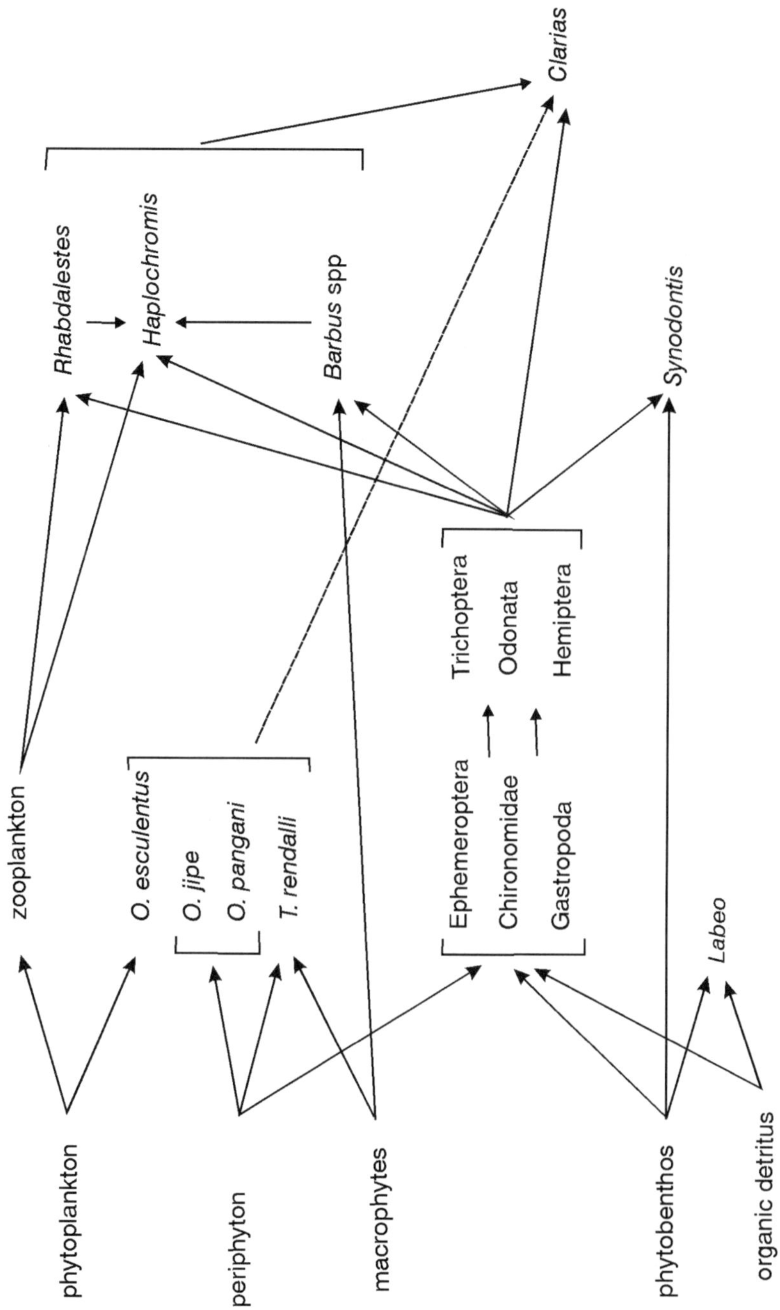

Fig. 5.4 The roles of tilapias in Nyumba ya Mungu reservoir, Tanzania, 1974 (simplified food web, redrawn from Bailey *et al.*, 1978).

5.6 THE ROLES OF TILAPIAS IN SPECIES FLOCKS

Lake Malawi

Lake Malawi is the only Great Lake to have a species flock of tilapia, a series of three species, *O. karongae* (syn. *saka*), *O. squamipinnis* and *O. lidole*, which have evolved within the lake where they have adapted to increasingly open water conditions through their schooling, feeding and breeding habits (Lowe, 1952, 1959). The substrate-spawning *Tilapia rendalli*, widely distributed in southern Africa, also occurs in Lake Malawi but is restricted to lagoons and amongst aquatic plants, on which they feed, leading a solitary existence. Another *Oreochromis*, *O. shiranus* is also caught in shallow areas, including Lake Malombe into which Lake Malawi flows; this species includes higher plants and detritus in its diet but feeds on phytoplankton when this is abundant. In the 1940s *O. shiranus* grew to 37 cm TL, the males growing larger than the females and developing enlarged lower jaws (as in other members of the *O. mossambicus* group to which this species belongs; Trewavas, 1983).

Lying well south of the equator (Fig. 5.1) limnological conditions are more seasonal in Lake Malawi than in the other great lakes, and breeding is seasonal amongst the Malawi tilapias, as described in an earlier chapter (see Turner and Robinson). In Lake Malawi tilapias are more abundant at the south end of the lake where the water is shallower (less than 40 m) and where the southwest trade-wind causes seasonal up-welling, which increases nutrients stimulating phytoplankton blooms. At the north end of the 600 km long lake tilapias are restricted to inshore waters.

The biology of each tilapia species was found to be related to its place, inhabiting inshore to more open water conditions. From the more inshore-living to open water-living species: (a) the food contained a higher proportion of phytoplankton instead of bottom diatoms and debris;(b) the tendency to school in small, close schools increases; (c) the amount of movement about the lake increases; *O. shiranus* populations appeared to be very localized and *O. lidole* is a far-ranging fish; (d) the young are mouth-brooded by the female to an increasingly large size, and (e) growth in weight for length decreases, so that open-water species are lighter for a given length; the big-headed *O. lidole* often resembled a 'hunger form'.

The balance between the numbers of the different species may depend on the lake level, which fluctuates considerably in cycles of years. The tilapia species that live in the shelter of weeds for much of the life cycle will suffer from a rising lake drowning weed beds. High lake levels may have enabled *O. lidole* to exploit new waters at the south end of the lake. In the 1990s Turner recorded brooding female *O. lidole* in Lake Malombe into which Malawi flows, which had become much more open than at the time of the initial surveys, when no brooding *O. lidole* were found there. Schooling in adult tilapia is unusual and seems to be a juvenile character, together with the black 'tilapia mark' on the hind end of the dorsal fin, retained in these species which have adopted an open water habit. *O.*

shiranus generally contained vegetable debris, probably picked off the bottom, though individuals were feeding on *Anabaena* when the maximum algal bloom was at its height in September 1946. *T. rendalli* contained *Ceratophyllum* leaves in a condition suggesting they were taken off the live plants. *O. lidole*, *O. squamipinnis* and *O. karongae* caught at same time and place in September 1946 had the same phytoplankton in stomach contents, indicating that all three species fed on whatever algae were most easily available. But in time of plankton scarcity *O. squamipinnis* and *O. karongae* tended to feed on bottom algae, whereas *O. lidole* appeared to range further over open water, notably losing condition when plankton was scarce. The young tilapia browsed on algae attached to rooted vegetation. Fry taken from the mother's mouth showed that feeding started as soon as the yolk-sac was involuted and before it was fully absorbed; they contained dinoflagellates and small *Melosira* and *Navicula* (Lowe, 1952). Stable isotope analyses of food partitioning among Lake Malawi near-shore fishes have confirmed that the more pelagic *O. lidole* relies on a phytoplankton-based food chain, whereas *O. squamipinnis* feeding on sandy sediments was also primarily an algae eater (Bootsma *et al.*, 1996). Quantitative field and laboratory studies have indicated that *O. lidole* and *O. karongae* exploit plankton resources better than other tilapia species (Hartnell, 1996).

In 1945–47 growth studies indicated that *O. karongae*, *O. squamipinis* and *O. lidole* all reached their minimum breeding sizes (about 27 cm TL, 24 cm and 28 cm, respectively) when just on 3 years old, the males and females growing at the same rate. *O. shiranus* females bred at 18 cm, males at 22 cm (Lowe, 1952). *O. shiranus* growth parameters have since been studied in detail in two Malawian reservoirs (Mattson and Kaunda, 1997).

Predators included *Bagrus meridionalis* and *Clarias* catfishes that preyed on adult tilapias, and numerous other cichlids fed on juveniles. Crocodiles and the immense numbers of fish-eating birds (including cormorants, pelicans, fish eagle, kingfishers, herons) took a heavy toll; reed cormorants (*Phalacrocorax africanus*) and *Anhinga* fed most heavily on young tilapia in the shallows (54 juveniles found in one *Anhinga*). Linn and Campbell (1986) estimated the c.10 000 white-breasted cormorants (*P. carbo*) utilized 261 g fish day^{-1}, thus catching c. 950 t yr^{-1}, mainly of small benthic cichlids, but very few tilapia, from water less than 12 m deep. But man became the chief predator. In the 1930/1940s fishermen caught tilapia mainly from traps set in reeds and small shore seines, hand-woven from plant fibre or string extracted from old car tyres. Industrial commercial fishing started late 1930s when Greek fishermen introduced large (40 m deep) ring-nets to surround tilapia schools in open water. Some gillnets caught *Bagrus* and *Clarias* and other species. Human population pressures, which have resulted in heavy exploitative fishing and the use of increasingly small-meshed (including illegal mosquito net-lined) seine nets, have led to the collapse of the tilapia fishery in Lake Malombe and the south end of the lake (FAO, 1993; Turner, 1995; Tweddle *et al.*, 1995, analysed by Palsson *et al.*, 1998).

Although the deep rift valley Lake Tanganyika greatly resembles Lake Malawi limnologically, this has only one endemic tilapia, *Oreochromis (Neotilapia) tanganicae*. This is restricted to inshore waters and is not important in the fisheries. For zoogeographical reasons (see Fig. 5.1) Lake Tanganyika, colonized from the faunal rich Zaire basin, has an open-water fauna of endemic clupeids and *Lates* which together support the main commercial fisheries.

O. niloticus, which originated from Lake Kivu, occurs in northern estuaries, and the vast (1100 km^2) swamps of the ancient Malagarasi affluent system have two endemic tilapias, *O. karomo* and *O. malagarasi* (miscalled *O. niloticus*). Living sympatrically, these breed at different times of year and eat different foods. *O. karomo* rasps algae off the abundant water-lily leaves. *O. malagarasi* feeds on soft flocculant bottom deposits of finely divided plant material with many protozoa and bacteria, much of which appeared to have passed through the guts of the cohabiting *Alestes* and *Distichodus* which fed on the water-lily leaves (Lowe, 1956b).

Tilapia species flocks in Cameroon crater lake ecosystems

The small Cameroon crater lake Barombi Mbo, only 2.5 km in diameter but over 110 m deep, has four endemic *Sarotherodon* species among its eleven endemic cichlid species (Trewavas *et al.*, 1972). These provide an example of ecological separation to exploit the meager resources of an oligotrophic lake. Barombi Mbo, which lies in a small forested crater with an outflow to the Mungo river, is structurally uniform with relatively few food resources: phytoplankton, organic debris, aquatic invertebrates, fishes and adult insects. The clear water permitted underwater observations of fish behaviour. Most of the cichlids were found to be clearly separated from one another in their distribution in space or time (feeding by day or night) or in their feeding habits.

Of the four endemic *Sarotherodon* species, two lived inshore and two offshore when adult. The two inshore species both fed mainly on organic debris, but one (*S. lohbergeri*) collected it together with aufwuchs from the surfaces of rock and plants, the other (*S. steinbachi*) mainly from sandy areas. Young individuals of the offshore tilapias (*S. linnellii* and *S. caroli*) fed close inshore in water too shallow for other fishes, taking flotsum as well as bottom aufwuchs and insect larvae. Phytoplankton formed the sole food of the adults, both species swimming a metre or so below the surface in mid-lake. The restricted space for spawning sites round the shore may be more limiting than food for these fishes, and there was some evidence that they breed asynchronously. *S. linnellii* is a mouth-brooder and the large and few ova ripening at a time suggested that the others are also mouth-brooders.

The suggested scenario was that a sand sifter *Sarotherodon galilaeus*-like form became more exclusively adapted to this habit (to become *S. steinbachi*). A second form became divided in response to increased opportunity and competition (especially for limited aufwuchs) into a specialized aufwuchs feeder (*S.*

lohbergeri) and two phytoplankton feeders (*S. linnellii* and *S. caroli*). Predation on invertebrates was shared by various non-tilapiine cichlids specializing on particular items. The factors that initiated speciation within the lake in the apparent absence of geographical barriers remained open to speculation. Trewavas (1972) pointed out that even the same stream species invading the lake in floods at successive time intervals would find conditions in the lake different at each time as niches were occupied by earlier invaders. Recent DNA analyses of these tilapias support the view that ecological diversification within the lake has been a key factor responsible for speciation after colonization (Schliewen *et al.*, 1994).

Stiassny *et al.* (1992) described a species flock of eight *Tilapia* spp. from another Cameroon crater lake, the 0.5 km^2 Lake Bermin. Here an endemic radiation of nine substratum-spawning tilapiine cichlids dominates the ichthyofauna. Eight of these, tentatively included in the genus *Tilapia*, show an impressive range of morphological variation, especially in dentition. Whilst most species feed primarily on organic debris and detrital plant material, at least two trophic specializations were recognized: a phytoplanktivore (*T. imbriferna*)and a sponge eater (*T. spongotroktis*). All the Bermin species are relatively small, *T. synderae* females attaining sexual maturity at 25 mm SL, making this the smallest tilapiine recorded. These Lake Bermin cichlids are the first example of a substrate-spawning tilapiine lacustrine radiation. They demonstrate that in the absence of haplochromine lineages tilapiines will speciate in lacustrine environments and that mouth-brooding is not a prerequisite for tilapiine lacustrine speciation; they also appear to have evolved sympatrically in Lake Bermin which is considered to be a relatively old lake. Another cichlid radiation, harboured in Ejagham Lake, also in the Cameroun Cross River drainage, has yet to be investigated.

5.7 TILAPIAS IN SPECIALIZED ECOSYSTEMS

Rift valley soda lakes and thermal springs

Thermal springs often carry populations of dwarf tilapia (e.g. those near Lake Turkana of *O. niloticus vulcani*). The very alkaline lakes with hot springs in the eastern rift valley are also inhabited by very small endemic tilapias. The ecology of *O. alcalicus grahami* in Lake Magadi (Kenya), a very alkaline lake, pH 10.5, 43°C, extremely rich in sodium salts, is best known (Trewavas, 1983). These fishes browse on cyanobacteria together with small copepods and dipteran larvae on the stony bottom in the warm shallows near the hot springs, moving into deeper water at night when the surface temperature begins to fall. High temperatures (>32°C) appear necessary for males to assume breeding colours, and they then breed almost continuously. Mouth-brooding females release the young in very shallow water where they escape predation by adult fish but are attacked by numerous kinds of birds, especially pelicans. There is also some cannabilism of eggs. The smallest brooding female was only 25 mm SL, the smallest ripe male 35 mm (see also Seegers and Tichy, 1999).

This species was introduced into Lake Nakuru c. 1960, where it altered the whole ecology of the lake (Vareschi, 1979). Fish distribution here was very patchy. At noon they concentrated near the shore, at night offshore, a pattern thought to reflect preference for high temperatures. Here they fed primarily on the very dense cyanophyte *Spirulina platensis*, 80% of the tilapia living and feeding in the top 100 cm. Nakuru waters are cooler than those of Lake Magadi from where the *O. a. grahami* were stocked. There was also more living space in Lake Nakuru and under these conditions the species grew to a larger size (127 mm SL). The total ichthyomass of *O. a. grahami* in this lake was estimated to increase from 90 t (dry weight) in 1972 to 400 t in 1973 (2.1 g m^{-2} to 10.2 g m^{-2}). The Lesser Flamingo (*Phoenicopterus minor*), which also fed on *Spirulina*, deserted the lake, but the main impact of the introduction was in extending food chains to fish-eating birds, especially pelicans and cormorants. Before the introduction these were occasional visitors, but after the introduction over 50 species of bird were recorded, of which the great white pelican (*Pelecanus onacrotalus roseum*) predominated. Adult pelican have been estimated to eat 1330 g d^{-1} (fresh weight) of tilapia, the young some 770 g d^{-1} during the period from hatching to fledging. This suggested pelicans were removing 16 000–20 000 kg fresh weight of fish per day from the lake. Since they nested on a neighbouring lake, taking fish there to feed their young, they were estimated to be exporting some 13 t phosphorus a year from the Nakuru system.

West African coastal lagoons

In West African coastal lagoons *Sarotherodon melanotheron*, a species that can withstand high and changing salinities, was studied as a possible candidate for culture with mullet (*Mugil* sp.) by Pauly (1976). This tilapia only grows to ~25 cm TL, the adults feeding on algae and detritus in the bottom mud (30% organic content, 1.2 kcal g^{-1} energy content). Daily food intake of a 20 g fish was estimated to be 1.5 g (dry weight) and the energy assimilation 900 kcal d^{-1}. Juveniles took some zooplankton and phytoplankton. *S. melanotheron* is a male mouth-brooder, whose behaviour has been much studied in aquaria under the names *T. macrocephala* and *T. heudelotii* (see Trewavas, 1983). In the field this species breeds throughout the year with equinoxial peaks. Parasitism was low: this was attributed to the varying salinity in the lagoon. Trewavas (1983) also described the ecology and behaviour of several subspecies (including *S. m. heudelotii* and *S. m. nigripinnis*) distributed in coastal lagoons along Africa's west coast and south to Namibia.

5.8 *O. MOSSAMBICUS* IN LAKE SIBAYA, SOUTHERN AFRICA

The ecology of *O. mossambicus* in Lake Sibaya where is it indigenous, is of particular interest as this species is now so widely distributed in other parts of the world and important for fish culture. This 77 km^2, 43 m deep, endorheic lake on the coastal plain near the Indian Ocean, was studied intensively as an IBP project looking at all trophic levels (Allanson, 1979; Bowen, 1979; Bruton, 1979). It has a slow-growing, early-maturing and very abundant population of *O. mossambicus* (Trewavas, 1983).

The fish fauna (18 species) was dominated by four species of cichlid, *O. mossambicus*, *T. sparrmanii*, T. *rendalli swierstrae*, and *Pseudocrenilabrus philander*. *O. mossambicus* nested on terraces in warm, more shallow water than the other cichlids which nested among the macrophytes, after intense intraspecific conflicts establishing territories in September when water temperature first exceeded 20°C. Mouth-brooding females released fry into water less than 10 cm deep. Juveniles occupied progressively deeper water as they grew larger. The fry had access to nutritious food in these shallows, but food quality was so poor offshore where the larger *O. mossambicus* lived that these became stunted. Males grew faster than females after the first year. Maximum age was c. 7–8 years; modal maturation sizes were: males 17 cm SL, females 11–12 cm. Females bred after 1 year, males after 1 or 2 years. The largest male was 29 cm TL, the largest female 18.5 cm. Poor food quality led to precocious breeding and growth decelerated after maturity earlier in the life cycle than in 'normal' populations; maturation length/maximum length 0.43 (lower than the characteristic cichlid value of c. 0.7).

Primary production is low in Lake Sibaya, with little phytoplankton. *O. mossambicus* feeds mainly on benthic diatoms and bacteria, but this is only sufficient to maintain good growth and condition in the juveniles which can feed on the terrace in well-lit warm water <1 m deep. The benthic algal populations available to the adult *O. mossambicus* had a low nutritive value and they therefore fed to a small extent on other algae, small fishes, aquatic insects, *Caridina nilotica*, gastropods and aquatic macrophyes. *T. rendalli* was here a specialist macrophytophage; *T. sparrmanii* was omnivorous. The predatory *Clarias gariepinus*, *Ctenopoma* and *Glossogobius* included small tilapia in a varied diet with many invertebrates.

The most abundant fishes in the lake were all eurytopic, adaptable species able to make use of the rich, but harsh and variable littoral zone. This appeared to be the key to the successful utilization of the lake. The high energetic cost of feeding at elevated temperatures was offset by a superabundance of food in marginal areas with high light intensities. In spite of it small size, *O. mossambicus* was one of the most abundant fishes in the lake and it was suggested that precocious breeding here is part of an adaptive strategy which enables this species to obtain maximum benefit from the rich resources in shallow water. By breeding precociously generation time is shortened and a greater intrinsic rate of population increase results, high recruitment rates balancing the high mortality rates from the drying up of marginal pools and heavy bird predation.

Precocious breeding here was considered an adaptation that maximizes reproductive output and makes best use of inshore feeding areas. 'Stunting' was not related to overtaxing the food supply (as in overstocked ponds), but with abundant food for immature fishes and early maturity. Utilization of primary productivity on the terrace (bacteria and diatoms) in the production of 20 g *O. mossambicus* was estimated to be 0.5–5% per day (Allanson, 1979).

5.9 THE ROLES OF TILAPIAS ESTABLISHED IN ECOSYSTEMS OUTSIDE AFRICA

In addition to numerous translocations of tilapia species within Africa, since 1940 some species, mainly *O. mossambicus* and *O. niloticus*, *T. rendalli* and various hydrids, have been widely distributed by man as food fishes and become naturalized in many tropical countries (see Philippart and Ruwet, 1982; Welcomme, 1988; Baluyut, 1992; Juarez-Palacios and Olmos-Tomassini, 1992; Lever, 1996). Lever, who devotes 40 pages to the distributions of naturalized tilapias and their ecological and socio-economic impacts on the indigenous fish faunas, should be consulted for details.

Tilapias established in natural waters, often as escapes from fish culture activities, have come to be regarded as either beneficial to local fisheries (and sometimes in control of mosquitoes or aquatic plants), or as pests (stunted populations, or thought to compete with or damage indigenous fish species), or their presence is controversial, some authors stressing the benefits, others the negative effects within that geographical area. In many cases they have become established from introductions of very few individuals, resulting in a very small genetic base. For example only 3 males and 1 female survived in the batch of *O. mossambicus* introduced to the Philippines from Thailand in 1949. The Papua New Guinea populations, which support an important fishery in the Sepik River basin, originated from 250 *O. mossambicus* fry sent to PNG from Malaysia in 1954. *O. mossambicus*, now widely dispersed on Pacific islands, appear to have originated from 14 survivors of a batch of 60 small *O. mossambicus* sent to Honolulu in 1951. Once established, tilapia have proved impossible to eradicate, as attempted on some islands where they are now living in tidal estuaries, reputedly to the detriment of local mullet populations.

Aquarists were responsible for the first introductions outside Africa. Lever quotes Tubb's (1954) report that *O. mossambicus* came to Hong Kong with aquarium imports pre 1940, from Singapore and the USA. This species, found in the wild in Java in 1938/39, was then widely dispersed as a food fish in Indonesia and Malaysia during the Second World War when it was difficult to get juvenile milkfish (*Chanos chanos*) to stock brackish fishponds. Postwar experiments (notably at Malacca Fish Culture station, Malaysia) to produce better-growing monosex (male) hybrids for pond culture led to imports of other tilapias, including *O. niloticus* and *O. urolepis hornorum* (the 'Zanzibar tilapia', see Trewavas, 1983). *O. aureus* was brought from the USA to the Philippines in 1977, where it hybridized with others already there (*O. mossambicus* from 1950, *O. niloticus* post 1970 from Thailand and Israel) and became established in lake and rivers. In the Philippines tilapias have enhanced fisheries for carnivorous fishes in three natural lakes; in three reservoirs they support fisheries as there were previously no endemic fish feeding at low trophic levels (Guerrero, 1999).

Tilapias were also stocked deliberately into hydroelectric impoundments in countries that lacked lacustrine fishes (e.g. Sri Lanka), and as cheap food for rural poor (the West Indies). Asian fish faunas (predominantly cyprinid) have few indigenous species which crop primary production; also the fisheries often utilize carnivorous species (e.g. *Ophiocephalus* spp., murrels) for which tilapia provide good forage, boosting the catches. Neotropical fish faunas have few herbivores, though they have many detritivorous species. Tilapias have also been introduced to control mosquitoes and midges (e.g. in Pacific islands). In Sri Lanka planktivorous *O. mossambicus* were found to create a new food source for indigenous species in the form of partially digested algae.

The macrophyte-feeding *T. rendalli* and *T. zillii* (often confused, and hybridizing if both stocked) have been used to control aquatic plants (e.g. in Puerto Rico). This can, however, have drastic results, as it did in Lake Kinkony, Madagascar, where within3 years *T. rendalli* devastated 3000 ha of *Ceratophyllum* and *Nymphaea* beds, resulting in the disappearance of valuable endemic cichlids. In Madagascar introduced *O. macrochir* are said to be established in mangrove swamps; in Lake Itasy the species hybridized with *O. niloticus* (producing the hybrid locally known as 'tilapia Trois Quarts'; Moreau *et al.*, 1988). Tilapias are now the main freshwater fish in Madagascan fish markets (personal observation) but have decimated the unique cichlids (M. Stiassny, personal communication).

Despite the benefits, some *O. mossambicus* populations exploded then declined (for reasons unknown) or 'stunted' and are now regarded as pests. *O. niloticus* are considered to grow better. Competition for food, spawning places, and removal of vegetation on which many cyprinids and other indigenous fish lay their eggs made tilapias unpopular in India, where they reputedly suppress production of major carps. In coastal areas of some countries they have interfered with *Chanos chanos* milkfish and *Mugil cephalus* striped mullet production. Tilapia introductions have also affected bird populations on some lakes (e.g. in Nicaragua and Solomon Islands). In the USA escaped tilapia are extending their ranges in warmer southern waters, to the detriment of indigenous centrarchids and other species.

The biology of O. *mossambicus* was studied in Plover Cove Reservoir, Hong Kong, a freshwater reservoir formed from a marine inlet (Man and Hodgkiss, 1977). The initial 20 species included marine species, which survived but did not breed, and some rice-field and riverine species, but it was *O. mossambicus* which escaped from an aquaculture centre that soon became the principal species in gill-net catches. The tilapia flourished by having the widest food spectrum; they also helped to control both chironomids and algae, and by removing nutrients from the mud indirectly assisted in reducing algal growth.

Analyses of niche partitioning of indigenous riverine carps and *O. mossambicus* in a shallow Sri Lankan reservoir (Tissawewa) showed that of the twelve common species, overlap in food and living space only occurred between *O. mossambicus* and the small cyprinid *Amblypharyngodon melettinus*. Both fed mainly on detritus and algae (Pet and Piet, 1993; Piet *et al.*, 1994). T. *rendalli* and *O. niloticus* here fed mainly on macrophytes and porifera.

Australian studies have indicated that competition between *O. mossambicus* and indigenous species for food and breeding territories could devastate indigenous fish populations in the Darling River system (Arthington and Bluhdorn, 1996). In Papua New Guinea, on the other hand, *O. mossambicus* supported the main fishery on the Sepik River floodplain without, it seems, detriment to the indigenous fishes (mainly ariid catfishes and gudgeons); Papua New Guinea, like Australia, lacks primary freshwater fishes (Coates, 1987).

5.10 DISCUSSION: THE SUCCESS OF TILAPIAS

Thus to sum up: the success of tilapias as widely dispersed and abundant fishes in African waters appears to be due mainly to (a) their use of plant food, including phytoplankton, which is rarely limiting, with detritus and some other foods - invertebrates, zooplankton - as a reserve, and (b) their flexibility in growth rate and maturation size according to the prevailing environmental conditions.

This phenonomen of 'stunting or dwarfing', an understanding of which is so important for tilapia culture, has been explored by Iles (1973), Noakes and Balon (1982), Pauly (1984), Kolding (1995) and elsewhere in this book (see Lorenzen). Iles's examples of species with dwarfed populations in natural waters included *O. rukwaensis, O. spilurus niger, O. grahami, O. mortimeri, O. macrochir, O. leucostictus* and *S. galilaeus*. Iles thought small size represented an adaptation towards survival under extreme physical conditions, for small tilapia are very hardy and have a higher temperature tolerance than do other small cichlids. However, breeding size is reduced to an even greater degree than final size, giving very low values for maturity/total length ratios. Furthermore stunted tilapia are precocious breeders (they can mature in 3 months compared with 2–4 years in lake populations), thereby speeding up the whole life cycle. Iles' view was that stunting represents an adaptive mechanism enabling tilapia populations to withstand extremely high mortality rates (such as from birds in drying up pools). He considered that stunted tilapia may be unique in the cumulative effect of a combination of changes: marked increase in relative growth rate, increase in brood frequency and decrease in egg size.

As already mentioned, Pauly (1984) took the view that we are not looking for a sudden 'switch', but that diminishing oxygen supply per unit of body weight serves in growing fish as the master factor in inducing the transition from juvenile to adult. As a mechanism, still to found, he suggested that a steady decrease in availability of oxygen per unit weight activates those parts of the brain capable of responding to stimuli for reproduction, and that any stress factors (such as high temperatures, crowding, osmotic costs) which raise the maintenance metabolism would result in a reduced final size of fish. The conditions under which dwarf populations of tilapias have been found, as discussed above, supports this view. The Lake Albert lagoon populations were very stressed. In Lake Turkana the crater lake *O. niloticus* appeared to have little food and space while those in Ferguson's Gulf which produced the enormous 1976 boom appeared very well fed (rich phytoplankton bloom from newly flooded land). However, Kolding (1993, 1995) postulated low, or rapidly fluctuating, oxygen conditions resulting from the bloom.

Tilapia species also vary very much in aggressiveness, *O. niloticus* being one of the most aggressive, their large males out-competing other species when spawning places are limited. This species is also very flexible in feeding and growth according to environmental conditions. They manage to coexist with large piscivorous *Lates*, with which they have co-evolved. When they occasionally hybridize with other species, as with *O. variabilis* in Lake Victoria and *O. macrochir* both in Madagascar and lac Ihema, the resulting generations resemble *O. niloticus*.

The success of tilapias is also shown by the readiness with which they have spread in ecosystems outside Africa, in many cases from the introduction of very few individuals. Two species, *O. niloticus* and *O. mossambicus*, are now particularly widely dispersed throughout the tropics (see Moreau *et al.*, 1988; Lever, 1996; Baluyut, 1992). Interactions with indigenous species in Sri Lanka have been studied in considerable detail by Pet and Piet (1993). To establish *O. mossambicus* in some Brazilian dams, it was sometimes necessary, however, to remove the piscivorous *Serrasalmus* species (M. Petrere, personal communication). In Guyana during trials with *O. mossambicus* in flood-fallow sugarcane fields many fell prey to the piscivorous characid *Hoplias malabaricus* (personal observations). They did, however, flourish in brackish water ponds in Guyana. In Trinidad *O. mossambicus* escaped and is now established in mangrove swamps, so invading yet another ecosystem.

REFERENCES

Agnèse, J.F. and Gourène, B.P. (1998) Natural hybridization between *Tilapia zillii* (Gervais, 1848) and *T. guineensis* (Bleeker, 1962) in Ivory Coast, in *Abstracts from the Symposium on African Fishes and Fisheries: Diversity and Utilization, Grahamstown, South Africa, 1998*, p. 14.

Allanson, B.R. (1979) Lake Sibaya, in *Lake Sibaya. Mongr. Biol.* 36, Dr W. Junk, The Hague, pp. 313–328.

Arthington, A.H. and Bluhdorn, D.R. (1996) The effects of species introductions resulting from aquaculture operations, in *Aquaculture and Water Resource Management* (eds D.J. Baird, M.C.M. Beveridge, L.A. Kelly and J.F. Muir), Blackwell, Oxford, pp. 114–139.

Bailey, R.G. (1966) The dam fisheries of Tanzania. *E. Afric. Agricult. Forest. J.* **32**(1), 1–15.

Bailey, R.G. (1988) Fish and fisheries, in *The Jonglei Canal* (eds P. Howell, M. Lock and S. Cobb), Cambridge University Press, Cambridge, pp. 328–349.

Bailey, R.G. (1996) Changes in the fish and fisheries ecology of a large man-made lake in Tanzania, 1965–94. *Fish. Manage. Ecol.* **3**, 251–260.

Bailey, R.G., Churchfield, S., Petr, T. and Pimm, R. (1978) The ecology of the fishes in Nyumba ya Mungu reservoir, Tanzania. *Biol. J. Linn. Soc.* **10**, 109–137.

Balirwa, J.S. (1998*) Lake Victoria Wetlands and the Ecology of the Nile tilapia* Oreochromis niloticus L., PhD dissertation, Agricultural University, Wageningen, Balkema, Rotterdam, the Netherlands.

Balogun, J.K. and Ibeun, M.O. (1995) *Additional Information on Fish Stocks and Fisheries of Lake Kaini (Nigeria).* CIFA Technical Paper, 30, FAO, Rome, pp. 19–40.

Baluyut, E.A. (ed.) (1992) *Papers Contributed to the Workshop on Tilapia in Capture and Culture-Enhanced Fisheries in the Indo-Pacific Fishery Commission Countries, Bogor, Indonesia, June 1991*. FAO Fish. Rep., 458, (Suppl.), FAO, Rome.

Beauchamp, R.S.A. (1964) The Rift Valley lakes of Africa. *Verh. Int. Verein. Limnol.* **15**, 91–99.

Bootsma, H.A., Hecky, R.E., Hesslein, R.H. and Turner. G.F. (1996) Food partitioning among Lake Malawi nearshore fishes as revealed by stable isotope analyses. *Ecology* **77**, 1286–1290.

Bowen, S.H. (1979) A nutritional constraint in detritivory by fishes: the stunted population of *Sarotherodon mossambicus* in Lake Sibaya, South Africa. *Ecol. Monogr.* **49**, 17–31.

Bowen, S.H. (1980) Detrital non-protein amino acids are the key to rapid growth of *Tilapia* in Lake Valencia, Venezuela. *Science* **207**, 1216–1218.

Bowen, S.H. (1984) Detritivory and herbivory, in *Biology and Ecology of African Freshwater Fishes* (eds C. Leveque, M.N. Bruton and G.W. Ssentongo), Travaux et Documents, 216, ORSTOM, Paris, pp. 243–247.

Braimah, L.I. (1995) Recent developments in the fisheries of Volta Lake (Ghana), *CIFA Technical Paper*, 30, FAO, Rome, pp. 111–134.

Bruton, M.N. (1979) The fishes of Lake Sibaya, in *Lake Sibaya* (ed. B.R. Allanson), *Mongr. Biol.* 36, Dr W. Junk, The Hague, pp. 162–245.

Burgis, M.J. and Dunn, I.G. (1978) Production in three contrasting ecosystems, in *Freshwater Fish Production* (ed. S.D. Gerking), Blackwell, Oxford, pp. 137–158.

Coates, D. (1987) Considerations of fish introductions into the Sepik River, Papua New Guinea. *Aquacult. Fish. Manage.* **18**, 231–241.

Crespi, V. and Ardizzone, G.D. (1995) Fishery resources and some economic aspects of four fishing villages on Lakes George and Edward in the Queen Elizabeth National Park, Uganda. *Afr. J. Trop. Hydrobiol. Fish* **6**, 11–20.

Crul, R. *et al.* (1995) *A Bibliography of Lake Victoria (East Africa).* IHP-IV Project M-5.1 SC-95/WS/16 UNESCO, Paris.

Duponchelle, F. and Panfili, J. (1998) Variations in age and size at maturity of female Nile tilapia *Oreochromis niloticus*, populations from man-made lakes of Cote d'Ivoire. *Env. Biol. Fishes* **52**, 453–465.

FAO (1993) *Fisheries Management in the Southeast Arm of Lake Malawi, the Upper Shire River and Lake Malombe, With Particular Reference to the Fisheries on Chambo* (Oreochromis *spp.*). CIFA Technical Paper, 21, FAO, Rome.

Fryer, G. (1961) Observations on the biology of the cichlid fish *Tilapia variabilis* in the northern waters of Lake Victoria (East Africa). *Rev. Zool. Bot. Afr.* **64**, 1–33.

Gophen, M., Pollingher, U. Ochumba, P.O.B. and Kaufman, L.S. (1993) Nile perch (*Lates niloticus*) invasion in Lake Victoria (East Africa). *Verh. Internat. Verein. Limnol.* **25**, 856–859.

Gourène, G., Teugels, G.G., Hugueny, B. and Thys van den Audenaerde, D.F.E. (1998) Evaluation of the icthyological diversity of a West African basin after the construction of a dam (Bia River, Ivory Coast and Ghana), in *Abstracts, Symposium on African Fishes and Fisheries, Grahamstown, South Africa, 1998*, p. 42.

Greenwood, P.H.G. and Lund, J.W.G. (1976) A discussion on the biology of an equatorial lake: Lake George, Uganda. *Proc. Roy. Soc. Lond.* **184**B, 227–346.

Guerrero, III, R.D. (1999) Impacts of tilapia introductions on the endemic fishes in some Philippine lakes and reservoirs, in *Fish and Fisheries of Lakes and Reservoirs in SE Asia and Africa* (eds W.L.T. van Densen and M.J. Morris), Westbury Academic and Scientific Publishing, Otley, UK, pp. 151–157.

Gwahaba, J.J. (1973) Effects of fishing on the *Tilapia nilotica* populations of Lake George, Uganda, over the past twenty years. *East Afr. Wild. J.* **11**, 317–328.

Hartnell, R.M. (1996) *Laboratory and Field Investigations into the Feeding Performance of Lake Malawi* Oreochromis, unpublished PhD Thesis, University of Wales, Bangor.

Hecky, R.E. (1993) The eutrophication of Lake Victoria. Kilham Memorial Lecture, 25th Congress of SIL, *Verh. Int. Verein. Limnol.* **25**, 39–48.

Hickley, P. and Bailey, R.G. (1987) Fish communities in the eastern seasonal-floodplain of the Sudd, Southern Sudan. *Hydrobiologia* **144**, 243–250.

Holden, M.J. (1963) *The Populations of Fish in Dry Season Pools of the River Sokoto.* Colonial Office Fishery Publications,19, Colonial Office, London.

Hopson, T. *et al.* (eds) (1982) *Lake Turkana. A Report on the Findings of the Lake Turkana.* Project 1972–1975, ODA, London.

Iles, T.D. (1973) Dwarfing or stunting in the genus *Tilapia* (Cichlidae) a possibly unique recruitment mechanism. *Rapp. Reun. Cons. Perm. Int. Explor. Mer.* **164**, 247–254.

Ita, E.O. (1984) Lake Kainji (Nigeria), in *Status of African Reservoir Fisheries* (eds J.M. Kapetsky and T. Petr), CIFA Technical Paper, 10, FAO, Rome, pp. 43–103.

Johnson, T.C., Scholz, C.A., Talbot, M.R., Kelts, K., Ricketts, R.D., Ngobi, G., Beuning, K., Ssemmanda, I. and McGill, J.W. (1996) Late Pleistocene desiccation of Lake Victoria and rapid evolution of cichlid fishes. *Science* **273**, 1091–1093.

Juarez-Palacios J.R. and Olmos-Tomassini, M.E. (1992) *Tilapia in Capture and Culture-enhanced Fisheries in Latin America.* FAO Fish, Technical Report 458, Suppl., pp. 244–273.

Kaufman, L. (1992) Catastropic change in species-rich freshwater ecosystem of Lake Victoria. *BioScience* **42**, 846–858.

Kolding, J. (1992) A summary of Lake Turkana: an ever-changing mixed environment. *Mitt. Verein. Internat. Limnol.* **23**, 25–35.

Kolding, J. (1993) Population dynamics and life-history styles of Nile tilapia, *Oreochromis niloticus*, in Ferguson's Gulf, Lake Turkana, Kenya. *Env. Biol. Fishes* **37**, 25–46.

Kolding, J. (1995) Changes in species composition and abundance of fish populations in L. Turkana, Kenya, in *The Impact of Species Changes in African Lakes* (eds T.J. Pitcher and P.J.B. Hart), Chapman and Hall, London, pp. 335–363.

Lauzanne, L. (1972) Regimes alimentaires principales especes de poissons de l'archipel oriental du lac Tchad. *Verh. Internat. Verein. Limnol.* **18**, 636–646.

Lehman, J.T. (ed.) (1998) *Environmental Change and Response in East African Lakes*, Kluwer Academic Press, Dordrecht.

Lever, C. (1996) *Naturalized Fishes of the World*, Academic Press, London.

Linn, I. J. and K.L.I. Campbell. (1986) *Cormorants and Fisheries. A Report on the Biology of the White-breasted Cormorant* Phalacrocorax carbo *as it Affects the Commercial Fisheries of Lake Malawi.* Overseas Development Administration, London.

Lowe, R.H. (1952) *Report on the Tilapia and Other Fish and Fisheries of Lake Nyasa 1945–47*, Colonial Office, London, Fishery Publ. **1**(2), 1–126.

Lowe, R.H. (1955a) New species of *Tilapia* (Pisces, Cichlidae) from Lake Jipe and the Pangani River, East Africa. *Bull. Brit. Mus. (Nat. Hist.)* **2**, 349–368.

Lowe (McConnell), R.H. (1955b) The fecundity of *Tilapia* species. *East Afr. Agric. J.* **11**, 45–52.

Lowe (McConnell), R.H. (1956a) Observations on the biology of *Tilapia* (Pisces-Cichlidae) in Lake Victoria, East Africa. *East Afr. Fish. Res. Org, Suppl. Pub.* **1**, 1–72.

Lowe (McConnell), R.H. (1956b) The breeding behaviour of *Tilapia* species in natural waters: observations on *T. karomo* Poll and *T. variabilis* Blgr. *Behaviour* **9**, 140–163.

Lowe (McConnell), R.H. (1957) Observations on the diagnosis and biology of *Tilapia leucosticta* Trewavas in East Africa. *Rev. Zool. Bot. Afr.* **55**, 353–373.

Lowe (McConnell), R.H. (1958) Observations on the biology of *Tilapia nilotica* Linne in East African waters. *Rev. Zool. Bot. Afr.* **57**, 129–170.

Lowe (McConnell), R.H. (1959) Breeding behaviour patterns and ecological differences between *Tilapia* species and their significance for evolution within the genus *Tilapia* (Pisces: Cichlidae). *Proc. Zool. Soc. (Lond)* **132**, 1–30.

Lowe-McConnell, R.H. (1982) Tilapias in fish communities, in *The Biology and Culture of Tilapias (ICLARM Conference Proceedings*, Vol. 7) (eds R.S.V. Pullin and R.H. Lowe-McConnell), ICLARM, Manila, pp. 83–114.

Lowe-McConnell, R.H. (1987) *Ecological studies of Tropical Fish Communities*, Cambridge University Press, Cambridge.

Man, H.S.H. and Hodgkiss, I.J. (1977) Studies on the ichthyofauna in Plover Cove Reservoir, Hong Kong: feeding and food relations. *J. Fish Biol.* **11**, 1–13.

Mattson, N.S. and Kaunda, E.K.W.H. (1997) Population dynamics of *Oreochromis shiranus* in two small water bodies in Malawi. *J. Fish Biol.* **50**, 592–607.

Moreau, J., Arrignon, J. and Jubb. R.A. (1988) Introduction of foreign fishes in African inland waters. Suitability and problems, in *Biology and Ecology of African Freshwater Fishes* (eds C. Leveque, M.N. Bruton and G.W. Ssentongo), Travaux et Documents No 216, ORSTOM, Paris, pp. 395–425.

Moreau, J., Christensen, V. and Pauly, D. (1993) A trophic ecosystem model of Lake George, Uganda, in *Ecotrophic Models of Aquatic Ecosystems (ICLARM Conference Proceedings*, Vol. 26) (eds V. Christensen and D. Pauly), ICLARM, Manila, pp, 124–129.

Moriarty, D.J.W., Darlington, J.P.E.C., Dunn, I.G., Moriarty, C.M and Tevlin, M.P. (1973) Feeding and grazing in Lake George, Uganda. *Proc. R. Soc. Lond.* **184**B, 299–319.

Muchiri, S.M., Hart, P.J.B.and Harper, D.M. (1995) The persistence of two introduced tilapia species in Lake Naivasha, Kenya, in the face of environmental variability and fishing pressure, in *The Impacts of Species Changes in African Lakes* (eds T. Pitcher and P.J.B. Hart), Chapman and Hall, London, pp. 291–319.

Mwanja, W. and Kaufman, L. (1995) A note on recent advances in the genetic characterization of Tilapiine stocks in Lake Victoria Region. *Afr. J. Trop. Hydrobiol. Fish.* **6**, 51–53.

Noakes, D.L.G. and Balon, E.K. (1982) Life histories of tilapias: an evolutionary perspective, in *The Biology and Culture of Tilapias (ICLARM Conference Proceedings*, Vol. 7) (eds R.S.V. Pullin and R.H. Lowe-McConnell), ICLARM, Manila, pp. 61–80.

Ogutu-Ohwayo, R. (1992) The purpose, costs and benefits of fish introductions; with specific reference to the Great Lakes of Africa. *Mitt. Internat. Verein. Limnol.* **23**, 37–44.

Ogutu-Ohwayo, R. 1995. Diversity and stability of fish stocks in Lake Victoria, Kyoga and Nabugabo after the establishment of introduced species, in *The Impacts of Species Changes in African Lakes* (eds T. Pitcher and P.J.B. Hart), Chapman and Hall, London, pp. 59–81.

Palsson, O.K., Bulirani, A. and Banda, M. (1998) A review of the population dynamics of chambo *Oreochromis* spp. (Cichlidae) in Lakes Malawi and Malombe, in *Abstracts from the Symposium on African Fishes and Fisheries: Diversity and Utilization, Grahamstown, South Africa, 1998*, p. 158.

Pauly, D. (1976) The biology, fishery and potential for aquaculture of *Tilapia melanotheron* in a small West African lagoon. *Aquaculture* **7**, 33–49.

Pauly, D. (1984) A mechanism for the juvenile-to-adult transition in fishes. *J. Cons. Int. Expl. Mer.* **41**, 280–284.

Pet, J.S. and Piet, G.J. (1993) The consequences of habitat occupation and habitat overlap of the introduced tilapia *O. mossambicus* and indigenous fish species in resource management in a Sri Lankan reservoir. *J. Fish Biol.* **43** (Suppl. A), 193–208.

Petr, T. (1967) Fish population changes in the Volta Lake in Ghana during its first sixteen months. *Hydrobiologia* **30**, 193–220.

Philippart, J.-Cl., and Ruwet, J.-Cl. (1982) Ecology and distribution of Tilapias, in *The Biology and Culture of Tilapias (ICLARM Conference Proceedings*, Vol. 7) (eds R.S.V. Pullin and R.H. Lowe-McConnell), ICLARM, Manila, pp. 15–59.

Piet, F.J., Pet, J.S. and Guruge, W.A.H.P. (1994) Niche partitioning of indigenous riverine carps and the exotic lacustrine tilapia (*Oreochromis mossambicus*) in a reservoir in SE Sri Lanka. *Verh. Internat. Verein. Limnol.* **25**, 2183–2187.

Pitcher, T.J. and Hart, P.J.B. (eds) (1995) *The Impact of Species Changes in African Lakes*, Chapman and Hall, London.

Plisnier, P.D, Micha, J.-C. and Frank, V. (1988) *Biologie et Exploitation des Poissons du Lac Ihema (bassin Akagera), Rwanda*, Presses Universitaires de Namur, Belgium.

Rashid, M.M. (1995) *Some Additional Information on Limnology and Fisheries of Lakes Nasser (Egypt) and Nubia (Sudan)*, CIFA Technical Paper, 30, FAO, Rome, pp. 81–109.

Schiemer, F. (1996) Significance of filter-feeding fish in tropical waters, in *Perspectives in Tropical Limnology* (eds F. Schiemer and K.T. Bolland), SPB Academic bv, Amsterdam, pp. 65–75.

Schliewen, U.K., Tauz, D. and Paabo, S. (1994) Sympatric speciation suggested by monophyly of crater lake cichlids. *Nature (Lond.)* **368**, 629–632.

Seegers, L. (1996) *The Fishes of the Lake Rukwa Drainage. Ann. Mus. r. Afr. Cent. Sci. Zool.* **278**, 407.

Seegers, L. and Tichy, H. (1999) The *Oreochromis alcalicus* flock from Lakes Natron and Magadi, Tanzania and Kenya, with descriptions of two new species. *Ichthyol. Explor. Freshwaters* **10**, 97–146.

Stiassny, M., Schliewen, U.K. and Dominey, W.J. (1992) A new species flock of cichlid fishes from Lake Bermin, Cameroon with a description of eight new species of *Tilapia* (Labroidei: Cichlidae). *Ichth. Expl. Freshwat.* **3**, 311–346.

Trewavas, E. (1983) *Tilapiine Fishes of the Genera* Sarotherodon, Oreochromis *and* Danakilia, British Museum (Natural History) London.

Trewavas, E., Green, J. and Corbet, S.A. (1972) Ecological studies on crater lakes in West Cameroon. Fishes of Barombi Mbo. *J. Zool. (Lond).* **167**, 41–95.

Tubb, J.A. (1954) Introduction of *Tilapia* to Hong Kong. *Hong Kong University Fisheries Journal* **1**, 63–64.

Turner, G.W. (1995) Management, conservation and species changes of exploited fish stocks in Lake Malawi, in *The Impacts of Species Changes in African Lakes* (eds T. Pitcher and P.J.B. Hart), Chapman and Hall, London, pp. 365–395.

Tweddle, D., Turner, G.W. and Seisay, M.B.D. (1995) Changes in species composition and abundance as a consequence of fishing in Lake Malombe, Malawi, in *The Impacts of Species Changes in African Lakes* (eds T. Pitcher and P.J.B. Hart), Chapman and Hall, London, pp. 413–424.

Twongo, T. (1995) Impact of fish species introductions on the tilapias of Lakes Victoria and Kyoga, in *The Impacts of Species Changes in African Lakes* (eds T. Pitcher and P.J.B. Hart), Chapman and Hall, London, pp. 45–56.

Vareschi, E. (1979) The ecology of Lake Nakuru (Kenya) II Biomass and spatial distribution of fish (*T. grahami*). *Oecologia (Berlin)* **37**, 321–335.

Welcomme, R.L. (1964a) Notes on the present distribution and habits of the non-endemic species of *Tilapia* which have been introduced into Lake Victoria. *E. Afr. Fish. Res. Org. Ann. Rept.* **1962/63**, 36–39.

Welcomme, R.L. (1964b) The habitats and habitat preferences of the young of the Lake Victoria *Tilapia* (Pisces, Cichlidae). *Rev. Zool. Bot. Afr.* **70**, 1–28.

Welcomme, R. L. (1966) Recent changes in the stocks of *Tilapia* in L. Victoria. *Nature* **212**, 52–54.

Welcomme, R.L. (1967b) The relationship between fecundity and fertility in the mouth-brooding cichlid fish *Tilapia leucosticta. J. Zool. (London)* **151**, 453–468.

Welcomme, R.L. (1968) Observations on the biology of the introduced species of tilapia in Lake Victoria. *Rev. Zool. Bot. Afr.* **76**, 249–279.

Welcomme, R.L. (1970) Studies on the effects of abnormally high waterlevels on the ecology of fish in certain shallow regions of L. Victoria. *J. Zool. (London)* **160**, 405–436.

Welcomme, R.L. (1988) *International Introductions of Inland Aquatic Species*. FAO Fish. Technical Paper, 294, FAO, Rome.

Witte, F., Goldschmidt, T. and Wanink, J.H. (1995) Dynamics of the haplochromine cichlid fauna and other ecological changes in the Mwanza Gulf of Lake Victoria in *The Impacts of Species Changes in African Lakes* (eds T. Pitcher and P.J.B. Hart), Chapman and Hall, London, pp. 83–135.

Worthington, E.B. (1937) On the evolution of fish in the Great Lakes of Africa. *Int. Rev. gesamt. Hydrobiol. Hydrog.* **35**, 304–317.

Worthington, E.B. (1954) Speciation of fishes in African lakes. *Nature* (Lond.) **173**, 1064.

Worthington, E.B. and Ricardo, C.K. (1936) Scientific Results of the Cambridge Expedition to the East African Lakes, 1930–1. No 15. The fish of Lake Rudolf and Lake Baringo. *J. Linn. Soc. Zool.* **39**, 353–389.

Chapter six

Population dynamics and management

K. Lorenzen

6.1 INTRODUCTION

The tilapias are of major economic and ecological importance, being exploited in capture fisheries and aquaculture systems at all levels of intensity, and having invaded tropical fresh and brackish water ecosystems on all continents as a result of intentional or accidental introductions. Perhaps the most striking ecological feature of the tilapias is their ability to reproduce readily in almost any environment, from lakes to aquaculture tanks or aquaria. This ability, together with a remarkable plasticity in life history traits, is seen as a reason for their successful colonization of such a wide range of habitats. It is also, however, a source of management problems in culture where early maturation and prolific breeding may lead to 'stunting', dense populations of small individuals below marketable size. Population dynamics, the study of the abundance and biomass dynamics of populations, their underlying mechanisms and their management consequences, is therefore relevant to all forms of tilapia exploitation. At the same time, the wide range of manipulations to which tilapia populations are being subjected provides a unique opportunity to investigate fundamental problems in fish population biology.

The phenomenon of 'stunting' or 'dwarfing' in tilapias has attracted much attention in the literature, not least because of the practical problem it presents in aquaculture. It has often been suggested that 'stunting' is a response, possibly unique to the tilapias, to conditions of stress and high mortality (e.g. Iles, 1973). However, a review of 'stunting' in wild and farmed tilapias (section 6.4) indicates that this view is misleading: 'stunting' is a collective term applied to different life

M.C.M. Beveridge and B.J. McAndrew (eds), Tilapias: Biology and Exploitation, 163–225.

histories linked in but one trait (low maximum size), it may occur under conditions of low as well as high mortality, and it is by no means unique to tilapias. Because the term 'stunting' is so poorly defined, it will not be used in this chapter until section 6.4 where the phenomenon is reviewed in depth.

The aims of this chapter are threefold: to provide an introduction to tilapia population dynamics, to show how population dynamics approaches can aid the management of wild as well as farmed stocks, and to explore the significance of life history variation to both areas. Significant advances in our understanding of tilapia population biology will require constructive interaction between experimentally orientated aquaculture, field studies of wild populations, and the development of a common theoretical framework. The chapter aims to foster this development by integrating across wild and farmed populations, and emphasizing the applicability of common concepts and approaches in aquaculture and fisheries ecology.

Because the aim of this chapter is to synthesize current understanding of tilapia population biology and its application to practical management, it relies heavily on the detailed information available for a few well-studied species, mainly of the genus *Oreochromis*. Although these species are not necessarily representative of the tilapias as a whole, they represent the most important exploited species and there is no reason to believe that their population biology is fundamentally atypical.

The chapter is organized as follows. The first section reviews information relating to the processes of individual growth, mortality and reproduction. This is followed by a discussion of the dynamics and management of tilapia populations, both in the wild and in culture, and the dynamics of tilapia invasions. The final section explores life history variation in the tilapias, with particular reference to the 'stunting' problem and its underlying mechanisms.

6.2 POPULATION PROCESSES

The present section deals individually with the processes of body growth, mortality and reproduction. To analyse each trait in isolation helps to identify the key proximate factors governing them. Ultimately, however, the traits are related through fitness trade-offs, which lead to the evolution of certain relationships between them (Beverton and Holt, 1959; Beverton, 1992; Roff, 1992; Charnov, 1993; Mangel, 1996; Jensen, 1996). Such relationships in the tilapias are explored further in section 6.4.

Growth

Individual growth is of key importance to the management of wild and cultured stocks, and has been studied extensively in tilapias. To allow a quantitative comparison of growth, the relevant patterns must be described mathematically.

Mathematical models of different degrees of complexity have been developed to describe growth in fish. Simple models such as the von Bertalanffy growth function describe growth by parameters that relate directly to key patterns (e.g. asymptotic size), and can be estimated relatively easily. Complex physiological and bio-energetic growth models have the potential to provide insights into the factors and mechanisms determining growth patterns, but require a larger number of parameters that describe physiological processes and are not in any obvious way related to particular growth patterns. Such models are outside the scope of this chapter, but interested readers are referred to Liu and Chang (1992), Van Dam and Pauly (1995), and Yi (1998).

In ecological studies the choice of growth model is usually determined by practical considerations, and how well alternative models fit the available data. In optimality models of life history theory, however, the choice of growth model may have far-reaching implications and must be considered carefully (Day and Taylor, 1997).

Describing growth: the VBGF

Throughout this chapter, growth is described by the conventional von Bertalanffy growth function (VBGF) or closely related models. Von Bertalanffy (1957) developed the VBGF on the basis of physiological considerations. The commonly used forms of the equation,

$$L_t = L_\infty\,(1 - e^{-k(t-t_0)}) \qquad \text{(Eq. 6.1a)}$$

for growth in length and

$$W_t = W_\infty\,(1 - e^{-k(t-t_0)})^3 \qquad \text{(Eq. 6.1b)}$$

for growth in weight were derived by Beverton and Holt (1957). In these equations, L_t and W_t are the length and weight at time t, and L_∞ and W_∞ are the asymptotic length and weight towards which the fish grow at rate K. The parameter t_0 defines the position of the growth curve along the time axis. Equation 6.1b assumes an isometric length-weight relationship (i.e. weight proportional to the cube of length). An empirically determined allometric exponent may be substituted for the third power in Equation 6.1b to derive a more general weight growth equation (see also Pauly, 1981).

In order to facilitate the comparison of overall growth performance between populations that differ in both asymptotic size and growth rate, it is desirable to define an index that conveys growth performance information in a single number. An obvious candidate for such an index is the maximum weight growth rate, which is attained at the inflexion point of Equation 6.1b:

$$(dW/dt)\text{max} = (4/9)\,KW_\infty \qquad \text{(Eq. 6.2)}$$

Pauly (1981) defined the growth performance index P, which is proportional to the logarithm of the maximum weight growth rate:

$$P = \log_{10}(KW_{\infty}) \qquad \text{(Eq. 6.3)}$$

Although other indices of growth performance have been proposed (Moreau *et al.*, 1986), P is the most readily interpretable in terms of the maximum growth rate. Jensen (1997) has added a further interpretation to the index, which is discussed below.

The physiological basis of the VBGF may be summarized as follows: growth is the net result of the opposing processes of anabolism and catabolism (metabolism). Anabolism is limited by the rate at which resources can be acquired by the organism, which is assumed to be approximately proportional to body length to the power of two. Different reasons have been advanced for this proportionality. The limiting effects of body surfaces on resource uptake in general, or of the gill surface area on oxygen uptake in particular, have been advanced by von Bertalanffy (1957) and Pauly (1981), while Walters and Post (1993) propose that the search rate and swimming speed of fishes scale with the second power of length. Catabolism is approximately proportional to body weight, i.e. the third power of length. Several authors have interpreted and extended the VBGF (Beverton and Holt, 1957; Pauly, 1981; Roff, 1983; Walters and Post, 1993; Lorenzen, 1996a; Jensen, 1997), and this has led to some generalizations with regards to the growth patterns that can be expected under particular circumstances: (1) Differences purely in the rate at which resources are acquired by the organism (relative to the requirement, proportional to length to the power of two) will lead to differences in asymptotic size (L_{∞} and W_{∞}), but not in the growth rate (K). (2) Differences in the allocation of resources to somatic growth and other body functions such as reproduction or predator avoidance will affect the growth rate (K) as well as the asymptotic size. The growth rate K may also be affected if the relative rate at which resources are acquired changes with the size of fish, for example because population biomass and therefore the degree of resource competition changes as the fish grow (e.g. single cohort aquaculture, Lorenzen, 1996a), or because competition between size groups is asymmetric (Walters and Post, 1993). Jensen (1997) shows that, for organisms that acquire resources at the same rate but allocate them differently, K and W_{∞} can be expected to be inversely proportional. This also implies that the maximum weight growth rates and growth performance index P of these organisms will be the same (cf. Eqs 6.2, 6.3), but will be achieved at different sizes and ages (see also Fig. 6.1).

Beverton and Holt (1957), Walters and Post (1993) and Lorenzen (1996a) have extended von Bertalanffy's theory of growth to account for density-dependence resulting from competition for food. The model by Lorenzen (1996a), which will be applied to tilapia growth later on, assumes a linear relationship between asymptotic length and population biomass and the equivalent relationship for asymptotic weight:

$$L_{\infty B} = (L_{\infty L} - g\,B) \qquad \text{(Eq. 6.4a)}$$
$$W_{\infty B} = (W_{\infty L}^{1/3} - c\,B)^3 \qquad \text{(Eq. 6.4b)}$$

where $L_{\infty L}$ and $W_{\infty L}$ are the (extrapolated) asymptotic length and weight in the absence of competition ($B \rightarrow 0$), and g and c are the competition coefficients for length and for weight. The model of Walters and Post (1993) was developed specifically to address competitive asymmetries in age/size structured populations, and allows for more complex relationships between the abundance of age/size groups and asymptotic size. A result of general interest is that competitive asymmetries are expected to result in apparent changes in the growth rate K with density. This suggests that simple models will describe density-dependent growth well in omnivorous species that do not show major ontogenic shifts in diet or competitive asymmetries (inequitable partitioning of resources between fish of different ages/sizes). This has indeed been shown for carp (Lorenzen, 1996a).

Comparative analysis of growth patterns

Extensive, comparative studies of tilapia growth performance in wild stocks and in aquaculture have been reported by Moreau *et al.* (1986) and Pauly *et al.* (1988). Tilapia growth patterns vary widely between and within species. In natural populations the highest asymptotic sizes of 500 mm in length or 7000 g in weight have been recorded for *O. niloticus*, while populations of *O. aureus* and *Tilapia zillii* rarely reach more than 250 mm in length or 600 g in weight. In aquaculture, asymptotic sizes of less than 200 mm in length or 500 g in weight are common regardless of the species used. Typical values for the growth rate K range from 0.2 to 0.8 y^{-1} in natural populations, but can be an order of magnitude higher in culture.

An overview of observed growth parameters in natural and cultured populations of *O. niloticus* is given in Fig. 6.1(a). There are two clearly separated clusters, one of wild populations with high W_∞ and low K, and one of cultured populations with low W_∞ and high K. The line of slope −1 indicated in Fig. 6.1(a) connects points of equal growth performance P (Eq. 6.3; Pauly, 1981). The average growth performance P and therefore maximum growth rates in natural and farmed populations of *O. niloticus* overlap widely, although on average P is higher in culture (see also Fig. 6.2). The different growth patterns implied by different points on the same isoline of P are illustrated on Fig. 6.1(b).

The question remains why values of K in aquaculture tend to be much higher than in natural stocks. The above discussion of VBGF patterns and their physiological interpretations suggests several, not mutually exclusive possibilities. Under culture conditions, fish may increase allocation of resources to maintenance metabolism (Pauly, 1981) and reproduction, and/or resource

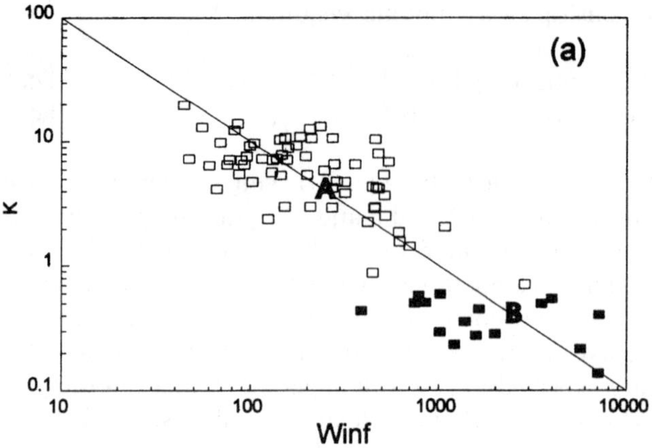

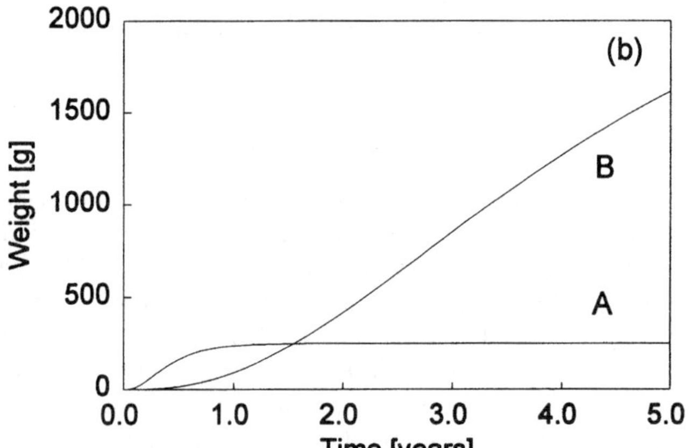

Fig. 6.1 Growth patterns in populations of *O. niloticus*. (a) VBGF parameters in wild (■) and farmed (□) populations. The solid line of slope −1 denotes equal growth performance index P. (b) Growth trajectories associated with points A and B. Both parameter combinations have the same growth performance index P (and therefore the same maximum growth rate), but lead to very different growth trajectories. Data from Moreau *et al.* (1986) and Pauly *et al.* (1988).

acquisition may decline relative to requirements as the population approaches the system's carrying capacity (Lorenzen, 1996a). The analysis of density-dependent growth in pond culture of *O. spilurus niger* (see below) suggests a combination of both mechanisms: density-dependence accounts for some apparent variation in K, but even the value of K estimated for the density-dependent VBGF is high relative to the values found in natural populations.

Pauly *et al.* (1988) compared overall growth performance of different tilapia species in the wild and in culture. Growth performance P measured in wild stocks overlapped widely among the species (Fig. 6.2). Median growth performance was higher in culture than in natural stocks for must species, with the differences in growth performance P evident in *O. aureus* and *T. zillii* (22 and 33% respectively). These conclusions are moderately sensitive to the choice of growth performance index: Pauly *et al.* (1988) based their final assessment on a different growth performance index (ϕ), which indicated a far greater magnitude of differences between cultured and wild stocks, but similar relative patterns.

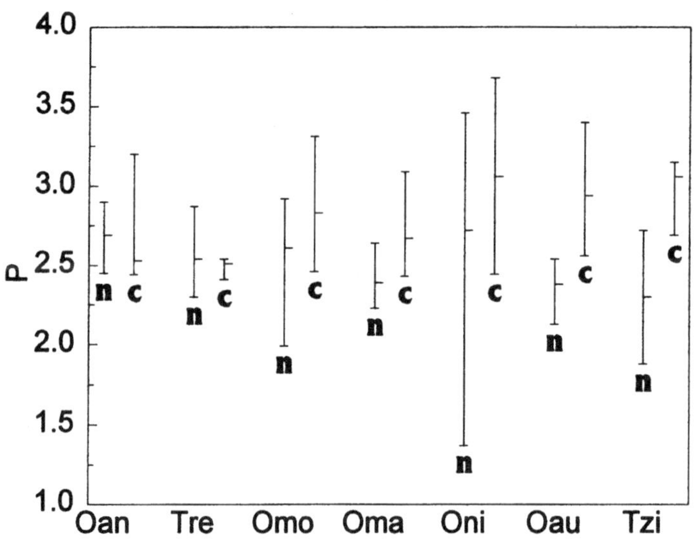

Fig. 6.2 Comparison of the growth performance index P between natural (n) and cultured (c) populations of *O. andersonii* (Oan), *T. rendalli* (Tre), *O. mossambicus* (Omo), *O. macrochir* (Oma), *O. niloticus* (Oni), *O. aureus* (Oau) and *T. zillii* (Tzi). Bars indicate the range and median. Based on data in Pauly *et al.* (1988).

The maximum or asymptotic size in tilapia populations tends to be positively related to the size of the water body in which they live. This has been noticed by Lowe (McConnell) (1958) for *O. niloticus*, and is also borne out in the *O. mossambicus* data assembled in Table 6.1. (Fig. 6.3). This suggests that the absolute allocation of resources to growth is higher in individuals living in larger water bodies, either because of higher acquisition or because of a higher relative allocation to growth (or a combination of the two).

Table 6.1 Life history parameters of certain tilapia populations. These data are used for exploratory analyses in various places in the chapter. Values in brackets have been inferred indirectly and are not use in quantitative analyses. L_∞ and L_m in total length

Species	Water body	Area [ha]	Z [/year]	K [/year]	L_∞ [mm]	L_m [mm]	L_m/L_∞	A_m [years]	Source
O. andersonii	Kafue Floodplain	430 000	0.8	0.21	434	276[c]	0.64	4.0_c	Kapetsky (1974); Dudley (1974)
O. esculentus	Lake Victoria	6 863 500	3.0	0.28	295	230	0.78	5.4[a]	Garrod (1959, 1963); Fryer and Iles (1972)
O. karongae	Lake Malawi	303 000	0.81	0.18	413	285	0.69	6.5[a]	FAO (1993)
O. lidole	Lake Malawi	303 000	0.85	0.18	432	285	0.66	6.0[a]	FAO (1993)
O. macrochir	Kafue Floodplain	430 000	0.7	0.31	308	211[c]	0.69	3.0[c]	Kapetsky (1974); Dudley (1974)
O. mossambicus	Tank	0.0001	0.5		224	(>120)[d]	(>0.53)[d]		Silliman (1975)
O. mossambicus	Rufanes Pool	0.12			265	114	0.43	1.0[b]	James and Bruton (1992)
O. mossambicus	Kowie	0.6			275	217	0.79	3.0[b]	James and Bruton (1992)
O. mossambicus	Bradshaw's Mill Dam	0.75	1.2		256	177	0.69	2.0[b]	James and Bruton (1992)
O. mossambicus	Mill Farm Dam	8	0.4		358	264	0.74	2.5[b]	James and Bruton (1992)
O. mossambicus	Pimburettwa	834	2.4	0.34	393	205	0.52	2.2[a]	Amarasinghe (1996); De Silva (1986)
O. mossambicus	Parakrama	2262	1.7	0.26	380	170	0.45	2.3[a]	Amarasinghe (1996); De Silva (1986)
O. mossambicus	Kaudulla	2537	1.6	0.52	437	210	0.48	1.3[a]	Amarasinghe (1996); De Silva (1986)
O. mossambicus	Minneriyaa	2560	3.1	0.45	450	195	0.43	1.3[a]	Amarasinghe (1996); De Silva (1986)

Table 6.1 (cont)

Species	Water body	Area [ha]	Z [/year]	K [/year]	L_∞ [mm]	L_m [mm]	L_m/L_∞	A_m [years]	Source
O. niloticus	Kaudulla	2537	1.9	0.34	545	225	0.41	1.6[a]	Amarasinghe (1996); Amarasinghe and De Silva (1992)
O. niloticus	Minneriyaa	2560	3.6	0.43	545	210	0.39	1.1[a]	Amarasinghe (1996); Amarasinghe and De Silva (1992)
O. niloticus	Lake George	27 000	3.2		450	205	0.46		Gwahaba (1973)
O. niloticus	Lake Tukana	854 700			610	390	0.64		Lowe (McConnell) (1958)
O. shiranus	Chisombezi	2.2	2.1	0.67	238	140	0.59	1.3[a]	Mattson and Kaunda (1997)
O. shiranus	Mbvoniha	3.6	5.8	1.08	304	150	0.49	0.6[a]	Mattson and Kaunda (1997)
O. squamipinnis	Lake Malawi	303 000	1.09	0.24	378	255	0.67	4.7[a]	FAO (1993)
S. melanotheron	Fosu Lagoon	61	5.0	0.82	161	60	0.37	0.6[a]	Blay and Asabere-Ameyaw (1993)
S. melanotheron	Tank	0.001	(0.2)[e]		86	50	0.58		Eyeson (1983)
T. rendalli	Kafue Floodplain	430 000	1.4	0.31	364	248[c]	0.68	3.0[c]	Kapetsky (1974); Dudley (1974)

[a]Calculated from L_m and growth parameters, assuming $t_0 = 0$; [b]determined by direct ageing; [c]approximate A_m, L_m calculated using growth parameters; [d]based on minimum length at which sexes can be distinguished externally; [e]average Z in pond and cage culture.

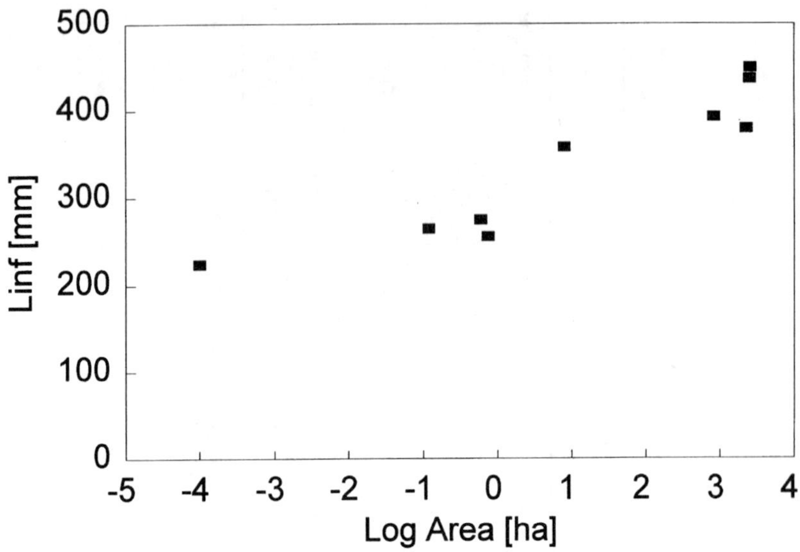

Fig. 6.3 Relationship between water body area and asymptotic size in populations of *O. mossambicus*. Data from Table 6.1.

Plasticity in growth

The growth of tilapias is highly plastic, responding immediately to changes in resource availability and other factors. Van Someren and Whitehead (1959) transferred mature males of *O. spilurus niger* from a densely populated pond where they were approaching their asymptotic weight to a less densely populated pond, where they resumed rapid growth towards a higher asymptotic weight. This demonstrates clearly that, rather than being determined early on in life, the growth of tilapias responds quickly to changing conditions throughout life.

An aspect of plasticity of particular interest to the management of populations and the 'stunting' problem is density-dependence. Perhaps the best experimental data on density-dependent growth in tilapias can be found in van Someren and Whitehead (1961), who stocked and harvested, in a staggered fashion, four cohorts of *O. spilurus niger* in a pond and monitored their growth. The resulting growth patterns are illustrated in Fig. 6.4(a), together with a fitted density-dependent VBGF model. Total biomass in the experimental pond increased rapidly over 6 months and declined thereafter, and this pattern is mirrored by changes in asymptotic weight (Fig. 6.4(b)). Density-dependence has generated a wide variety of growth patterns. The most striking differences are apparent between the first stocked cohort, growing to a low asymptotic weight at a high apparent growth rate as biomass increases, and the last stocked cohort growing

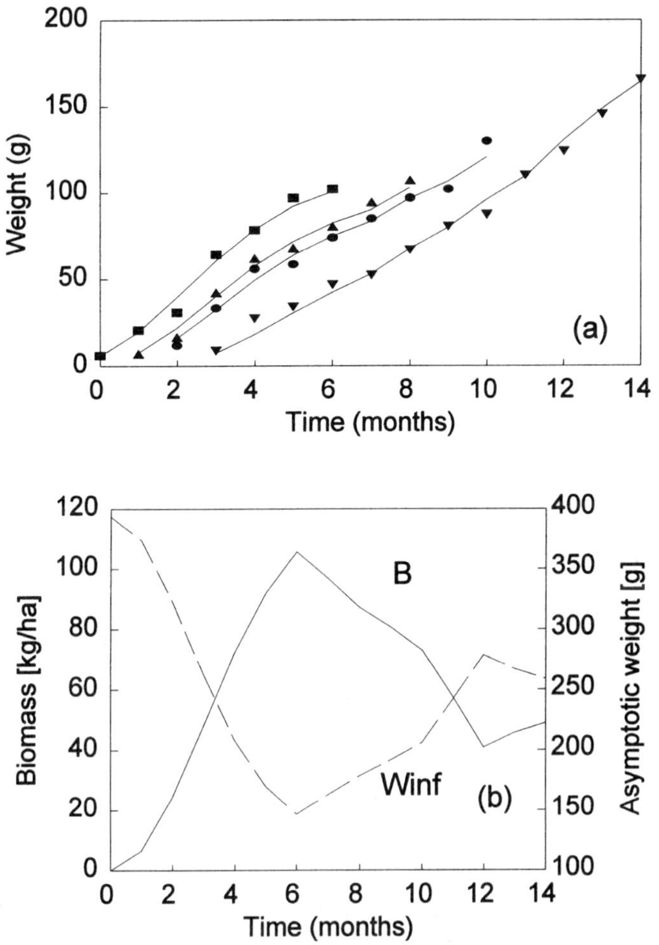

Fig. 6.4 Density-dependent growth of *O. spilurus niger* in extensive pond culture (without feeding). Four all-male cohorts were stocked and harvested in a staggered fashion. (a) Growth trajectories of the four cohorts, symbols denote observed weight and lines indicate predictions from a density-dependent VBGF (Eq. 6.4b) with parameters $W_{\infty L}$ = 394 g, K = 2.1/year, and c = 0.019 $g^{1/3}$ ha/kg. (b) Population biomass B and corresponding asymptotic weight $W_{\infty B}$. Data from Van Someren and Whitehead (1961).

almost linearly towards a large size as biomass is gradually reduced. Using a conventional (non density-dependent) VGBF (Eq. 6.1b), growth in the first cohort is described by W_∞ = 132 g and K = 4.1/year, while growth in the last stocked cohort is described by W_∞ = 349 g and K = 1.3/year. The full range of patterns is well described by a simple density-dependent model with constant K and asymptotic size as function of total biomass (Eq. 6.8, parameter values: $W_{\infty L}$ = 394 g, K = 2.1/year, and c = 0.019 $g^{1/3}$ ha/kg). The estimated value of K = 2.1/year is somewhat higher than the values estimated for conventional VBGF models applied to wild populations (Pauly *et al.*, 1988; see also Fig. 6.1(a)). This suggests that the high values of K estimated in single cohort aquaculture (Pauly *et al.*, 1988) result from a combination of increasing resource competition as the culture cycle progresses, and a higher allocation to functions other than growth.

The primary factor mediating density dependence in growth in natural and extensively farmed fish is competition for food (e.g. Lorenzen, 1996a). Food availability in itself, of course, has a strong effect on tilapia growth (see Diana, 1997, for a review). Nutritional constraints in particular ontogenic stages have been linked to restricted growth in the *O. mossambicus* population of Lake Sibaya (Bowen, 1979), and similarly 'trophic bottlenecks' have been implicated in growth depression in non-tilapiines (Heath and Roff, 1996).

Apart from population density and food supply in general, a range of other environmental factors influence growth in tilapias. A frequently cited pond culture experiment by Chen and Prowse (1964) suggests that the size of the pond in which the population lives may affect the growth of tilapias even when density is held constant. However, this result rests on the markedly higher growth observed in a single, large pond and has not been replicated.

Changes in habitat availability have been shown to affect growth in resident tilapia populations. Dudley (1974) found a strong positive correlation between growth increments in three tilapia species and the extent of seasonal flooding in the Kafue floodplain, Zambia. The damming of the Zambezi to form Lake Kariba has led to a marked increase in the growth of the endemic *O. mortimeri* (Harding, 1965).

The relationship between oxygen availability and growth is of practical as well as theoretical interest. Pauly (1981) hypothesized that oxygen availability was the primary factor limiting the growth of fishes and developed a 'generalized' version of the VBGF to account for the limitations to growth arising from geometric constraints on gill surface area. Although this hypothesis has not been tested explicitly, there is experimental evidence for an effect of oxygen availability on growth in tilapias (Tsadik and Kutty, 1987; Papotsoglou and Tziha, 1996). Van Dam and Pauly (1995) incorporated an oxygen limitation term based on Pauly's theory into a dynamic simulation model for fish growth, and showed that the model can reproduce the effects of low oxygen saturation on final weights of *O. niloticus* observed by Tsadik and Kutty (1987). This does not, however, imply that oxygen availability is the principal factor limiting fish growth under natural or culture conditions. Many studies point to an overwhelming effect of food supply (e.g. Prein, 1993a; Lorenzen, 1996a; Diana 1997), but oxygen may become limiting under very high densities and rates of feed or fertilizer supply (Teichert-Coddington and Green, 1993; see also Ross, this volume).

Sexual growth dimorphism

It has long been noted that in many tilapia species, males tend to grow at a higher rate and/or to a larger asymptotic size than females (e.g. Fryer and Iles, 1972). This pattern is unusual within the fishes, where females often grow larger than males. Among the reasons advanced for male growth 'superiority' in tilapias are a sex-specific physiological growth capacity, cessation of feeding and growth during female mouth brooding, and behavioural interactions. Schreiber *et al.* (1998) demonstrate that in individually reared *O. niloticus*, males show a lower growth rate and feed conversion efficiency than females. This indicates that the physiological growth capacity is in fact higher in females than in males, and the opposite relationship commonly observed in group rearing must be due to either the effect of reproduction or behavioural factors. Empirical evidence with regards to the latter two possibilities is somewhat inconclusive. A number of studies indicate marked growth differences between monosex populations of the same species (*O. spilurus*: Van Someren and Whitehead, 1960; *O. niloticus*: Pauly *et al.*, 1988), implying that the differences persist even when no breeding takes place. Conversely, Bolivar *et al.* (1993) show that in some strains of *O. niloticus* at least, non-breeding females have growth patterns similar to those of males. Bramick *et al.* (1995) studied triploid *O. niloticus* and found no difference in growth between diploid and triploid sibs up to the onset of reproduction in diploids. Thereafter, growth in diploid females declined immediately relative to their later maturing triploid sibs. Growth in diploid males only declined after large numbers of fry occurred in the experimental ponds, suggesting that the decline in growth was linked to increased competition rather than the energetic costs of reproduction.

Individual variability and genetics

Individual variability in growth may have important implications for the dynamics and life history evolution of populations, and consequently for their management (Lomnicki, 1988; De Angelis and Gross, 1992; Tyler and Rose, 1994; Mangel, 1996; Stamps *et al.*, 1998). Variability in growth may be due to phenotypic plasticity as well as genetically determined growth potential. In laboratory situations, tilapias may exhibit a high degree of individual variability in growth linked to hierarchical interactions. Fernandes and Volpatro (1993) show how hierarchical interactions lead to increased heterogeneity in growth, related largely to metabolic differences between dominant and subordinate fish. The question to what extent individual variability in growth has a genetic basis is of obvious relevance for selective breeding programmes, as well as the selective exploitation of natural populations. A comparative study of the growth of different strains of *O. niloticus* under a range of farming conditions has shown that strain effects were significant, but only accounted for a low proportion (1.5%) of the overall observed variation in growth rates (Eknath *et al.*, 1993).

The proportion of variation explained by genetic differences is, of course, expected to be higher if only populations within a single environment are considered. Silliman (1975) subjected laboratory populations of *O. mossambicus* to selective and unselective fishing over a period of about four generations, and subjected groups of mature fish from both treatments to growth experiments under similar conditions. He found no difference in the growth of females, while males from the unselectively fished population grew larger than those from the selectively fished one. The result is therefore somewhat inconclusive, as well as of questionable validity. The growth tests have been carried out using large fish that had experienced different conditions during husbandry prior to the test, rather than individuals of the following generation reared under identical conditions.

Mortality

Despite of its obvious importance to life history evolution, ecology and management, mortality has been studied far less than growth or reproduction, being more difficult to measure in field situations and considered less controllable in culture.

Describing mortality

Traditionally, mortality is described as proportional survival over a culture period in aquaculture, and as an instantaneous rate in fisheries ecology (for a discussion see Hopkins and Pauly, 1993). The development of population numbers over time under the influence of a constant instantaneous mortality rate is described by

$$N_t = N_0 \, e^{-Zt} \qquad \text{(Eq. 6.5)}$$

where N_t is the population abundance at time t, N_0 is the initial number at time t = 0, and Z is the instantaneous mortality rate. Proportional survival S from t = 0 to t is defined by

$$S = N_t/N_0 \qquad \text{(Eq. 6.6)}$$

Consequently, S and Z are related by

$$S = e^{-Zt} \qquad \text{(Eq. 6.7)}$$

Among the advantages of using instantaneous mortality are that they are independent of the study period and can therefore be compared easily, and that they are additive. In fisheries ecology, the total mortality rate Z is often written as the sum of the mortality rate F caused by fishing, and the natural mortality rate M due to all other causes (e.g. predation, parasitism, or adverse environmental conditions):

$$Z = M + F \qquad \text{(Eq. 6.8)}$$

Comparative information on mortality rates

An overview of the range and median of total mortality rates determined for some tilapia populations in different habitats is given in Fig. 6.5. Data for natural populations are from Table 6.1, and reflect mortality due to fishing as well as natural causes. Data for pond and cage culture have been extracted from Lorenzen (1996b), and reflect natural mortality during the on-growing period in culture, i.e. excluding the final, catastrophic event of fishing mortality characteristic of stocks in single cohort aquaculture. Total mortality rates in the natural populations range from 0.4 to 5.8, with median values of 1.8/year in lakes (>100 ha), 0.8/year in rivers, and 2.1/year in small water bodies of less than 100 ha in area. Sample sizes in all categories are too small to provide anything more than a rough indication, but its is worth noting that some populations suffer extremely high total mortality rates, and that there is no obvious relationship between (natural) habitat type and total mortality rate. The highest degree of variation is evident in small water bodies, which account for the lowest and well as the highest recorded values. Mortality rates in pond and cage culture are comparatively low, ranging from 0.01/year to 1.4/year, with a median of 0.2/year. This category includes many ponds that are not unlike natural small water bodies in terms of fluctuations of physico-chemical parameters, etc. but there are no instances of particularly high mortality. Mortality rates in small water bodies are particularly variable, but there is no evidence to suggest that they are different on average from those in larger water bodies.

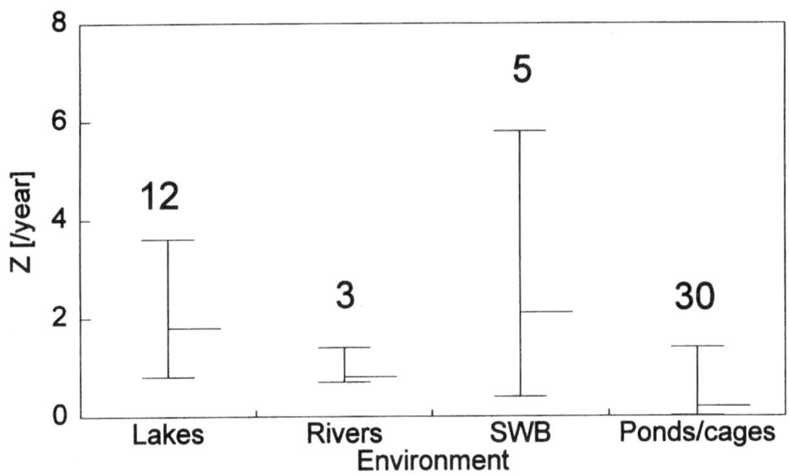

Fig. 6.5 Range and median of total mortality rates (Z) measured for tilapia populations in different environments: lakes (>100 ha); rivers; small water bodies (SWB, <100 ha); and pond and cage aquaculture. Data from Table 6.1; and Lorenzen (1996b). Numbers indicate sample size.

Certain methods of fisheries assessment (dynamic pool models, see p. 194) require the separation of total mortality into its components due to fishing and due to natural causes. Unfortunately, the natural mortality rate of an exploited population is difficult to assess, and there is only one well-documented estimate of natural mortality in a wild tilapia population, for *O. esculentus* in Lake Victoria in the late 1950s (Garrod, 1963). Pauly (1980) developed a multiple regression model predicting natural mortality from growth parameters and mean environmental temperature, which has been widely used to provide estimates of natural mortality in assessments of tilapia (as well as many other) stocks. Pauly's original data set included Garrod's natural mortality estimate for *O. esculentus* as the only estimate for tilapiines.

Size dependence

Natural mortality rates in fish are strongly size-dependent, and can be described by a power function of the form

$$M_W = M_u W^b \qquad \text{(Eq. 6.9)}$$

where M_W is the natural mortality at weight W, M_u is the natural mortality at unit weight, and b is the allometric exponent of the relationship. A comparative analysis of mortality-weight relationships across a wide range fish species (Lorenzen, 1996b) showed that natural ecosystems are characterized by a common natural mortality-weight relationship with an allometric exponent of $b = -0.29$. Natural mortality in aquaculture systems was lower overall than in natural ecosystems, but more strongly weight-dependent with an allometric exponent of $b = -0.43$.

The study by Lorenzen (1996b) included 36 natural mortality estimates for tilapias in pond and cage culture, spanning a range of mean weights from 0.45 to 330 g. These data are shown in Fig. 6.6, together with the corresponding mortality-weight relationship estimated by non-parametric (Theil) regression (Sprent, 1989). The parameters (with 90% CI) of the relationship for tilapia pond and cage culture are $b = -0.60$ [−0.67, −0.45] and $M_u = 2.9$ [2.3, 3.7]/year. These parameters are not significantly different from those estimated for the full set of pond and cage culture data, including or excluding the tilapia data analysed here. This suggests that in tilapias, too, natural mortality rates in aquaculture are more strongly size dependent than in wild stocks. Consequently, natural mortality rates of large/mature are far lower in culture than in natural stocks. There are no data on mortality-weight relationships in wild tilapia populations, but these relationships have been found to be quite consistent among other populations (Lorenzen, 1996b) and there is no reason to expect very different patterns in the tilapias.

Size-dependence in mortality has important ecological and life history implications, because it provides a direct link between growth and survival. In

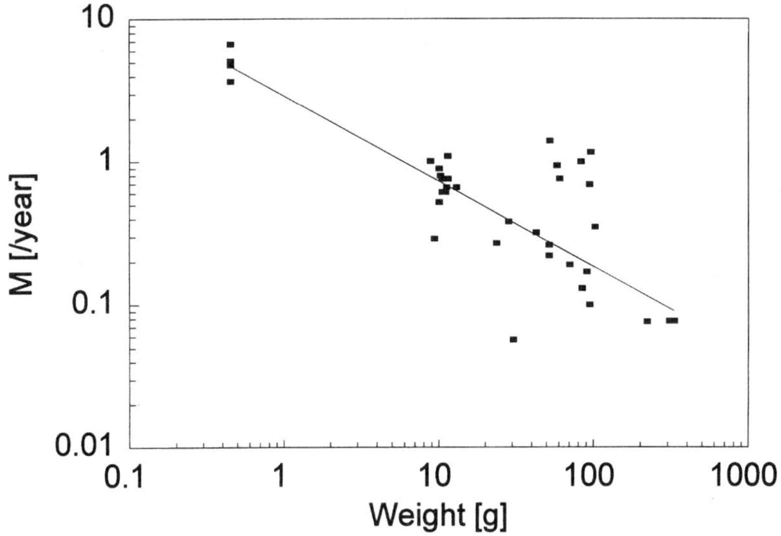

Fig. 6.6 Relationship between natural mortality and body weight in tilapia pond and cage culture. The solid line indicates a relationship (Eq. 6.9) determined by non-parametric (Theil) regression. The parameters b = –0.60 (90% CI [–0.67, –0.45]) and Mu = 2.9 (90% CI [2.3, 3.7]) are not significantly different from the overall means for pond and cage culture determined by Lorenzen (1996b). Data compiled by Lorenzen (1996b) from information in Inland Fisheries Project (1977); Williams *et al.* (1987); AIT (1990); Costa-Pierce and Hadikusuma (1990); Cruz and Ridha (1991) and Diana and Lin (1991).

the context of the 'stunting' problem, it is worth noting that populations characterized by a low maximum size can be expected to show reduced longevity compared to large growing populations, simply as a result of size-dependent mortality.

Factors determining mortality rates

Mortality rates are determined by a combination of environmental factors (e.g. temperature, presence of predators), and factors rooted in the behaviour, physiology and genetics of the individual organism (e.g. choice of habitat, predator avoidance behaviour). These factors and their interaction are poorly understood, and have been little studied in the tilapias. However, a study on the influence of salinity on mortality in cultured tilapia illustrates the potential usefulness of systematic mortality studies to management. Hopkins and Pauly (1993) showed that in pure *O. niloticus* stocks, mortality increases with increasing salinity, while *O. niloticus* × *O. mossambicus* hybrid stocks show the opposite trend. Further quantitative studies on mortality in culture and in wild populations could lead to significant improvements in management, and may also provide insights into fundamental ecological questions (see e.g. the discussion on predation mortality in Lorenzen, 1996b).

Reproduction

The tilapias have certain reproductive strategies that set them apart from the vast majority of fish species. All tilapias normally build and spawn in nests, around which they defend territories. Spawning is followed by guarding of the brood by both parents in the Genus *Tilapia*, female mouthbrooding of eggs and fry in *Oreochromis*, and bi-parental mouthbrooding of eggs only in *Sarotherodon*. Relative gonad weight in tilapias rarely exceeds 6% of body weight, and is therefore near the lower limit observed for teleost fish (Peters, 1963; Iles, 1973). Overall, tilapias are characterized by a very high fertility: 150 adult fish have been reported to have produced 15 000 fry in less than 4 months (Hickling, 1963). Fecundity (the number of eggs produced) is approximately proportional to weight, or the third power of length, in the substrate spawning tilapias, but scales with the second power of length in mouthbrooders (Welcomme, 1967; Jalabert and Zohar, 1982).

Reproductive biology is covered in depth by Turner and Robinson (this volume), and the discussion here focuses on aspects of particular interest to population dynamics and life history analysis: size and age at maturity.

Describing size and age at maturity

In physiological terms, the onset of the maturation process may be distinguished from a phase of final maturation, and the two phases appear to be influenced by different cues (Brummett, 1995). The maturation process may be described in terms of maturity stages, based on gonadal morphology (e.g. Nikolsky, 1969), and 'late' stages may be distinguished form 'early' stages in order to classify organisms as mature or immature. The age and length at maturity is usually expressed as the age/length at which 50% of the population are mature (i.e. in late stages of gonadal development), determined by fitting a logistic function to proportional maturity data (King, 1995). Where information based on gonadal morphology is not available, other indicators of maturity may be used as a proxy, for example coloration (e.g. Lowe (McConnell), 1958).

Variation in reproductive traits

The tilapias are renowned for their variability in reproductive traits, in particular the age and size at maturity. Length at maturity L_m has been shown to vary from 35 mm (Hecht and Zway, 1984) to 275 mm (James and Bruton, 1992) in *O. mossambicus*, or from 100 mm (Bolivar *et al.*, 1993) to 390 mm (Lowe (McConnell), 1958) in *O. niloticus*. Information on age at maturity is less easily available, but *O. niloticus* have been shown to mature at an age of 3 months in tanks (Bolivar *et al.*, 1993), and possibly over 2 years in large lakes

(Gwahaba, 1973; Kolding, 1993). In general, size and age at maturity are correlated, i.e. tilapias mature either early and small, or late and large (Duponchelle and Panfili, 1998; section 6.4 this chapter). However, there are also examples of populations with a low maximum size where maturation appears to be delayed or suppressed completely (Silliman, 1975; Eyeson, 1983). In such populations, small mature fish may be quite old. Different sizes at maturity are associated with differences in fecundity and brood frequency, with small maturing individuals being relatively more fecund and breeding more frequently than large maturing ones (Iles, 1973).

The observation that the largest maturing tilapia populations are found in large lakes and river systems suggests that length at maturity may be related to the size of the water body in which a population lives. Empirical relationships between length at maturity and water body area have been established by Lowe (McConnell) (1958) for African lakes, by De Silva (1986) for Sri Lankan reservoirs, and by Duponchelle and Panfili (1998) for reservoirs in Ivory Coast. However, a plot of length at maturity in *O. mossambicus* populations against water body size, covering several orders of magnitude, suggests that such relationships exist within sets of similar water bodies, but not globally (Fig. 6.7).

A comparative analysis across African lakes also suggests a negative correlation between the condition factor of fish and the onset of maturation, i.e. populations in poor condition appear to mature at a smaller size than those in a good condition (Lowe (McConnell), 1958). There are a number of possible

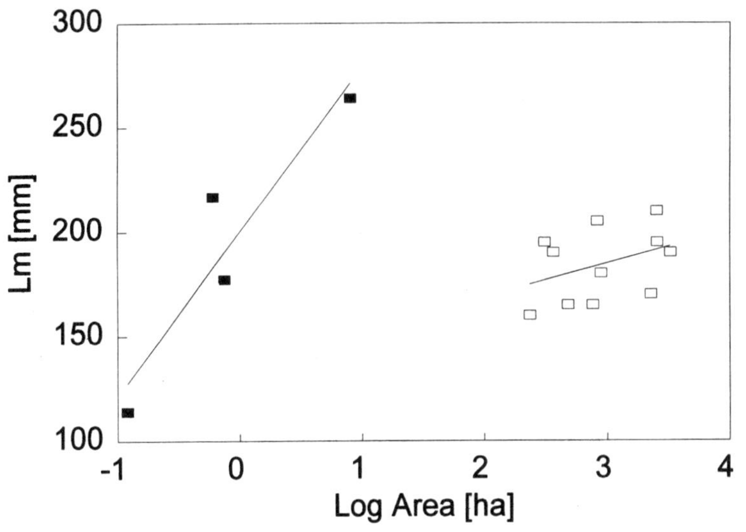

Fig. 6.7 Relationship between water body area and length at maturity in populations of *O. mossambicus* in South African (■) and Sri Lankan (□) reservoirs (data from De Silva, 1986; James and Bruton, 1992).

explanations for this correlation, e.g. poor condition causes maturation at small size, maturation at small size causes poor condition, or poor condition limits somatic growth so that fish mature small but not necessarily early. Additional data on growth are required to resolve this problem.

Mechanisms underlying variation in reproductive traits

Is the observed variation in reproductive traits a reflection of phenotypic plasticity, or the result of genetic selection? Many observations, such as life history comparisons between long established populations, are indeed consistent with either mechanism, but there is compulsive evidence from natural environments, transfer experiments and population genetics that the observed differences are primarily phenotypic in origin. Lowe (McConnell) (1958) and Welcomme (1970) observed maturation at small size in temporarily isolated lagoons of large lakes. In the early days of tilapia culture, transfers of specimens from large-maturing African lake populations to ponds or aquaria were common, and led to immediate reductions in size at maturity to values around 100 mm, far outside the range (200–300 mm) observed in the lake populations. Although these experiments are not well documented, they are regularly referred to in the earlier literature (see e.g. Cridland, 1961; Fryer and Iles, 1972; Iles, 1973). Such drastic and immediate changes in maturity traits can not be explained by genetic selection and must be phenotypic in origin. In a recent study, Duponchelle *et al.* (1998) subjected groups of *O. niloticus* from two reservoir populations with very different reproductive characteristics to a common environment, and found that the differences disappeared within 5 months. Further evidence for the phenotypic nature of variation in size at maturity is the failure of population genetics studies to detect any differences between small and large maturing feral *O. mossambicus* populations of single origin (Bluhdorn and Arthington, 1990).

Having established that the observed variation in age and size at maturity in tilapias is primarily phenotypic in nature, the question remains: how is the expression of this plasticity controlled? There are two main lines of thought in this respect, one that interprets plasticity in terms of the individual organism's physiological responses to environmental stimuli, and another that sees intraspecific interactions as a key factor.

The main physiological hypotheses focus on (1) unspecific stress, (2) the oxygen regime, and (3) other physical or chemical cues. Several authors have linked early maturation in tilapias to unspecific stress, related to factors such as strong fluctuations in physico-chemical parameters, high rates of mortality due to catastrophic events or predation, and extreme reduction in living space in periods of drought (Lowe-McConnell, 1982, also this volume; Kolding, 1993). However, the stress hypothesis is difficult to evaluate because stress is defined in terms of consequences, not causes. Indeed, it is not uncommon in the literature to find descriptions of early maturing populations that are considered stressed simply because they mature early. Unless the level of stress can be established independently (e.g. by physiological indicators such as levels of stress hormones), the stress hypothesis is little more than a circular argument. Even if a correlation between independent indicators of stress and early maturation could be established, this would be of limited practical use unless the underlying causes of stress could be isolated.

The relationship between mortality and reproductive traits is explored in section 6.4. In established, self-sustaining populations age at maturity is inversely related to the mortality rate, while the relationship between size at maturity and mortality is more complex. However, in situations of demographic dis-equilibrium, such as aquaculture facilities stocked only with juveniles, maturation may occur very early even though mortality rates are extremely low (cf. p. 177).

Pauly (1984) proposed a theory that the transition to maturity is induced by declining oxygen availability, due to the geometric constraints on gill surface area as fish grow. He hypothesized further that the (as yet to be discovered) physiological mechanism inducing maturation in response to oxygen levels could be triggered off prematurely under conditions of fluctuating ambient oxygen concentrations, or stress-related increases in oxygen consumption within the organisms itself. The latter point would provide a mechanistic basis for the stress hypothesis, including a potentially measurable stress indicator (elevated oxygen consumption). Empirical evidence for a link between low or fluctuating oxygen levels and early maturation at a small size is anecdotal. Undoubtedly, early maturation has been observed under conditions associated with low or strongly fluctuating oxygen levels, such as in shallow lagoons or highly productive ponds (see also Lowe-McConnell, this volume). However, there is no evidence that high and stable oxygen saturation alone would cause a substantial delay in maturation (see e.g. Horstgen-Schwark and Langholz, 1998, who maintained oxygen saturation within narrow limits, close to 100%). Also, there is evidence for suppressed or delayed reproduction in some populations maintained under a lower degree of saturation. In Silliman's 1975 study, little reproduction was observed in the *O. mossambicus* population over a long period while oxygen saturation barely exceeded 50%. Also, Eyesons's (1983) *S. melanotheron* population showed low reproductive activity while maintained in a shallow outdoor tank, a habitat most likely associated with low and fluctuating oxygen saturation.

Environmental factors or 'cues' affecting tilapia reproduction have recently been reviewed by Brummett (1995), who identified temperature, light intensity, quantity and quality of food, and certain water quality parameters as important environmental cues. Food supply in particular may be expected to have a strong influence on reproduction. In experiments with *O. niloticus* × *O. aureus* hybrids, low feeding levels reduced the proportion of mature fish, weight at maturity, and fecundity (Siddiqui *et al.*, 1997). However, fecundity per unit weight of females was highest in the fish fed the lowest ration. Coward and Bromage (1999) show that under prolonged food restriction, *T. zillii* will sacrifice somatic growth to maintain reproductive investment. These observations suggest a complex relationship between food supply and reproductive traits, possibly involving a high degree of individual variability in responses. This area calls for further investigation. Manipulations in food supply have been successful in influencing maturation in other species, e.g. salmonids (Bagenal, 1969; Thorpe *et al.*, 1990).

Intraspecific interactions may have a strong influence on final maturation and reproduction in tilapias. Extreme crowding may completely suppress final maturation and reproduction (Coward *et al.*, 1998). Under moderately crowded conditions, dominance hierarchies and territorial behaviour may restrict reproduction to a small fraction of the population. Van Someren and Whitehead (1959) describe aquarium experiments in which only a limited number of males show breeding coloration at any one time, and if these males are removed others will take their place. Behavioural studies of tilapias and other cichlids show that body size is a key determinant of the outcomes of conflicts about territories (Bearends and Baerends-van Roon, 1950; Oliveira and Almada, 1996), although this may be modulated by other factors such as gonad size (Neat *et al.*, 1998). These observations suggest competition for breeding territories, with small/inferior individuals queuing for territories, as a mechanism influencing maturation size. This hypothesis has been clearly formulated by Moriarty (1982). Deferred maturation as a consequence of territoriality has been described in many bird populations (Sutherland, 1996), and has been analysed theoretically by Ens *et al.* (1995).

The stress and oxygen hypotheses have been formulated to explain the occurrence of small and early maturation, while large and late maturation has been considered 'normal' and therefore not in need of an explanation. The opposite view that early maturation is the norm, but may be delayed if the fish is unable to acquire the resources necessary for successful reproduction, is embodied e.g. in the territorial behaviour hypothesis. While there is no *a priori* reason to prefer one perspective over the other, the latter perspective is supported by the fact that early maturation has been observed under a very wide range of conditions (e.g. high as well as low mortality, low and fluctuating as well as high and fairly stable levels if oxygen saturation). The delay perspective is also supported by studies in other species (e.g. Policansky, 1983; Thorpe, 1994) and has been applied fruitfully in the analysis of life history plasticity (Thorpe *et al.*, 1998). The interpretation of observed reproductive patterns is complicated by the fact that maturation takes several months to complete (Jalabert and Zohar, 1982), which implies that conditions preceding actual maturation by at least a few months are likely to be of key importance.

In conclusion, the mechanisms controlling maturation and reproduction in tilapias are still poorly understood, and there is no clear support for any of the main hypotheses advanced in this respect. The stress hypothesis is poorly defined and therefore not readily testable. A link does exist between mortality and maturation, but this link is more complex than is generally appreciated (see section 6.4). The oxygen hypothesis is consistent with field data in the sense that early maturation has been observed under conditions of low or fluctuating oxygen availability, but this could be purely coincidental and there is no indication that high/stable oxygen availability alone is sufficient to induce late maturation. This hypothesis is, however, eminently testable. Likewise, the degree to which maturation can be controlled by behavioural interactions should be explored further in experiments. Given the ubiquitous occurrence of early maturation under culture conditions, convincing evidence for control of maturation by a particular factor would require a demonstration that the factor in question can substantially delay maturation.

Individual variability and genetics

There is evidence of considerable within-population variation in age and size at maturity. This is obvious from maturity-at-age or at-length curves derived for natural populations (e.g. De Silva, 1986; Duponchelle and Panfili, 1998). In a long-term experiment with *O. niloticus*, 28% of the females matured within 60 days and a further 15% between 61 and 210 days, while 57% of the females had not matured by the end of the experiment at 210 days (Bolivar *et al.*, 1993). The basis of this variation is has not been evaluated; possibilities include genetic, as well as phenotypic factors such as behavioural interactions.

Whereas a high degree of phenotypic plasticity is evident in reproductive traits, this does not imply the absence of a genetic component: several studies have indicated a high degree of heritability (around 0.5) for age and/or size at maturity (Lester *et al.*, 1988; Kronert *et al.*, 1989; Oldorf·*et al.*, 1989). This suggests that, although the observed variation between populations is primarily phenotypic in origin, selection for late maturity may bring about a delay sufficient to allow completion of a culture cycle of 4–6 months without reproduction.

6.3 DYNAMICS AND MANAGEMENT OF POPULATIONS

The section starts with a brief overview of the structural diversity of tilapia populations and their responses to exploitation. This is followed by a review of the dynamics and management of wild and farmed populations, and the dynamics of invasions. Farmed populations are defined here as those subject to management intervention other than harvesting. The degree of intervention defines culture intensity as extensive (stocking and harvesting only, production based on natural aquatic productivity), semi-intensive (natural productivity enhanced by fertilization and some feeding), and intensive (production based mainly on complete feeds). However, it must be remembered that even farmed tilapia populations may show many of the dynamic features of wild populations and that on the other hand, many small-scale tilapia aquaculture systems entail little manipulation other than harvesting of a largely self-sustaining pond population.

The diversity of tilapia populations

Many tilapia species can form breeding populations in habitats ranging from large lakes to small aquaria, and may be exploited and manipulated at wide range of intensities. Different conditions give rise to different patterns of growth, mortality and reproduction, which have been reviewed in the previous section. These patterns translate into populations of widely different age and size structure.

A further aspect of population structure that deserves attention for its dynamic and management implications is spatial structure. Large water bodies usually offer a diversity of habitats that differ in terms of the physical environment (e.g. temperature, turbidity), the resources available (e.g. types and abundance of food, spawning substrate), and the intensity of intra and interspecific interactions (e.g. competition for food, predation risk). Individuals choose habitats on the basis of their requirements and competitive abilities, and it has been shown that in general, the resulting pattern of habitat use maximizes the individual's fitness (see Tyler and Rose, 1994). Different life history stages of fishes differ in their requirements and competitive abilities, and therefore tend to occupy different habitats. In tilapia populations in larger water bodies, juveniles remain in near-shore areas at first and gradually move towards deeper, open waters as they grow (Fryer and Iles, 1972). Large individuals appear to be the most cosmopolitan, occupying open waters as well as the shallow near-shore areas. Breeding is generally limited to particular areas of relatively shallow water. This general pattern of spatial structure seems to hold even in relatively small water bodies, as demonstrated in a small (200 ha) reservoir in Sri Lanka (Pet and Piet, 1993). However, spatial structure necessarily becomes less pronounced in very small water bodies such as ponds that offer little habitat diversity. A similar situation can arise when shallow areas of larger water bodies are cut off from the main water body, and all life stages are forced to share the same habitat (Welcomme, 1970). Breeding for example is no longer confined to particular areas in uniform water bodies, but occurs over the whole bottom area (see e.g. Huet, 1978). Most tilapias are flexible in their habitat requirements and able to complete their life cycle within a single habitat, even if it lacks any of the features normally preferred by different stages such as cover or substrate suitable for building nests. It is clear, however, that confinement to a single, uniform habitat greatly increases the interaction between size/age groups, and reduces the opportunity for individuals to escape limiting conditions by moving to a different habitat. The implications of differences in habitat diversity for the dynamics of tilapia populations have received little explicit attention, but this area may well be of key importance to the understanding of life history variation (section 6.4). Tyler and Rose (1994) review approaches to the modelling of spatial heterogeneity in fish populations. The diversity of tilapia populations makes them ideally suited for an empirical analysis, and the testing of models that may help to unravel the proximate basis of habitat choice and the resulting population dynamics.

Population regulation and responses to exploitation

Population regulation by density-dependent processes is an important factor in the long-term persistence of populations, and the regulatory processes also determine the levels of exploitation to which a population can be subjected on a sustainable basis. It is therefore important to understand the main mechanisms by which tilapia populations are regulated and respond to exploitation.

Several potential regulatory mechanisms have been identified in the previous section. Density-dependence in growth, mediated by competition for resources, is well-established (p. 172). Combined with size dependent mortality (p. 178), density-dependence in growth also generates density dependence in mortality. Competition for breeding territories and suppression of maturation in fish unable to defend territories (p. 184) suggests a mechanism for density-dependent, behavioural control of reproduction. Observations of density-dependent responses in wild populations are limited, not least because environmental variability often obscures density-dependent responses. However, it is worth noting Gwahaba's (1973) observation that increasing exploitation (as suggested by increasing yields) of the *O. niloticus* stock in Lake George was associated with declining length at maturity. A comparative analysis of life history responses to exploitation is given in section 6.4.

Perhaps the most detailed information on responses to exploitation in a tilapia stock has been obtained in a long-term laboratory experiment. Silliman (1975) maintained a population of *O. mossambicus* for more than 6 years, in an 850 l tank under constant feed supply and environmental conditions, and subjected the stock to different levels of harvesting. The pattern of total (natural and fishing) mortality, and the resulting patterns of numerical abundance, biomass, maximum individual length, and recruitment are shown in Fig. 6.8. Natural mortality in the experiment can be assessed as total mortality prior to exploitation, i.e. M = Z = 0.1/year. The population was then subjected to fishing mortality rates of F = 0.5/year and F = 1.1/year in addition to the natural mortality, leading to the total mortality schedule shown in Fig. 6.8(a). Population abundance (Fig. 6.8(b)) was stable at around 240 individuals in the beginning, but dropped to a lower level of about 180 as a result of a catastrophic mortality event linked to equipment failure (indicated by an arrow in Figs 6.8(b) and (c)) at the end of year 2. The onset of exploitation at F = 0.5/year in year 3 did not lead to an immediate change in abundance, but there was a recruitment-related increase later in the year followed by continuous decline to about 120 at the beginning of year 6. The higher level of exploitation at F = 1.1/year from the beginning of year 6 was again linked to initial increases in abundance as a result of recruitment, followed by a decline to the lowest level of 100 at the end of the experiment. Population biomass (Fig. 6.8(c)) increased sharply as result of individual growth until the end of year 2 and remained at a stable level of about 10 kg until the middle of year 4, but declined subsequently under the influence of exploitation. Maximum individual length (Fig. 6.8(d)) in the catch was stable

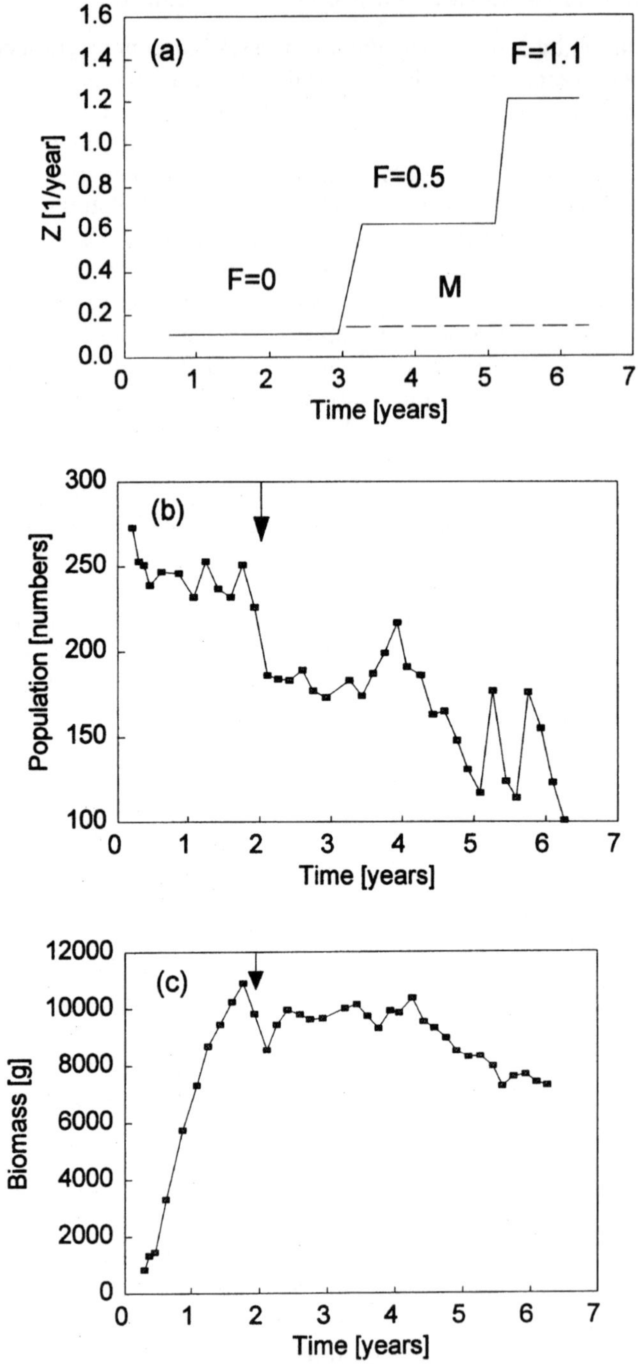
(a)
F=1.1
F=0.5
F=0
M
Z [1/year]
Time [years]
(b)
Population [numbers]
Time [years]
(c)
Biomass [g]
Time [years]

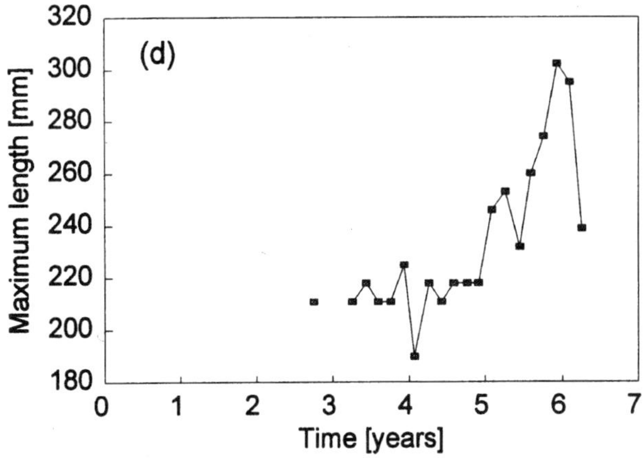

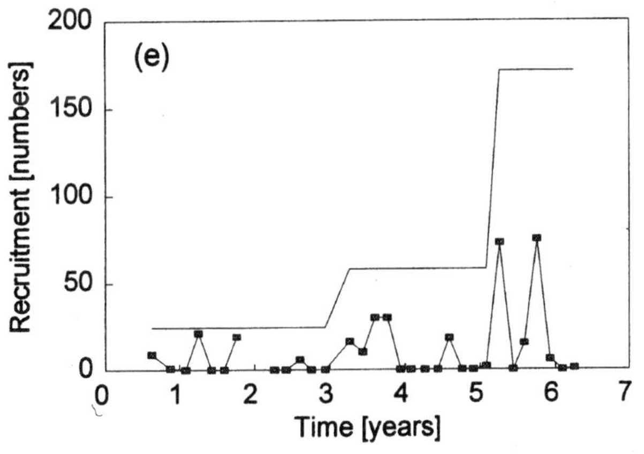

Fig. 6.8 (opposite and above) Dynamics of an experimentally exploited population of *O. mossambicus*, maintained for over six years in a 850 l tank at constant feed supply (Silliman, 1975). (a) Total mortality rate; (b) population; (c) biomass; (d) maximum length in catch; and (e) recruitment (squares: actual recruits per sampling period, line: cumulative recruits per year). Arrows indicate a catastrophic mortality event due to equipment failure.

at about 220 mm until the end of year 4, but increased markedly thereafter as population abundance and biomass declined. Recruitment (Fig. 6.8(e)) increased markedly with increasing levels of exploitation and the associated reduction in abundance and biomass. The evidently high degree of variability in recruitment per sampling period is a consequence of the demographic stochasticity inherent in a very small population, but a strong and consistent increase in the level of recruitment with increasing exploitation becomes evident when recruitment is aggregated over the experimental phases. The experimental population was tightly regulated, there is clear evidence for density dependence in the level of recruitment, as well as in the somatic growth of adults.

Unfortunately, Silliman (1975) reported no detailed observations on reproductive traits in the experimental population, so that it is impossible to ascertain the mechanisms involved in the regulation of recruitment. However, Eyeson (1983) studied a similarly confined tank population of *S. melanotheron* and found reproductive activity to be suppressed to a very low level. Coward *et al.* (1998) show that crowding can lead to complete suppression of spawning, associated with low levels of sex steroids, in *T. zillii*.

Fishery yields and ecosystem properties

Environmental conditions such as ecosystem productivity define fundamental limits to the productivity of populations and the yields obtainable from their exploitation. This is evident in the strong correlations found between indicators of productivity/trophic levels and average tilapia yields in different water bodies. In Sri Lankan reservoirs, for example, commercial fish yields (dominated by the introduced *O. mossambicus*) are strongly correlated with the concentration of chlorophyll a, an indicator of primary production (Fig. 6.9; from Moreau and De Silva, 1991). The tilapias generally occupy low trophic levels in the food webs of tropical aquatic ecosystems, and are correspondingly efficient at transforming primary production into fish yield. Turner (1995) compared transfer efficiencies of primary production into fish yield in twenty African lakes, and found that the tilapia-dominated lakes in the set had significantly higher transfer efficiencies than lakes with fisheries dominated by the predatory *Lates*. More detailed trophic ecosystem models have been constructed for some ecosystems with significant tilapia stocks, e.g. Lake George (Moreau *et al.*, 1993) and Lake Kinneret (Walline *et al.*, 1993). Trophic ecosystem models provide a framework to synthesize ecological data and evaluate the relationship between ecosystem properties and the yield and biomass of populations (Christensen and Pauly, 1993).

Relationships between average yields and ecosystem properties can provide a deceptively static picture of populations that are in fact highly dynamic. Yields from many tilapia fisheries have been highly variable in the medium term, due to environmental factors as well as exploitation. Examples of yield variation in tilapia fisheries from three African lakes are shown in Fig. 6.10. In Lake Naivasha, a high altitude Rift Valley lake, *O. leucosticus* and *T. zillii* have been introduced in the 1960. Following an initial surge with a peak yield of over 1000 t in the late 1960s, annual yield of the combined tilapia fishery has varied between 20 and 530 t from 1970 to 1996. Yield variation in the established fishery has been related primarily to fluctuations in the water level, the extent of

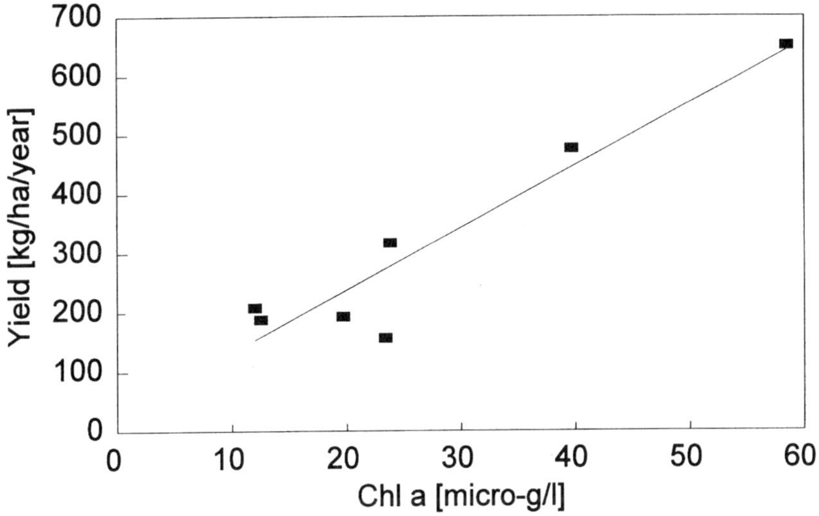

Fig. 6.9 Relationship between the trophic status indicator chlorophyll a and fish yield in *O. mossambicus*-dominated fisheries in Sri Lankan reservoirs (Moreau and De Silva, 1991).

macrophyte vegetation, and high but variable fishing effort in the 1980s (Muchiri and Hickley, 1991; Muchiri, 1997). Lake Malombe is a large, shallow inundation of the Shire River in southern Malawi. Yield from the indigenous 'Chambo' (*Oreochromis* sp.) fishery was high and only moderately variable between 1976 and 1986, but declined drastically thereafter. The rapid decline in the Chambo fishery has been attributed to the development of a seine net fishery for small haplochromines, with a large bycatch of juvenile *Oreochromis* sp. (FAO, 1993). Lake Turkana is a large rift valley lake that supports a productive fishery based primarily on *O. niloticus*. The variation in *O. niloticus* yield has been linked primarily to change in water levels, although changes in exploitation may also play a role (Kolding, 1993; Muchiri, 1997). These examples illustrate the high degree of variation experienced in tilapia fisheries even in large water bodies.

Quantitative assessment of exploitation

The aim of quantitative stock assessment is to evaluate effects of alternative exploitation regimes on variables of management interest such as yield or catch rates. The quantitative approaches that have been used in the assessment of tilapia fisheries fall into three broad categories: empirical yield models, production models, and dynamic pool models.

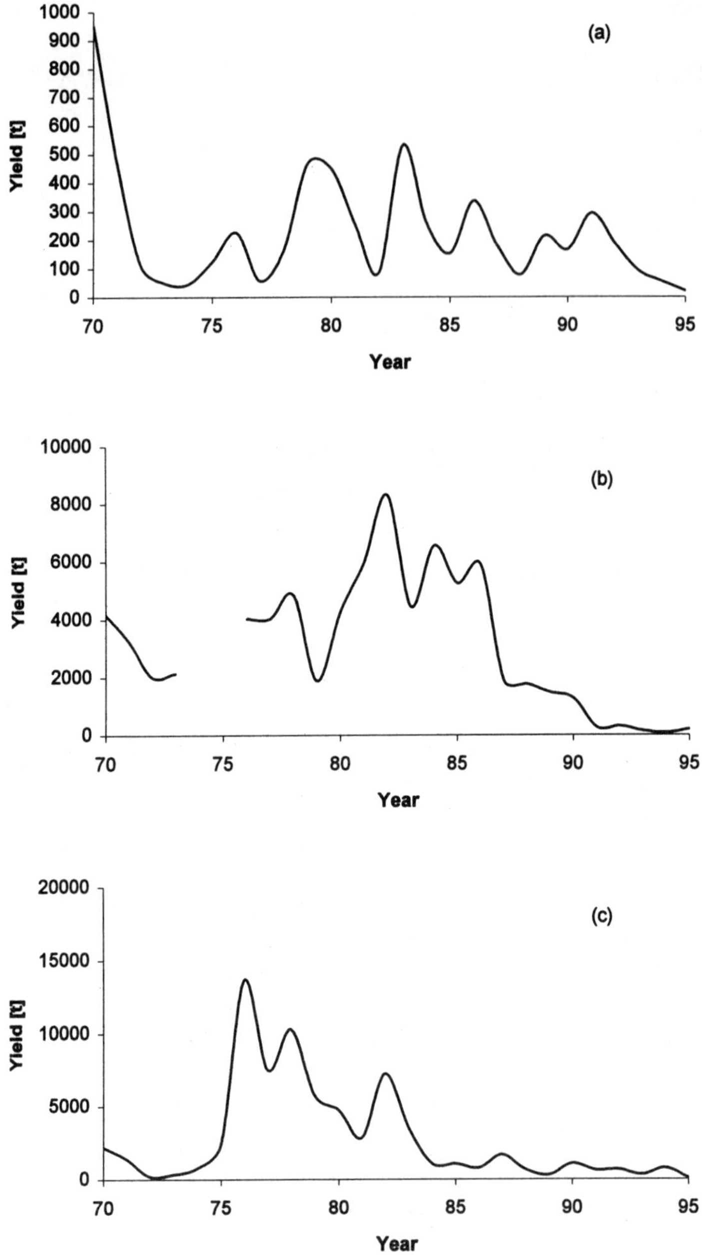

Fig. 6.10 Variation in Yield in three tilapia fisheries: (a) Lake Naivasha (*O. leucostictus* and *T. zillii*), (b) Lake Malombe (*Oreochromis* sp., 'Chambo'), and (c) Lake Turkana (mainly *O. niloticus*). Data from Muchiri (1997).

Empirical yield models

Empirical yield models are single or multiple regression models relating yield to, for example, indicators of productivity (e.g. the morphoedaphic index, chlorophyll a concentration, primary production), fishing effort, and/or water level. While no empirical models have been developed explicitly for tilapia fisheries, a number of such models have been developed for fisheries in which tilapias account for a large proportion of the catch. Empirical models for Sri Lankan reservoirs, where commercial catches are dominated by introduced *O. mossambicus*, predict yields from the ratio of catchment to reservoir area, chlorophyll a concentration, and fishing effort (Moreau and De Silva, 1991). Empirical yield models make no assumptions regarding the mechanisms underlying patterns in yields, and their predictions should not be extrapolated beyond the range for which data are available. Empirical models are most useful in situations where data are available for a wide range of conditions, such as for example yield, effort and trophic status indicators for a number of different water bodies. This makes empirical models ideal tools for the assessment of small water body fisheries with their high potential for replication. However, there are problems and potential pitfalls in the assessment of exploitation regimes using empirical yield models. Often, for example, water body productivity and fishing effort are correlated (the more productive water bodies attract more fishers). This can give rise to an apparently linear relationship between yield and effort, which however does not imply that the fisheries are underexploited but merely reflects underlying differences in water body productivity.

Production models

Production models are based on elementary ideas of production ecology, and are commonly used to estimate optimal effort levels from time series of catch and effort data (overview in Hilborn and Walters, 1992; Hilborn and Mangel, 1997). Such data are relatively easy to collect at least in principle, and production models therefore hold good potential for the assessment of fisheries for which more detailed information is lacking. However, the measurement of fishing effort can be a major problem where many different gear types are used and fishing methods change over time. Production models have been used to assess the Chambo (*Oreochromis* sp.) fishery of Lake Malawi (Alimoso, 1991), and the *O. leucosticus* dominated gill net fishery of Lake Naivasha (Muchiri and Hickley, 1991). In both cases the traditional, equilibrium method of fitting the production model has been used. The method assumes that each annual yield estimate represents the equilibrium yield corresponding to the fishing effort exerted in that year. This assumption is clearly questionable in fisheries undergoing rapid change, and its violation can lead to severe overestimation of the sustainable level of fishing effort (see e.g. Hilborn and Walters, 1992). It is therefore strongly recommended to fit production models using dynamic (non-equilibrium) approaches. This may appear more mathematically demanding than the equilibrium approach at first, but can in fact be accomplished relatively easily in computer spreadsheets (see Hilborn and Mangel, 1997). The dynamic approach is also true to the ecological production concept underlying the formulation of production models, whereas equilibrium fitting is not principally different from the use of empirical yield (regression) models.

Dynamic pool models

Dynamic pool approaches to stock assessment break down abundance and biomass dynamics into the component processes of growth, mortality and reproduction, and allow the assessment of management options formulated at a high level of technical detail (e.g. mesh size regulations). A problem with the application of dynamic pool approaches to tilapias and other fish in the tropics species is that these fish are difficult to age, making it difficult to analyse age-dependent aspects such as growth or mortality. This has prompted the development of a variety of approaches to estimating growth and mortality parameters from length composition data (Pauly and Morgan, 1987; Rosenberg and Beddington, 1988). Most of the commonly used methods establish a growth function (usually the VBGF, see p. 165) to relate length to age, and then proceed to conduct further analysis in an essentially age-based framework. Growth studies of tilapias using length composition data include those of Getabu (1987), Amarasinghe and De Silva (1992), and Blay and Asabere-Ameyaw (1993).

Most tilapia fisheries are exploited predominantly by gill nets. The size-selectivity of gill nets implies that, unless a very wide range of mesh sizes is used, the length structure of catches does not reflect that of the population. This makes gill net catch data difficult to interpret, as illustrated by Amarasinghe and De Silva (1992): their use of catch length distributions obtained from relatively unselective seine nets resulted in a drastic re-assessment of growth patterns determined in an earlier study using gill nets only (De Silva and Senaratne, 1988). The problem of gill net selectivity can be addressed by using a wide range of mesh sizes, and correcting catch length distributions for gill net selectivity. Gill net selectivity has been estimated for several tilapia species, e.g. *O. esculentus* (Garrod, 1961); *O. mossambicus* (De Silva and Sirisena, 1987; Pet *et al.*, 1995); *O. niloticus* (Pet *et al.*, 1995); and *O. shiranus* (Mattson, 1994). The selectivity models developed in these studies may be used to calculate approximate selectivity curves for tilapia fisheries where no specific experiments have yet been carried out, and may indicate to what extent fishery samples are representative of the population.

Conventional dynamic pool models are based on the assumption that fish populations are regulated at the pre-recruit stage, while growth and mortality in the adult population are considered to be constant (density-independent). This implies that the optimal exploitation of the of the growth potential of a recruited cohort is independent of the level of recruitment, and can be assessed in a yield-per-recruit analysis even if the level of recruitment is not known. Yield per recruit analysis identifies optimal combinations of gear selectivity and fishing effort to exploit stocks of known growth and natural mortality parameters. Such analyses have been used for example in Sri Lanka to determine the optimal size of first capture in the commercial fisheries for *O. mossambicus* (Amarasinghe and De Silva, 1992). As mentioned earlier (and again in section 6.4), there is strong evidence for density-dependence in individual growth in tilapias, and in principle this violates a key assumption of conventional yield-per-recruit analyses. The implication of density-dependent growth is that, due to its compensatory nature, changes in exploitation will lead to smaller changes in yield than predicted from analyses assuming constant growth.

The second element of dynamic pool models, apart from yield-per-recruit analysis, is the analysis of the relationship between spawning stock and recruitment. In general, such studies are based on analyses of catch-at-age data to reconstruct past cohort size and thereby recruitment as well as stock biomass. Unfortunately, both recruitment and stock size are difficult to measure in tilapias and other tropical species that reproduce over an extended period and are difficult to age. Consequently, there are no tilapia stocks for which a stock recruitment relationship has been determined. The tank experiment using *O. mossambicus* reviewed earlier (p. 187) shows that recruitment in tilapias can be very strongly density-dependent, with very low recruitment when stock levels are high. This suggests a Ricker-type stock-recruitment relationship with a clear maximum at intermediate stock sizes.

The spatial structure of tilapia populations can have important implications for management and may need to be considered explicitly in certain situations. The tendency of young *O. mossambicus* to stay in near-shore areas while larger size groups venture into midwater implies that small mesh gill nets can be used in midwater to catch small species without affecting the tilapia stock. This was first realized and investigated by De Silva and Sirisena (1987). A dynamic pool model for a Sri Lankan reservoir fishery with explicit consideration of size-dependent habitat occupation has been developed by Pet *et al.* (1996b). The model confirmed that a subsidiary gill net fishery for the small cyprinids could be established in open waters if small mesh gill nets of about 15 mm were used, while larger mesh nets of 30 mm would cause major damage to the *O. mossambicus* fishery.

Accounting for environmental variation and life history plasticity

The various conventional methods of fish stock assessment applied to tilapia stocks have clearly provided useful information for management, and in some cases this information has been used to identify management measures such as mesh size limits (e.g. in Sri Lankan reservoirs). However, certain features of tilapia fisheries, such as large environmental variations and a high degree of plasticity in maturation and growth pose limitations to the use of conventional models. In spite of the wide recognition of these factors in the literature, little attempt has been made at incorporating them into quantitative assessments. This is a key area for future research in stock assessment, of tilapias as well as other species.

Culture-enhanced fisheries

Culture-enhanced fisheries, i.e. fisheries based at least in part on the regular stocking of farm-produced (or occasionally wild caught) seed fish, are at the interface between capture fisheries and aquaculture (Lorenzen, 1995; see also Welcomme and Bartley, 1998, for a review of enhancements). Culture-enhanced fisheries should be distinguished from other forms of management that may also involve stocking, for example for introductions with the aim of establishing a self-sustaining population (see Cowx, 1999, for a review of stocking strategies). Tilapia culture-enhanced fisheries are widespread in the tropics (Sugunan, 1997), although their precise extent is difficult to determine due to their position between the statistical reporting categories. Also, due to the tendency of tilapias to form breeding populations within a few months of stocking, the distinction between culture-enhanced fisheries and introductions is somewhat difficult. The direct contribution of stocking to yields is most obvious in seasonally dry water bodies, but is difficult to assess, and may possibly be negligible in permanent water bodies where breeding populations of tilapias exist. Yields from tilapia culture-enhanced fisheries can be up to 3 t/ha/year in productive water bodies (Lorenzen *et al.*, 1998b).

Although culture-enhanced fisheries may generate benefits even in the absence of any systematic management, benefits are maximized if stocking and harvesting regimes are carefully co-ordinated and adapted to the natural conditions of the water body. The most comprehensive analysis of management options is possible with population dynamics models, while empirical yield models often allow a rapid assessment of key variables using comparative data from sets if culture fisheries. The population dynamics of fisheries based entirely on stocking, with density-dependent growth and size-dependent mortality of the type demonstrated for tilapias in section 6.2, has been studied theoretically by Lorenzen (1995), and an example of a practical assessment using the methodology is given in Lorenzen *et al.* (1997). However, the tilapias' tendency to reproduce in almost any environment means that, in contrast e.g. to the major carps, pure culture-based fisheries for tilapias are rare and limited to the stocking of monosex fry (see below). There is as yet no model to adequately capture the dynamics of the more common culture-enhanced tilapia fisheries where natural reproduction as well as stocking must be considered. Empirical models of culture fisheries involving tilapias have been derived by Lorenzen *et al.*, (1998b) for village fisheries in NE Thailand; and Quiros (1998) for larger reservoirs in Latin America. Lorenzen *et al.* (1998b) show that if stocked in combination with major carp species, *O. niloticus* develop dominant breeding populations in highly productive water bodies, while the stocked carp will dominate water bodies of low to intermediate productivity. Although tilapia-dominated culture-enhanced fisheries therefore appear to be more productive than carp-dominated ones, this may reflect the underlying productivity of the water body more than an inherently higher trophic efficiency of tilapias.

Van der Mheen (1994) discusses practical aspects of stocking small water bodies with wild caught, mixed tilapia seed and also derives some empirical models to predict catches from stocking densities. Stocking of wild caught tilapia seed has been carried out on a large scale and with formal government support in e.g. in Zimbabwe, but such transfers are also common in rural areas of Asia, often on an entirely informal basis (Lorenzen, personal observation).

Culture-enhanced fisheries constitute an investment into fish stocks, often exploited as common pool resources. Such investment may precipitate changes in the access regime to, and exploitation of the fishery. In small water bodies in Laos for example, tilapia stocking appeared to facilitate the development of community management systems with active effort regulation. A comparative study by Lorenzen *et al.* (1998a) indicates that this management regime resulted in substantial standing stocks and yields of tilapias (based on a combination of regular stocking and natural recruitment), and also had a positive effect on the abundance of indigeneous stocks. The latter impact was unexpected and is rooted in changes to the exploitation regime precipitated by culture fisheries development. This illustrates that there is more to culture fisheries than stocking, and that prediction of outcomes of culture fisheries development requires an approach integrating the analysis of biological and institutional dynamics with a strong element of adaptive learning (Lorenzen and Garaway, 1998).

Dynamics and management of isolated cohorts

In commercial tilapia culture, reproduction is controlled as far as possible, and the different stages of the rearing cycle are separated in hatcheries, nurseries and grow-out (as is common in the culture of most non-tilapiine species). The isolated cohort is therefore the dominant population structure in commercial growout, and the management of such populations deserves explicit attention.

Annual yields from isolated cohorts can be in the same order as those from culture fisheries in extensive systems without fertilization or feeding, but can be far higher in semi-intensive and intensive systems (see the respective chapters in this book). Empirical analyses of yield in relation to management inputs using multivariate statistical methods or multiple regression (Draper and Smith, 1981; Manly, 1994) can provide useful information for management without the need to obtain detailed data on population processes such as growth or mortality. In an analysis of commercial polyculture systems in Israel, Milstein (1995) shows that the highest yields are achieved in systems where tilapia (*O. niloticus* × *O. aureus* hybrids) is the main species, and also evaluated the relationship between input levels and yields.

Cohort growth and mortality models allow a more detailed evaluation of management options than empirical yield models, and are increasingly being used in bio-economic analyses of fish farms (for an introduction see Bjorndal, 1990). Springborn *et al.* (1992) used a population model based on the VBGF and an exponential mortality equation to predict the economically optimal harvesting time for *O. niloticus* cohorts in pond culture under two different fertilization regimes. The two regimes analysed gave rise to very different growth patterns, with consequent large differences in the optimal harvesting time of 105 vs. 191 days. The analysis did not, however, account for the rotation problem: in order to maximize returns per unit time it is usually optimal to harvest before the time that maximizes the production or profit from the individual cohort, because the cohort can be replaced with faster growing young fish (see e.g. Bjorndal, 1990).

A constraint on the use of simple models such as the conventional VBGF and constant mortality is that the parameters for these models must be estimated separately from experiments for each different level of management inputs. It is therefore desirable to develop growth and possibly mortality models that incorporate explicit terms for management inputs. For example, the density-dependent growth model and the mortality-weight relationship discussed in section 6.2 can be used to evaluate stocking density and size as well as harvesting regimes in extensive culture (Lorenzen, 1995; Lorenzen *et al.*, 1997). The multiple regression models of fish growth described by Pauly *et al.* (1993) and Prein and Pauly (1993) allow analysis of a wide range of management impacts on growth, and may aid the improvement of semi-intensive and intensive husbandry systems. Prein (1993a,b) describe the application of these models to the analysis of tilapia growth in pond culture. For integrated farms in the Philippines, for example, the analysis showed positive effects of solar radiation, pond area, and manure loading rate, and negative effects of stocking density and early morning oxygen saturation on growth. Oxygen saturation was negatively correlated with pond productivity, and overall the positive effect of productivity overrides the negative effect of low morning oxygen saturation. This example illustrates the potential to derive useful generalizations from the analyses of experimental and commercial data using growth models that can account for management impacts. However, there are also potential pitfalls associated with the statistical design of experiments that have not been rigorously planned to provide data adequate for the purpose of model development (see e.g. Riley and Darmi, 1995).

The potential of population and related modelling approaches to aid planning, and substantially reduce the experimental effort required to improve culture systems, has not been fully appreciated among aquaculture practitioners. The further development of modelling approaches and a more rigorous application of statistical principles in experimental design and analysis (Riley and Darmi,

1995; Knud-Hansen, 1997) are likely to yield increasingly more useful and reliable generalizations. This will require an increasingly close collaboration between practitioners, experimentalists and modellers.

Controlling reproduction and 'stunting' in culture

Although most commercial culture systems are designed to raise single, isolated cohorts, the tendency of tilapias to breed within a few months of stocking means that this ideal is not easily achieved. Reproduction can quickly turn fast growing single cohorts into dense populations dominated by small recruits, with consequent decline in the growth of stocked fish. The absolute production of such 'stunted' pond populations may be extremely high, but there is often a lower limit to the size at which tilapias can be marketed profitably, and therefore 'stunting' is considered a problem in commercial farms. However, uncontrolled reproduction in culture is not universally a problem: recent research shows that subsistence-oriented fish farmers may value reproduction as positive, and that markets for 'undersized' tilapias exist (Little and Hulata, this volume; Cheftel and Lorenzen, 1999).

Approaches to the problems of reproduction and 'stunting'

Four different approaches to the problems arising from early maturation in cultured tilapias can be distinguished: (1) complete control of reproduction through culture of mono-sex or sterile populations; (2) stock or environmental manipulations aimed at delaying maturation; (3) the use of physical means or predators to minimize survival of offspring; or (4) increased feeding to minimize the impact of recruits on the growth of the parent stock. Approach 1 is perhaps the most widely used in commercial culture of tilapias, and methods for the production of mono-sex stocks are discussed elsewhere in this book. Attempts at delaying maturation by means of stock or environmental manipulations (approach 2) have not so far been very successful, and no such manipulations are routinely used in culture. Approach 3 is associated with two main methods: cage culture and predator stocking. Cage culture does not prevent breeding altogether, but the eggs are lost from the cage at the time spawning (Pagan-Font, 1975). Predator stocking is discussed in more detail below. Approach 4, increasing food supply to offset consumption by new recruits, is likely to be a widespread response by farmers, and has been shown to be potentially effective (De Graaf *et al.*, 1996). Although it has been argued that increased feeding exacerbates the underlying problem of early maturation and prolific breeding (e.g. Noakes and Balon, 1982), experimental evidence reviewed in pp. 182–184 suggests that tilapias tend to respond to resource scarcity with increased relative allocation to reproduction, and that therefore reproductive traits are not very sensitive to manipulations of feed supply within the constraints set by practical fish culture. The economics of feeding decisions in this context are, however, intriguing, and worthy of further analysis.

Predator–prey culture

A complex approach of particular interest in the context of population dynamics is the stocking of predators to control recruitment. This approach probably originated with the experiments of Swingle (1960), who extended his previous work on the creation of predator–prey systems in American farm ponds to tilapias. Predator stocking has since been practised and reported widely throughout the world (Dunseth and Bayne, 1978; McGinty, 1984, 1985; Bedavi, 1985; Ofori, 1988; El Gamal, 1992; Fisher and Grant, 1994; Lazard and Oswald, 1995; DeGraaf *et al.*, 1996). Predator stocking undoubtedly works well in principle, and the approach has often succeeded in producing market-size fish. However, careful selection of the species, size and number of predators stocked is essential to produce the desired results. If predators are too small, too few or feed on a wide spectrum of fish and invertebrate prey, recruitment control may be ineffective and stunting may result despite the presence of predators (Lazard and Oswald, 1995). Conversely, if predators are too large and numerous, they may decimate the stocked fish and cause loss of production. Predator stocking is thus complex, given that both the stocked tilapia and the predator grow in the course of the culture cycle and adequate size ratios and predation pressures must be maintained throughout the cycle. Some predators have themselves a tendency to reproduce and thus create a mixed-size predator population (e.g. Dunseth and Bayne, 1978). Even once these complexities have been mastered, the system remains sensitive to variations in the growth of either species, which may often be impossible to avoid. Overall this has given the predator approach to recruitment control the reputation to be somewhat unreliable, if effective in principle (Huet, 1978). The development of population models for predator–prey systems in pond culture has great potential to advance the understanding and aid the management of such systems, as illustrated by Fisher and Grant's (1994) model of a Nile tilapia (*O. niloticus*)-tucunare (*Cichla ocellaris*) system.

Avian predators are often considered a nuisance if not a major problem by fish culturists. In the case of tilapia farming, these predators may contribute to limiting unwanted recruitment and therefore have a beneficial effect. This was suggested in a study by Ashkenazi and Yomtov (1996), who argue that stocked fingerlings in the size range susceptible to heron predation are available in Israeli fish farms for only short period (6 weeks), and that the largest part of heron predation occurs on unwanted recruits.

Tilapias as invaders

Originating from Africa, tilapias are now found in the tropical and sub-tropical regions of all continents. Two species, *O. mossambicus* and *O. niloticus*, in particular must be regarded as some of the most successful invaders among the fishes (Welcomme, 1988; FAO, 1998). The following, brief review of tilapias as invaders is structured according to the conceptual framework of Williamson

(1996), recognizing the attributes of arrival and establishment, spread, and equilibrium and effects.

Arrival and establishment

Virtually all recent arrivals of tilapias in new habitats can be traced back to human importations, mostly for the purpose of fish culture but also for the enhancement of capture fisheries (Welcomme, 1988). Studies of insect, bird and mammalian introductions have shown the probability of establishment to be strongly related to the number of arrivals of that species in a habitat (Beirne, 1975; Newsome and Noble, 1986; Griffith *et al.*, 1989). Williamson (1996) postulates as a general rule that propagule pressure, the number of arrivals, has a strong influence on establishment success. Most tilapia introductions have been conducted on a large scale either through the establishment of breeding centres distributing fry to a large number of farms, or through extensive movement of live fry between countries. This implies widespread and repeated arrivals of tilapias in the geographic range where they have been introduced, and may account at least in part for the apparently high degree of success in establishment. There are also, however, cases where successful establishment has occurred following accidental releases, for example in Australia where the escape of aquarium fish is though to be responsible for all invasions by *O. mossambicus* (Arthington and Bluhdorn, 1994).

A number of species characteristics have been suggested as possible predictors of invasion success and will be reviewed briefly in the context of tilapias: high rate of population growth, wide distribution in native habitat, and the existence of a 'vacant niche' (Williamson, 1996). Tilapias are capable of achieving very high rates of population growth, due to the reproductive traits discussed in section 6.2. For populations to persist, equilibrium rates of population growth must equal mortality rates, and both are related to body size (Beverton and Holt, 1959; Fenchel, 1974; Pauly, 1980; Lorenzen, 1996b). Analysing a large set of data on individual introductions, Pullin *et al.* (1997) showed that establishment success in fish was indeed inversely related to their maximum length. About 80% of introductions of fish with a maximum length of 10 cm lead to establishment, compared to only 40% in fish of 110 cm. Tilapias occupy an intermediate place in the maximum length range. The more commonly introduced tilapia species show establishment success rates between 60 and 95%, with the most successful being *O. mossambicus* at about 95%.

The most widely introduced tilpias, *O. mossambicus* and *O. niloticus*, are widely distributed in their native habitats (see e.g. Philippart and Ruwet, 1982), and thus conform to Williamson's (1996) second above mentioned criterion for successful invaders.

Some tilapia introductions, particularly for fisheries enhancement, have been conducted to fill perceived 'vacant niches' in aquatic ecosystems (Fernando and Holcik, 1991; Coates, 1993). The concept of the 'vacant niche' itself has been the subject of controversy. Herbold and Moyle (1986) have argued that the concept was fundamentally flawed, because the niche is a property of the organism, not the ecosystem and a niche cannot therefore be vacant. Instead, they argued that introductions invariably compress the realized niches of existing species. While this is true in principle, many ecologists find the 'vacant niche' concept useful and apply a slightly different, if possibly less rigorous definition. Williamson (1996) summarizes this view as 'a species occupying a vacant niche is one that plays an entirely new functional role in a community, not one that does not use any resources previously used by other species'. In fisheries ecology, a similarly broad but more explicitly utilitarian definition has been used. Fernando and Holcik (1991) and Coates (1993) identify 'vacant niches' in Sri Lankan reservoirs and the Sepik river system (Papua New Guinea) using two criteria: a comparatively low fish yield, and a relatively low diversity of indigenous species due to the absence of truly lacustrine species in Sri Lanka, and the generally low number of freshwater species in the geologically young Sepik-Ramu basin. Introductions of tilapias in these systems where 'vacant niches' had been identified have indeed led to successful establishment.

Tilapias have invaded highly diverse fish communities such as that of Lake Victoria (Twongo, 1995), as well as low-diversity communities in river systems of New Guinea and Australia (Coates, 1993; Arthington and Bluhdorn, 1994). This seems to suggest that the invasibility of communities by tilapias is not strongly related to diversity, although no quantitative assessment has been conducted. The question whether communities differ in their invasibility has generated much ecological debate, but the overall conclusion appears to be that there is no relation (Williamson, 1996), and this is consistent with observations on tilapias.

Spread

The spread of tilapias within a habitat has rarely been monitored, and the patterns of natural spread may be difficult to disentangle from those linked to repeated arrivals. An exception is the spread of *O. mossambicus* through the Gascoyne River system of Western Australia: the species is believed to have been released accidentally near the mouth of the river, and has colonized the approximately 800 km long river system within 10 years (Arthington and Bluhdorn, 1994).

Equilibrium and effects

The 'equilibrium' state and effects of fish introductions must be evaluated in terms of both, the fishery and the wider ecosystem. Many tilapia introductions (particularly of *O. mossambicus* and *O. niloticus*) have led to the development of commercial fisheries (De Silva, 1988; De Silva and Senaratne, 1988; Fernando and Holcik, 1991; Sreenivasan, 1991; Coates, 1993; Muchiri *et al.*, 1995;

Twongo, 1995). Increases in commercial yields and the rapid expansion of some commercial fisheries following tilapia introductions (e.g. De Silva, 1988) may be regarded as indicators of the economic success of introductions. However, increases in commercial fish yields may greatly overestimate ecological production benefits from introductions: in a Sri Lankan reservoir for example, the introduced tilapias accounted for over 80% of the commercial yield, but barely 10% of the ecological fish production (Pet *et al.*, 1996a). Owing to its tendency to form dense, stunted populations *O. mossambicus* is regarded as a pest in many of the countries were it was intentionally introduced (Welcomme, 1988), but this view is by no means universal (De Silva and Senaratne, 1988).

The major ecological impacts of fish species introductions may involve genetic effects such as hybridization with indigenous species or between exotic species and strains, habitat alterations, competition, predation and the introduction of exotic parasites and pathogens (Courtenay and Stauffer, 1984; Moyle *et al.*, 1986; Arthington, 1991; Carvalho and Hauser, 1995). Hybridization may occur between closely related species, and is therefore most likely in habitats that already contain endemic or previously introduced tilapias. Ogutu-Ohwayo and Hecky (1991) suggest that hybridization with introduced species was at least partly responsible for the disappearances of *O. spilurus* from Lake Naivasha, and *O. esculentus* and *O. variabilis* from Lake Victoria. However, in the absence of genetic investigations, the relative importance of hybridization and competition is difficult to ascertain. Hybridization is also common between *O. mossambicus* and *O. niloticus* where both species have been introduced (Amarasinghe and De Silva, 1996).

Habitat alterations and competition are key mechanisms by which some introduced tilapias have compressed the realized niches of existing species. Habitat alterations by tilapia have been reported primarily for the macrophytophagous *O. aureus*, *T. rendalli* and *T. zillii*, which have sometimes proved highly destructive to established vegetation (Philippart and Ruwet, 1982). Competition for spawning habitat has been reported between introduced *T. zillii* and the endemic *O. variabilis* in Lake Victoria, to the detriment of the latter (Welcomme, 1988).

A survey of perceived impacts of *O. niloticus* introductions in several Asian countries suggests no overly destructive effects (Pullin *et al.*, 1997). However, there is surprisingly little reliable information on ecological impacts of tilapia introductions other than those in the African lakes. Ecological impacts of introductions are often difficult to assess due to a lack of pre-introduction data and the effects of confounding factors. This problem can be overcome through spatially replicated studies, particularly in small and medium sized water bodies where sufficient replicates and controls can usually be identified. This approach was used by Lorenzen *et al.* (1998a) to assess the effects of tilapia stocking and fishing restrictions on small water body fisheries in Laos. The combination of *O. niloticus* stocking and fishing restrictions significantly increased total standing stocks, while *O. niloticus* stocking in itself had no impact on the standing stocks of wild fish. Possible effects of *O. niloticus* stocking on wild fish diversity could not be assessed conclusively, due to a high degree of variation between replicates. Further studies with larger numbers of replicates are urgently required to provide solid baseline data on the ecological effects of the most widely introduced tilapias.

It is now widely accepted that proposals for fish introductions should be evaluated very carefully, particularly where new introductions into a watershed are concerned (it may be rather impractical, at least in developing countries, to evaluate every single transfer within a watershed!). Coates (1998) reviews experiences with the use of codes of practice for introduction proposals. An example of the application of the EIFAC/ICES code of practice for introductions (Turner, 1988) is the evaluation of introductions to the Sepik River basin in Papua New Guinea (Coates, 1993, 1995).

6.4 LIFE HISTORY VARIATION AND 'STUNTING'

The tilapias display considerable variation in life-history traits between species, between populations of the same species, and between individuals within populations. Variation in individual traits has been explored in section 6.2. The present section focuses on relationships between life history traits, illustrating at first the existence of some broad patterns of symmetry in tilapia life histories, before considering the nature of 'stunting', and finally outlining approaches towards a quantitative understanding and prediction of life history variation.

Tilapia life history variation

Relationships between life history traits arise because of fitness trade-offs. In the fishes, such relationships have been analysed extensively using both, comparative empirical approaches and theoretical optimality models (e.g. Beverton and Holt, 1959; Roff, 1992; Beverton, 1992; Charnov, 1993; Mangel, 1996). The aims of these studies were both theoretical and practical: to understand the evolution of life histories, and to establish relationships that can be used to estimate population parameters that are difficult to measure (e.g. the natural mortality rate M) from more easily measured parameters (e.g. asymptotic length) (e.g. Pauly, 1980).

Most empirical studies have been conducted on populations close to their natural, undisturbed (unexploited) state in order to relate life history patterns to the conditions under which they have evolved through natural selection (Beverton, 1992). The empirical analysis of life history patterns in populations that have undergone recent changes in mortality or other environmental factors is complicated by the fact that observed patterns reflect immediate phenotypic responses within the genetic constraints shaped over evolutionary time scales. Such is the case in the following exploration of tilapia life histories, because most data are from populations that are heavily exploited and/or have only recently been introduced into new habitats. Moreover, information on mortality rates is limited to total mortality.

Some relationships between life history parameters among the tilapia populations included in Table 6.1 are explored in Fig. 6.11. Note logarithmic scales on both axes. Power relationships of the form $\log Y = \beta_0 + \beta_1 \log X$ have been estimated using non-parametric (Theil) regression; relationships with exponents β_1 of 1 or –1 imply proportional or inversely proportional relationships between the untransformed variables. There is a significant and approximately linear (β_1 not significantly different from 1) relationship between L_m and L_∞ (Fig. 6.11(a)). Although there is an apparent tendency for K to decline with increasing L_∞, the relationship is not significant (Fig. 6.11(b)). The relationship between K and Z is significant and positive, but a slope β_1 significantly less than 1 indicates a non-linear relationship, with K increasing less than in proportion to Z (Fig. 6.11(c)). Age at maturity A_m is inversely related to Z (β_1 not significantly different from –1, Fig. 6.11(d)). These patterns are broadly consistent with patterns of symmetry known as the 'Beverton and Holt invariants' (Charnov, 1993), which have been found to exist in many taxonomic groups of fishes (Beverton and Holt, 1959; Beverton, 1992; Charnov, 1993) and have been investigated theoretically by Charnov (1993), Mangel (1996), and Jensen (1996). The invariants are: the ratio between length at maturity and asymptotic length ($L_m/L_\infty = C_1$), the ratio between natural mortality and VBGF growth rate K ($M/K = C_2$); and age at maturity A_m is inversely proportional to the adult natural mortality rate M ($A_mM = C_3$). The L_m/L_∞ and A_mM invariants appear to hold in exploited tilapia populations (substituting Z for M), although it will be shown later that L_m/L_∞ is not strictly speaking an invariant, but depends strongly on Z. The M/K invariant does not hold for total mortality in exploited populations, because K does not increase as strongly with Z mortality as would be expected from comparative studies of unexploited populations or ultimate life history models.

'Stunting'

'To stunt', according to Chamber's Twentieth Century Dictionary, means 'to hinder for growth, dwarf'. By definition, therefore, 'stunted' fish populations are made up of individuals that are small relative to their conspecifics elsewhere. 'Stunting' is a common phenomenon in freshwater fish and has been studied extensively for more than 50 years, but its mechanisms are still poorly understood. Unfortunately, studies on the 'stunting' of tilapias have been little informed by the work conducted on other populations of freshwater fish, or by comparative studies on larger sets of tilapia life history data. This has led to the perception that stunting in tilapias is a special or even unique phenomenon, and given rise to some misleading interpretations. Iles (1973) suggested that tilapia stunting may be understood as a unique response to conditions of high

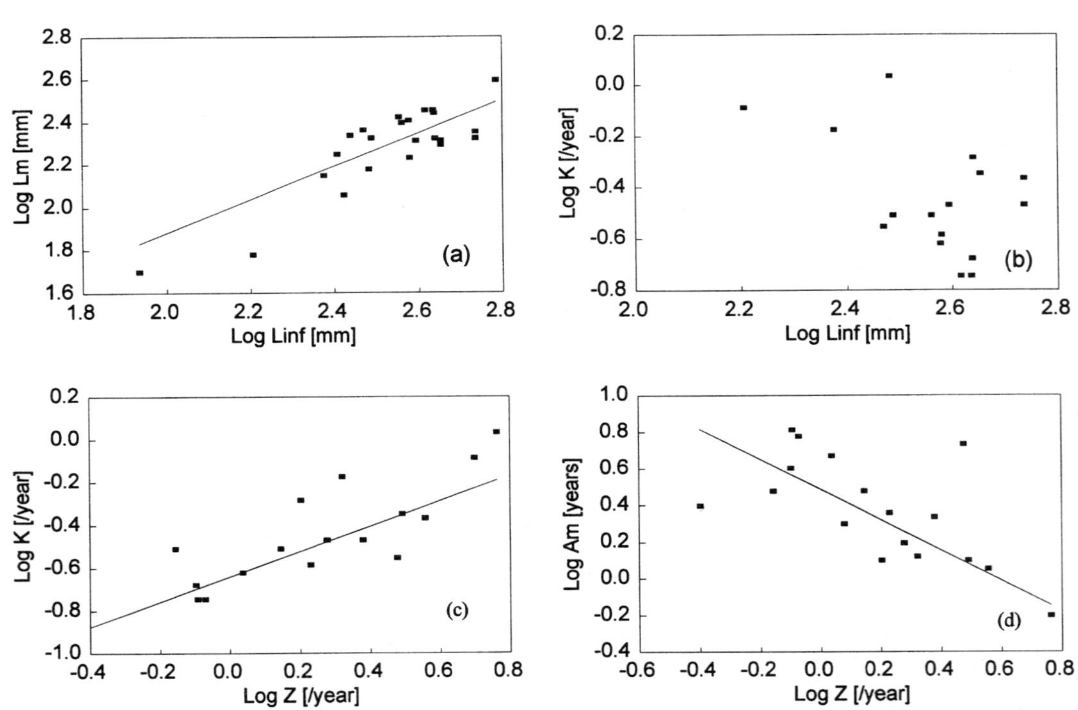

Fig. 6.11 Relationships between life history traits among the tilapia populations listed in Table 6.1. Where significant power relationships of the form log Y = $\beta_0 + \beta_1$ log X exist, parameters are given with 90% confidence intervals. (a) Length at maturity L_m versus asymptotic length L_∞; $\beta_0 = 0.33$ [–0.19, 1.19] and $\beta_1 = 0.78$ [0.44, 1.00]. (b) VBGF growth rate K versus L_∞; no significant relationship. (c) VBGF growth rate K versus total mortality rate Z; $\beta_0 = -0.64$ [–0.68, –0.60] and $\beta_1 = 0.59$ [0.38, 0.84]. (d) Age at maturity Am versus total mortality rate Z; $\beta_0 = 0.49$ [0.38, 0.58] and $\beta_1 = -0.83$ [–1.13, –0.50]. All relationships estimated by non-parametric (Theil) regression.

mortality, and identified early maturation was key mechanism responsible for stunting. He further suggested that stunted populations were characterized by early maturation at a small size even relative to the asymptotic size, and consequently identified the L_m/L_∞ ratio as an indicator of 'stunting' (with values below 0.7 characterizing a 'stunted' population). However, the following review shows that (a) the patterns of stunting in tilapias are not unlike those found in other species, and that (b) different life histories may be associated with the 'stunting', and the phenomenon can not be interpreted simply as a response to high mortality rates.

Life history variation within species

The data assembled in Table 6.1 contain sufficient information on life history variation in *O. mossambicus* to merit an exploratory analysis of 'stunting' in this species. The relationship between L_m and L_∞ is shown in Fig. 6.12(a), where the solid line indicates the median L_m/L_∞ ratio of 0.54. The populations represent a wide range of L_m and L_∞. Overall, L_m and L_∞ are correlated, and there is no obvious relationship between L_∞ and the ratio of L_m/L_∞: 'stunted' populations with a L_∞ below 300 mm show a similar range of L_m/L_∞ values as those growing to 400 mm or more. Hence the L_m/L_∞ ratio is not an indicator of 'stunting'. However, both the L_m/L_∞ ratio and the age at maturity A_m do decline with increasing total mortality rate (Fig. 6.12 (b,c)), as predicted by Iles (1973) and more recent life history models (see Roff, 1992). A better understanding of the responses involved can be obtained by looking at the relationships of L_m and L_∞ to Z individually (rather than their ratio). As shown in Fig. 6.12 (d), L_m declines slightly with increasing Z, but there is also an apprently more pronounced increase in L_∞. Hence the decline in L_m/L_∞ with Z reflects primarily an increase in growth combined with a decline in L_m and A_m. This shows that L_m/L_∞ ratios can not be understood in terms of plasticity of L_m alone, but must be considered in context of two highly plastic life history traits. Further comparative analyses of tilapia life histories are required in order to assess whether the patterns apparent in the limited set of *O. mossambicus* data are generalizable. It is interesting to note, however, that the patterns are not inconsistent with responses to exploitation observed in marine fish stocks. In a comparatitive study of 77 stocks, Rochet (1998) concludes that exploitation leads to earlier age and increased size at maturity.

The occurrence of stunted populations has been reported for many families of fish other than the cichlids, for example percids (Eschmeyer, 1936; Alm, 1946; Jansen, 1996), cyprinids (Burrough and Kennedy, 1979; Linfield, 1979; Piironen and Holopainen, 1988; Holopainen *et al.*, 1997), coregonids (Healey, 1980; Ridgeway and Chapleau, 1994), salmonids (Alm, 1959; Leggett and Power, 1969; Hindar and Jonsson, 1982; Donald and Alger, 1989; Amundsen *et al.*, 1993). Alm (1959) pointed out that the low maximum size characteristic of stunted populations may be combined with either early (see also e.g.

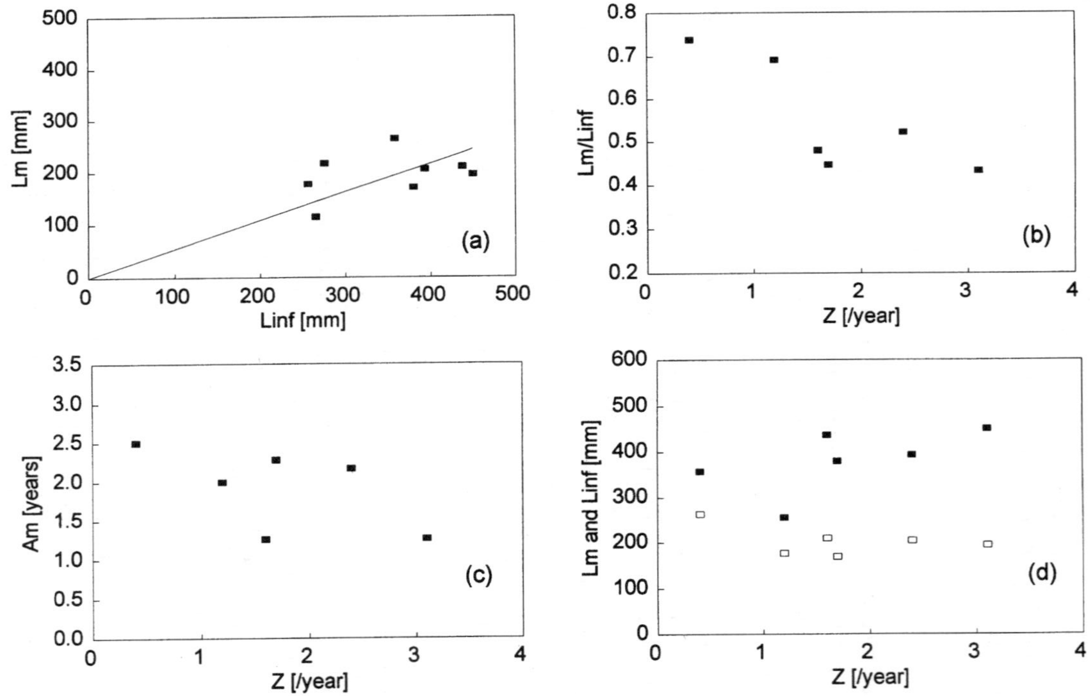

Fig 6.12 Relationship between life history traits in *O. mossambicus* populations. (a) Length at maturity L_m versus asymptotic length L_∞; the line indicates the median L_m/L_∞ ratio of 0.53. (b) L_m/L_∞ ratio versus total mortality rate Z. (c) Age at maturity A_m versus Z. (d) L_m (open squares) L_∞ (solid squares) versus Z. All data from Table 6.1.

Ridgeway and Chapleau, 1994; Jansen, 1996; Doyon *et al.*, 1998), or late maturation (see also e.g. Donald and Alger, 1989). The same is true for tilapias: while the occurrence of early maturation has been widely reported (e.g. in Iles, 1973), there are also examples of stunted populations where maturation appears to be delayed or suppressed, e.g. in Silliman (1975) or Eyeson (1983). Note that the populations in the latter studies were also characterized by comparatively high L_m/L_∞ ratios (Table 6.1). Stunted populations are characterized by a low maximum size and will necessarily show a low absolute size at maturity as well, even if maturity is delayed relative to 'normal' populations. The relationship between 'stunting' and mortality is complex. Stunted populations may be characterized by reduced longevity (i.e. higher mortality) compared to non-stunted populations of the same species (Roff, 1992; Jansen, 1996). This increased mortality suffered by some 'stunted' populations may well be a consequence of reduced growth and size-dependent mortality, rather than a cause of 'stunting'. Further increases in mortality through experimental exploitation have often been shown to increase growth and 'rehabilitate' stunted populations (Donald and Alger, 1989; Amundsen *et al.*, 1993).

If the patterns of life history variation and 'stunting' in tilapias are not substantially different from those exhibited by other species, is the degree of variation exceptionally great? Table 6.2 summarizes the coefficients of variation of asymptotic length, length at maturity and their ratio for populations of *O. mossambicus* and *O. niloticus*, as well as for four non-tilapiines, herring (*Clupea harengus*), cod (*Gadus morhua*), walleye (*Stizostedion vitreum*) and lake trout (*Salvelinus namaycush*). The tilapiines do indeed show relatively high degree of life history variation, which is however on par with that found in cod and walleye. It appears, then, that tilapia life history variation is neither qualitatively nor quantitatively unique.

Table 6.2 Comparison of coefficient of variation in length at maturity, asymptotic length, and their ratio between populations of tilapias and other fish species

Species	CV L_∞	L_m	L_m/L_∞	N	Source
O. mossambicus	21	21	25	8	Table 6.1
O. niloticus	11	30	21	4	Table 6.1
Clupea harengus	15	16	8	12	Beverton (1992)
Gadus morhua	20	28	20	17	Beverton (1992)
Stizostedion vitreum	22	9	25	13	Colby *et al.* (1979); Colby and Nepszy(1981)
Salvelinus namaycush	9	9	5	6	Shuter *et al.* (1998); Trippel (1993)

Phenotypic response or genetic selection?

It has already been established in section 6.2 that the growth and reproductive traits implicated in 'stunting' are highly plastic, and this suggests that tilapia 'stunting' is also primarily a phenotypic response. This conclusion is further supported by the rapid emergence of 'stunted' phenotypes following isolation of sub-populations from non-stunted populations (e.g. Welcomme, 1970). Also, a transfer experiment (Duponchelle et *al.*, 1998) and a population genetics study (Bluhdorn and Arthington, 1990) failed to provide evidence for genetic differentiation between 'stunted' and 'normal' populations of tilapias.

Many studies on 'stunting' in other species have likewise pointed to a phenotypic response, reversible by experimental manipulations such as transfers or reductions in density (e.g. Heath and Roff, 1987; Donald and Alger, 1989; Amundsen *et al.*, 1993). However, there are also examples where a degree of genetic differentiation is apparent (Svedaeng, 1990; Taylor and Bentzen, 1993; Szczerbowskie *et al.*, 1997). Genetic differentiation is likely to occur when 'stunted' and 'normal' forms persists in isolation for prolonged periods of time. Hence the occurrence of genetic differentiation between the life history forms is likely to reflect local conditions more than a property of particular species.

Environmental characteristics associated with 'stunting'

A survey of the literature points to certain environmental conditions that are often associated with 'stunting', even though the phenomenon itself is not identified with a uniform life history strategy. The conditions that seem most regularly associated with stunting in tilapias and in other fishes are: small water bodies, isolated populations, and a low diversity of the local fish community. None of these conditions necessarily leads to stunting, they merely characterize situations where stunting is likely to occur and may aid the identification of underlying mechanism.

The association of 'stunting' with small water bodies has often been made in the literature on tilapias as well as other species (see e.g. this chapter for tilapias; Holopainen *et al.*, 1997 for crucian carp). Many records of stunted populations are from ponds or small lakes (e.g. Pechlaner, 1984; Ridgeway and Chapleau, 1994; Holopainen *et al.*, 1997). However, not all populations of species known to 'stunt' necessarily do so in small water bodies: there are examples of 'stunted' and 'normal' populations of the same species in water bodies of similar size, or even the same water body (e.g. Fenderson, 1964; Doyon *et al.*, 1998). Neither is the occurrence of 'stunting' limited to the smallest of water bodies: 'stunted' populations have been reported from lakes of more than 1000 ha area (e.g. Amundsen, 1988; Amundsen *et al.*, 1993), although 'stunting' in such large lakes tends to be less severe than in smaller water bodies.

Isolation of a population is a factor that often coincides with its occurrence in a small water body. However, there are situations where isolation can be identified as a key factor in its own right, for example in landlocked populations of anadromous species such as salmon (e.g. Leggett and Power, 1969; Gibson *et al.*, 1996). There is an obvious similarity between isolation in anadromous species and populations in small water bodies in that in both cases, fish are prevented from migrating into open waters to escape the intense competition characteristic of shallow littoral areas (Hanson and Leggett, 1985), or extremes of physical conditions. However, isolated populations may also show 'normal' growth patterns, either naturally or in response to experimental exploitation (see e.g. Pechlaner, 1984; Donald and Alger, 1989).

Another factor often associated with stunting is a low diversity of the resident fish community, sometimes to the extent that the 'community' is monospecific (Welcomme, 1970; Johnson, 1983; Ridgeway and Chapleau, 1994; Holopainen *et al.*, 1997). A low diversity is in itself a characteristic of small water bodies (e.g. Minns, 1989), and the independent significance of this factor is therefore not easy to assess. A very low diversity is certainly not a necessary condition for 'stunting', as the occurrence of 'stunted' populations in lakes of moderate diversity shows (e.g. Heath and Roff, 1996).

Conclusion

The term 'stunting' has long been used, and will likely continue to be used as a descriptor of populations characterized by a low maximum size. However, 'stunting' is not identified with a particular life history strategy, but rather describes a variety of life histories united in but one trait: a low maximum size. The perception that stunting in tilapias is unique and does represent a particular life history strategy (Iles, 1973; Noakes and Balon, 1982) is not borne out in a comparative perspective. The 'stunting indicator' proposed in this context, the L_m/L_∞ ratio, is unrelated to 'stunting' as it is commonly understood (as a low L_∞). Use of this indicator is, in fact, misleading as the following examples illustrate. Mattson and Kaunda (1997) compared two populations of *O. shiranus* in small waterbodies, with L_m/L_∞ ratios of 0.59 and 0.48 (see also Table 6.1). While the indicator would suggest that the latter population was the more 'stunted', it was actually characterized by much higher growth and overall productivity than the former. Another example is given by the two *S. melanotheron* populations included in Table 6.1. The tank population (Eyeson, 1983) with an L_∞ of 86 mm is clearly 'stunted', but its L_m/L_∞ ratio of 0.58 is far higher than the value of 0.37 found in the larger growing (L_∞ = 161 mm) lagoon population. In both these examples, as in the *O. mossambicus* data explored before, L_m varied less between populations than L_∞.

If 'stunting' is not synonymous with a particular life history strategy, but rather a collective descriptor for a variety of strategies, then it is unlikely that 'stunting' can be understood in terms of a single factor such as 'a high mortality rate'. As discussed elsewhere in this chapter, such 'blanket' explanations appear to fail even when 'stunting' is re-defined to describe a particular life history strategy or individual trait. The problem is thus one of understanding life history plasticity in general, rather than the ill-defined phenomenon of 'stunting'. This is likely to require a far more detailed analysis of responses in individual traits and their interactions than has so far been attempted in the search of explanations for 'stunting'.

Towards a mechanistic understanding of life history plasticity

There is little doubt that the remarkable life history plasticity of tilapias is the legacy of their evolution in variable habitats (Lowe-McConnell, 1982; Noakes and Balon, 1982). What is less clear, but of immense theoretical and practical importance, is the question: how is the expression of this plasticity controlled?

Examination of life history data from established tilapia populations shows patterns of symmetry consistent with ultimate considerations of life history theory, i.e. maximization of some measure of fitness. However, an understanding the proximate basis life history plasticity, i.e. the mechanisms that animals use to achieve optimum life history patterns (Thorpe *et. al.*, 1998) is required in order to manage this plasticity in cultured and exploited stocks. The challenge this involves is illustrated in the example of *O. mossambicus* life history plasticity discussed in section 6.4. Comparative examination of the data shows a decline in the L_m/L_∞ ratio and in A_m with increasing Z, a pattern consistent with predictions from ultimate life history models for optimal age and size at maturity (e.g. Roff, 1992). However, this response in involves plasticity in several traits (primarily an increase in L_∞), which is difficult to accommodate with the common life history models predicting optimal age and size at maturity from externally determined growth and mortality patterns (Roff, 1992; Stearns, 1992; but see Stamps *et al.*, 1998 for an alternative approach). The problem is that there are several highly plastic traits, and it is not clear *a priori* which of these traits would respond and to what extent. This problem can not be solved by re-writing ultimate models, for example to predict optimal L_∞ subject to L_m. A dynamic framework is required that allows to analyse and predict development paths on the basis of state dependent, proximate responses to environmental factors.

Such a dynamic framework must accommodate processes that are more or less continuous in nature (e.g. growth, mortality), as well as others that involve discrete 'switches' (e.g. maturation, ontogenic shifts in habitat use). Continuous processes tend to be easier to understand and model; for example growth may be modelled on the basis of the acquisition of resources and their allocation to maintenance metabolism, somatic growth and reproduction. Processes involving switches are less well understood and more difficult to model. Perhaps the most promising approach is to interpret switches as determined by the organism's present physiological state and rate of change, in relation to a genetically determined threshold (Roff, 1996; Thorpe *et al.*, 1998). Thorpe *et al.* (1998) have used this approach in the development of a first, proximate life history model to predict quantitatively the occurrence of smolting and maturation in Atlantic salmon.

A comparative analysis of the wealth of information now available from field studies on tilapia populations could generate a matrix of conditions associated with alternative life histories, which may be used to test the predictions from proximate models. Much of the detailed information on state-dependent responses in development paths required to construct such models in the first place, however, will have to be gained from experimentally manipulated populations, not least in aquaculture.

6.5 CONCLUDING REMARKS

Tilapia populations are being exploited and otherwise manipulated in an extraordinarily wide range of habitats, and have been studied extensively under a wide range of conditions. This has resulted in a wealth of information pertinent to their population dynamics and life histories, the significance of which is most fully appreciated in a comparative perspective integrating across wild and cultured populations. This chapter aims to develop such an integrated perspective, but much remains to be done.

The analysis of growth and reproductive traits has shown a high degree phenotypic plasticity in both. The mechanisms governing plasticity in growth are relatively well understood at least in qualitative terms, and through a set of different but largely complementary quantitative models. By contrast, little is known about the mechanisms governing plasticity in reproductive traits, and the existing hypotheses are either inconsistent with the data or too loosely defined to be of real value. Recent developments in evolutionary ecology, analysing maturation in terms of the organism's status relative to genetically defined thresholds, may be promising to explore. Patterns of mortality have been relatively little explored, in spite of their importance to management, ecology and evolution.

Population dynamics approaches have substantial potential to aid the management of wild as well as farmed tilapia stocks, as illustrated in a range of examples throughout this chapter. This potential has not been realized as much as it deserves to be, but it is hoped that this chapter will provide a new impetus. Many of the approaches and methods discussed are easily transferable. However, most of the studies reviewed use conventional stock assessment models, which do not account for the plasticity of life history traits found in the tilapias (as well as many other species of fish). This indicates a need to develop new approaches that can account for such plasticity, particularly in the assessment of management strategies for populations that can be subjected to a high level of manipulation. Density-dependent growth appears to be a particularly pervasive aspect of plasticity, and is relatively easy to incorporate into assessment models.

'Stunting' in tilapias is a reflection of the plasticity in life history traits already discussed. However, 'stunting' in tilapias is neither qualitatively nor quantitatively unique: similar patterns are found in many species of freshwater fish. 'Stunting' is a collective term applied to a variety of life history forms characterized by a common trait, a low maximum size. The variety of life histories associated with a low maximum size and of the underlying causes for their expression defies a simple definition and causal explanation. In particular, the widespread use of the L_m/L_∞ ratio as stunting indicator in tilapias is misleading and should be abandoned. While 'stunting' may be used as a descriptor of populations characterized by a low maximum size, an understanding of the underlying mechanisms will most likely emerge from an appreciation of the underlying variety of life histories. Understanding the proximate causes of life history variation is perhaps the greatest challenge in tilapia population biology. This understanding will most likely emerge from a combination of quantitative modelling with laboratory and field experiments: it will rely on a close integration of the study of wild and farmed stocks, and ultimately benefit the management of both.

Comparative information suggests that life history variation in tilapias is not as unique as it has often been considered to be, and this implies that tilapias may serve as useful model to explore life history plasticity in fish in general.

ACKNOWLEDGEMENTS

Comments on a draft version of this chapter by Malcolm Beveridge, Marc Mangel, Sampsa Vilhunen and Robin Welcomme have substantially improved the manuscript. Needless to say, the author takes full responsibility for any remaining errors of fact or interpretation. Work on this chapter was partially supported by the Nuffield Foundation (NUF-NAL), and by the UK Department for International Development (DFID, Fisheries Management Science Programme). The views expressed are not necessarily those of the DFID.

REFERENCES

AIT (1990) Assessment and trial of alternative feed options for project fry nursing recommendations in nylon hapas. *AIT Aquaculture Outreach Report.*

Alimoso, S.B. (1991) Catch effort data and their use in the management of fisheries in Malawi, in *Catch Effort Sampling Strategies* (ed. I.G. Cowx), Fishing News Books, Oxford, pp. 393–403.

Alm, G. (1946) Reasons for the occurrence of stunted fish populations. *Reports of the Institute of Freshwater Research Drottningholm* **25**, 1–146.

Alm, G. (1959) Connection between maturity, size and age in fishes. *Reports of the Institute of Freshwater Research Drottningholm* **40**, 1–145.

Amarasinghe, U.S. (1996) Stock assessment in Sri Lankan reservoirs, in *Stock Assessment in Inland Fisheries* (ed. I.G. Cowx), Fishing News Books, Oxford, pp. 345–356.

Amarasinghe, U.S. and De Silva, S.S. (1992) Population dynamics of *Oreochromis mossambicus* and *O. niloticus* (Cichlidae) in two reservoirs in Sri Lanka. *Asian Fisheries Science* **5**, 37–61.

Amarasinghe, U.S. and De Silva, S.S. (1996) Impact of *Oreochromis mossambicus* × *O. niloticus* (Pisces: Cichlidae) hybridization on population reproductive potential and long-term influence on a reservoir fishery. *Fisheries Management and Ecology* **3**, 239–250.

Amundsen, P.A. (1988) Effects of an intensive fishing programme on age structure, growth and parasite infection of stunted whitefish (*Coregonus lavaretus* L.) in Lake Stuorajarvi, northern Norway. *Finnish Fisheries Research* **9**, 425–434.

Amundsen, P.A., Klemetsen, A. and Grotnes, P.E. (1993) Rehabilitation of a stunted population of Arctic char by intensive fishing. *North American Journal of Fisheries Management* **13**, 483–491.

Arthington, A.H. (1991) Ecological and genetic impacts of introdced and translocated freshwater fishes in Australia. *Canadian Journal of Fisheries and Aquatic Sciences* **48** (Supplement 1), 33–43.

Arthington, A.H. and Bluhdorn, D.R. (1994) Distribution, genetics, ecology and status of the introduced cichlid, *Oreochromis mossambicus*, in Australia. *Mitteilungen des Internationalen Vereins fuer Limnologie* **24**, 53–62.

Ashkenazi, S. and Yomtov, Y. (1996) Herons and fish farming in the Huleh valley, Israel – conflict or mutual benefit? *Colonial Waterbirds* **19**, 143–151.

Baerends, G.R. and Baerends-Van Roon, J.M. (1950) An introduction to the study of ethology of cichlid fishes. *Behaviour* **1** (Supplement), 1–242.

Bagenal, T.B. (1969) The relationship between food supply and fecundity in brown trout *Salmo trutta* L. *Journal of Fish Biology* **1**, 167–182.

Bedavi, R.M. (1985) Recruitment control and production of market-size *Oreochromis niloticus* with predator *Lates niloticus* in Sudan. *Journal of Fish Biology* **26**, 459–464.

Beirne, B.P. (1975) Biological control attempts by introductions against pest insects in the field in Canada. *Canadian Entomologist* **107**, 225–236.

Beverton, R.J.H. (1992) Patterns of reproductive strategy and parameters in some marine teleost fishes. *Journal of Fish Biology* **41** (Supplement B), 136–160.

Beverton, R.J.H. and Holt, S.J. (1957) On the dynamics of exploited fish populations. Fishery Investigations, Series II. Ministry of Agriculture, Fisheries and Food. HMSO, London.

Beverton, R.J.H. and Holt, S.J. (1959) A review of the lifespans and mortality rates of fish in nature, and their relation to growth and other physiological characteristics, in *Ciba Foundation Colloquia on Ageing*, Vol. 5 (eds G.E.W. Wolstenholme and M. O'Connor), Churchill, London, pp. 142–177.

Bjorndal, T. (1990) *The Economics of Salmon Aquaculture*, Blackwell, London.

Blay, J. and Asabere-Ameyaw, E. (1993) Assessment of the fishery of a stunted population of the cichlid, *Sarotherodon melanotheron* (Rueppel), in a 'closed' lagoon in Ghana. *Journal of Applied Ichthyology* **9**, 1–11.

Bluhdorn, D.R and Arthington, A. (1990) The incidence of stunting in Australian populations of the introduced Cichlid *Oreochromis mossambicus* (Peters), in *The Second Asian Fisheries Forum* (eds R. Hirano and I. Hanyu), Asian Fisheries Society, Manila, pp. 41–44.

Bolivar, R.B., Eknath, A.E., Bolivar, H.L. and Abella, T.A. (1993) Growth and reproduction in individually tagged Nile tilapia (*Oreochromis niloticus*) of different strains. *Aquaculture* **111**, 159–169.

Bowen, S.H. (1979) A nutritional constraint on detritivory by fishes: the stunted population of *Sarotherodon mossambicus* in Lake Sibaya, South Africa. *Ecological Monographs* **49**, 17–31.

Bramick, U., Puckhaber, B., Langholz, H.J. and Horstgen-Schwank, G. (1995) Testing of triploid tilapia (*Oreochromis niloticus*) under tropical pond conditions. *Aquaculture* **137**, 343–352.

Brummett R.E. (1995) Environmental regulation of sexual maturation and reproduction in tilapia. *Reviews in Fisheries Science* **3**, 231–248.

Burrough, R.J. and Kennedy, C.R. (1979) The occurrence and natural alleviation of stunting in a population of roach, *Rutilus rutilus* (L.). *Journal of Fish Biology* **15**, 93–109.

Carvalho, G.R. and Hauser, L. (1995) Genetic impacts of fish introductions: a perspective on African lakes, in *The Impact of Species Changes in African Lakes* (eds T.J. Pitcher and P.J.B. Hart), Chapman and Hall, London, pp. 457–493.

Charnov, E.L. (1993) *Life History Invariants*, Oxford University Press, Oxford.

Cheftel, J. and Lorenzen, K. (1999) An appraisal of the potential uptake of genetically male tilapia by small-scale fish farmers in Nueva Ecija, Central Luzon, Philippines. Research Report, Huxley School, Imperial College, London, 86 pp.

Chen, F.Y. and Prowse G.A. (1964) The effect of living space on the growth rate of fishes. *Ichthyologica* **3**, 11–20.

Christensen, V. and Pauly, D. (eds) (1993) *Trophic Models of Aquatic Ecosystems*, ICLARM, Manila.

Coates, D. (1993) Environmental management implications of aquatic species introductions: a case study of fish introductions into the Sepik-Ramu Basin, Papua, New Guinea. *Asian Journal of Environmental Management* **1**, 39–49.

Coates, D. (1995) Implementing the EIFAC/ICES code of practice: experiences with the evaluation of international fish transfers into the Sepik River Basin, Papua New Guinea, in *Proceedings of the World Fisheries Congress* (ed. C.W. Voigtlander), Oxford and IBH Publishing, New Delhi, pp, 159–173.

Coates, D. (1998) Codes of practice for the stocking and introduction of fish, in *Stocking and Introduction of Fish* (ed. I.G. Cowx), Fishing News Books, Oxford, pp. 383–396.

Colby, P.J. and Nepszy, S.J. (1981) Variations among stocks of walleye (*Stizostedion vitreum vitreum*): management implications. *Canadian Journal of Fisheries and Aquatic Science* **38**, 1814–1822.

Colby, P.J., McNicol, R.E. and Ryder, R.A. (1979) Synopsis of biological data on the walleye *Stizostedion v. vitreum* (Mitchell 1818). *FAO Fish Synopsis* 119.

Costa-Pierce, B.A. and Hadikusumah, H.Y. (1990). Research on cage aquaculture systems in the Saguling Reservoir, West Java, Indonesia, in *Reservoir Fisheries and Aquaculture Development for Resettlement in Indonesia* (ed. B.A. Costa-Pierce), ICLARM Technical Report 23, pp. 112–217.

Courtenay, W.R. and Stauffer, J.R. (eds) (1984) *Distribution, Biology, and Management of Exotic Fishes*, Johns Hopkins University Press, Baltimore.

Coward, K. and Bromage, N.R. (1999) Spawning frequency, fecundity, egg size and ovarian histology in groups of *Tilapia zillii* maintained upon two distinct food ration sizes from first-feeding to sexual maturity. *Aquatic Living Resources* **12**, 11–22.

Coward, K., Bromage, N.R. and Little, D.C. (1998) Inhibition of spawning and associated suppression of sex steroid levels during confinement in the substrate-spawning *Tilapia zillii*. *Journal of Fish Biology* **52**, 152–165.

Cowx, I.G. (1999) An appraisal of stocking strategies in the light of developing country constraints. *Fisheries Management and Ecology* **6**, 21–34.

Cridland, C.C. (1961) Laboratory experiments on the growth of *Tilapia* spp. The reproduction of *Tilapia esculenta* under artificial conditions. *Hydrobiologia* **18**, 177–184.

Cruz, E.M. and Ridha, M. (1991). Production of the tilapia *Oreochromis spilurus* Guenther stocked at different densities in sea cages. *Aquaculture* **99**, 95–103.

Day, T. and Taylor, P.D. (1997) Von Bertalanffy's equation should not be used to model age and size at maturity. *American Naturalist* **149**, 381–393.

DeAngelis, D.L. and Gross, L.J. (1992) (eds) *Individual-based Models and Approaches in Ecology: Populations, Communities and Ecosystems*, Chapman and Hall, London.

DeGraaf, G., Galemoni, F. and Banzoussi, B (1996) Recruitment control of Nile tilapia, *Oreochromis niloticus*, by the African catfish, *Clarias gariepinus* (Burchell 1822), and the African snakehead, *Ophiocephalus obscuris*. 1. A biological analysis. *Aquaculture* **146**, 85–100.

De Silva, S.S. (1986) Reproductive biology of *Oreochromis mossambicus* populations of man-made lakes in Sri Lanka: a comparative study. *Aquaculture and Fisheries Management* **17**, 31–47.

De Silva, S.S. (1988) Reservoirs in Sri Lanka and their fisheries. *FAO Fisheries Technical Paper* 298.

De Silva, S.S. and Senaratne, K.A.D.W (1988) *Oreochromis mossambicus* is not universally a nuisance species: the Sri Lankan experience, in *The Second International Symposium on Tilapia in Aquaculture* (eds R.S.V. Pullin, T. Bhukaswan, K Tonguthai and J.L. MacLean), ICLARM, Manila, pp. 445–450.

De Silva, S.S. and Sirisena, H.K.G. (1987) New fish resources in Sri Lanka: feasibility of introduction of a subsidiary gillnet fishery for minor cyprinids. *Fisheries Research* **6**, 17–34.

Diana, J.S. (1997) Feeding strategies, in *Dynamics of Pond Aquaculture* (eds H.S. Egna and C.E. Boyd), CRC, Boca Baton, pp. 245–262

Diana, J.S. and Lin, C.K. (1991). Tilapia and *Clarias* polyculture, in *Aquaculture/Pond Dynamics Collaborative Research Support Program: Ninth Annual Administrative Report*, Oregon State University, Corvallis, pp. 45–48.

Donald, D.B. and Alger, D.J. (1989) Evaluation of exploitation as a means of improving growth in a stunted population of brook trout. *North American Journal of Fisheries Management* **9**, 177–183.

Doyon, J.F., Schetagne, R. and Verdon, R. (1998) Different mercury bioaccumulation rates between sympatric populations of dwarf and normal lake whitefish (*Coregonus clupeaformis*) in the La Grande complex watershed, James Bay, Quebec. *Biogeochemistry* **40**, 203–216.

Draper, N.R. and Smith, H. (1981) *Applied Regression Analysis*, 2nd edn, Wiley, New York.

Dudley, R.G. (1974) Growth of *Tilapia* on the Kafue floodplain Zambia: predicted effects of the Kafue Gorge Dam. *Transactions of the American Fisheries Society* **103**, 281–291.

Dunseth, D.R. and Bayne, D.R. (1978) Recruitment control and production of *Tilapia aurea* (Steindachner) with the predator *Cichlasoma managuense* (Guenther). *Aquaculture* **14**, 383–390.

Duponchelle, F. and Panfili, J. (1998) Variations in age and size at maturity of female Nile tilapia, *Oreochromis niloticus*, populations from man-made lakes of Cote d'Ivoire. *Environmental Biology of Fishes* **52**, 453–465.

Duponchelle, F., Pouyaud, L. and Legendre, M. (1998) Evidence of environmental effects on reproductive characteristics of Nile tilapia (*Oreochromis niloticus*) populations from man-made lakes of Ivory Coast. *Aquatic Living Resources* **11**, 137–144

El Gamal, A.A. (1992) Predation by Nile perch *Lates niloticus* (L.) on *Oreochromis niloticus* (L.), *Cyprinus carpio* (L.), *Mugil* sp. and its role in controlling tilapia recruitment in Egypt. *Journal of Fish Biology* **40**, 351–358.

Eknath, A.E., Tayamen, M.M. Palada-de Vera, M.S., Danting, J.C., Reyes, R.A. Dionisio, E.E., Capili, J.B., Bolivar, H.L., Abella, T.A., Circa, A.V., Bentsen, H.B., Gjerde, B., Gjedrem, T. and Pullin, R.S.V.P. (1993) Genetic improvement of farmed tilapias: the growth performance of eight strains of *Oreochromis niloticus* tested in different farm environments. *Aquaculture* **111**, 171–188.

Ens, B.J., Weissing, F.J. and Drent, R.H. (1995) The despotic distribution and deferred maturity: two sides of the same coin. *American Naturalist* **146**, 625–650.

Eschmeyer, R.W. (1936) Some characteristics of a population of stunted perch. *Michigan Academy of Science, Arts and Letters* **22**, 613–678.

Eyeson, K.N. (1983) Stunting and reproduction in pond reared *Sarotherodon melanotheron*. *Aquaculture* **31**, 257–267.

FAO (1993) Fisheries management in the south-east arm of Lake Malawi, the Upper Shire River and Lake Malombe, with particular reference to the fisheries on chambo (*Oreochromis* spp.) *CIFA Technical Paper* 21, FAO, Rome.

FAO (1998) Database on introductions of aquatic species. Accessible via the world wide web: http://www.fao.org/WAICENT/FAOINFO/FISHERY/statist/fisoft/dias/index.htm.

Fenchel, T. (1974) Intrinsic rate of natural increase: the relationship with body size. *Oecologia* **14**, 317–326.

Fenderson, O.C. (1964) Evidence of sub-populations of lake Whitefish, *Coregonus clupeaformis*, involving a dwarf form. *Transactions of the American Fisheries Society* **93**, 77–94.

Fernandes, M.D and Volpatro, G.L. (1993) Heterogeneous growth in the Nile tilapia-social stress and carbohydrate-metabolism. *Physiology and Behaviour* **54**, 319–323.

Fernando, C.H. and Holcik, J. (1991) Some impacts of fish introductions into tropical freshwaters, in *Ecology of Biological Invasions in the Tropics* (ed. P.S. Ramakrishnan), International Science Publications, New Delhi, pp. 103–129.

Fisher, G.W. and Grant, W.E. (1994) Use of a native predator to control overcrowding in warm-water polyculture ponds: simulation of a tucunare (*Cichla monoculus*) – tilapia (*Oreochromis niloticus*) system. *Ecological Modelling* **72**, 205–227.

Fryer, G. and Iles D.T. (1972) *The Cichlid Fishes of the Great Lakes of Africa*, Oliver and Boyd, Edinburgh.

Garrod, D.J. (1959) The growth of *Tilapia esculenta* Graham in Lake Victoria. *Hydrobiologia* **12**, 268–298.

Garrod, D.J. (1961) The selection characteristics of nylon gill nets to *T. esculenta* Graham. *Journal du Conseil Permanent International pour l'Exploration de la Mer* **26**, 191–203.

Garrod, D.J. (1963) An estimation of the mortality rates in a population of *Tilapia esculenta* Graham (Pisces, Cichlidae) in Lake Victoria, East Africa. *Journal of the Fisheries Research Board of Canada* **20**, 195–227.

Getabu, A. (1987) Aspects of Lake Victoria Fisheries with emphasis on *Oreochromis niloticus* and Alestes sadleri from Nyanza Gulf. *FAO Fisheries Report* 389, 416–431.

Gibson, R.J., Williams, D.D., McGowan, C. and Davidson, W.S. (1996) The ecology of dwarf fluvial Atlantic salmon, *Salmo salar* L., cohabiting with brook trout, *Salvelinus fontinalis* (Mitchill), in southeastern Newfoundland. *Polskie Archiwum Hydrobiologii* **43**, 145–168.

Griffith, B., Scott, M.J., Carpenter, J.W., and Reed C. (1989) Translocation as a species conservation tool: status and strategy. *Science* **245**, 477–480.

Gwahaba, J.J. (1973) Effects of fishing on the *Tilapia nilotica* population in L. George, Uganda, over the past twenty years. *East African Wildlife Journal* **11**, 317–328.

Hanson, J.M. and Leggett, W.C. (1985) Experimental and field evidence for inter- and intraspecific competition in two freshwater fishes. *Canadian Journal of Fisheries and Aquatic Sciences* **42**, 280–286.

Harding, D. (1965) Research on Lake Kariba, 1962–1963. *Fisheries Research Bulletin Zambia* 1962/63, 38–55.

Healy, M.C. (1980) Growth and recruitment in experimentally exploited lake whitefish (*Coregonus clupeaformis*) populations. *Journal of the Fisheries Research Board of Canada* **37**, 255–267.

Heath, D. and Roff, D.A. (1987) A test of genetic differentiation in growth of stunted and non-stunted populations of perch and pumpkinseed. *Transactions of the American Fisheries Society* **116**, 98–102.

Heath, D.D and Roff, D.A. (1996) The role of trophic bottlenecks in stunting: a field test of an allocation model for growth and reproduction in yellow perch, *Perca flavescens*. *Environmental Biology of Fishes* **45**, 53–63.

Hecht, T. and Zway, P. (1984) On the stunted Mocambique tilapia *Oreochromis mossambicus* (Peters, 1852) (Pisces: Cichlidae) of the Matiovila Hot Spring, Kruger National Park. *Koedoe* **27**, 25–38.

Herbold, B. and Moyle, P.B. (1986) Introduced species and vacant niches. *American Naturalist* **128**, 751–760.

Hickling, C.F. (1963) The cultivation of tilapia. *Scientific American* **208**, 143–152.

Hilborn, R. and Mangel, M. (1997) *The Ecological Detective: Confronting Models with Data*, Princeton University Press, Princeton.

Hilborn, R. and Walters, C.J. (1992) *Quantitative Fisheries Stock Assessment: Choice, Dynamics and Uncertainty*, Chapman and Hall, London.

Hindar, K. and Jonsson, B. (1982) Habitat and food segregation of the dwarf and normal arctic char (*Salvelinus alpinus*) from Vangsvtnet Lake, Western Norway. *Canadian Journal of Fisheries and Aquatic Sciences* **93**, 1030–1045.

Holopainen, I. J., Tonn, W.M. and Paszkowski, C.A. (1997) Tales of two fish: the dichotomous biology of crucian carp (*Carassius carassius* (L.)) in northern Europe. *Annales Zoologici Fennici* **34**, 1–22.

Hopkins, K.D. and Pauly, D. (1993) Instantaneous mortalities and multivariate models: application to tilapia culture in saline waters, in *Multivariate Methods in Aquaculture Research: Case Studies of Tilapias in Experimental and Commercial Systems* (eds M. Prein, G. Hulata and D. Pauly), ICLARM, Manila.

Horstgen-Schwark, G. and Langholz, H.J. (1998) Prospects of selecting for late maturity in tilapia (*Oreochromis niloticus*) III. A selection experiment under laboratory conditions. *Aquaculture* **167**, 123–133.

Huet, M. (1978) *Textbook of Fish Culture*, 2nd edn, Fishing News Books, Oxford.

Iles, T.D. (1973) Dwarfing or stunting in the genus Tilapia (Cichlidae): a possibly unique recruitment mechanism. *Rapports et proces-verbaux de la Conseil Internatinal d'Exploration de la Mer* **164**, 247–254.

Inland Fisheries Project (1977). *Progress Report* (Report No. 10). Diliman: College of Fisheries, University of the Philippines.

Jalabert, B. and Zohar, Y. (1982) reproductive physiology in cichlid fishes, with particular reference to *Tilapia* and *Sarotherodon*, in *The Biology and Culture of Tilapias* (eds R.S.V. Pullin and R.H. Lowe-McConnell), ICLARM, Manila, 129–140.

James, N.P.E. and Bruton, M.N. (1992) Alternative life-history traits associated with reproduction in *Oreochromis mossambicus* (Pisces: Cichlidae) in small water bodies of the eastern Cape, South Africa. *Environmental Biology of Fishes* **34**, 379–392.

Jansen, W.A. (1996) Plasticity of maturity and fecundity of yellow perch, *Perca flavescens* (Mitchill): comparison of stunted and normal-growing populations. *Annales Zoologici Fennici* **33**, 403–415.

Jensen, A.L. (1996) Beverton and Holt life history invariants result from optimal trade-offs of reproduction and survival. *Canadian Journal of Fisheries and Aquatic Sciences* **53**, 820–822.

Jensen, A.L. (1997) Origin of the relation between K and Linf and synthesis of relations among life history parameters. *Canadian Journal of Fisheries and Aquatic Sciences* **54**, 987–989.

Johnson, L. (1983) Homeostatic characteristics of single species fish stocks in Arctic lakes. *Canadian Journal of Fisheries and Aquatic Sciences* **40**, 987–1024.

Kapetsky, J.M. (1974) *Growth, Mortality, and Reproduction of Five Fish Species of the Kafue River Floodplain*, Zambia, PhD thesis, University of Michigan.

King, M. (1995) *Fisheries Biology, Assessment and Management*, Fishing News Books, Oxford.

Knut-Hansen, C.F. (1997) Experimental design and analysis in aquaculture, in *Dynamics of Pond Aquaculture* (eds H.S. Egna and C.E. Boyd), CRC Press, Boca Baton, pp. 325–376.

Kolding, J. (1993) Population dynamics and life-history styles of Nile tilapia, *Oreochromis niloticus*, in Ferguson's Gulf, Lake Tukana, Kenya. *Environmental Biology of Fishes* **37**, 25–46.

Kronert, U., Horstgen-Schwark, G. and Langholz, H.J. (1989) Prospects of selecting for late maturity in tilapia (*Oreochromis niloticus*) I: family studies under laboratory conditions. *Aquaculture* **77**, 113–121.

Lazard, J. & Oswald, M. (1995) Association silure africain tilapia: polyculture ou controle de la reproduction? *Living Aquatic Resources* **8**, 445–463.

Leggett, W.J. and Power, G. (1969) Differences between two populations of landlocked Atlantic salmon (*Salmo salar*) in Newfoundland. *Journal of the Fisheries Research Board of Canada* **35**, 1585–1596.

Lester, L.T., Aballa, T.A., Palada, M.S. and Keus, H.J. (1988) Genetic variation in size and sexual maturation of Oreochromis niloticus under hapa and cage culture conditions, in *The Second International Symposium on Tilapia in Aquaculture* (eds R.S.V. Pullin, T. Bhukaswan, K Tonguthai and J.L. MacLean), ICLARM, Manila, pp. 223–230.

Linfield, R.S. (1979) Changes in the rate of growth in a stunted roach, *Rutilus rutilus*, population. *Journal of Fish Biology* **15**, 275–289.

Liu, K.M. and Chang, W.Y.B. (1992) Bioenergetic modelling of effects of fertilisation, stocking density, and spawning on the growth of the Nile tilapia, *Oreochromis nilotica*. *Aquaculture and Fisheries Management* **22**, 291–301.

Lomnicki, A. (1988) *Population Ecology of Individuals*, Princeton University Press, Princeton.

Lorenzen, K. (1995) Population dynamics and management of culture-based fisheries. *Fisheries Management and Ecology* **2**, 61–73.

Lorenzen, K. (1996a) A simple von Bertalanffy model for density-dependent growth in extensive aquaculture, with an application to common carp (*Cyprinus carpio*). *Aquaculture* **142**, 192–205.

Lorenzen, K. (1996b) The relationship between body weight and natural mortality in juvenile and adult fish: a comparison of natural ecosystems and aquaculture. *Journal of Fish Biology* **49**, 627–647.

Lorenzen, K. and Garaway, C.J. (1998) How predictable is the outcome of stocking? *FAO Fisheries Technical Paper* 347, 133–152.

Lorenzen, K., Garaway, C.J., Chamsingh, B. and Warren, T.J. (1998a) Effects of access restrictions and stocking on small water body fisheries in Laos. *Journal of Fish Biology* **53** (Supplement 1) 345–357.

Lorenzen, K., Juntana, J., Bundit, J. and Tourongruang, D. (1998b) Assessing culture fisheries practices in small water bodies: a study of village fisheries in Northeast Thailand. *Aquaculture Research* **29**, 211–224.

Lorenzen, K., Xu, G., Cao, F., Ye, J. and Hu, T. (1997) Analysing extensive fish culture systems by transparent population modelling: bighead carp, *Aristichthys nobilis* (Richardson 1845), culture in a Chinese reservoir. *Aquaculture Research* **28**, 867–880.

Lowe (McConnell), R.H. (1958) Observations on the biology of *Tilapia nilotica* in L. Victoria, E. Africa. *East African Fisheries Research Organisation Supplementary Publication* **1**, 1–72.

Lowe-McConnell, R.H. (1982) Tilapias in fish communities, in *The Biology and Culture of Tilapias* (eds R.S.V. Pullin and R.H. Lowe-McConnell), ICLARM, Manila, pp. 83–113.

McGinty, A.S. (1984) Effects of predation by tucunare (*Cichla ocellaris*) on *Tilapia nilotica* in ponds. *Journal of the Agricultural University of Puerto Rico* **68**, 101–105.

McGinty, A.S. (1985) Effects of predation by largemouth bass in fish production ponds strocked with *Tilapia nilotica*. *Aquaculture* **46**, 269–274.

Mangel, M. (1996) Life history invariants, age at maturity and the ferox trout. *Evolutionary Ecology* **10**, 249–263.

Manly, B.F.J. (1994) *Multivariate Statistical Methods: A Primer*, 2nd edn, Chapman and Hall, London.

Mattson, N.S. (1994) Direct estimates of multi-mesh gillnet selectivity to *Oreochromis shiranus chilwae*. *Journal of Fish Biology* **45**, 997–1012.

Mattson, N.S. and Kaunda, E.K.W.H. (1997) Population dynamics of *Oreochromis shiranus* in two small water bodies in Malawi. *Journal of Fish Biology* **50**, 592–607.

Milstein, A. (1995) Fish-management relationships in Israeli commercial fish farming. *Aquaculture International* **3**, 292–314.

Minns, C.K. (1989) Factors affecting fish species richness in Ontario lakes. *Transactions of the American Fisheries Society* **118**, 533–545.

Moreau, J. and De Silva, S.S. (1991) Predictive yield models for lakes and reservoirs of the Philippines, Sri Lanka and Thailand. *FAO Fisheries Technical Paper* 319.

Moreau, J., Bambino, C. and Pauly, D. (1986) A comparison of four indices of overall fish growth performance, based on 100 tilapia populations (Fam. Cichlidae), in *The First Asian Fisheries Forum*, Asian Fisheries Society, Manila, pp. 201–206.

Moreau, J., Christensen, V. and Pauly, D. (1993) A trophic ecosystem model of Lake George, Uganda, in *Trophic Models of Aquatic Ecosystems* (eds V. Christensen and D. Pauly), ICLARM Conference Proceedings 26, pp. 124–129.

Moriarty, D.J.W. (1982) Physiology: chairman's overview, in *The Biology and Culture of Tilapias* (eds R.S.V. Pullin and R.H. Lowe-McConnell), ICLARM, Manila, pp. 115–117.

Moyle, P.B., Li, H.W. and Barton, B.A. (1986) The Frankenstein effect: impact of introduced fishes on native fishes in North America, in *Fish Culture in Fisheries Management* (ed. R.H. Stroud), American Fisheries Society, Bethesda, pp. 415–426.

Muchiri, S.M. (1997) Fluctuations in the fisheries of Kenya's Rift Valley Lakes: causes and prospects for the future, in *Developing and Sustaining World Fisheries Resources, 2nd World Fisheries Congress* (eds D.A. Hancock, D.C. Smith, A. Grant and J.P. Beumer), CSIRO, Collingwood, pp. 90–97.

Muchiri, S.M. and Hickley, P. (1991) The fishery of Lake Naivasha, Kenya, in *Catch Effort Sampling Strategies* (ed. I.G. Cowx), Fishing News Books, Oxford, pp. 382–392.

Muchiri, S.M., Hart, P.J.B. and Harper, D.M. (1995) The persistence of two introduced tilapia species in Lake Naivasha, Kenya, in the face of environmental variability and fishing pressure, in *The Impact of Species Changes in African Lakes* (eds T. Pitcher and P.J.B. Hart), Chapman and Hall, London, pp. 299–319.

Neat, F.C., Huntigford, F.A. and Beveridge, M.C.M. (1998) Fighting and assessment in male cichlid fish: the effects of asymmetries in gonadal state and body size. *Animal Behaviour* **55**, 883–891.

Newsome, A.E. and Noble, I.R. (1986) Ecological and physiological characteristics of invading species, in *Ecology of Biological Invasions: an Australian Perspective* (eds R.H. Groves and J.J. Burdon), Australian Academy of Science, Canberra, pp. 1–20.

Nikolsky, G.V. (1969) *Fish Population Dynamics as the Biological Background for the Rational Exploitation and Management of Fishery Resources*, Oliver and Boyd, Edinburgh.

Noakes, D.L.G and Balon, E.K. (1982) Life histories of tilapias: an evolutionary perspective, in *The Biology and Culture of Tilapias* (eds R.S.V. Pullin and R.H. Lowe-McConnell), ICLARM, Manila, pp. 61–82.

Ofori, J.K. (1988) The effect of predation by *Lates niloticus* on overpopulation and stunting in mixed sex culture of tilapia species in ponds, in *The Second International Symposium on Tilapia in Aquaculture* (eds R.S.V. Pullin, T. Bhukaswan, K. Tonguthai and J.L. McLean), ICLARM, Manila, pp. 69–73.

Oldorf, W., Kronert, U., Balarin, J., Haller, R., Horstgen-Schwark, G. and Langholz, H.J. (1989) Prospects of selecting for late maturity in tilapia (*Oreochromis niloticus*) II. Strain comparisons under laboratory and field conditions. *Aquaculture* **77**, 123–133.

Oliveira, R.F. and Almada, V.C. (1996) On the (in)stability of dominance hierarchies in the cichlid fish *Oreochromis mossambicus. Aggressive Behaviour* **22**, 37–45.

Ogutu-Ohwayo, R. and Hecky, R.E. (1991) Fish introductions in African and some of their implications. *Canadian Journal of Fisheries and Aquatic Sciences* **48** (Supplement 1), 8–12.

Pagan-Font, F.A. (1975) Cage culture as mechanical method for controlling reproduction of *Tilapia aurea* (Steindachner). *Aquaculture* **6**, 243–247.

Papotsoglou, S.E. and Tziha, G. (1996) Blue tilapia (*Oreochromis aureus*) growth rate in relation to dissolved oxygen concentration under recirculated water conditions. *Aquacultural Engineering* **15**, 181–192.

Pauly, D. (1980) On the interrelationship between natural mortality, growth parameters, and mean environmental temperature in 175 fish stocks. *Journal du Conseil Permanent internationale pour l'Explorartion de la Mer* **39**, 175–192.

Pauly, D. (1981) The relationship between gill surface area and growth performance in fish: a generalisation of von Bertalanffy's theory of growth. *Berichte der Deutschen Wissenschaftlichen Kommission fuer Meeresforschung* **28**, 251–282.

Pauly, D. (1984) A mechanism for the juvenile to adult transition in fishes. *Journal du Conseil Permanent internationale pour l'Explorartion de la Mer* **41**, 280–284.

Pauly, D. and Morgan, G.P. (eds) (1987) *The Theory and Application of Length-Based Methods of Stock Assessement*, ICLARM, Manila.

Pauly, D., Moreau, J. and Prein, M. (1988) A comparison of overall growth performance of tilapia in open water and in aquaculture, in *The Second International Symposium on Tilapia in Aquaculture* (eds R.S.V. Pullin, T. Bhukaswan, K. Tonguthai and J.L. McLean), ICLARM, Manila, pp. 469–479.

Pauly, D., Prein, M. and Hopkins, K.D. (1993) Multiple regression analysis of aquaculture experiments based on the 'extended Gulland and Holt plot', in *Multivariate Methods in Aquaculture Research: Case Studies of Tilapias in Experimental and Commercial Systems* (eds M. Prein, G. Hulata and D. Pauly), ICLARM, Manila, pp. 13–23.

Pechlaner, R. (1984) Dwarf populations of Arctic charr in high-mountain lakes of the Alps resulting from under-exploitation, in *Biology of Arctic charr* (eds L. Johnson and B.L. Burns), University of Manitoba Press, Winnipeg, pp. 319–327.

Pet, J.S. and Piet, G.J. (1993) The consequences of habitat occupation and habitat overlap of the introduced tilapia *Oreochromis mossambicus* and indigenous fish species for fishery management in a Sri Lankan reservoir. *Journal of Fish Biology* **43** (Supplement A), 193–208.

Pet, J.S., Pet-Soede, C. and Densen, W.L.T. van (1995) Comparison of methods for the estimation of gillnet selectivity to tilapia, cyprinids, and other species of Sri Lankan reservoir fish. *Fisheries Research* **24**, 141–164.

Pet, J.S., Gevers, G.J.M., Densen, W.L.T. van and Vijverberg, J. (1996a) Management options for a more complete utilisation of the biological production in Sri Lankan reservoirs. *Ecology of Freshwater Fish* **5**, 1–14.

Pet, J.S., Machiels, M.A.M. and Densen, W.L.T. van (1996b) A size-structured simulation model for evaluating management strategies in gillnet fisheries exploiting spatially differentiated populations. *Ecological Modelling* **88**, 195–214.

Peters, H.M. (1963) Eizahl, Eigewicht und Gelegeentwicklung in der Gattung *Tilapia* (Cichlidae, Teleostei). *Internationale Revue der gesamten Hydrobiologie* **48**, 547–576.

Philippart, J.C. and Ruwet, J.C. (1982) Ecology and distribution of tilapias, in *The Biology and Culture of Tilapias* (eds R.S.V. Pullin and R.H. Lowe-McConnell), ICLARM, Manila, pp. 15–59.

Piironen, J. and Holopainen, I.J. (1988) Length structure and reproductive potential of crucian carp (*Carassius carassius* (L.)) populations in some small forest ponds. *Annales Zoologici Fennici* **25**, 203–208.

Policansky, D. (1983) Size, age and demography of metamorphosis and sexual maturation in fishes. *American Zoologist* **23**, 57–63.

Prein, M. (1993a) Multiple regression and path analysis of Nile tilapia growth in in integrated livestock-fish culture in the Philippines, in *Multivariate Methods in Aquaculture Research: Case Studies of Tilapias in Experimental and Commercial Systems* (eds M. Prein, G. Hulata and D. Pauly), ICLARM, Manila, pp. 50–66.

Prein, M. (1993b) Multivariate analysis of tilapia growth experiments in Israel, Zambia and Peru, in *Multivariate Methods in Aquaculture Research: Case Studies of Tilapias in Experimental and Commercial Systems* (eds M. Prein, G. Hulata and D. Pauly), ICLARM, Manila, pp. 75–87.

Prein, M. and Pauly D. (1993) Two new approaches to examining multivariate aquaculture growth data: the 'extended Bayley plot' and path analysis, in *Multivariate Methods in Aquaculture Research: Case Studies of Tilapias in Experimental and Commercial Systems* (eds M. Prein, G. Hulata and D. Pauly), ICLARM, Manila, pp, 32–49.

Pullin, R.S.V, Palomares, M.L., Casal, C.V., Dey, M.M. and Pauly, D. (1997) Environmental impacts of tilapias, in *Proceedings of the Fourth International Symposium on Tilapia in Aquaculture*, Vol. 2 (ed. K. Fitzsimmons), Northeast Regional Agricultural Engineering Service Cooperative Extension, Ithaca, pp. 554–570.

Quiros, R. (1998) Reservoir stocking in Latin America, an evaluation. *FAO Fisheries Technical Paper* 347, 91–117.

Ridgeway, L.L. and Chapleau, F. (1994) Study of a stunted population of yellow perch (*Perca flavescens*) in a monospecific lake in Gatineau Park, Quebec. *Canadian Journal of Zoology* **72**, 1576–1582.

Riley, J. and Darmi, M. (1995) A survey of statistical use in aquaculture research. *Aquaculture Research* **26**, 95–101.

Rochet, M.J. (1998) Short-term effects of fishing on life history traits of fishes. *ICES Journal of Marine Science* **55**, 371–391.

Roff, D.A. (1983) An allocation model of growth and reproduction in fish. *Canadian Journal of Fisheries and Aquatic Sciences* **40**, 1395–1404.

Roff, D.A. (1992) *The Evolution of Life Histories: Theory and Analysis*, Chapman and Hall, New York.

Roff, D.A. (1996) The evolution of threshold traits in animals. *Quarterly Review of Biology* **71**, 3–35.

Rosenberg, A.A. and Beddington, J.R. (1988) Length-based methods of fish stock assessment, in *Fish Population Dynamics* (ed. J.A. Gulland), Wiley, Chichester, pp. 83–103.

Schreiber, S., Focken, U. and Becker, K. (1998) Individually reared female Nile tilapia (*Oreochromis niloticus*) can grow faster than males. *Journal of Applied Ichthyology* **14**, 43–47.

Shuter, B.J., Jones, M.L., Korver, R.M. and Lester, N.P. (1998) A general, life history based model for regional management of fish stocks: the inland lake trout (*Salvelinus namaycush*) fisheries of Ontario. *Canadian Journal of Fisheries and Aquatic Sciences* **55**, 2161–2177.

Siddiqui, A.Q., Al-Harbi, A.H. and Al-Hafedh, Y.S. (1997) Effects of food supply on size at first maturity, fecundity and growth of hybrid tilapia, *Oreochromis niloticus* (L.) × *Oreochrimis aureus* (Steindachner), in outdoor concrete tanks in Saudi Arabia. *Aquaculture Research* **28**, 341–349.

Silliman, R.P. (1975) Selective and unselective exploitation of experimental populations of *Tilapia mossambica*. *Fishery Bulletin* **73**, 495–507.

Sprent, P. (1989) *Applied Nonparametric Statistical Methods*. Chapman and Hall, London.

Springborn, R.R., Jensen, A.L., Chang, W.Y.B., and Engle, C. (1992) Optimum harvest time in aquaculture: an application of economic principles to a Nile tilapia, *Oreochromis niloticus* (L.), growth model. *Aquaculture and Fisheries Management* **23**, 639–647.

Sreenivasan, A. (1991) Transfers of freshwater fishes into India, in *Ecology of Biological Invasions in the Tropics* (ed. P.S. Ramakrishnan), International Science Publications, New Delhi, pp. 131–138.

Stamps, J.A., Mangel, M. and Phillips, J.A. (1998) A new look at the relationship between size at maturity and asymptotic size. *The American Naturalist* **152**, 470–479.

Stearns, S.C. (1992) *The Evolution of Life Histories*, Oxford University Press, Oxford.

Sugunan, V.V. (1997) Fisheries management of small water bodies in seven countries in Africa, Asia and Latin America. *FAO Fisheries Circular* 933.

Sutherland, W.J. (1996) *From Individual Behaviour to Population Ecology*, Oxford University Press, Oxford.

Svedaeng, H. (1990) Genetic basis of life-history variation of dwarf and normal Arctic charr, *Salvelinus alpinus* (L.), in Stora Roesjoen, central Sweden. *Journal of Fish Biology* **36**, 917–932

Swingle, H.S. (1960) Comparative evaluation of two tilapias as pond fishes in Alabama. *Transactions of the American Fisheries Society* **89**, 142–148.

Szczerbowski, A., Zakes, Z., Luczynski, M.J. and Szkudlarek, M. (1997) Maturation and growth of a stunted form of crucian carp *Carassius carassius* (L.) in natural and controlled conditions. *Polskie Archiwum Hydrobiologii* **44**, 171–180.

Taylor, E.B. and Bentzen, P. (1993) Molecular genetic evidence for reproductive isolation between sympatric populations of smelt *Osmerus* in Lake Utopia, south-western New Brunswick, Canada. *Molecular Ecology* **2**, 345–357.

Teichert-Coddington, D. and Green, B.W. (1993) Tilapia yield improvement through maintenance of minimal oxygen concentrations in experimental grow-out ponds in Honduras. *Aquaculture* **118**, 63–71.

Thorpe, J.E. (1994) Reproductive strategies in Atlantic salmon, *Salmo salar* (L.). *Aquaculture and Fisheries Management* **25**, 77–87.

Thorpe, J.E., Talbot, C., Miles, M.S. and Keay, D.S. (1990) Control of maturation in cultured salmon, *Salmo salar*, in pumped seawater tanks, by restricting food intake. *Aquaculture* **86**, 315–326.

Thorpe, J.E., Mangel, M., Metcalfe, N.B. and Huntingford, F.A. (1998) Modelling the proximate basis of salmonid life-history variation, with application to Atlantic salmon, *Salmo salar* L. *Evolutionary Ecology* **12**, 581–599

Trippel, E.A. (1993) Relations of fecundity, maturation, and body size of lake trout and implications for management in northwestern Ontario lakes. *North American Journal of Fisheries Management* **13**, 64–72.

Tsadik, G.G. and Kutty, M.N. (1987) Influence of ambient oxygen on feeding and growth of the tilatia, *Oreochromis niloticus* (Linnaeus). Working Paper ARAC/87WP/10. African Regional Aquaculture Centre, Port Harcourt, Nigeria.

Turner, G.E. (1988) Codes of practice and manual of procedures for consideration of introductions and transfers of marine and freshwater organisms. *EIFAC Occasional Paper* 23, FAO, Rome.

Turner, G.F. (1995) Maximisation of yields from African lakes, in *Stock Assessment in Inland Fisheries* (ed. I.G. Cowx), Fishing News Books, Oxford, pp. 465–481.

Twongo, T. (1995) Impact of fish species introductions on the tilapias of Lakes Victoria and Kyoga, in: *The Impact of Species Changes in African Lakes* (eds T.J. Pitcher and P.J.B. Hart), Chapman and Hall, London, pp. 457–493.

Tyler, J.A. and Rose, K.A. (1994) Individual variability and spatial heterogeneity in fish population models. *Reviews in Fish Biology and Fisheries* **4**, 91–123.

van Dam, A.A. and Pauly, D. (1995) Simulation of the effects of oxygen on food consumption and growth of the Nile tilapia, *Oreochromis niloticus* (L.). *Aquaculture Research* **26**, 427–440.

Van der Mheen, H. (1994) Practical aspects of stocking in small water bodies: an example from Zimbabwe. *CIFA Technical Paper* 28, FAO, Rome.

van Someren, V.D. and Whitehead, P.J. (1959) The culture of *Tilapia nigra* (Guenther) in ponds. I. Growth after maturity in male *T. nigra*. *East African Agriculture and Forestry Journal* **25**, 42–46.

van Someren, V.D. and Whitehead, P.J. (1960) The culture of *Tilapia nigra* (Guenther) in ponds. III. Early growth of males and females at comparable stocking rates, and the length-weight relationship. *East African Agriculture and Forestry Journal* **25**, 169–173.

van Someren, V.D. and Whitehead, P.J. (1961) The culture of *Tilapia nigra* (Guenther) in ponds. V. The effect of progressive alterations in stocking density on the growth of male *T. nigra*. *East African Agriculture and Forestry Journal* **26**, 145–155.

Von Bertalanffy, L. (1957) Quantitative laws in metabolism and growth. *Quarterly Reviews in Biology* **32**, 217–231.

Walline, P.D., Pisanty, S., Gophen, M. and Berman, P. (1993) The ecosystem of Lake Kinneret, Israel, in *Trophic Models of Aquatic Ecosystems* (eds V. Christensen and D. Pauly), ICLARM, Manila, pp. 103–109.

Walters, C.J. and Post, J.R. (1993) Density-dependent growth and competitive asymmetries in size-structured fish populations: a theoretical model and recommendations for field experiments. *Transactions of the American Fisheries Society* **122**, 34–45.

Welcomme, R.L. (1967) The relationship between fecundity and fertility in the mouth-brooding cichlid fish *Tilapia leucostica*. *Journal of Zoology, London* **151**, 453–468.

Welcomme, R.L. (1970) Studies on the effects of abnormally high water levels on the ecology of fish in certain shallow regions of Lake Victoria. *Journal of Zoology, London* **160**, 405–436.

Welcomme, R.L. (1988) International introductions of inland aquatic species. *FAO Fisheries Technical Paper* 294.

Welcomme, R.L. and Bartley, D.M. (1998) Current approaches to the enhancement of fisheries. *Fisheries Management and Ecology* **5**, 351–382.

Williams, K., Gebhard, G.E. and Maughan, O.E. (1987) Enhanced growth of cage cultured channel catfish through polyculture with blue tilapia. *Aquaculture* **62**, 207–214.

Williamson, M. (1996) *Biological Invasions*, Chapman and Hall, London.

Yi, Y. (1998) A bioenergetics growth model for Nile tilapia (*Oreochromis niloticus*) based on limiting nutrients and fish standing crop in fertilized ponds. *Aquacultural Engineering* **18**, 157–173.

Chapter seven

Genetics for the management and improvement of cultured tilapias

D. J. Penman and B. J. McAndrew

7.1 INTRODUCTION

Tilapia were first farmed experimentally in Kenya in 1924 (*Oreochromis spilurus niger*) and in 1937 in Zaire (*Oreochromis nyasalapia macrochir* and *Tilapia rendalli*) (reviewed by Chimits, 1955, 1957). The main period of expansion in Africa was in the 1950s and 1960s as the full potential of this group of fish for natural stocking and culture was realized (reviewed by Balarin and Hatton, 1979). At about the same time, tilapia were gaining importance in Asia after the introduction of *O. mossambicus* into Java (Indonesia) pre-1939 (Chimits, 1955) and subsequent introductions from there to many other countries (reviewed by Pullin, 1988). In Africa, farming of tilapia developed but despite the many beneficial characteristics, problems associated with excessive reproduction in ponds and the small size of the fish produced resulted in disillusionment. In Asia and Israel farmers persisted with species more suited for culture. With increased understanding of their biology and genetics, allied with improvements in management, tilapia have become very important farmed species. *O. niloticus* and certain hybrids have now become the major cultured tilapia.

Historically, the first application of genetics to tilapia culture was probably the inadvertent hybridization between *O. mossambicus* and *O. urolepsis hornorum*, the 'Malacca hybrids' (Hickling, 1960). Fry from these crosses were all males, and this stimulated considerable further research on hybridization, sex determi-

M.C.M. Beveridge and B.J. McAndrew (eds), Tilapias: Biology and Exploitation, 227–266.

nation and sex ratio control. In this chapter we do not attempt to present the application of genetics in tilapia culture from a historical perspective, but rather to organize the chapter by beginning with the initial choice of species, strains and hybrids for different culture environments, then by progressing through the various applications of genetics to tilapia culture and describing some of the associated research.

7.2 CHOICE OF SPECIES AND STRAINS

Genetic techniques for species identification

The most commonly accepted classification of tilapia recognizes four genera, *Oreochromis*, *Sarotherodon*, *Tilapia* and *Danakilia* (Trewavas, 1983). The first three of these are important in aquaculture, with species from the genus *Oreochromis* accounting for the bulk of world cultured tilapia production (FAO, 1998). The phylogenetics and zoogeography of tilapias are dealt with in detail earlier in this book (Chapter 1).

The widespread movement and mixing of species in Africa has resulted in hybridization between tilapia species in many rivers and lakes (see below and Wohlfarth and Hulata, 1983). The adoption of hybridization for sex control and to improve the performance of red tilapia strains used in aquaculture (see below) and the subsequent poor management of these fish has resulted in breeding between hybrids and their parental stocks and other farmed strains, resulting in serious genetic introgression (Macaranas *et al.*, 1986). There has been a need for methodologies that can unequivocally identify the purity, specific status and genetic potential of an individual or strain. A key based on morphological traits, breeding colour or behaviour will be unreliable in many circumstances because of the high incidence of hybrid introgression and the fact that these species are now grown under a wide range of differing environments outside the species' natural range.

Species discrimination techniques have been developed which rely on detecting variation between individuals and species at the protein and DNA level. These traits have the advantage that they are relatively stable: they are generally not influenced by environmental factors and are inherited in highly predictable ways. The earliest developments in general protein separation and identification were applied to the problem of species identification and discrimination in this group (reviewed by McAndrew and Sodsuk, in press). Initial studies used general proteins, haemoglobin and easily stained enzyme loci (Iles and Howlett, 1967; Chen and Tsuyuki, 1970; Avtalion and Wojdani, 1971; Basasibwaki, 1975; Avtalion *et al.*, 1975, 1976; Avtalion, 1982). Further development of protein- and enzyme-specific stains rapidly expanded the number of gene loci that could be rapidly and cost-effectively studied. McAndrew and Majumdar (1983) and Sodsuk and McAndrew (1991) identified 25 and 40 allozyme loci, respectively, that could be used in both species identification and phylogenetic studies. Table 7.1 shows the monomorphic diagnostic loci that can be used to discriminate between the four most commonly farmed (and hybridized) *Oreochromis* species.

Table 7.1 Diagnostic loci (fixed allele differences) between pairs of the most commonly farmed *Oreochromis* species (after Sodsuk, 1993). The locus designation and preferred buffer systems can be obtained from Sodsuk *et al.* (1995)

	O. aureus	*O. u. hornorum*	*O. mossambicus*
O. niloticus	AAT-3* ADA* ADH* FH-1* G6PDH-2* MDH-3*	ADA* AH-2* EST-1* FBALD-2* FH-1* G6PDH-1* MDH-3* SOD*	ADA* CK* DDH-1* FBALD-2* FH-1* G6PDH-2* IDDH* MDH-3* SOD*
O. aureus		AAT-3* ADA* ADH* AH-2* EST-1* GDA* G6PDH-1* G6PDH-2* SOD*	AAT-3* ADA* ADH* ALAT* CK* DDH-1* GDA* IDDH* SOD*
O. u. hornorum			AH-2* ALAT* CK* DDH-1* EST-1* G6PDH-1* G6PDH-2* IDDH*

One criticism of the protein gel electrophoresis approach is the general need to kill fish to obtain different tissues and the need for relatively sophisticated laboratory facilities. Some work has been done to try to develop field testing kits based on a suite of blood agglutination and immunological tests, in which each species would give a specific pattern (Falk *et al.*, 1996). However, gel electrophoresis of blood proteins (paralbumins and haemoglobins) was more successful for hybrid identification in this study. The agglutination technique, although possible in the field with a minimum of equipment, appeared to lack the specificity required to identify hybrid individuals and would also be limited by the size at which blood samples could be removed from small fish without killing them.

Macaranas *et al.* (1996) showed that *O. mossambicus*, *O. aureus* and *O. niloticus* could be distinguished on the basis of haemoglobin electrophoresis patterns and that the Kenyan subspecies of *O. niloticus* (*O. n. vulcani*) could be separated from the other *O. niloticus* strains.

Species identification has proved fairly straightforward in tilapia using protein electrophoresis but this technique is not generally suitable for discriminating between populations or subspecies because of the relatively low levels of genetic variation detected. Seyoum and Kornfield (1992) showed that all seven subspecies of *O. niloticus* could be distinguished using mtDNA RFLP, which relies on the visualization of the cleavage fragments produced when purified mtDNA is cut using restriction endonucleases.and separated by gel electrophoresis. However this technique is not suitable in its present form for routine screening of farmed stock as it requires quite large amounts of mtDNA which normally requires the animal to be killed. The development of the polymerase chain reaction (PCR) for the analysis of variation at the DNA level now means that small amounts of tissue are all that are required in many analyses. If subspecies-specific variation can be identified to a short sequence of mitochondrial or nuclear DNA it should prove amenable to this technology. At this stage in the development of PCR-based DNA techniques in tilapia, few species- or subspecies-diagnostic markers have been identified. Bardakci and Skibinski (1994, 1999) and Dinesh *et al.* (1996) showed that RAPD markers could be used to distinguish between tilapia species (*O. niloticus*, *O. aureus* and *O. mossambicus*). Bardakci and Skibinski (1994) and Naish *et al.* (1995) also found differences between strains of *O. niloticus* with some strain-specific DNA bands.

The application of microsatellite DNA loci, which is also PCR-based, to species identification may be hindered by the nature of the loci identified. Microsatellite DNA loci are identified by the presence of simple repeats in cloned DNA, then amplified using primers which hybridize to the unique DNA sequences flanking such repeats. Some primer sites mutate, resulting in the appearance of null alleles or a complete lack of amplification in different species or genera. In the case of a complete lack of amplification, it is not possible to tell if the locus is absent or is present but with mutations in both primer sites which prevent amplification. As more microsatellite loci are identified and more cross-species amplification is attempted, it should not be long before a reliable set of such markers can be developed for species identification purposes. Multiple diagnostic loci increase the probability of identifying hybrid introgression in wild and farmed strains even after several generations of subsequent breeding. Sobolewska (personal communication) identified a suite of four microsatellite loci, originally isolated from *O. niloticus* (Lee and Kocher, 1996) that reliably amplify across all Tilapiine genera and could form the basis of a simple multilocus species diagnosis test.

Choice of pure species and hybrids for aquaculture in different environments

Much of the genetic improvement in tilapia to date has been through the application of one-off strategies to improve yield such as single sex populations or hybridization for the development of some red tilapia strains. However, the correct choice of species is a critical initial factor at the start of any tilapia aquaculture enterprise.

Despite the large number of different tilapia species available, aquaculture has tended to become focused on relatively few species. In the early days in Africa a large number of local species were farmed (Balarin and Hatton, 1979). To some extent this is still the case today, particularly where there are issues or worries about biodiversity (Maluwa and Dickson, 1996). In Asia the early industry was based on a chance discovery of *O. mossambicus* in Java (Schuster, 1952). The source of this introduction is unknown but presumably it was a direct introduction from southern Africa. This has proved to be a particularly unsuitable species in freshwater aquaculture.

In tropical freshwater *O. niloticus* or one of its subspecies now dominate, whereas in subtropical freshwater *O. aureus*, which has increased cold tolerance, is often substituted for *O. niloticus* or used to produce a hybrid. In brackish or saline conditions a number of species can be used but the preference, particularly in the Caribbean or Latin America, has been for one of the hybrid red tilapia strains based on the Taiwanese or Florida strains. These strains are red *O. mossambicus* that has been hybridized to pure or hybrid *O. niloticus*, *O. aureus* or *O. hornorum* strains (reviewed by McAndrew and Wohlfarth, in press). In higher salinity and temperature conditions such as in the Red Sea, species such as *O. spilurus spilurus* and *O. s. niger* have been assessed (Cruz *et al.*, 1990). In most Asian countries tilapia farmers have changed from using *O. mossambicus* or *O. mossambicus*/*O. hornorum* hybrids to *O. niloticus* or *O. aureus*. A catalogue of the introductions and their known movements are given by Pullin and Capili (1988).

There has been extensive hybridization, both intentional and unintentional, between different tilapia species. The reports on hybridization in the literature can be divided into three different categories: (i) observations of hybridization in natural or disturbed populations; (ii) hybridization to study the genetics of reproductive behaviour in the different tilapia genera (Heinrich, 1967; Bauer, 1968); and (iii) aquaculture-related (see Wohlfarth and Hulata, 1983; Majumdar and McAndrew, 1983; McAndrew, 1993). Aquaculture-related studies include both practical studies on single sex production and basic research into sex-determination mechanisms (see section 7.5) and strain improvement (particularly of red tilapia strains), in addition to inadvertent introgression between species in culture.

Hybridization in disturbed natural populations has been widely recorded (Wohlfarth and Hulata, 1983; Trewavas, 1983). The expansion of modern agriculture techniques in Africa during the various colonial periods required the building of dams and reservoirs for domestic and irrigation purposes: many of these were stocked with fish caught from the wild. This resulted in the movement of species between drainages and the breakdown of geographical isolation. The artificial conditions within these water bodies also resulted in the breakdown of reproductive barriers between species. These problems were exacerbated by the lack of information on the biology of the species and their taxonomic relationships. This period resulted in the displacement of species sometimes on quite a large scale e.g. *Oreochromis esculentus* in Lake Victoria disappeared after they had hybridized with the introduced *O. niloticus* (Welcomme, 1967), which has been subsequently confirmed using species-diagnostic genetic markers (Mwanza and Kaufman, 1995). The disappearance of the other Lake Victoria endemic Tilapiine *O. variabilis* is also believed to have been the result of hybridization (Welcomme, 1964). Today these species are only present in a few satellite lakes around Lake Victoria and are in serious risk of extinction (Agnese *et al.*, 1999). Hybridization and apparent loss of pure species occurred after *O. niloticus* was introduced into Lake Itasy in Madagascar, which already had an introduced population *of O. ny. macrochir*, which apparently was well established. Both typical forms of the two species gradually disappeared in less than 13 years (Daget and Moreau, 1981) through a process of hybrid introgression.

Among the many studies on hybridization for aquaculture purposes, there have been few attempts to study or quantify the economic benefit in terms of important traits such as growth performance. Anecdotal stories and the pervading impression in the industry is that heterosis is common in fish hybrids. This impression comes from the widespread and significant improvements that have arisen from intraspecific hybridization between inbred strains in crop and animal production. At the level of interspecific or intergeneric hybridization in fish there are few examples of hybrid vigour *per se* (Chevassus, 1983). Studies that have produced interspecific and intergeneric hybrids and have compared growth rates of the hybrid against both parental species under standardized conditions are few (McAndrew and Majumdar, 1988; Yapi-Gnaore, 1996; Rana *et al.*, 1996) and do not suggest that heterosis is a widespread or significant phenomenon in tilapia.

Some interspecific hybrids may exhibit a trait or combination of traits that endow a management or commercial advantage. This is exemplified in Israel where advantageous traits from two species are combined: fast growth in *O. niloticus* and increased hardiness and cold tolerance from *O. aureus*, with the added advantage of a sex ratio highly skewed towards male, often reinforced with hormone sex-reversal. The hybrid is seen as advantageous over either of the pure species because of the need to overwinter fish to achieve the premium market size (Wohlfarth, 1994).

One of the few studies to report positive heterosis in intraspecific crosses between *O. niloticus* strains (Jayaprakas *et al.*, 1988) used strains that allozyme variation studies (Brummett *et al.*, 1988) suggest had been highly inbred. A large scale cross breeding experiment between eight strains of *O. niloticus*, as part of the GIFT programme, suggested that any heterosis in crosses between the different wild and domestic strains studied was less than 6% (Bentsen *et al.*, 1998) and was not worth exploiting directly.

The studies on aquaculture-related hybridization reported above largely concern intentionally produced F1 hybrids at the interspecific or intraspecific level. Introgression between species through several generations of hybridization, either intentional or unintentional, has also been widespread in tilapia. Intentional hybridization between *O. niloticus* and *O. aureus* in Israel has been used extensively but poor management enabled hybrids to breed with the pure species and so reduce the percentage of males in the hybrid fry. In order to correct this problem molecular markers had to be used to test the purity of all broodstock being used in Israeli hatcheries (Avtalion, 1982; Wohlfarth, 1994).

Unintentional hybridization and subsequent introgression caused by feral populations of *O. mossambicus* entering ponds and breeding with imported farmed strains of *O. niloticus* and *O. aureus* in the Philippines has been shown to have dramatically reduced the commercial performance of these strains (Taniguchi *et al.*, 1985; Macaranas *et al.*, 1995). This was one of the factors which gave rise to the Genetic Improvement of Farmed Tilapia (GIFT) programme, which introduced new wild isolates of *O. niloticus* from Africa as the basis of a breeding programme (Eknath *et al.*, 1991).

Red tilapia

Strain improvement by hybridization has however been achieved in the development of red tilapia strains from red mutants that originally occurred in *O. mossambicus*. A number of different red variant *O. mossambicus* strains have occurred independently in Asia and the USA. The light body colour and the absence of a black mesentery in these strains were seen as having a commercial advantage over the black colour that was usual in this species. The red and wild type *O. mossambicus* strains generally performed poorly in comparison with the faster growing *O. niloticus* and *O. aureus* strains that were quickly replacing them in many countries. A number of hybrid breeding programmes have crossed the red *O. mossambicus* with faster growing pure or hybrid strains of *O. niloticus* or *O. aureus* (Sipe, 1979; Behrends *et al.*, 1982; Liao and Chen, 1983; Behrends and Smitherman, 1984a,b; Kuo, 1988; reviewed by McAndrew and Wohlfarth, in press).

This has resulted in a large number of different strains that vary in species composition, colour inheritance, growth rates, salinity tolerances, fecundity and other commercially important traits. The development of the more widespread red hybrids strains is complicated and even within the named strains (Taiwanese, Philippine, Florida) various isolates have been modified by selection and further hybridization as they have been transferred from farm to farm. The more structured crossing programmes have shown that the hybrid red strains show significant improvements in growth or other traits such as cold tolerance over the red *O. mossambicus* progenitor but are not necessarily better than the *O. niloticus* and/or *O. aureus* strains used in the hybridization (Behrends *et al.*, 1984b).

The genetic mechanisms controlling the red coloration in the different strains also appear to vary, often between isolates of the same nominal strain. The red coloration is a highly variable trait: the term red can refer to an individual fish or a strain in which the normal wild-type coloration is replaced with various single or multicolour patterns which can include varying amounts of red, yellow, orange, pink and white which may be also be covered with black blotches. Skin colour pattern in fish is controlled by a number of pigment cells generically called chromatophores. In red tilapia it is usually the absence of the dendritic melanophores which contain the black melanin and dark brown eumelanin pigment that enable the red coloured erythrophores and yellow xanthophores to dominate the skin colour (McAndrew *et al.*, 1988).

The inheritance of red colour has been reviewed by Wohlfarth *et al.* (1990) and McAndrew and Wohlfarth (in press). It appears that a number of independent red mutations that have occurred in Asia and the USA have been used as the basis for developing these red strains. Subsequent developments have often involved the mixing of different red stains with different underlying mutations and different species compositions. The different genes and genetic backgrounds in these hybrid strains has made it very difficult to generate pure red breeding strains for aquaculture as many different colour phenotypes including wild type will segregate each generation. This has to be compared to the relatively simple inheritance observed in pure *O. mossambicus* (Tave *et al.*, 1989) and *O. niloticus* (McAndrew *et al.*, 1988) red strains.

In recent years developments in red tilapia have followed two separate directions with the development of salt-tolerant strains better suited to brackish and seawater aquaculture as well as those needed for freshwater conditions. The salt-tolerant strains have mostly been developed from the Florida red tilapia which has an *O. mossambicus* and *O. hornorum* parentage (Watanabe, 1991). Freshwater performance of the early red strains was always inferior to the *O. niloticus* and *O. aureus* wild type strains, hence the need for hybridization. Today there is a need to evaluate the commercial value of the various hybrid and pure strains that have been developed for growth and colour segregation. Hulata *et al.* (1995) have undertaken comparative commercial trials on red tilapia in freshwater comparing Philippine red tilapia (PRT), Stirling Red Niloticus (SRN) with wildtype *O. niloticus* and *O. aureus* and a number of hybrid crosses between these strains. Hybrids between wildtype *O. niloticus* and SRN and PRT × *O. aureus* gave the best performance but the homogeneous red colour produced by the SRN × *O. niloticus* cross compared to the 60% segregation of the less favoured bronze phenotype in the PRT cross made the former the preferred combination.

Performance differences between strains within species

In the early stages of the GIFT programme, four existing Asian farmed strains and imported fish from four wild populations from across the natural range of *O. niloticus* were compared in eleven different aquaculture environments commonly found in the Philippines (Eknath *et al.*, 1993). This experiment looked at individually tagged pure strains grown communally in all environments. In a second experiment fish from a complete 8 × 8 diallele cross between all strains were evaluated in eight test environments (Bentsen *et al.*, 1998). The results of the first trial showed that apart from the Ghana strain the wild strains outperformed the established farmed *O. niloticus* strains and that the Nile (Egypt)/Rift valley (Kenya) strains were consistently better than the West African strains across most test environments (see Chapter 1). The results of the diallel crossing experiment agreed well with the first trial (Bentsen *et al.*, 1998). The overall ranking of the different pure strains was not changed in that the Kenyan and Nile strains were the best and all wild strains apart from Ghana were better than the farm strains. The experiments showed that although there was some heterosis it was small and not consistent across test environments. The highest per cent heterosis (39%) was observed in crosses (Singapore × Thailand) which still performed more poorly than the best wild strain (Kenya). The best performance overall was in the Kenya × Egypt which was about 11% better than the pure Kenya strain.

The overall conclusion from this study was that any heterosis was low and while it could contribute approximately 10% improvement over the best pure strain this would require the development of specialized sire and dam lines to exploit it to the full. Even then this heterosis was not evenly expressed in all culture environments and thus might require particular combinations of strains for specific environments. This particular approach was rejected because of the logistics of establishing and maintaining a crossbreeding programme at this early stage in the tilapia improvement programme. The preferred path was selective breeding to improve performance generation on generation. This simplifies the dissemination in that only a single improved strain needs to be given to the farmers. Selective improvement was considered to be likely to exceed that obtained in a crossbreeding programme after very few generations. This is the only published large-scale study comparing different strains of one tilapia species in such detail.

7.3 SELECTIVE BREEDING

The application of quantitative genetics to the improvement of tilapia strains began slowly because of the diversity of the culture systems and the difficulties the developing tilapia aquaculture sector was having in trying to overcome basic husbandry and technical problems. The earliest studies were also not very encouraging as they showed relatively low heritability estimates for growth. Thien (1971) showed heritability estimates of 0.10 and 0.16 for growth in male and female *O. mossambicus* respectively. Tave and Smitherman (1980) estimated heritabilities of 0.10 in the Ivory Coast strain of *O. niloticus* which subsequently yielded a negative realized heritability for growth (Teichert-Coddington and Smitherman, 1988). It is very likely that the genetic variation in the strains used in these early Auburn experiments had been seriously reduced by genetic bottlenecking (see Brummet *et al.*, 1988). Uraiwan (1988) observed low heritabilities for age and size at maturity in both males (0.10 ± 0.04 and 0.01 ± 0.03) and females (0.0 and 0.02 ± 0.0) in a Thai (Chitralada) strain of *O. niloticus*.

An attempt to improve the growth rate of *O. niloticus* (Ghana strain) by a mass selection approach (Hulata *et al.*, 1986) also failed to obtain any significant improvement over an unselected control over two generations of selection. Huang and Liao (1990) also failed to obtain a significant response to mass selection for weight in an *O. niloticus* strain imported from Japan in 1966. In both of these studies inbreeding of the strains was given as the most likely cause of failure to obtain a response. Lester *et al.* (1988) studied growth and reproductive performance in four Philippine *O. niloticus* strains. Although the heritability values they obtained were generally higher, standard errors were large because of the experimental design and the limited number of families studied.

Bondari *et al.* (1983) obtained a realized heritability of 0.23 ± 0.05 for body weight in a Tifton *O. aureus* strain. Other early studies which showed that there was potential for selective improvement in tilapia utilized strains that had come directly from the wild and been managed to maintain original levels of genetic variation as far as possible. Kronert *et al.* (1987, 1989) and Oldorf *et al.* (1989) studied selection for late maturation in various *O. niloticus* strains under laboratory and farm conditions. Kronert *et al.* (1987, 1989) showed medium to high heritability values and good genetic variability for a range of gonadal development traits and growth, initially in the Stirling *O. niloticus* strain (from Lake Manzala in Egypt). These results suggested that selection for growth and late maturity should be very promising. This work was repeated using a number of *O. niloticus* strains and comparing full sib families under laboratory and farm environments (Oldorf *et al.*, 1989). Despite a drop in growth performance and the obvious technical problems of working in a farm environment, overall the heritability values remained close to those observed in the laboratory. The conclusions were that the underlying mechanisms controlling these traits in the different environments were similar and that accurate heritability estimates determined under laboratory conditions could be used for an actual breeding programme until more specific estimates of comparable accuracy become available. They also noted that there was significant inter-strain variability for late maturation and that this should be recognized in identifying possible founder stocks for any breeding programme.

Technically, tilapia are quite a difficult species to selectively improve as spawning activity is difficult to synchronize in large populations and results in mixed aged fry. There are also divergent growth patterns between male and female fish as well as differences between individuals related to the onset of sexual maturation. All of these can complicate experimental design and subsequent genetic analysis. Various innovative solutions were proposed by Uraiwan and Doyle (1986), Villegas and Doyle (1986), Basiao and Doyle (1990), and Brzeski and Doyle (1995) that would enable selective improvement to be undertaken in relatively small hatcheries and farms, particularly in Asia. These smaller scale programmes would help to increase or maintain diversity in local farmed strains, which is likely to be lost with the distribution of fish from national or international breeding programmes (Doyle *et al.*, 1991).

Basiao and Doyle (1999) recently reported a 3% improvement over the controls and a realized heritability of 0.16 after one generation of mass selection for body length in a net cage population in the Philippines. This utilized the collimation method (Doyle and Talbot, 1986), i.e. removal of larger fry and smaller fry at 21 days to generate a population that has an increased ratio of genetic to phenotypic variation. Non-genetic or environmentally-induced size variation is a particular problem for pond produced tilapia fry (generally as a result of asynchronous spawning), making selection based on large phenotypic size often a poor indicator of breeding value under these conditions. A group of 1500 fry were removed from the middle 16–17 mm grading and were ongrown for 8 weeks, when the largest 300 individuals where selected. An unselected group (fish near to the mean) was also chosen. The fish were marked and ongrown communally for a further 12 weeks. At this point 50 of the largest males and females were chosen as the selected broodfish and 50 males and females near the mean of the unselected line were chosen as the control broodstock. Both populations were then mass-spawned and their fry were marked and divided into 10 separate lots, in which the selected and control fry were size-matched (± 1 mm), mixed and grown communally in 10 separate cages under normal artisanal farm conditions. The least square means for body length after 3 months were 98.2 ± 0.59 mm and 95.3 ± 0.59 mm for the selected and control groups respectively. Sanchez *et al.* (1995) reported a realized heritability value of 0.19 over five generations of selection for weight in *O. aureus* in Cuba. Scientifically based selection programmes in tilapia are still fairly rare.

The Genetically Improved Farmed Tilapia programme (GIFT), based in the Philippines, was the first to develop a selective breeding programme for *O. niloticus* at a national and international level (Eknath *et al.*, 1993). The GIFT project was well funded and designed to overcome many of the shortfalls that had hindered previous improvement programmes in this species (Eknath *et al.*, 1991).

The GIFT programme concentrated on growth performance to the end of a 120 day grow-out period. A synthetic base population was constructed from the original wild and cultured strains initially tested (see section 7.2). The contribution of each founder strain to the synthetic base population was in relation to its performance in the strain comparisons: better performing strains contributed more. Cultured *O. niloticus* strains in Asia have been shown to be introgressed with *O. mossambicus* (Macaranas *et al.*, 1986, 1995). These Asian cultured strains contributed to the synthetic base population (Circa *et al.*, 1995), so it is likely that this synthetic population contained a small proportion of *O. mossambicus* genes. Three generations of selection were completed by 1993 (Longalong *et al.*, 1999). Subsequent information on the progress of the GIFT selection programme for growth performance has not yet been published and no accurate estimates of genetic gain have been released. However, an evaluation of the selected fish in comparison to locally cultured strains in several Asian countries has been carried out ('Distribution and Evaluation of Genetically Improved Tilapia in Asia', or DEGITA). Mazid *et al.* (1996) reported that the selected fish grew 40–60% faster than the local tilapia strain in Bangladesh. Comparisons with local strains in other countries, in contrast, showed much smaller differences in growth rate, probably due to being compared to better local strains. Survival rates were also reported to be higher in the GIFT strain.

The GIFT programme also studied the genetic parameters of other economically important traits, such as age at maturation (Eknath *et al.*, 1995; Longalong *et al.*, 1999), carcass traits (Velesco *et al.*, 1995), survival and resistance to disease, for inclusion in a multi-trait selection programme in the future (Chiayvareesajja *et al.*, 1999).

A GIFT study, the first on survival and disease resistance parameters, looked at lysozyme activity and spontaneous haemolytic (SH) activity of blood in 42 full-sib groups within 21 paternal half-sib groups. Chiayvareesajja *et al.* (1999) estimated relatively high heritabilities (0.6–0.7) for lysozyme activity at 30°C and intermediate levels (0.3) at 15°C. There was also a significant negative genetic correlation between serum lysozyme activity and survival rate. It was suggested that serum lysozyme activity may be a promising candidate trait for indirect selection to improve survival rate. Heritability estimates for SH activity were zero for the sire component but 0.7 for the dam component. However, there was no correlation with survival or body weight at harvest and this was not seen as a useful indicator for increased survival.

In an attempt to overcome prolific spawning in production ponds GIFT studied the timing of female maturation in 42 full-sib families within 21 randomly chosen half-sib families. Female maturity was recorded in the families 4 weeks after the first swim-up fry were observed in the ponds containing the communally grown families. Potential breeders were chosen from the families with high (>75%) and low (<20%) proportions of mature females at this stage. The offspring from 25 progeny families of the high and low selected parents were nursed and then grown communally in ponds. The response to selection was highly significant but the results showed a significant correlated response in that the earlier maturing fish were also the largest or faster growing. Because of this negative genetic association, the authors suggested that combined selection for the two traits should be carried out (Longalong *et al.*, 1999).

It is now clear that there is substantial additive genetic variation present in wild and properly managed tilapia strains. It is also clear that strain selection can have profound effects e.g. Nile/Rift valley populations of *O. niloticus* are faster growing compared to those obtained from west Africa. Strain differences for other commercially important traits may also exist making the conservation of these wild populations critical for future developments in strain improvement.

There have been rapid advances in molecular technologies that will also assist in the management and genetic improvement of farmed tilapia strains. The rapid development of highly variable genetic markers and there integration into selective improvement programmes should help geneticists to maximize genetic gains.

One future approach will be the use of marker-assisted selection and the identification of candidate genes. The first requirement for this is the availability of a large number of highly variable and informative markers (Lee and Kocher, 1996). The first tilapia genetic linkage map to be published, for *O. niloticus*, used a combination of microsatellite loci and Amplified Fragment Length Polymorphisms (AFLP) markers (Kocher *et al.*, 1998). Practical programmes aimed at determining quantitative trait loci (QTLs) for cold and salinity tolerance using linkage mapping are presently underway in Israel (Hulata, personal communication).

7.4 CHROMOSOME SET AND PLOIDY MANIPULATIONS

Techniques

During meiosis, the diploid set of chromosomes duplicates, recombines (exchanges pieces of the chromatids between non-sister chromatids) and then goes through two reductional divisions (from four to two copies of the chromosomes in the first meiotic division and then from two sets to one in the second meiotic division) to produce the final haploid gamete. The joining of the haploid sets of chromosomes from an egg and a sperm then restores the zygote to diploid status. At the time of ovulation, an unfertilized fish egg still contains two sets of chromosomes as it has yet to undergo the second meiotic division.

In the normal course of events, the second meiotic division takes place before fusion of the male and female pronuclei, resulting in a diploid zygotic nucleus with one set of chromosomes from each parent and the extra maternal set of chromosomes expelled as the second polar body. The single zygotic cell will subsequently undergo mitosis (duplication of the chromosomes and separation into two identical diploid sets) and cell division to give two cells. This process is then repeated many times to give the multicellular embryo.

Application of a temperature or pressure shock or treatment with certain chemicals shortly after fertilization can prevent the formation and release of the polar body, resulting in a zygotic nucleus which contains three sets of chromosomes (both of the maternal sets and the single paternal set). This will give rise to a triploid fish. Triploidy has been induced in at least three different species of tilapia in the genus *Oreochromis* (*O. niloticus*, *O. aureus* and *O. mossambicus*) using heat, cold or pressure shocks (Chourrout and Itzkovich, 1983; Don and Avtalion, 1986, 1988; Penman *et al.*, 1987b; Pandian and Varadaraj, 1988; Hussain *et al.*, 1991). Hussain *et al.* (1991) describe a detailed comparison of the efficiency of heat, cold and pressure in inducing triploidy in *O. niloticus*. It was concluded that pressure shocks gave the most consistently high yield of triploids.

Similar shocks can be used to interfere with first mitosis, which results in duplication of the diploid chromosome set but no cell division (sometimes referred to as 'endomitosis', Chourrout, 1987). This produces a tetraploid embryo. Although tetraploid embryos have been produced in fairly large numbers in three species of tilapia they appear to have very low viability (Myers, 1986; Don and Avtalion, 1988; Mair, 1988) and rarely survive beyond hatching.

Appropriate ultraviolet (UV) irradiation of sperm before fertilization will result in a sperm which is still capable of fertilizing the egg but which carries an inviable paternal genome. This results in a haploid gynogenetic embryo, with only one set of maternal chromosomes. Many such embryos will survive to hatch but very few will survive to first feeding. There is a report of haploid gynogenetic *O. mossambicus* which survived to the age at which the diploid controls were mature (Varadaraj, 1993) but this appears to be the only case in the literature of haploid fish surviving for so long.

Combining UV irradiation of sperm with an 'early' shock to prevent second polar body extrusion results in a diploid gynogenetic fish, which carries the two sets of maternal chromosomes which were present in the unfertilized egg (see above). This is often referred to as 'meiotic' gynogenesis, to distinguish it from 'mitotic' gynogenesis, in which fertilization of an egg with UV-irradiated sperm is followed by a 'late' shock, which causes endomitosis and restores diploidy. Mitotic gynogenetics are completely homozygous, due to duplication of a single chromosome set, while meiotic gynogenetics are partially heterozygous, depending on the level of heterozygosity in the mother and the degree of recombination during meiosis. Both types of gynogenetics have been produced in tilapia species (Chourrout and Itzkovich, 1983; Penman *et al.*, 1987a; Mair *et al.*, 1987; reviewed by Mair, 1993). The enzyme locus ADA* can be used to distinguish between the two types of gynogenetics: meiotic gynogenetics from a heterozygous female will all be heterozygous due to the high level of recombination observed for this locus, while mitotic gynogenetics will be homozygous for one or other of the ADA* alleles (Hussain *et al.*, 1993).

Irradiation of unfertilized eggs followed by fertilization with normal sperm will result in haploid androgenetic embryos. Late shocks will restore diploidy, giving completely homozygous androgenetic diploids. The only part of the genome of such animals which is not of paternal origin is the mitochondrial DNA, which appears to be able to escape the effects of irradiation at doses which inactivate the maternal nuclear genome. Haploid and diploid androgenetic *O. niloticus* were first produced by Myers *et al.* (1995). Marengoni and Onoue (1998) produced diploid androgenetics using eggs from *O. niloticus* and sperm from either *O. niloticus* or *O. aureus*.

One generation of mitotic gynogenesis or androgenesis from outbred female or male broodstock respectively will generate completely homozygous individuals. There will still be genetic variation between such fish from a single outbred parent, due to the way in which the heterozygosity across the genome segregates into individual gametes which are then duplicated by mitotic gynogenesis or androgenesis. From a homozygous first-generation gynogenetic or androgenetic, another generation of gynogenetic or androgenetic reproduction will produce completely identical individuals. These can be regarded as clones, although they are genetically more equivalent to fully inbred lines. Unlike conventional techniques for producing highly inbred lines, which would involve many generations of sib mating, this has been achieved in two generations (Fig. 7.1). Such lines have been produced in *O. niloticus* (Müller-Belecke and Hörstgen Schwark, 1995; Hussain *et al.*, 1998; Sarder *et al.*, 1999).

Applications

The main potential applied interest in triploidy in tilapia is in controlling reproduction during culture. This aspect is dealt with in section 7.5.

Gynogenesis and androgenesis have proven to be very useful techniques in the elucidation of sex determining systems in tilapia (section 7.5). For example, gynogenesis was used to confirm female homogamety in *O. niloticus* and *O. mossambicus* and to detect gene-centromere recombination for the sex determining loci in *O. aureus* and *O. niloticus* (Penman *et al.*, 1987a; Avtalion and Don, 1990; Mair *et al.*, 1991a,b).

Gynogenesis also has application in gene mapping: haploid gynogenetic embryos were used to construct the first linkage map of *O. niloticus* (Kocher *et al.*, 1998) and meiotic gynogenesis, which has been used to estimate gene-centromere recombination distances for various types of loci (Hussain *et al.*, 1994; Markert *et al.*, 1999), can be used to 'anchor' such linkage groups to centromeres.

Fully inbred tilapia clonal lines can be used to study the genetic basis of many traits, as such lines have proven useful in laboratory organisms such as mice. They provide large numbers of genetically identical animals within a clone and a diversity of genotypes between clones. Segregation of the genes responsible for traits of interest can be studied by making crosses between lines of extreme phenotypes for these traits. Sarder *et al.* (in press) demonstrated significant differences between *O. niloticus* clonal lines for rates of infection and mortality following an artificial challenge with *Aeromonas hydrophila*. These lines also varied significantly for non-specific immune response traits (lysozyme activity and macrophage activity): macrophage activity appeared to be associated with

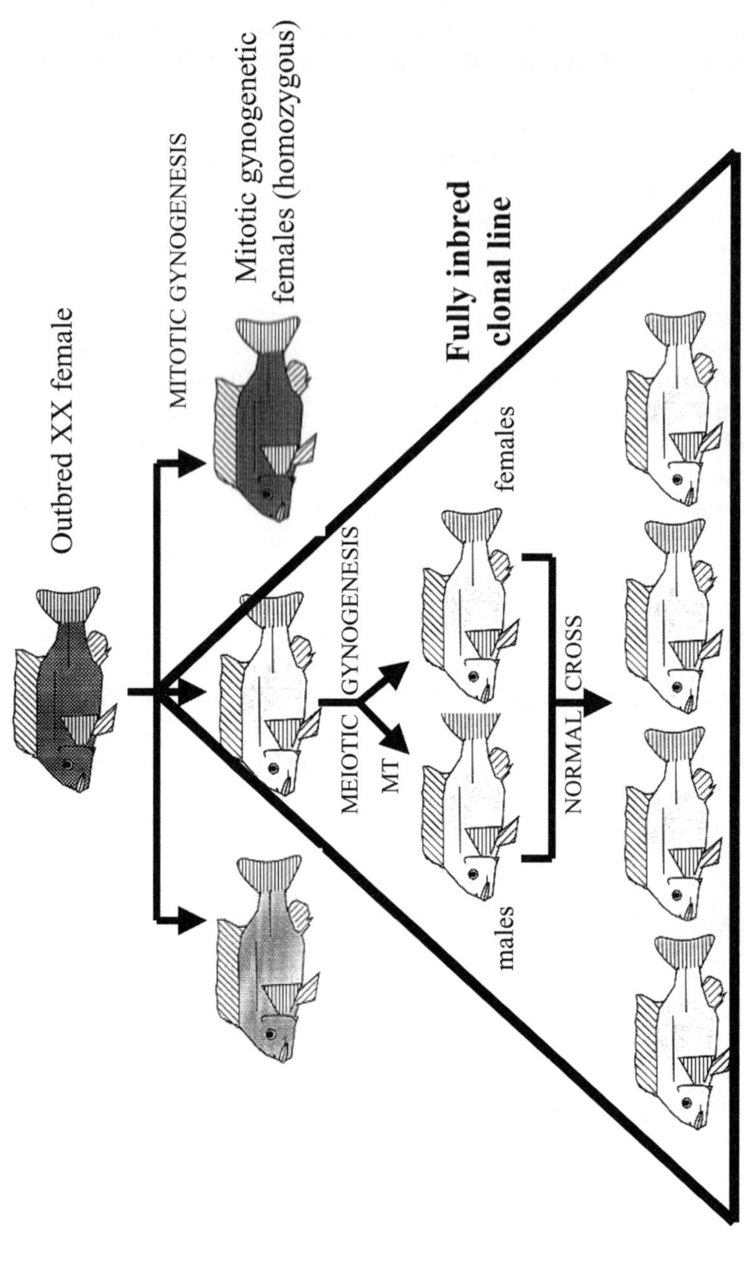

Fig. 7.1 Production of gynogenetic fully inbred clonal lines in *O. niloticus*. All of the fish shown within the triangle are completely homozygous and genetically identical to each other. See text for further details. MT = masculinization by treatment with 17α-methyltestosterone.

infection rate. These lines could also be used in studies on other aspects of disease resistance, toxicology, nutrition, etc.

Androgenesis can also be used in *ex situ* gene banking of fish. The major problem here is that although fish sperm can be cryopreserved routinely, neither eggs or embryos have been successfully cryopreserved (McAndrew *et al.*, 1993). Androgenesis can be used to recover organisms with diploid genotypes from sperm gene banks, by fertilizing UV-irradiated donor eggs with thawed sperm then using a heat shock to diploidize the haploid sperm genome. While not a perfect solution to the problem, due to low survival rates and the high degree of inbreeding, the technique has been demonstrated in the laboratory (Penman *et al.*, 1997).

7.5 CONTROL OF REPRODUCTION

The problems associated with growing tilapia in ponds stimulated research to find better husbandry methods and species for aquaculture. Much of the initial research focused on methods for the control of reproduction to avoid the serious problems of breeding and overcrowding that resulted when mixed sex fry were used to stock production ponds. Several techniques have been used to try to remove fry as they are produced, by the use of predators, or to establish single sex male populations so that reproduction is avoided during the ongrowing phase. Some of these techniques have become more sophisticated as our understanding of the problem and the biology of the species has improved, while others are now seldom used. The major genetic and non-genetic techniques are described here for comparison.

Predator control

By introducing predatory fish into ponds along with stocked tilapia fingerlings, the final size of the tilapia produced can be increased. Various predator species have been used depending on availability. In Thailand *Ophicephalus* and *Clarias* species were used along with *O. mossambicus* (Chimits, 1957). In Africa a number of species have been tried, including *Lates niloticus* (Meschkat, 1967), and *Micropterus salmoides* (Meschkat, 1967). In Central America *Cichlasoma managuense* (Dunseth and Bayne, 1978) or *Cichla ocellaris* (McGinty, 1983) have also been used for this purpose. All of the predator species impose other restrictions and require optimization of predator: prey ratios to achieve a balance between the fecundity of the tilapia species and the voracity of the predator. The timing of the introduction and the size of predator have also to be considered. This technique was only really successful in countries with the management ability to be able to breed or collect predators of the correct size and in the numbers needed at the correct time of the production cycle to be successful. This technique was also relatively inefficient because of the overall losses in growth rate associated with maturation and reproduction in the tilapia and the recycling of this through another species. This is the reason that there has been a move towards the production of single sex all-male fish for commercial production.

Hand sexing

Tilapia show a large degree of sexual dimorphism compared to most other fish species used in aquaculture. The males and females can be identified on the basis of external genital morphology (see p. 117 in Balarin and Hatton, 1979) at approximately 10 g. Although wasteful in labour and fish, because the females are usually discarded, this technique was widely applied (Hickling, 1963; Meschkat, 1967; Shell, 1967). The first successful application was described by van Someran and Whitehead (1959) on *O. spilurus niger* in Kenya. The process was labour intensive (approximately 2000 fish per person per day) and would take an average of 25 person-days to stock a 5 hectare pond which would result in the wastage of 250 kg of females, assuming an average weight of 10 g (Lovshin and DeSilva, 1975). The technique requires the operator to assess the number of openings in the urinogenital papillae of each fish. Males have a single urogenital aperture, while females have two separate apertures. Hand sexing is more accurate for larger fish. At best the technique is 80 to 90% accurate. Trying to hand sex fish of less than 10 g will increase the proportion of females in the 'male' group. Today this technique is rarely used but may occasionally prove useful to save a production cycle in a poorly managed pond in which there has been extensive uncontrolled reproduction, by eliminating females and small, unsexable fry then restocking only the males.

Monosex fry by hybridization

In Asia the tilapia culture was initially based on *O. mossambicus* that had been spread from Indonesia to many other countries in that area. In the late 1950s C.F. Hickling, working at the Malacca Station in Malaysia, undertook a programme to develop and improve the performance of the introduced *O. mossambicus*. In an attempt to produce a sterile fish by hybridization he imported a new '*O. mossambicus*' strain from Zanzibar. Experimentation showed that in crosses between the females of the existing strain and the Zanzibar males all the fry produced were male, while the reciprocal cross produced 75% males (3 male : 1 female) (Hickling, 1960). The relatively simple technology involved in producing an all-male fry population appeared to solve the majority of the problems associated with unwanted reproduction in tilapia culture. Subsequent analysis of the Zanzibar strain (Trewavas, 1968) showed that it was in fact *O. urolepis hornorum*. The success of this work resulted in the widespread dispersal of these strains throughout the world (Pullin and Capili, 1988).

This apparently simple solution to the major problem of fry production in ponds stimulated research into finding other species with better commercial characteristics that would also hybridize and produce all-male fry. *O. mossambicus* and *O. u. hornorum* males and their F1 hybrids mature into fish with large heads and black skin which fetched a low market price, particularly in Asia. Other crosses were reported that gave all or nearly all-male hybrid fry: *O. niloticus* × *O. u. hornorum* (Pruginin, 1967), *O. niloticus* × *O. nyasalapia macrochir* (Jalabert *et al.*, 1971), *O. niloticus* × *O. aureus* (Fishelson, 1962). The results of this work showed that there were relatively few species of commercial value that would consistently produce F1 sex ratios close to all-male (Wohlfarth and Hulata, 1983; Majumdar and McAndrew, 1983). Of these it was the *O. niloticus* × *O. aureus* cross that stimulated the most interest and resulted in the widespread dispersal of these two species to establish new hatchery stocks, mostly from Israeli sources (Pullin and Capili, 1988).

The widespread dispersal of these species resulted in a revival of tilapia culture but inconsistencies in hybrid sex ratios began to appear as poor management resulted in the hybrid fry entering the pure species broodstock population, causing hybrid introgression. Today very few examples of commercial F1 hybridization still exist. The problems of identifying, obtaining and monitoring pure strains and the added management problems and reduced yields compared to single species culture have made this technique uneconomic. Wohlfarth (1994) reviewed some of the problems and missed opportunities encountered in a number of countries where hybridization programmes were initiated. The most widespread use of hybrid tilapia is in Israel where the cross between *O. niloticus* × *O. aureus* gives a high percentage of male fry with a greater tolerance to cold. The hybrid survives over-wintering in Israel better than pure *O. niloticus*. This has made it the fish of choice even though the percentage of males has declined, because of introgression between the parental species, and hormone sex reversal is needed to achieve all-male populations.

Monosex male production by hormones

With the discover by Yamamoto (1969) that steroids can be used to change the phenotypic sex of fish, androgens have been used widely to produce all-male fry populations in a number of different tilapia species (reviewed by McAndrew, 1993; Macintosh and Little, 1995; Green *et al.*, 1997). Today a significant proportion of the tilapia industry is using a protocol based on the use of the synthetic androgen 17α-methyltestosterone incorporated into the fry diet at between 40 and 60 ppm. It is important that any sex-reversal treatment should begin at or shortly (a few days) after the first feeding stage and last for around 30 days. Therefore in order to obtain the highest possible sex-reversal rates the hatchery must be capable of producing large numbers of fry of known age, rather than simply harvesting small fry of unknown age from breeding ponds. This imposes design and management demands on such hatcheries (see Little and Hulata, this volume).

Direct sex-reversal of tilapia fry is used for most tilapia production in the USA and the Caribbean, and a smaller but significant proportion in some Asian countries, but it has come under increasing criticism because of the potential for the hormone entering the human food chain or the environment. In the USA sex-reversal of tilapia is a licensed procedure (USDEA, 1997) while in Europe the direct use of hormones is banned (Council of the European Communities, 1996). Research has identified the labile period for treatment and has shown that extending the treatment period beyond 30 days and dosages beyond the optimum levels of 30–60 ppm does not increase the efficiency of the treatment but may cause increased mortality and paradoxical sex-reversal (Green *et al.*, 1997). Therefore there is little to be gained by farmers from exceeding the standardized protocols. Research has also shown that when the optimal dosages are given the whole body levels of hormone are reduced to $<5\%$ in fry 100 h after withdrawal (Johnstone *et al.*, 1983) and $<1\%$ in 21days (Goudie *et al.*, 1986). Although oral administration techniques are apparently safe for the consumer, the environmental impact from uneaten food and metabolites may be a greater problem. Short periods of immersion in steroids have been shown to be successful in salmonids (Donaldson and Hunter, 1982) and there has been some preliminary success with tilapias (Gale *et al.*, 1995). Immersion should use only a relatively small volume of water, which could be treated to ensure the hormone does not enter the environment (Green *et al.*, 1997).

Sex determination systems in tilapias and monosex male production by genetic manipulation

With the development of hormonal sex-reversal it became possible to invert the phenotypic sex of tilapia and produce functional neomales[1] (genetically female but phenotypically male) and neofemales (genetically male but phenotypically female). In combination with chromosome-set manipulations (gynogenesis and androgenesis), this was used by a number of research groups to develop not only our understanding of the sex-determination mechanisms in tilapia but also techniques for the production of genetically male tilapia (GMT; Mair *et al.*, 1997).

---

[1]The prefixes neo, Δ- and pseudo- have been used to avoid ambiguity about the phenotypic and genotypic sex of sex-reversed fish (Chevassus, 1987; Mair *et al.*, 1987; Melard, 1995). All of these prefixes have a common meaning, indicating that the sex indicated is the phenotypic sex, which is the opposite of the fish's genotypic sex. For example, a neomale is a fish which is phenotypically male, generally through treatment with an androgen, but genetically female. In this chapter we use the neo-prefix due to its chronological precedence.

The sex determination systems of three *Oreochromis* species (*O. niloticus*, *O. aureus* and *O. mossambicus*) are now moderately well understood, while information on other tilapia species is limited at best and mostly comes from studies on sex ratios in hybrid crosses (Hickling, 1960; Chen, 1969; Avtalion and Hammerman, 1978; Hammerman and Avtalion, 1979).

O. niloticus has an XX female/XY male system with additional influence on sex ratios coming from autosomal loci and temperature. Female homogamety was demonstrated by crosses between neomales and ordinary females (Jalabert *et al.*, 1974) and by gynogenesis (Penman *et al.*, 1987a; Mair *et al.*, 1991a). YY males have been produced (Scott *et al.*, 1989; Mair *et al.*, 1991a). The sex determining region has been localized to a distal region of a chromosome arm (Mair *et al.*, 1991a), which has been shown to be on the largest chromosome by analysis of the synaptonemal complex (Foresti *et al.*, 1993; Carrasco *et al.*, 1999) (Fig. 7.2). The presence of autosomal genes which affect sex ratios in this species has been suggested by variable sex ratios in hybrid crosses (e.g. Majumdar and McAndrew, 1983) and the presence of males in crosses between XX neomales and ordinary XX females (Hanson *et al.*, 1983; Baroiller, 1996), in gynogenetics from ordinary females (Mair *et al.*, 1991a; Hussain *et al.*, 1994; Müller-Belecke and Hörstgen-Schwark, 1995) and in one fully inbred clonal line (Sarder *et al.*, 1999). Baroiller *et al.* (1995a,b) and Abucay *et al.* (1999) demonstrated that temperatures of around 35°C or greater can modify sex ratios. Interestingly, high temperatures caused sex reversal in both directions: the strongest effect was masculinisation of XX fry but feminisation of both XY and YY fry was also observed. Abucay *et al.* (1999) showed that there was a correlation between the percentage of males at 28°C and 35°C among batches of fry split and reared at these two temperatures, suggesting that the effects of autosomal genes and temperature on sex ratio may be connected.

Kwon *et al.* (2000) showed that treatment of YY male *O. niloticus* fry with Fadrazole, a chemical inhibitor of cytochrome P450 aromatase, could suppress high temperature feminization. Aromatase is a membrane-bound enzyme which is responsible for the conversation of androgens to estrogens (principally testosterone to estrogen). These data suggest that high temperature may result in unusually high levels of aromatase expression, the effects of which are suppressed by treatment with Fadrazole. Taken together with the masculinizing effects of Fadrazole when administered to genetically female (XX) *O. niloticus* fry during sexual differentiation (Kwon *et al.*, in press), aromatase is implicated as a key factor in sexual differentiation in the Nile tilapia, as also appears to be the case in other teleost species (e.g. Piferrer *et al.*, 1994; Kitano *et al.*, 1999). However, it is likely that it is some way down the cascade from the primary sex determining gene(s).

O. aureus has a WZ female/ZZ male sex determination system which can be influenced by autosomal genes and temperature. Female heterogamety was demonstrated by Guerrero (1975), Liu (1977: cited by Jensen and Shelton, 1979) and Mair *et al.* (1991b) and male homogamety was demonstrated by Liu (1977:

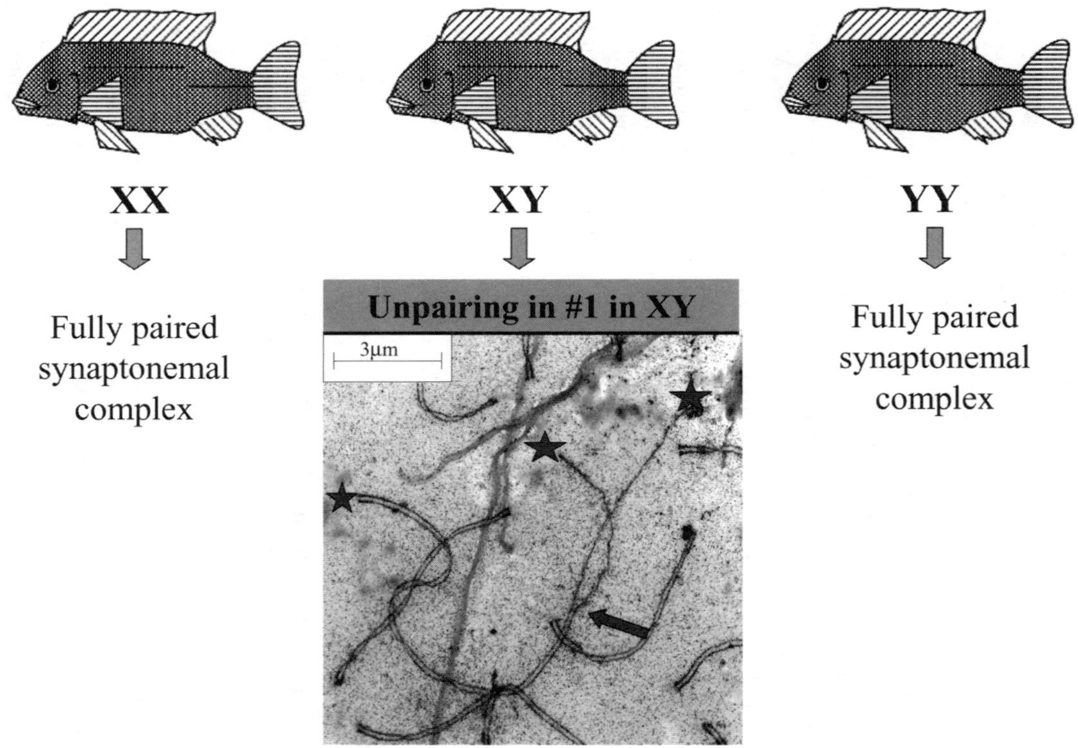

Fig. 7.2 Analysis of the synaptonemal complex in three different genotypes (XX, XY and YY) of *O. niloticus*. Both of the homogametic genotypes (XX and YY) show completely paired bivalents, whereas the heterogametic genotype (XY) shows an unpaired region in the longest bivalent (chromosome 1). The ends of the longest bivalent are indicated with asterisks and the end of the unpaired region is indicated by an arrow. EM photograph provided by Luis Carrasco.

cited by Jensen and Shelton, 1979), Hopkins (1979), Mair *et al.* (1991b) and Melard (1995). Like *O. niloticus*, the sex determining region or gene appears to be located distally to the centromere (Penman *et al.*, 1987a; Avtalion and Don, 1990), but the sex chromosome has not yet been identified. Autosomal genes appear to be able to modify sex ratios (Shelton *et al.*, 1983; Mair *et al.*, 1991b; Melard, 1995) and high temperature (34°C) has been shown to cause female to male sex reversal (Desprez and Melard, 1998).

O. mossambicus has an XX/XY sex determination system. Female homogamety was demonstrated by Clemens and Inslee (1968), Penman *et al.* (1987a) and Pandian and Varadaraj (1990). There is no good evidence for autosomal influence on sex ratios. The data of Mair *et al.* (1990) suggested that temperature can affect sex ratios, but unlike in *O. niloticus* and *O. aureus*, in this case lower temperatures (19 or 20°C) caused an increase in the proportion of males. Varadaraj and Pandian (1989) produced meiotic gynogenetic *O. mossambicus* from XY females. These gynogenetics had a sex ratio near to 1:1 and progeny testing demonstrated that these consisted of XX females and YY males. This suggests that very little recombination takes place during meiosis in the interval between the *O. mossambicus* major sex determining gene and its centromere (otherwise recombinant XY gynogenetics would also have been found, skewing the sex ratio of the gynogenetic groups towards a majority of males). This contrasts with the situation in *O. niloticus* and *O. aureus*, where evidence from similar experiments, and from synaptonemal complex analysis, suggests the major sex determining region or gene is located on a chromosome arm in a distal position from the centromere.

In *O. niloticus*, crosses between an XY neofemale and a normal male (XY) segregate for the following genotypes: 1 XX female: 2 XY males: 1 YY male. Through progeny testing it has been shown that YY males are viable and in crosses with XX females produced fry (XY) which were all males or had at most low percentages of females (Scott *et al.*,1989; Mair *et al.*, 1991b). Commercial-scale production of YY males for GMT production has been achieved by several generations of breeding, sex reversal and progeny testing (Mair and Little, 1991; Mair *et al.*, 1997). Sex reversal of some YY offspring using estrogens produces YY neofemales (Mair *et al.*, 1997). Once these fish are available then normal crosses between YY males and YY neofemales produce 100% YY male offspring. Mair *et al.* (1997) describe the development of this system in *O. niloticus*. To some extent this technology will suffer from the need to maintain the purity of the YY males to avoid contamination by XY males. However in countries such as the Philippines which has a well-defined hatchery sector licensed producers are achieving consistently higher percentages of males compared to those relying on hormonal sex-reversal (G.C. Mair, personal communication). It is necessary to obtain >95% males to reduce fry production to insignificant levels (Mair and Little, 1991): deviations from 100% males in GMT production appear to be due to the influence of autosomal loci and temperature (see above). The fry being produced by this process also have the added advantage of all being genetically male fish which reduces the size dimorphism compared to that observed in mixed sex culture and in phenotypically all-male hormone sex-reversed batches from mixed sex fry sources (Mair *et al.*, 1997).

The number of generations required to produce YY males can be shortened by the use of chromosome set manipulations, although this has not yet been demonstrated on a commercial scale. Myers *et al.* (1995b) showed that YY males can be produced directly from normal (XY) male *O. niloticus* by androgenesis and Sarder *et al.* (1999) showed that such males can be reproduced clonally using androgenesis. The combination of sex-reversal and chromosome-set manipulations has enabled single sex populations of both male and female phenotypes to be produced routinely with any genotype (XX, XY or YY) in *O. niloticus*.

Genetically male tilapia have also been produced in *O. mossambicus* (Varadaraj and Pandian, 1989). Here the YY males were produced by gynogenesis from XY neofemales. In *O. aureus*, Jensen and Shelton (1979) suggested a breeding scheme in which ZZ neofemales, produced by hormonal feminization and crossed with normal males (ZZ), would give all male or nearly all male fry. This has been demonstrated (Liu ,1977: cited by Jensen and Shelton, 1979; Hopkins, 1979; Mair *et al.*, 1991b; Melard, 1995).

Sex-linked DNA markers, such as those which have been found in chinook salmon *Onchorynchus tshawytscha* (Devlin *et al.*, 1991, 1994) would be very useful in basic research on sex determination as well as in the management of GMT production. Ultimately it would be desirable to have markers for both the major sex determining systems (XX/XY or WZ/ZZ) and for autosomal sex modifying genes which appear to have a significant influence at least in some studied populations. However, to date there are no confirmed reports of such markers.

Triploidy-induced sterility

Triploids have been produced in many different species of fish in an attempt to reduce the effects of sexual maturation in farmed populations, as a means of avoiding the risks of hybrid introgression or for the control of exotic introductions (Chourrout, 1987). In common with triploids in most other fish species, female tilapia are sterile in that they do not appear to be able to produce viable gametes, while at least some males do produce sperm which will fertilize eggs and give rise to aneuploid embryos which die before yolk sac resorption (Penman *et al.*, 1987b; Pandian and Varadaraj, 1988; Varadaraj and Pandian, 1990; Hussain *et al.*, 1996).

Hussain *et al.* (1995) compared the performance of three groups of triploid *O. niloticus* (produced by heat, cold or pressure shocks) against their diploid sibs in replicated communal stocking in 1 m^2 fibreglass tanks. In this study the method of triploid production did not effect subsequent performance, after any initial differences in treatment mortality. Both triploid sexes grow as well as their diploid sibs under communal conditions. Proximate analysis of carcasses showed significant differences in crude protein, percentage crude lipid and percentage moisture between the sexes but no significant interaction between sex and ploidy was observed. The gonadal weight and gonadosomatic index (GSI) of the triploid females were significantly smaller than those of the diploid females: the triploid ovaries appeared string-like and contained only small numbers of oogonia or previtellogenic oocytes. In males testes weight or GSI did not differ between the ploidy groups but the triploid testes, although full of seminal fluid, contained few spermatozoa. Serum calcium level, a good predictor of vitellogenin level, remained low in the triploid females and did not rise above that observed in the males, suggesting no exogenous vitellogenesis was occurring in these animals and probably accounting for the small ovaries and the large visceral lipid stores seen in the triploid females.

Serum concentrations of the steroid hormones testosterone and 17β-oestradiol were significantly lower in the triploid females than in the diploid females. This was correlated with the greatly reduced population of follicular cells in the ovaries of the triploid females. These normally surround the developing oocytes and are a major site of steroid production. The consequence of this was that the triploid females displayed few secondary sexual characteristics, such as changes in breeding colour or swelling of the urogenital papilla. There was no significant difference in the steroid levels or secondary sexual characteristics between the male ploidy groups. Even though the triploid males produced few spermatozoa the testes contained ample steroidogenic cells.

Carrasco (1998) studied chromosome pairing during meiosis in diploid and triploid *O. niloticus*. To obtain synaptonemal complex spreads in genetic females (XXX), it was necessary to treat fry with 17α-methyltestosterone and study these as phenotypic males, since synaptonemal complex spreads could not be prepared from ovaries. Significant differences were found between comparable genetic female (XXX^{MT}) and male (XXY^{MT}) triploids in chromosome pairing during meiosis. In the XXX^{MT} fish, multiple synapses were observed, resulting in spreads with a complex, tangled appearance. In the XXY^{MT} fish, however, relatively high numbers of fully-paired bivalents were observed, often with the third unpaired axial element lying parallel to the bivalent but at some distance from it. This difference could be one of the critical factors in determining that gametogenesis progresses further in triploid males than triploid females.

While most studies on triploid tilapia have been carried out in laboratories, Brämick *et al.* (1995) compared the performance of triploid and diploid *O. niloticus* in earthen ponds in Nigeria. Higher harvest weights were obtained for both triploid males and females compared to the diploids (66% and 90% larger, respectively). Fry production was high in the diploid ponds but very low or absent in the triploid ponds. Testing of the ploidy status of the fish stocked in the triploid groups revealed the presence of 3% diploids, which were presumed to be responsible for the very low level of fry production in the three out of five 'triploid' ponds where fry were observed. Ripe eggs and unexpectedly high GSI values were found in 15% of the triploid females, although eggs taken from the mouths of six triploid females failed to develop into viable larvae. Such advanced maturity in triploid females was in contrast to most of the laboratory studies, although Penman *et al.* (1987b) did observe large oocytes and high GSI values in the oldest group of triploid *O. aureus* studied.

It is clear that triploidy would help reduce precocious maturity and excessive fry production in ponds and the study of Brämick *et al.* (1995) also suggests that harvest weights could be increased. The challenge lies in trying to find a cost-effective way of producing the large numbers of fry needed. The direct handling of individual batches of eggs and the accurate timing of the shocks required to produce high triploid yields would make the process uneconomic in most commercial tilapia hatcheries. A possible solution would be via tetraploid male x diploid female crosses which would generate triploid offspring without the need for any physical intervention. This has been successfully carried out in the rainbow trout (e.g. Myers and Hershberger, 1991) but in many other species of fish, including tilapia, tetraploids appear to have zero or very low viability.

Control of reproduction summary

At this point hormonal sex reversal is still the most widely used technique, although the production of genetically male tilapia (GMT) has reached large-scale production in some countries recently. To be successful in terms of preventing reproduction, a minimum of 95% males is required. Hormonal sex reversal can fail to achieve this if the treatment is not adequately controlled, while genetic and environmental factors can prevent GMT from reaching this threshold. Tilapia culture is very diverse, and there are situations where control of reproduction is not required (e.g. intensive cage culture) or is even undesirable (e.g. backyard ponds where fish are left to breed and only enough are harvested for immediate household consumption) but there are also many situations where suppression of reproduction is highly desirable.

7.6 TRANSGENIC TILAPIA

Individual genes can be cloned and, if desired, molecularly rearranged to bring together for example the promoter from one gene with the coding sequences from another. Such DNA constructs can be introduced into the genome of target organisms to confer new properties or to alter existing ones, e.g. to cause a protein to be expressed at higher levels or in a broader range of cell or tissue types. Organisms with such DNA sequences stably integrated into their genomes are called transgenic organisms. According to the definition adopted by the EU (Council of the European Communities, 1990), transgenics come under the definition of genetically modified organisms (GMOs), while techniques such as triploidy and GMT (despite the unfortunate similarity in the acronyms!) do not. However, other organizations have adopted a broader definition (Pullin, 1994).

A variety of techniques have been developed for the introduction of DNA into the genome of fish and other vertebrates, including electroporation of eggs or sperm and particle bombardment, but microinjection into the cytoplasm of eggs shortly after fertilization is still the most efficient technique in most cases (Penman *et al.*, 1995). Targeted knockout or replacement of specific genes is now routinely carried out in mice (e.g. Gonzalez and Kimura, 1999; Picciotto, 1999; Eynon and Flavell, 1999) but this has not yet been achieved in fish, where gene transfer still relies on apparently random integration.

Apart from basic aspects of gene transfer such as ensuring reasonably efficient integration and expression of the introduced DNA in the target organism, the main objective of gene transfer in tilapia to date has been growth enhancement. Brem *et al.* (1988) introduced a DNA construct into *O. niloticus* eggs which consisted of mammalian promoter sequences and growth hormone coding sequences. No growth enhancement was obtained. Indig and Moav (1988), Maclean *et al.* (1992) and Rahman and Maclean (1992) describe expression of various reporter genes in tilapia. Alam *et al.* (1996a) were able to demonstrate germline transmission and expression of a lacZ containing transgene in the same species (lazZ is a reporter gene often used in model studies as expression is easy to detect). Alam *et al.* (1996b) suggested that DNA constructs which are as homologous as possible to the DNA of the host organism are more effective, as higher expression levels were found from a DNA construct with a carp β-actin regulatory sequence than from one with the equivalent rat regulatory sequence. Rahman *et al.* (1997) used a coinjection strategy to improve the poor integration efficiency of a growth hormone gene construct in *O. niloticus*. The growth hormone construct was injected along with a reporter gene construct which had a higher integration efficiency. From these founders, they developed lines of transgenic fish which expressed the growth hormone.

Martinez *et al.* (1999) and Rahman and Maclean (1999) were able to demonstrate significantly enhanced growth in transgenic *O. hornorum* and *O. niloticus* respectively. Martinez *et al.* (1999) used a DNA construct expressing tilapia growth hormone cDNA under control of human cytomegalovirus regulatory sequences. Their results suggested a transgene-dosage effect, in that homozygous transgenics (carrying two copies of the transgene) grew more slowly than hemizygous ones (carrying only one copy). The transgenic tilapia described by Rahman and Maclean (1999), containing an 'all-fish' construct expressing chinook salmon growth hormone under the control of an ocean pout (*Macrozoarces americanus*) antifreeze regulatory sequence, were approximately three times the size of non-transgenic siblings. The results of Engidaw (1998) and Rahman *et al.* (1998) suggested that the transgenic tilapia carrying the ocean pout antifreeze/chinook salmon GH construct utilized diets more efficiently than non-transgenic controls.

Little research on modifying other traits in tilapia has been published, although Rahman and Maclean (1999) stated that their long-term goal is to produce entirely sterile growth enhanced tilapia, which would presumably minimize environmental impacts. In other species of fish, research has been directed to altering metabolic capabilities with a view to being able to use alternative food sources or eliminate certain components from the diet. Pitkänen *et al.* (1999) describe transfer and expression of glucose transporter and hexokinase genes in salmonids, with the aim of improving carbohydrate metabolism and allowing a partial substitution of fish oil in the diet. Toyohara *et al.* (1996) transferred the gene for L-gulono-γ-lactone oxidase (GLO) into medaka (*Oryzias latipes*), which lacks this enzyme, to see if this could enable the medaka to complete the biosynthesis of L-ascorbic acid. GLO activity was detected in offspring from founder transgenics, although L-ascorbic acid content in transgenics was not significantly different from that of non-transgenics. A detailed understanding of the molecular genetics, biochemistry, physiology, etc. of potential target traits is clearly necessary for successful transgenic modification.

Beardmore (1997) and Rahman and Maclean (1999), among others, have suggested that transgenic organisms in which the transgene construct is as homologous as possible to the host genome (ideally conspecific) are likely to have greater consumer acceptance. However, given the spreading negative reaction to genetically modified organisms at the time of writing, it is difficult to predict whether transgenic tilapia will find acceptance in aquaculture. Consumer acceptance may vary considerably from country to country.

Possible environmental impacts of escaped or deliberately released transgenic fish are very hard to predict. Hallerman *et al.* (1999) have developed a decision support software designed to guide risk assessment and risk management for proposed experiments on aquatic genetically modified organisms. Hopefully the use of such tools could help to identify potential risks and minimize the likelihood of environmental impacts during research on such organisms. Muir and Howard (1999) modelled the potential environmental impacts of transgenic medaka (*Oryzias latipes*) expressing an introduced growth hormone gene. They suggested that if this transgene was introduced into a natural population, it might spread due to higher mating success of transgenic fish (due to larger body size) but that reduced viability of transgenic offspring might lead to extinction of the population. While the reduced viability of offspring was actually taken from experimental data, increased mating success of transgenics was assumed from observations that larger wild type medaka had higher mating success than smaller ones. It is of course possible that the diversion of more energy into growth in growth-enhanced transgenics might result in lower mating success, and Rahman and Maclean (1999) observed that growth-enhanced male transgenic Nile tilapia showed low, and in some cases zero, sperm production.

Modelling of potential environmental impacts of transgenic tilapia (and other fish), incorporating relevant data from laboratory experiments, is highly desirable.

7.7 SUMMARY

The farming of tilapia has developed rapidly over the last 10 years, with production growing from 295 000 t in 1987 to 801 000 t in 1996 (FAO, 1998). Much of this increase has come from the widespread use of *O. niloticus* and improvements in hatchery production, husbandry and applied genetics, enabling this species to express its potential in aquaculture. There is still major scope to further improve tilapia genetic resources for aquaculture. The spread of improved strains from programmes such as GIFT will help to improve the basic stocks in many countries. However, without knowledge of how to manage such stocks to maintain and continue their improvement while avoiding other problems such as hybrid introgression, this effort will be wasted.

Recent years have also seen the application of sophisticated molecular genetics techniques in research on tilapia, which will hopefully translate into applications in aquaculture in the near future.

REFERENCES

Abucay, J.S., Mair, G.C., Skibinski, D.O.F. and Beardmore, J.A. (1999) Environmental sex determination: the effect of temperature and salinity on sex ratio in *Oreochromis niloticus* L. *Aquaculture* **173**, 219–234.

Agnèse, J-F. Adépo-Gourène, B. Owino, J. Pouyaud, L. and Aman, R. (1999) Genetic characterisation of a pure relict population of *Oreochromis esculentus*, an endangered tilapia. *Journal of Fish Biology* **54**, 1119–1123.

Alam, M.S., Popplewell, A. and Maclean, N. (1996a) Germ line transmission and expression of a lacZ containing transgene in tilapia (*Oreochromis niloticus*). *Transgenic Research* **5**, 87–95.

Alam, M.S., Lavender, F.L., Iyengar, A., Rahman, M.A., Ayad, H.H., Lathe, R., Morley, S.D. and Maclean, N. (1996b) Comparison of the activity of carp and rat β-actin gene regulatory sequences in tilapia and rainbow trout embryos. *Molecular Reproduction and Development* **45**, 117–122.

Avtalion, R.R. (1982) Genetic markers in *Sarotherodon* and their use for sex and species identification, in *The Biology and Culture of Tilapia* (eds R.S.V. Pullin and R.H. Lowe-McConnell), ICLARM Conference Proceedings 7, pp. 269–277.

Avtalion, R.R. and Don, J. (1990) Sex-determining genes in tilapia: a model of genetic recombination emerging from sex ratio results of three generations of diploid gynogenetic *Oreochromis aureus*. *J. Fish. Biol.* **37**, 167–173.

Avtalion, R.R. and Hammerman, I.S. (1978) Sex determination in *Sarotherodon (Tilapia)*. I. Introduction to a theory of autosomal influence. *Bamidgeh* **30**, 110–115.

Avtalion R.R. and Wojdani, A. (1971) Electrophoresis and immunoelectrophoresis of sera from known F1 hybrids of *Tilapia*. *Bamidgeh* **23**, 117–124.

Avtalion, R.R., Pruginin, Y. and Rothbard, S. (1975) Determination of allogenic and xenogenic markers in the genus Tilapia: I. Identification of sex and hybrids in tilapia by electrophoretic analysis of serum proteins. *Bamidgeh* **27**, 8–13.

Avtalion, R.R., Duczyminer, M., Wojdani, A. and Pruginin, Y. (1976) Determination of allogeneic and xenogeneic markers in the genus of Tilapia. II. Identification of *T. aurea, T. vulcani* and *T. nilotica* by electrophoresis analysis of their serum proteins. *Aquaculture* **7**, 255–265.

Balarin, J.D and Hatton, J.P. (1979) *Tilapia: A Guide to their Biology and Culture in Africa*, University of Stirling, 174p.

Bardakci, F. and Skibinski, D.O.F. (1994) Application of the RAPD technique in tilapia fish species and subspecies identification. *Heredity* **73**, 117–123.

Bardakci, F. and Skibinski, D.O.F. (1999) A polymorphic SCAR-RAPD marker between species of tilapia (Pisces: Cichlidae). *Animal Genetics* **30**, 78–79.

Baroiller, J.F. (1996) Significant proportions of unexpected males in progenies from single pair matings with sibling sex reversed males of *Oreochromis niloticus*, in *The Third International Symposium on Tilapia in Aquaculture* (eds R.S.V. Pullin, J. Lazard, M. Legendre, J.B. Amon Kothias and D. Pauly), ICLARM Conference Proceedings 41, pp. 229–237.

Baroiller, J.F., Clota, F. and Geraz, E. (1995a) Temperature sex determination in two tilapia, *Oreochromis niloticus* and the red tilapia (Red Florida strain): effect of high or low temperature, in *Proceedings of the Fifth International Symposium on the Reproductive Physiology of Fish* (eds F.W. Goetz and P. Thomas), Fish Symposium 95, Austin, Texas, USA, pp. 158–160.

Baroiller, J.F., Chourrout, D., Fostier, A. and Jalabert, B. (1995b) Temperature and sex chromosomes govern sex ratios of the mouthbrooding cichlid fish *Oreochromis niloticus*. *J. Exp. Zool.* **272**, 213–223.

Baroiller, J.F., Nakayama, I., Foresti, F. and Chourrout, D. (1996) Sex determination studies in two species of teleost fish, *Oreochromis niloticus* and *Leporinus elongatus*. *Zool. Studies* **35**, 279–285.

Basasibwaki., P. (1975) Comparative electrophoretic patterns of lactate dehydrogenase and malate dehydrogenase in five Lake Victoria cichlid species. *African Journal of Tropical Hydrobiology and Fisheries* **4**, 21–26.

Basiao, Z.U. and Doyle, R.W. (1990) Interactions between test and reference populations when tilapia strains are compared by the 'internal control' technique. *Aquaculture* **85**, 207–214.

Basiao, Z.U. and Doyle, R.W. (1999) Test of size specific mass selection for Nile tilapia, *Oreochromis niloticus* L., cage farming in the Philippines. *Aquaculture Research* **10**, 373–378.

Bauer, J. (1968) Vergleichende untersuchungen zum kontakverhalten verschiedener arten der gattung *Tilapia* (Cichlidae, Pisces) und ihrer bastarde. *Z. Tierpsychol.* **25**, 22–70.

Beardmore, J.A. (1997) Transgenics: autotransgenics and allotransgenics. *Transgenic Research* **6**, 107–108.

Behrends, L.L. and Smitherman, R.O. (1984a) Genetics and breeding of red tilapia, in *Proceedings Auburn Symposium on Fisheries and Aquaculture* (eds R.O. Smitherman and D. Tave), Auburn University, Alabama, USA, pp.19–26.

Behrends, L.L. and Smitherman, R.O. (1984b) Development of a cold tolerant population of red tilapia through introgressive hybridization. *J. World Aqua. Soc.* **15**, 172–178.

Behrends, L.L., Nelson, R.G., Smitherman, R.O. and Stone, N.M. (1982) Breeding and culture of the red-gold color phase of tilapia. *J. World Maricult. Soc.* **13**, 210–220.

Behrends, L.L., Kingsley, J.B. and Price III, A.H. (1990) Two stage hybridisation and introgression for improving production traits of red tilapia, in *Genetics in Aquaculture: Proceedings of the Sixteenth US-Japan Meeting on Aquaculture* (ed. R.S. Svrjcek), NOAA Technical Report NMES 92, pp. 77–81.

Bentsen, H.B., Eknath, A.E., Palada-de Vera, M.S., Danting, J.C., Bolivar, H.L., Reyes, R.A., Dionisio, J.C., Longalong, F.M., Circa, A.V., Tayamen, M.M., Gjerde, B. (1998) Genetic improvement of farmed tilapias: growth performance in a complete diallel cross experiment with eight strains of *Oreochromis niloticus*. *Aquaculture* **160**, 145–173.

Bondari, K., Dunham, R.A., Smitherman, R.O. Joyce, J.A. and Castillo, S. (1983) Response to bidirectional selection for body weight in blue tilapia, in *International Symposium on Tilapia in Aquaculture* (eds L. Fishelson and Z. Yaron), Tel Aviv University, Israel, pp. 300–310.

Brämick, U., Puckhaber, B., Langholz, H.-J. and Hörstgen-Schwark, G. (1995) Testing of triploid tilapia (*Oreochromis niloticus*) under tropical pond conditions. *Aquaculture* **137**, 343–353.

Brem, G., Brenig, B., Hörstgen-Schwark, G. and Winnacker, E.L. (1988) Gene transfer in tilapia (*Oreochromis niloticus*). *Aquaculture* **68**, 209–219.

Brummett, R.E., Halstrom, M.L. Dunham, R.A. and Smitherman, R.O. (1988) Development of biochemical dichotomous keys for the identification of American populations *of Oreochromis aureus, O. mossambicus, O. niloticus, O. urolepis hornorum* and red tilapia, in *The Second International Symposium on Tilapia in Aquaculture* (eds Pullin, Bhukasawan, Tonguthai and Maclean), ICLARM Conference Proceedings 15, pp. 135–141.

Brzeski, V.J. and Doyle, R.W. (1995) A test of an on-farm selection procedures for tilapia growth in Indonesia. *Aquaculture* **137**, 219–230.

Carrasco, L.A.P. (1998) *The Effects of Induced Triploidy on the Reproduction of the Rainbow Trout* (Oncorhynchus mykiss) *and the Nile tilapia* (Oreochromis niloticus), PhD dissertation, University of Stirling.

Carrasco, L.A.P., Penman, D.J. and Bromage, N. (1999) Evidences for the presence of sex chromosomes in the Nile tilapia (*Oreochromis niloticus*) from synaptonemal complex analysis of XX, XY and YY genotypes. *Aquaculture* **173**, 207–218.

Chen, F.Y. (1969) Preliminary studies on the sex-determining mechanism of *Tilapia mossambica* Peters and *T. hornorum* Trewavas. *Verh. Internat. Verein. Limnol.* **17**, 719–724.

Chen, F-Y. and Tsuyuki, M. (1970) Zone electrophoretic studies on the proteins of *Tilapia mossambica* and *T. hornorum* and their F1 hybrids, *T. zillii* and *T. melanopleura*. *J. Fish. Res. Bd. Can.* **27**, 2167–2177.

Chevassus, B. (1983) Hybridization in fish. *Aquaculture* **33**, 245–262.

Chevassus, B. (1987) Caractéristiques et performances des lignées uniparentales et des polyploides chez les poissons d'eau froides, in *Proceedings of a World Symposium on Selection, Hybridization and Genetic Engineering in Aquaculture* (ed. K. Tiews), Vol II. Heeneman, Berlin, pp. 145–161.

Chiayvareesajja, J., Røed, K.H. Eknath, A.E., Danting, J.C., De Vera, M.P. and Bentsen, H.B. (1999) Genetic variation in lytic activities of blood serum from Nile tilapia and genetic associations with survival and body weight. *Aquaculture* **175**, 49–62.

Chimits, P. (1955) The *Tilapia* and its culture: a preliminary bibliography. *FAO Fish. Bull.* **8**(1), 1–33.

Chimits, P. (1957) The *Tilapia* and their culture, a second review and bibliography. *FAO Fish. Bull.* **10**(1), 1–24.

Chourrout, D. (1987) Genetic manipulations in fish: review of methods, in *Proceedings of a World Symposium on Selection, Hybridization and Genetic Engineering in Aquaculture* (ed. K. Tiews), Vol II, Heeneman, Berlin, pp. 111–126.

Chourrout, D. and Itzkovich, J. (1983) Three manipulations permitted by artificial insemination in tilapia: induced diploid gynogenesis, production of all triploid populations and intergeneric hybridization, in *International Symposium on Tilapia in Aquaculture* (eds L. Fishelson and Z. Yaron), Tel Aviv University, Israel, pp. 246–255.

Circa, A.V., Eknath, A.E. and Taduan, A.G. (1995) Genetic improvement of farmed tilapias: the growth performance of the GIFT strain of Nile tilapia (*Oreochromis niloticus*) in rice-fish environments. *Aquaculture* **137**, 329–330.

Clemens, H.P. and Inslee, T. (1968) The production of unisexual broods by *Tilapia mossambica* sex-reversed with methyltestosterone. *Trans. Am. Fish. Soc.* **97**, 18–21.

Council of the European Communities (1990) Council Directive 90/220/EC on the deliberate release into the environment of genetically modified organisms. *Official Journal* No. L117(8/5/90), 15–27.

Council of the European Communities (1996) Council Directive 96/23/EC on measures to monitor certain substances and residues thereof in live animals. *Official Journal* L125(23/5/96), 10–32.

Cruz, E.M., Ridha, M. and Abdullah, M.S. (1990) Production of the African fresh-water tilapia *Oreochromis spilurus* (Gunther) in seawater. *Aquaculture* **84**, 41–48.

Daget, J and Moreau, J. (1981) Hybridation introgressive entre deux espèces de *Sarotherodon* (Pisces, Cichlidae) dans un lac de Madagascar. *Bull. Mus. Natn. Hist. Nat Paris 4e sér, section* A, n° 2 **3**, 689–703.

Desprez, D. and Melard, C. (1998) Effect of ambient water temperature on sex determinism in the blue tilapia *Oreochromis aureus*. *Aquaculture* **162**, 79–84.

Devlin, R.H., McNeil, B.K., Groves, T.D.D. and Donaldson, E.M. (1991) Isolation of Y-chromosomal DNA probe capable of determining genetic sex in chinook salmon (*Onchorynchus tshawytscha*). *Can. J. Fish. Aquat. Sci.* **48**, 1606–1612.

Devlin, R.H., McNeil, B.K., Solar, I.I. and Donaldson, E.M. (1994) A rapid PCR-based test for Y-chromosomal DNA allows simple production of all-female strains of chinook salmon. *Aquaculture* **128**, 211–220.

Dinesh, K.R., Lim, T.M., Chan, W.K. and Phang, V.P.E. (1996) Genetic variation inferred from RAPD fingerprinting in three species of tilapia. *Aquaculture International* **4**, 19–30.

Don, J. and Avtalion, R.R. (1986) The induction of triploidy in *Oreochromis aureus* by heat shock. *Theor. Appl. Genet.* **72**, 186–192.

Don, J. and Avtalion, R.R. (1988) Comparative study on the induction of triploidy in tilapias, using cold- and heat-shock techniques. *J. Fish Biol.* **32**, 665–672.

Donaldson, E.M. and Hunter, G.A. (1982) Sex control in fish with particular reference to salmonids. *Can. J. Fish. Aquat. Sci.* **39**, 99–110.

Doyle, R.W. and Talbot, A.J. (1986) Effective population size and selection in variable aquaculture stocks. *Aquaculture* **57**, 27–35.

Doyle, R.W., Shackel, N.L., Basiao, Z.U., Uraiwan, S., Matricia, T. and Talbot, A.J. (1991) Selective diversification of aquaculture stocks a proposal for economically sustainable genetic conservation. *Can. J. Fish. Aquat. Sci.* **48**, 148–154.

Dunseth, D.R. and Bayne, D.R. (1978) Recruitment control and production of *Tilapia aurea* (Steindachner) with the predator *Cichlosoma managuense* (Günther). *Aquaculture* **14**, 383–390.

Eknath, A.E., Bentsen, H.B., Gjerde, B., Tayamen, M.M., Abella, T.A., Gjedrem, T. and Pullin, R.S.V. (1991) Approaches to national fish breeding programs: pointers from a tilapia pilot study. *NAGA, The ICLARM Quarterly* **14**(2), 10–12.

Eknath, A.E., Tayamen, M.M., Palada-de Vera, M.S., Danting, J.C., Reyes, R.A., Capili, J.B., Bolivar, H.L., Abella, T.A., Circa, A.V., Bentsen, H.B., Gjerde, B., Gjedrem, T. and Pullin, R.S.V. (1993) Genetic improvement of farmed tilapias: the growth performance of eight strains of *Oreochromis niloticus* tested in different farm environments. *Aquaculture* **111**, 171–188.

Eknath, A.E., Reyes, R.A., Bolivar, H.L., Palada-de Vera, M.S., Danting, J.B., Dionisio, E.E., and Longalong, F.M. (1995) Genetic improvement of farmed tilapias: estimation of genetic variation and heritability for age and size at first spawning in Nile tilapia (*Oreochromis niloticus*). *Aquaculture* **137**, 279–280.

Engidaw, B.Z. (1998) *Growth and Nutrition Studies on Transgenic Nile Tilapia* (Oreochromis niloticus *L.) Carrying a Growth Hormone Construct*, MSc dissertation, University of Stirling.

Eynon, E.E. and Flavell, R.A. (1999) Walking through the forest of transgenic models of human disease. *Immunological Reviews* **169**, 5–10.

FAO (1998) *Aquaculture production statistics 1987–1996*. FAO Fisheries circular No 815, Rev 10, Rome, 197p.

Falk, T.M., Abban, E.K., Oberst, S., Villwock,W., Pullin, R.S.V. and Renwrantz, L. (1996) *A Biochemical Laboratory Manual for Species Characterisation of Some Tilapiine Fishes*, ICLARM Educ. Ser. 17, 93p.

Fishelson, L. (1962) Hybrids of two species fishes of the genus *Tilapia* (Cichlidae, Telecosti). *Fisherman's Bulletin* **4**(2), 14–19. (In Hebrew with English abstract.)

Foresti, F., Oliveira, C, Galetti, P.M. and Almeida-Toledo, L.F. (1993) Synaptonemal complex analysis in spermatocytes of tilapia *Oreochromis niloticus* (Pisces, Cichlidae). *Genome* **36**, 1124–1128.

Gale, W.L., Fitzpatrick, M.S. and Schreck, C.B. (1995) Immersion of Nile tilapia (*Oreochromis niloticus*) in 17α-methyltestosterone and mestalone for the production of all-male populations, in *Proceedings of the Fifth International Symposium on the Reproductive Physiology of Fish* (eds F.W. Goetz and P. Thomas), Fish Symposium 95, Austin, Texas, USA, p. 117.

Gonzalez, F.J. and Kimura, S. (1999) Role of gene knockout mice in understanding the mechanisms of chemical toxicity and carcinogenesis. *Cancer Letters* **143**, 199–204.

Goudie, C.A., Shelton, W.L. and Parker, N.C. (1986) Tissue distribution and elimination of radiolabelled methyltestosterone fed to sexually undifferentiated blue tilapia. *Aquaculture* **58**, 215–226.

Green, B.W., Verrica, K.L. and Fitzpatrick, M.S. (1997) Fry and fingerling production, in *Dynamics of Pond Aquaculture* (eds H.S. Egna and C.E. Boyd), CRC Press, Boca Raton, USA, pp. 215–243.

Guerrero, R.D. III. (1975) Use of oral androgens for the production of all male *Tilapia aurea* (Steindachner). *Trans. Am. Fish. Soc.* **104**, 342–348.

Hallerman, E., King, D. and Kapuscinski, A. (1999) A decision support software for safely conducting research with genetically modified fish and shellfish. *Aquaculture* **173**, 309–318.

Hammerman, I.S. and Avtalion, R.R. (1979) Sex determination in *Sarotherodon* (*Tilapia*). Part 2: The sex ratio as a tool for the determination of the genotype a model of autosomal and gonosomal influence. *Theor. Appl. Genet.* **55**, 177–187.

Hanson, T.R., Smitherman, R.O., Shelton, W.L. and Dunham, R.A. (1983) Growth comparisons of monosex tilapia produced by separation of sexes, hybridization and sex reversal, in *International Symposium on Tilapia in Aquaculture* (eds L. Fishelson and Z. Yaron), Tel Aviv University, Israel, pp. 570–579.

Heinrich, W. (1967) Untersuchungen zum sexualverhalten in der gattung *Tilapia* (Cichlidae, Teleostei) und bei artbastarden. *Z. Tierpsychol.* **24**, 684–754.

Hickling, C.F. (1960) The Malacca tilapia hybrids. *Genetics* **57**, 1–10.

Hickling, C.F. (1963) The cultivation of tilapia *Sci. Amer.* **208**, 143–152.

Hopkins, K.D. (1979) *Production of Monosex Tilapia Fry by Breeding Sex-reversed Fish*, PhD thesis, Auburn University, USA.

Huang, C-M. and Liao, I-C. (1990) Response to mass selection for growth rate in *Oreochromis niloticus*. *Aquaculture* **85**, 199–205.

Hulata, G., Wohlfarth, G.W. and Halvey, A. (1986) Mass selection for growth rate in the Nile tilapia (*Oreochromis niloticus*). *Aquaculture* **57**, 177–184.

Hulata, G., Karplus, I. and Harpaz, S. (1995) Evaluation of some red tilapia strains for aquaculture: growth and colour segregation in hybrid progeny. *Aquaculture Research* **26**, 765–772.

Hussain, M.G., Chatterji, A., McAndrew,B.J. and Johnstone, R. (1991) Triploidy induction in Nile tilapia, *Oreochromis niloticus* L., using pressure, heat and cold shock. *Theor. Appl. Genet.* **81**, 6–12.

Hussain, M.G., Penman, D.J., McAndrew, B.J. and Johnstone, R. (1993) Suppression of first cleavage in the Nile tilapia, *Oreochromis niloticus* L. – a comparison of pressure and heat shocks. *Aquaculture* **111**, 263–270.

Hussain, M.G., McAndrew, B.J., Penman, D.J. and Sodsuk, P. (1994) Estimating gene-centromere recombination frequencies in gynogenetic diploids of *Oreochromis niloticus* L., using allozymes, skin colour and a putative sex-determining locus (*SDL-2*), in *Genetics and Evolution of Aquatic Organisms* (ed. A.R. Beaumont), Chapman and Hall, London, pp. 502–509.

Hussain, M.G., Rao, G.P.S., Humayan, N.M., Randall, C.F., Penman, D.J., Kime, D., Bromage, N.R., Myers, J.M. and McAndrew, B.J. (1995) Comparative performance of growth, biochemical composition and endocrine profiles in diploid and triploid *Oreochromis niloticus* L. *Aquaculture* **138**, 87–98.

Hussain, M.G. Penman, D.J. and McAndrew, B.J. (1996) Effects of triploidy on sexual maturation and reproduction in Nile tilapia *Oreochromis niloticus*. L, in *The Third International Symposium on Tilapia in Aquaculture* (eds R.S.V. Pullin, J. Lazard, M. Legendre, J.B. Amon Kothias and D. Pauly), ICLARM Conference Proceedings 41, pp. 320–325.

Hussain, M.G., Penman, D.J. and McAndrew, B.J. (1998) Production of heterozygous and homozygous clones in the Nile tilapia. *Aquaculture International* **6**, 197–205.

Iles, T.D. and Howlett, C.J. (1967) Electrophoretic analysis of blood of *Tilapia leucosticta* Trewavas and *Tilapia zillii* (Gervais) from Lake Victoria. *East African Freshwater Fisheries Research Organisation Annual Report for 1967*, pp. 64-72.

Indig, F.E. and Moav, B. (1988) A prokaryotic gene is expressed in fish cells and persists in tilapia embryos following microinjection through the micropyle. *Colloq. Inst. Natl. Rech. Agron.* **44**, 221–225.

Jalabert, B., Kammacher, P., Lessent, P. (1971) Determinisme du sexe chez les hybrides entre *Tilapia macrochir* et *Tilapia nilotica*. Etude de la sex ratio dans les recroisement des hybrides de premiere generation par les especes parentes. *Ann. Biol. Anim. Bioch. Biophy.* **11**, 155–165.

Jalabert, B., Moreau, J., Planquette, P. and Billard, R. (1974) Déterminisme du sexe chez *Tilapia macrochir* et *Tilapia nilotica*: action de la méthyltestostérone dans l'alimentation des alevins sur la différentiation sexuelle; proportion des sexes dans la descendance des mâles 'inversés'. *Ann. Biol. Anim. Bioch. Biophys.* **14**, 729–739.

Jayaprakas, V., Tave, D. and Smitherman, R.O. (1988) Growth of two strains of *Oreochromis niloticus* and their F1 and F2 and backcross hybrids, in *The Second International Symposium on Tilapia in Aquaculture* (eds R.S.V. Pullin, T. Bhukaswan, K. Tonguthai and J. Maclean), ICLARM Conference Proceedings 15, pp. 197–202.

Jensen, G.L. and Shelton, W.L. (1979) Effects of estrogens on *Tilapia aurea*: implications for production of monosex genetic male tilapia. *Aquaculture* **16**, 233–242.

Johnstone, R., Macintosh, D.J. and Wright, R.S. (1983) Elimination of orally-administered 17-alpha methyltestosterone by *Oreochromis mossambicus* (tilapia) and *Salmo gairdneri* (rainbow trout) juveniles. *Aquaculture* **35**, 249–257.

Kitano, T., Takamune, K., Kobayashi, T., Nagahama, Y. and Abe, S.-I. (1999) Suppression of P450 aromatase gene expression in sex-reversed males produced by rearing genetically female larvae at a high water temperature during a period of sex differentiation in the Japanese flounder (*Paralichthys olivaceus*). *J. Mol. Endocrinol.* **23**, 167–176.

Kocher, T.D., Lee, W.-J., Sobolewska, H., Penman, D. and McAndrew, B. (1998) A genetic linkage map of a cichlid fish, the tilapia (*Oreochromis niloticus*). *Genetics* **148**, 1225–1232.

Kronert, U., Hörstgen-Schwark, G. and Langholz, H.J. (1987) Investigations on selection of tilapia for late maturity, in *Proceedings of a World Symposium on Selection,*

Hybridisation and Genetic Engineering in Aquaculture (ed. K. Tiews) Vol. I, Heeneman, Berlin, pp. 345–352.

Kronert, U., Hörstgen-Schwark, G. and Langholz, H.J. (1989) Prospects of selecting for late maturation in tilapia (*Oreochromis niloticus*). 1. Family studies under laboratory conditions. *Aquaculture* **77**, 113–121.

Kuo, H. (1988) Progress in genetic improvement of red hybrid tilapia in Taiwan, in *The Second International Symposium on Tilapia in Aquaculture* (eds R.S.V. Pullin, T. Bhukaswan, K. Tonguthai and J. Maclean), ICLARM Conference Proceedings 15, pp. 219–222.

Kwon, J.Y., McAndrew, B.J. and Penman, D.J. (2000) Inhibition of aromatase activity suppresses high-temperature feminisation of genetic male Nile tilapia, *Oreochromis niloticus*, in *Proceedings of the Sixth International Symposium on the Reproductive Physiology of Fish, Bergen, 4–9 July 1999* (eds B. Norberg, O.S. Kjesbu, G.L. Taranger, E. Andersson and S.O. Stefansson), p. 268.

Kwon, J.Y., Haghpanah, V., Kogson-Hurtado, L.M., McAndrew, B.J. and Penman, D.J. (in press) Masculinisation of genetic female Nile tilapia (*Oreochromis niloticus*) by dietary administration of an aromatase inhibitor during sexual differentiation. *J. Exp. Zool.*

Lee, W.-J. and Kocher, T.D. (1996) Microsatellite DNA markers for genetic mapping in the tilapia *Oreochromis niloticus*. *J. Fish Biol.* **49**, 169–171.

Lester, L.J., Abella, T.A., Palada, M.S., and Keus, H.J. (1988) Genetic variation in size and sexual maturation of *Oreochromis niloticus* under hapa and cage culture conditions, in *The Second International Symposium on Tilapia in Aquaculture* (eds R.S.V. Pullin, T. Bhukaswan, K. Tonguthai and J.L.Maclean), ICLARM Conference Proceedings 15, pp. 223–230.

Liao, I.C. and Chen, T.P. (1983) Status and prospects of tilapia culture in Taiwan, in *International Symposium on Tilapia in Aquaculture* (eds L. Fishelson and Z. Yaron), Tel Aviv University, Israel, pp. 588–598.

Longalong, F.M. Eknath, A.E. and Bentsen, H.B. (1999) Response to bi-directional selection for frequency of early maturing females in Nile tilapia (*Oreochromis niloticus*). *Aquaculture* **178**, 13–25.

Lovshin, L.L. and da Silva, A.B. (1975) Culture of monosex hybrid tilapias. *FAO/CIFA Symposium on Aquaculture in Africa*, CIFA/75/SR9, 16p.

McAndrew, B.J. (1993) Sex control in Tilapiines, in *Recent Advances in Aquaculture IV* (eds J.F. Muir and R.J. Roberts), Blackwell, London, pp. 87–98.

McAndrew, B.J. and Majumdar, K.C. (1983) Tilapia stock identification using electrophoretic markers. *Aquaculture* **30**, 249–261.

McAndrew, B.J. and Majumdar, K.C. (1988) Growth studies in juvenile tilapia using pure species, hormone treated and nine interspecific hybrids. *Aquaculture and Fisheries Management* **20**, 35–47.

McAndrew B.J. and Sodsuk P. (in press) Application of allozyme analysis in population genetics, stock management and taxonomy of tilapia, in *Applied Genetics of Tilapias* (ed. Mair, G.C.), DFID/ICLARM.

McAndrew B.J. and Wohlfarth G.W. (in press) Qualitative Phenotypes – Colour Varieties, in *Applied Genetics of Tilapias* (ed. G.C. Mair), DFID/ICLARM.

McAndrew, B.J., Roubal, F.R., Roberts, R.J., Bullock, A.M. and McEwen, I.M. (1988) The genetics and histology of red, blond and associated colour variants in *Oreochromis niloticus*. *Genetica* **76**, 127–137.

McAndrew, B.J., Rana, K.J. and Penman, D.J. (1993) Conservation and preservation of genetic variation in aquatic organisms, in *Recent Advances in Aquaculture* (eds J.F. Muir and R.J. Roberts), Vol IV, Croom Helm, pp. 295–336.

Mc Ginty, A.S. (1983) Southern region co-operative research. Project S-168 annual report 1982, warm water aquaculture Alabama Agricultural Experimental Station, Auburn University Alabama, USA.

Macintosh, D.J. and Little, D.C. (1995) Nile tilapia (*Oreochromis niloticus*), in *Broodstock Management and Egg and Larval Quality* (eds N.R. Bromage and R.J. Roberts), Blackwell Science, Oxford, pp. 277–320.

Maclean, N., Iyengar, A., Rahman, M.A., Sulaiman, Z. and Penman, D.J. (1992) Transgene transmission and expression in rainbow trout and tilapia. *Mol. Mar. Biol. Biotech.* **1**, 355–365.

Macaranas, J.M., Taniguchi, N., Pante, M.J.R., Capili, J.B. and Pullin, R.S.V.P. (1986) Electrophoretic evidence for extensive hybrid gene introgression into commercial *Oreochromis niloticus* (L.) in the Philippines. *Aquacult.Fish. Mgmt.* **17**, 249–258.

Macaranas, J.M., Agustin, L.Q., Ablan, M.C.A., Pante, M.J.R., Eknath, A.E. and Pullin, R.S.V. (1995) Genetic improvement of farmed tilapias: biochemical characterization of strain differences in Nile tilapia. *Aquaculture International* **3**, 43–54.

Macaranas, J.M., Agustin, L.Q. and Eknath, A.E. (1996) Multiple haemoglobins in three tilapiine species of the genus *Oreochromis* and in eight strains of *O. niloticus* (L.). *Aquaculture Research* **27**, 597–601.

Mair, G.C. (1988). *Studies on Sex Determining Mechanisms in* Oreochromis *Species*, PhD dissertation, University of Wales.

Mair, G.C. (1993) Chromosome-set manipulations in tilapia – techniques, problems and prospects. *Aquaculture* **111**, 227–244.

Mair, G.C. and Little, D.C. (1991) Population control in farmed tilapias. *NAGA, the ICLARM Quarterly* **14**, 8–13.

Mair, G.C., Penman, D.J., Scott, A., Skibinski, D.O.F. and Beardmore, J.A. (1987) Hormonal sex reversal and the mechanisms of sex determination in *Oreochromis*, in *Proceedings of a World Symposium on Selection, Hybridization and Genetic Engineering in Aquaculture* (ed. K. Tiews), Vol II, Heeneman, Berlin, pp. 301–312.

Mair, G.C., Beardmore, J.A. and Skibinski, D.O.F. (1990) Experimental evidence for environmental sex determination in *Oreochromis* species, in *Proceedings of the Second Asian Fisheries Forum* (eds R. Hirano and I. Hanyu), The Asian Fisheries Society, Manila, pp. 555–558.

Mair, G.C., Scott, A.G., Penman, D.J., Beardmore, J.A. and Skibinski, D.O.F. (1991a) Sex determination in the genus *Oreochromis*. 1. Sex reversal, gynogenesis and triploidy in *O. niloticus*. *Theor. Appl. Genet.* **82**, 144–152.

Mair, G.C., Scott, A.G., Penman, D.J., Skibinski, D.O.F. and Beardmore, J.A. (1991b) Sex determination in the genus *Oreochromis*. 2. Sex reversal, hybridisation, gynogenesis and triploidy in *O. aureus* Steindachner. *Theor. Appl. Genet.* **82**, 153–160.

Mair, G.C., Abucay, J.S., Skibinski, D.O.F., Abella, T.A. and Beardmore, J.A. (1997) Genetic manipulation of sex ratio for the large-scale production of all-male tilapia, *Oreochromis niloticus*. *Can. J. Fish. Aquat. Sci.* **54**, 396–404.

Majumdar, K.C and McAndrew, B.J. (1983) Sex ratios from interspecific crosses within the tilapias, in *International Symposium on Tilapia in Aquaculture* (eds L. Fishelson and Z. Yaron), Tel Aviv University, Israel, pp. 261–269.

Maluwa, A.O. and Dickson, M.W. (1996). Comparison of the growth of *Oreochromis karongae* and *O. shiranus* in fishponds in Malawi, in *The Third International Symposium on Tilapia in Aquaculture* (eds R.S.V. Pullin, J. Lazard, M. Legendre, J.B. Amon Kothias and D. Pauly), ICLARM Conference Proceedings 41, pp. 108–111.

Marengoni, N.G. and Onoue, Y. (1998) Ultraviolet-induced androgenesis in Nile tilapia, *Oreochromis niloticus* (L.) and hybrid Nile × blue tilapia, *O. aureus* (Steindachner). *Aquaculture Research* **29**, 359–366.

Markert, J.A., Kocher, T.D., Penman, D.J. and Karayucel, I. (1999) Gene-centromere distances in tilapia, in *Plant and Animal Genome VII*, San Diego, USA: abstracts, p. 33.

Martinez, R., Arenal, A., Estrada, M.P., Herrera, F., Huerta, V., Vazquez, J., Sanchez, T. and de la Fuente, J. (1999) Mendelian transmission, transgene dosage and growth phenotype in transgenic tilapia (*Oreochromis hornorum*) showing ectopic expression of homologous growth hormone. *Aquaculture* **173**, 271–283.

Mazid, M.A., Hussain, M.G. and Alam, M.J. (eds) (1996) *Proceedings of a Workshop on Current Status and Future Strategy for Dissemination of Genetically Improved Farmed Tilapia in Bangladesh.* Fisheries Research Institute, Mymensingh, Bangladesh, 50 p.

Melard, C. (1995) Production of a high percentage of male offspring with 17α-ethynylestradiol sex-reversed *Oreochromis aureus*. I. Estrogen sex-reversal and production of F2 pseudofemales. *Aquaculture* **130**, 25–34.

Meschkat, A. (1967) The status of warm-water fish culture in Africa. *FAO Fish. Rep.* **44**(2), 88–122.

Muir, W.M. and Howard, R.D. (1999) Possible ecological risks of transgenic organism release when transgenes affect mating success: Sexual selection and the Trojan gene hypothesis. *PNAS (USA)* **96**(24), 13853–13856.

Müller-Belecke, A. and Hörstgen-Schwark, G. (1995) Sex determination in tilapia (*Oreochromis niloticus*) sex ratios in homozygous gynogenetic progeny and their offspring. *Aquaculture* **137**, 57–65.

Myers, J.M. (1986) Tetraploid induction in *Oreochromis* spp. *Aquaculture* **57**, 281–287.

Myers, J.M. and Hershberger, W.K. (1991) Early growth and survival of heat-shocked and tetraploid-derived triploid rainbow trout (*Oncorhynchus mykiss*). *Aquaculture* **96**, 97–107.

Myers, J.M., Penman, D.J., Basavaraju, Y., Powell, S.F., Baoprasertkul, P., Rana, K.J., Bromage, N. and McAndrew, B.J. (1995a) Induction of diploid androgenetic and mitotic gynogenetic Nile tilapia (*Oreochromis niloticus*). *Theor. Appl. Genet.* **90**, 205–210

Myers, J.M., Penman, D.J., Rana, K.J., Bromage, N., Powell, S.F. and McAndrew, B.J. (1995b) Applications of induced androgenesis with tilapia. *Aquaculture* **137**, 150.

Mwanza, W. and Kaufman, L. (1995) A note on recent advances in genetic characterisation of tilapia stocks in Lake Victoria region. *African Journal of Tropical Hydrobiology and Fisheries* **6**, 51–53.

Naish, K.A., Warren, M., Bardakci, F., Skibinski, D.O.F., Carvalho, G.R. and Mair, G.C. (1995) Multilocus DNA-fingerprinting and RAPD reveal similar genetic relationships between strains of *Oreochromis niloticus* (Pisces, Cichlidae). *Molecular Ecology* **4**, 271–274.

Oldorf, W., Kronert, U., Balarin, J., Haller, R. Hörstgen-Schwark, G. and Langholz, H.J. (1989) Prospects of selecting for late maturation in tilapia (*Oreochromis niloticus*). II. Strain comparisons under laboratory and field conditions. *Aquaculture* **77**, 123–133.

Pandian, T.J. and Varadaraj, K. (1988) Techniques for producing all-male and all-triploid *Oreochromis mossambicus*, in *The Second International Symposium on Tilapia in Aquaculture* (eds R.S.V. Pullin, T. Bhukaswan, K. Tonguthai and J. Maclean), ICLARM Conference Proceedings 15, pp. 243–249.

Pandian, T.J. and Varadaraj, K. (1990) Development of monosex female *Oreochromis mossambicus* broodstock by integrating gynogenetic technique with endocrine sex reversal. *J. Exp. Zool.* **255**, 88–96.

Penman, D.J., Shah, M.S., Beardmore, J.A. and Skibinski, D.O.F. (1987a) Sex ratios of gynogenetic and triploid tilapia, in *Proceedings of the World Symposium on Selection, Hybridization and Genetic Engineering in Aquaculture* (ed. K. Tiews), Vol II, Heeneman, Berlin, pp. 268–276.

Penman, D.J., Skibinski, D.O.F. and Beardmore, J.A. (1987b) Survival, growth rate and maturity in triploid tilapia, in *Proceedings of the World Symposium on Selection,*

Hybridization and Genetic Engineering in Aquaculture (ed. K. Tiews), Vol. II, Heeneman, Berlin, pp. 277–287.

Penman, D.J., Woodwark, M. and McAndrew, B.J. (1995) Genetically modified fish populations, in *Environmental Impacts of Aquatic Biotechnology*, OECD, Paris, pp. 22–27.

Penman, D.J., Myers, J.M. and McAndrew, B.J. (1997) Restoration of diploid genotypes by androgenesis, in *Sciences et Techniques du Froids. Actes du Colloques Refrigeration and Aquaculture, Bordeaux Aquaculture '96*. International Institute of Refrigeration, Paris, pp. 469–473.

Picciotto, M.R. (1999) Knock-out mouse models used to study neurobiological systems. *Crit. Rev. Neurobiol.* **13**, 103–149.

Piferrer, F., Zanuy, S., Carrillo, M., Solar, I.I., Devlin, R.H. and Donaldson, E.M. (1994) Brief treatment with an aromatase inhibitor during sex differentiation causes chromosomally female salmon to develop as normal, functional males. *J. Exp. Zool.* 270, 255–262.

Pitkänen, T., Krasnov, A., Reinisalo, M. and Mölsä, H. (1999) Transfer and expression of glucose transporter and hexokinase genes in salmonid fish. *Aquaculture* **173**, 319–332.

Pruginin, Y. (1967) Report to the Government of Uganda on the experimental fish culture project in Uganda, 1965–66. FAO/UNDP (Technical Assistance). Reports on Fisheries. TA Reports 2446, FAO, Rome, 19 p.

Pullin, R.S.V. (ed.) (1988) *Tilapia Genetic Resources for Aquaculture*, ICLARM Conference Proceedings 16, 108p.

Pullin, R.S.V. (1994) Exotic species and genetically modified organisms in aquaculture and enhanced fisheries: ICLARM's position. *Naga, the ICLARM Quarterly* **17**(4), 19–24.

Pullin, R.S.V. and Capili, J. (1988) Genetic improvement of tilapias: problems and prospects, in *The Second International Symposium on Tilapia in Aquaculture* (eds R.S.V. Pullin, T. Bhukaswan, K. Tonguthai and J. Maclean), ICLARM Conference Proceedings 15, pp. 259–266.

Rahman, M.A. and Maclean, N. (1992) Production of transgenic tilapia (*Oreochromis niloticus*) by one-cell-stage microinjection. *Aquaculture* **105**, 219–232.

Rahman, M.A. and Maclean, N. (1999) Growth performance of transgenic tilapia containing an exogenous piscine growth hormone gene. *Aquaculture* **173**, 333–346.

Rahman, M.A., Iyengar, A. and Maclean, N. (1997) Co-injection strategy improves integration efficiency of a growth hormone gene construct, resulting in lines of transgenic tilapia (*Oreochromis niloticus*) expressing an exogenous growth hormone gene. *Transgenic Research* **6**, 369–378.

Rahman, M.A., Mak, R., Ayad, H., Smith, A. and Maclean, N. (1998) Expression of a novel piscine growth hormone gene results in growth enhancement in transgenic tilapia (*Oreochromis niloticus*). *Transgenic Research* **7**, 357–369.

Rana, K.J., McAndrew, B.J. Wohlfarth, G. and MacGowan, I. (1996) Observations on intergeneric hybrids in tilapias, in *The Third International Symposium on Tilapia in Aquaculture* (eds R.S.V. Pullin, J. Lazard, M. Legendre, J.B. Amon Kothias and D. Pauly), ICLARM Conference Proceedings 41, pp. 391–397.

Sanchez, T., Ponce de Leon, R., Aguilar, M., Vazquez, J. and McAndrew, B. (1995) Response to selection and heritability for weight in *Oreochromis aureus* Steindachner after five generations of selection. *Aquaculture* **137**, 271.

Sarder, M.R.I., Penman, D.J., Myers, J.M. and McAndrew, B.J. (1999) Production and propagation of fully inbred clonal lines in the Nile tilapia (*Oreochromis niloticus* L.). *J. Exp. Zool.* **284**, 675–685.

Sarder, M.R.I., Thompson, K.D., Penman, D.J. and McAndrew, B.J. (in press) Immune responses of Nile tilapia (*Oreochromis niloticus* L.) clones: I. Non-specific responses. *Dev. Comp. Immunology.*

Schuster, W.H. (1952) Fish-culture in brackish-water ponds of Java. *Indo-Pacific Fisheries Council Special Publication*, No. 1, 1–25.

Scott, A.G., Penman, D.J., Beardmore, J.A.B. and Skibinski, D.O.F. (1989) The 'YY' supermale in *Oreochromis niloticus* (L.) and its potential in aquaculture. *Aquaculture* **78**, 237–251.

Seyoum, S. and Kornfield, I. (1992) Identification of the subspecies of *Oreochromis niloticus* (Pisces, Cichlidae) using restriction endonuclease analysis of mitochondrial DNA. *Aquaculture* **102**, 29–42.

Shell, E.W. (1967) Relationship between rate of feeding, rate of growth and rate of conversion in feeding trials with two species of tilapia *Tilapia mossambica* Peters and *Tilapia nilotica* Linnaeus. *FAO Fish Rep.* **44**(3), 411–415.

Shelton, W.L., Meriwether, F.H., Semmens, K.J. and Calhoun, W.E. (1983) Progeny sex ratios from intraspecific pair spawnings of *Tilapia aurea* and *Tilapia nilotica*, in *International Symposium on Tilapia in Aquaculture* (eds L. Fishelson and Z. Yaron), Tel Aviv University, Israel, pp. 270–280.

Sipe, M. (1979) Golden perch. *Commercial Fish Farmer Aquaculture News* **5**(5), 56.

Sodsuk, P.K. (1993) *Molecular Genetics and Systematics of Tilapiine Cichlids using Allozymes and Morphological Characters*, PhD dissertation, University of Stirling.

Sodsuk, P. and McAndrew, B.J. (1991) Molecular systematics of three tilapiine genera *Tilapia, Sarotherodon* and *Oreochromis* using allozyme data. *J. Fish Biol.* **39** (Suppl. A), 301–308.

Sodsuk , P.K., McAndrew, B.J. and Turner G. (1995) Evolutionary relationships of the Lake Malawi *Oreochromis* species: evidence from allozymes. *Journal of Fish Biology* **47**, 321–333.

Taniguchi, N., Macaranas, J.M. and Pullin, R.S.V. (1985) Introgressive hybridisation in cultured tilapia stocks in the Philippines. *Bull. Jap. Soc. Fisheries* **51**(8), 1219–1224.

Tave, D. and Smitherman, R.O. (1980) Predicted response to selection for early growth in *Tilapia nilotica*. *Trans. Am. Fish. Soc.* **109**, 439–445.

Tave, D., Rezk, M. and Smitherman, R.O. (1989) Genetics of body color in *Tilapia mossambica*. *J. World Aqua. Soc.* **20**, 214–222.

Teichert-Coddington, D.R. and Smitherman, R.O. (1988) Lack of response by *Tilapia nilotica* to mass selection for rapid early growth. *Trans. Am. Fish. Soc.* **117**(3), 297–300.

Thien, C.M. (1971) Estimation of realised heritability in tilapia (*Tilapia mossambica* Peters) *Genetika* **7**(12), 53–59. (in Russian with English abstract) (English translation in: *Sov Genet.* **7**, 1550–1554).

Toyohara, H., Nakata, T., Touhata, K., Hashimoto, H., Kinoshita, M., Sakaguchi, M., Nishikimi, M., Yagi, K., Wakamatsu, Y. and Ozato, K. (1996) Transgenic expression of L-gulono-γ-lactone oxidase in medaka (*Oryzias latipes*), a teleost fish that lacks this enzyme necessary fo L-ascorbic acid biosynthesis. *Biochem. Biophys. Res. Comm.* **223**, 650–653.

Trewavas, E. (1968) The name and natural distribution of the *Tilapia* from Zanzibar (Pisces Cichlidae). *Proc World Symp. Warm-Water Pond Fish Culture*, FAO Fish. Rep. **44**(5), 246–254.

Trewavas, E. (1983) *Tilapiine fishes of the genera* Sarotherodon, Oreochromis *and* Danakilia. British Museum (Natural History) London.

Uraiwan, S. (1988) Direct and indirect responses to selection for age at first maturation of *Oreochromis niloticus*, in *The Second International Symposium on Tilapia in Aquaculture* (eds R.S.V. Pullin, T. Bhukaswan, K. Tonguthai and J. Maclean), ICLARM Conference Proceedings 15, pp. 295–300.

Uraiwan, S. and Doyle, R.W. (1986) Replicate variance and choice of selection procedure for tilapia (*Oreochromis nilotica*) stock improvement in Thailand. *Aquaculture* **57**, 93–98.

USDEA (1997) DEA No. 1601 Schedules of controlled substances: Exempt Anabolic Steroid Products. Drug Enforcement Administration drug and chemical evaluation section. Dept of Justice, Washington DC 20537.

Van Someren, V.D. and Whitehead, P.J. (1959) The culture of *Tilapia nigra* (Günther) in ponds. 1. Growth after maturity in male *T. nigra. East Afr. Agric. J.* **25**, 42–46.

Varadaraj, K. (1993) Production of viable haploid *Oreochromis mossambicus* gynogens using UV-irradiated sperm. *J. Exp. Zool.* **267**, 460–467.

Varadaraj, K. and Pandian, T.J. (1989) First report on production of supermale tilapia by integrating endocrine sex reversal with gynogenetic technique. *Curr. Sci.* **58**, 434–441.

Varadaraj, K. and Pandian, T.J. (1990) Production of all-female sterile-triploid *Oreochromis mossambicus. Aquaculture* **84**, 117–123.

Velesco, R.R., Janagap, C.C., Palaido de Vera, M., Afan, L.B., Reyes, R.A. and Eknath, A.E. (1995) Genetic improvement of farmed tilapias: estimation of heritability of body and carcass traits of Nile tilapia (*Oreochromis niloticus*). *Aquaculture* **137**, 280–281.

Villegas, C.T. and Doyle, R.W. (1986) Duration of feeding and indirect selection for growth of tilapia. *Aquaculture* **57**, 89–92.

Watanabe, W.O. 1991. Saltwater culture of tilapia in the Caribbean. *World Aquaculture* **22**(1), 49–54.

Welcomme, R.L. (1964) Notes on the present distribution and habits of the nonendemic species of tilapia which have been introduced into Lake Victoria. *Rep. E. Afr. Freshwater. Fish Res. Org.***1962/63**, 36–39.

Welcomme, R.L. (1967) Observations on the biology of the introduced species of tilapia in Lake Victoria. *Revue. de Zoologie et de Botanique Africaine* **86**, 249–279.

Wohlfarth, G.W. (1994) The unexploited potential of tilapia hybrids in aquaculture. *Aquaculture and Fisheries Management* **25**, 781–788.

Wohlfarth, G.W. and Hulata, G. (1983) *Applied Genetics of Tilapia* (second edition), ICLARM Studies and Reviews 6, 26p.

Wohlfarth, G.W., Rothbard, S., Hulata, G. and Szweigman, D. (1990) Inheritance of red body coloration in Taiwanese tilapia and in *Oreochromis mossambicus. Aquaculture* **84**, 219–234.

Yamamoto, T. (1969) Sex differentiation, in *Fish Physiology* Vol. 3 (eds W.S. Hoar and D.J. Randall), Academic Press, New York, pp. 117–175.

Yapi-Gnaore, C.V. (1996) Estimation of additive an nonadditive genetic parameters in the growth of fry of three strains of *Oreochromis* spp, in *The Third International Symposium on Tilapia in Aquaculture* (eds R.S.V. Pullin, J. Lazard, M. Legendre, J.B. Amon Kothias and D. Pauly), ICLARM Conference Proceedings 41, pp. 426–432.

Chapter eight

Strategies for tilapia seed production

D. C. Little and G. Hulata

8.1 INTRODUCTION

Tilapias stand alone among the important cultured fish in their singular ability to breed without any special inducement or modification to their environment. But an analysis of the global status of cultured tilapias suggests that the lack of an effective strategy for seed production can both weaken efforts to support current farmers raising tilapias and jeopardize successful adoption of improved systems and strains.

Biological and evolutionary aspects of tilapia reproduction have been described in a number of publications, notably in Lowe McConnell (1959), Fryer and Iles (1972), Noakes and Balon (1982), Trewavas (1983) and Turner and Robinson (this volume). Reproductive processes, including maturation, and egg and hatchling development of tilapias, were reviewed by Rana (1988), and more specifically for Nile tilapia (*Oreochromis niloticus*) by Macintosh and Little (1995). Production of all-male Nile tilapia fry using sex inversion techniques and YY males has been reviewed recently by Green *et al.* (1997) and by Penman and McAndrew, this volume.

Procedures for seed production in tilapias can be classified according to various criteria, such as the mode of harvesting, the age of harvested seed, the genetic characteristics of seed, the physical nature of the hatchery facilities used and their labour requirements. In view of the rather good recent coverage of the broad topic we have decided to focus in the current chapter on hatchery strategies that have developed to supply tilapia seed which can be defined generically as earthen pond, tank and hapa. We first consider the major factors

M.C.M. Beveridge and B.J. McAndrew (eds), Tilapias: Biology and Exploitation, 267–326.

that affect production of tilapia fry before an interpretation of the evolution of the systems in use today. A comparison of the efficiency of systems and their management over a spectrum of culture intensity will then be made and related to the criteria listed above.

8.2 DEVELOPMENT OF TILAPIA SEED PRODUCTION

Sources of tilapia seed

Farmers obtain tilapia seed for stocking (Fig. 8.1) from a variety of sources. Fish seed may be caught from wild or feral stocks in natural waters, collected ('saved') as juveniles spawned by fish growing in culture systems, purchased from hatcheries or produced on-site in a seed production system. The potential for control of breeding of stocked fish is a major advantage of hatchery-produced seed over wild or saved seed. The possible methods of population control are however highly dependent on the design and management of the hatchery process.

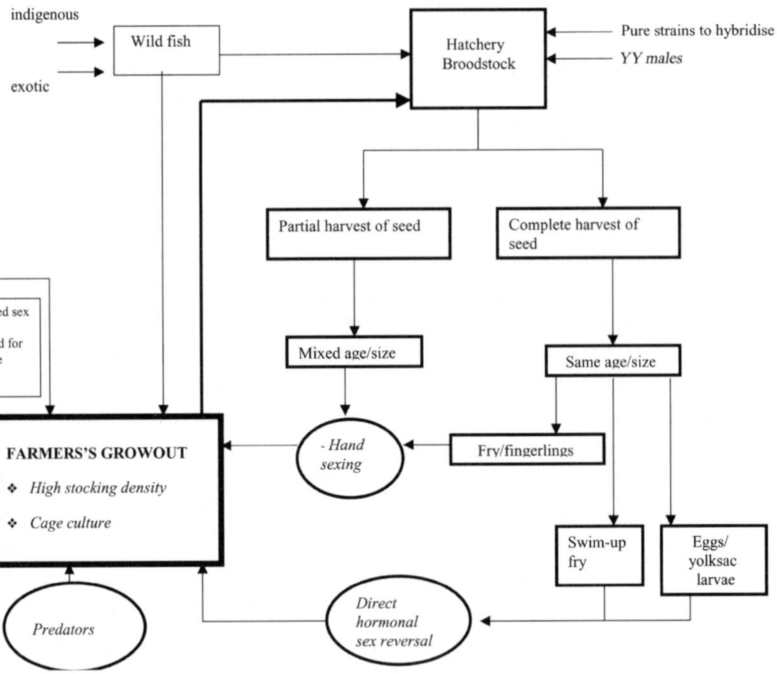

Fig. 8.1 Schema showing potential hatchery strategies for tilapias and opportunities for population control (italics).

Tilapia seed has been harvested from 'wild' populations both within and outside their natural range. The practice does not appear to have reached the levels found with carp in Asia where until recently traditional culture of riverine carps was based on seasonal supplies of wild-caught seed. The fact that tilapias reproduce without inducement in a variety of systems may explain why lack of seed has rarely constrained the beginnings of tilapia culture. However, the timely availability of high quality seed has become an issue and a major constraint to the further development of tilapia in many places.

In attempts to promote fish culture in Africa during the colonial era, seed was obtained from indigenous populations for stocking dams and from government experimental fish stations (R.H. Lowe-McConnell, personal communication). Supplementation of fry produced by government hatcheries with fish from feral populations, often mixed species, is still common in Egypt (B.W. Green, personal communication). 'Chambo', or *O. karongae*, fingerlings harvested by beach seine from Lake Malawi for stocking in ponds are difficult to differentiate from other species and in practice are often a mixture of *O. karongae*, *O. lidole* and *O. squamipinnis* (Brooks and Maluwa, 1997a). Hatchery management is haphazard in many developing countries and hatchery seed may be simply unavailable at the time of stocking or not produced in the quantities required. In Africa, almost everybody stocks wild fish which may out-perform poorly managed hatchery fish (R.E. Brummett, ICLARM, personal communication).

The breeding of tilapias within the grow-out system, although undermining the popularity of the fish where only large individuals are marketable, has been a major spur to their dissemination and adoption as cultured fish. Quite apart from ensuring a local supply of large seed, it can improve the returns from the fish enterprise as a whole. Engle (1997) found that when both Nile tilapia fingerlings and food fish could be sold in Rwanda, aquaculture became a much more attractive activity compared to alternative crops. Depending on a range of factors (age and density of stocked fish; productivity of the system), the seed produced can exceed that stocked in fertilized earthen ponds by a factor of 1.25 to more than 10 (AIT, 1986). This 'seed saving' can both sustain production without repeated introductions of hatchery-produced fish and result in the gradual spread, or farmer-to farmer adoption, of tilapia culture under the right conditions. Typically, however, backyard tilapia production is constrained by erratic quantities of low quality seed produced in this way. In three districts of Bangladesh, the inadequate supply of fingerlings was the most commonly reported difficulty among farmers raising Nile tilapia (Gupta *et al.*, 1992). Although only around 15% of farmers stocking tilapia in this study harvested and sold tilapia seed as a by-product of culture, the numbers of seed produced met from 38 to more than 200% of the estimated requirements for restocking in the areas concerned. However, a perennial lack of water is likely to reduce the actual availability of seed at the beginning of the new season.

The ability to produce self-sustaining crops may be an important consideration where hatcheries are undeveloped and also will affect species choice. In identifying suitable species for Malawian smallholders, Dickson and Brooks (1997) suggested that even though *O. karongae* had greater growth potential than *O. shiranus*, the latter was more fecund under low input conditions and more likely to sustain production of isolated smallholders.

Seasonal drought or predatory fish can reduce the availability of seed from one year to the next, necessitating restocking from external sources. Wild, predatory fish such as the snakehead (*Channa striata*) can be effective at removing any juvenile tilapias in mixed sex tilapia ponds (Edwards *et al.*, 1994). Low temperatures can also reduce the possibility of saving seed as a strategy for sustaining production. At higher, cooler elevations in Rwanda, Nile tilapia fingerlings produced as a by-product of culture were often insufficient for restocking (Hanson *et al.*, 1988).

Deterioration in the quality of tilapia strains is a common but poorly understood phenomenon. In the Philippines a major cause is genetic introgression (Macaranas *et al.*, 1986) through contamination with feral fish and poor management. Repeated use of recruits from small, isolated populations of pure species may lead to deterioration through indirect selection, genetic drift and inbreeding, especially if the original transfers were of small numbers of founder broodfish. This may be more problematic for some species, such as *O. mossambicus*, than the Nile tilapia; the apparently fast growing stocks such of the Chitralada strain of Nile tilapia in Thailand were based on small numbers of surviving individuals. Until the recent introduction of improved strains of Nile tilapia to northern Vietnam, the proportion of tilapia stocked in polycultures with carps in the Tranh Tri sewage-based culture systems around Hanoi appears to have declined substantially. The inability of the local tilapia strain to reach marketable size (100 g) within the normal culture period is understood to be the cause of this decline. This problem, while contributing to the poor acceptability of tilapia among those selling fish to market, may be an important part of its appeal to poorer people. In northwest Bangladesh, nearly half of backyard ponds and ditches contain small, stunted tilapias, often in polyculture with stocked carps and small indigenous 'weed' fish. As intensification has reduced availability of weed fish in common property waterbodies, such tilapia are believed to be important to women who have access to them for feeding the household. Plans to replace current strains with 'improved' strains have raised the possibility that these will become a commodity, like the carps, and be increasingly traded in the market. Potentially such 'improved' strains risk negative impacts on family nutrition in poor households and this should be recognized in dissemination efforts (Barman *et al.*, 1998).

Population control

Reproduction within the culture system has been a major constraint to the adoption of tilapia culture and has limited the popularity of tilapia as a cultured fish in places where it is established. The problem is especially acute where temperatures remain high year-round, supporting continuous breeding and recruitment in mixed sex populations. Simple hatchery systems that produce large, mixed sex fry necessitate some form of predator control and/or hand sexing to control the amount of recruitment and number of breeding females, respectively.

The stocking of predators to control tilapia breeding has been advocated as a practical method to raise larger individual fish (Guerrero, 1982), but the effectiveness of the strategy may be limited by availability of suitable species of the right size and at the right time. Effective control of recruitment, while preventing overpopulation, does not reduce other reproductive losses, especially to the production of females, and to courtship, spawning and parental care. Both predators and hand sexing to remove females impose serious management constraints but are used independently, or together, to control population in systems stocked with mixed sex and nearly all-male culture.

The type of culture systems also affects the necessity for complete control of spawning. While even a small percentage of females will result in uncontrolled populations in undrainable earthen ponds (Mair and Van Dam, 1996), courtship and spawning is inhibited at high densities. Breeding behaviour and the production of viable juveniles has less impact in intensive culture systems, i.e. systems in which fish are stocked at high density in tanks, raceways, cages or ponds with high water exchange. Spawning, recruitment and its effects on the growth of the stocked fish is typically negligible under such conditions and evidence of spawning in intensive systems indicates that fish densities are too low.

Considerable research has been conducted on production of all-male populations. Males grow faster than females, especially in some strains of Nile tilapia (Capili, 1995). Hatchery methods and scale determine possible approaches. If broodfish capable of producing all-male progeny directly are used, a range of hatchery options is possible. Theoretically, all or nearly all-male seed can be produced by crossing pure species broodfish to produce hybrids (Lovshin, 1982; McAndrew and Majumdar, 1983; McAndrew, 1993, this volume). The use of hybrids for population control has largely been discontinued or supplemented with other measures. The complexity of hatchery management required and the difficulties of keeping pure stocks apart have reduced the practical application of this research. Similar constraints face any widespread dissemination of the YY male technology. YY males, when bred with normal females (XX), produce genetically male stocks (XY) (Mair *et al.*, 1997). This technology was field tested in hatcheries in the Philippines (Mair, 1997) and now is being commercialized. The process to develop YY male broodfish requires several generations, with progeny testing at each stage, which in itself will limit the process to institutions with considerable research capacity. Also, unless control of contamination by wild tilapia can be built into the hatchery process, the predictability of sex ratio, and performance in farmers' ponds, will not meet expectations.

Direct hormone-in-feed treatment of sexually undifferentiated fry with androgens, usually 17α-methyltesteosterone (MT), is the main technique used commercially for producing nearly all-male stocks. Its successful adoption, however, has been uneven despite considerable research effort (see Green *et al.*, 1997, for review). The type and stage of seed harvested, the quality of feed and its management, and a range of environmental factors, including level of natural feed and temperature appear to affect success (Fig. 8.2).

Commercial hatchery production of tilapias suitable for hormonal sex reversal is based on use of earthen ponds, hapas suspended in either ponds or larger waterbodies, or tanks. Management intensity determines both productivity and suitability of fry for hormone treatment. Harvest method (complete or partial removal of fry at one time) and the type of seed harvested (swim-up fry or maternal-dependent seed, i.e. eggs/yolk-sac fry) appears to affect the consistency of results. In particular, the treatment of fry removed from the edge of ponds, hapas and tanks after grading to the same size appears to give unpredictable sex ratios under many conditions (Meyer *et al.*, 1997; Mair, 1997).

The production of seed suitable for direct hormone treatment, i.e. young and homogeneous, may also have positive effects on overall hatchery production. Early harvesting of seed and the raising of similar-sized larvae and fry in batches improves individual female productivity and reduces losses to cannibalism, respectively (Fig. 8.2). The possibility of using immersion of same-age fry in hormone solutions has been researched (see Yang Yi, 1992; Green *et al.*, 1997) but is constrained practically by a need for temperature control.

The techniques that have been used to produce monosex populations or to control recruitment in mixed sex culture have been reviewed by Mair and Little (1991).

Evolution of tilapia hatchery systems

Hatchery systems for tilapia have developed in different directions over the last two decades as demand for fry has increased. The centres of hatchery development have been in Israel and Asia, and both commercial operators and scientists have shaped the methods that are now routinely used. Methods developed at Auburn University also are widely used, especially in parts of Latin America.

The nature of demand, both for fry and marketable food fish, has shaped the development of tilapia hatcheries. The availability of scarce resources, particularly water, land, labour and nutrients, also has affected how hatchery systems have developed.

Mixed sex tilapia production continues to dominate in most of Asia. Where tilapia have become best established in markets and peoples' eating habits, small wild fish are typically staple foods, particularly among rural or poorer consumers. Current systems for producing mixed sex fish are based mainly on the fertilization of earthen ponds with low cost organic wastes, although more intensive methods are now developing in response to increased demand and higher prices. In the main areas of production, shallow ponds were developed by

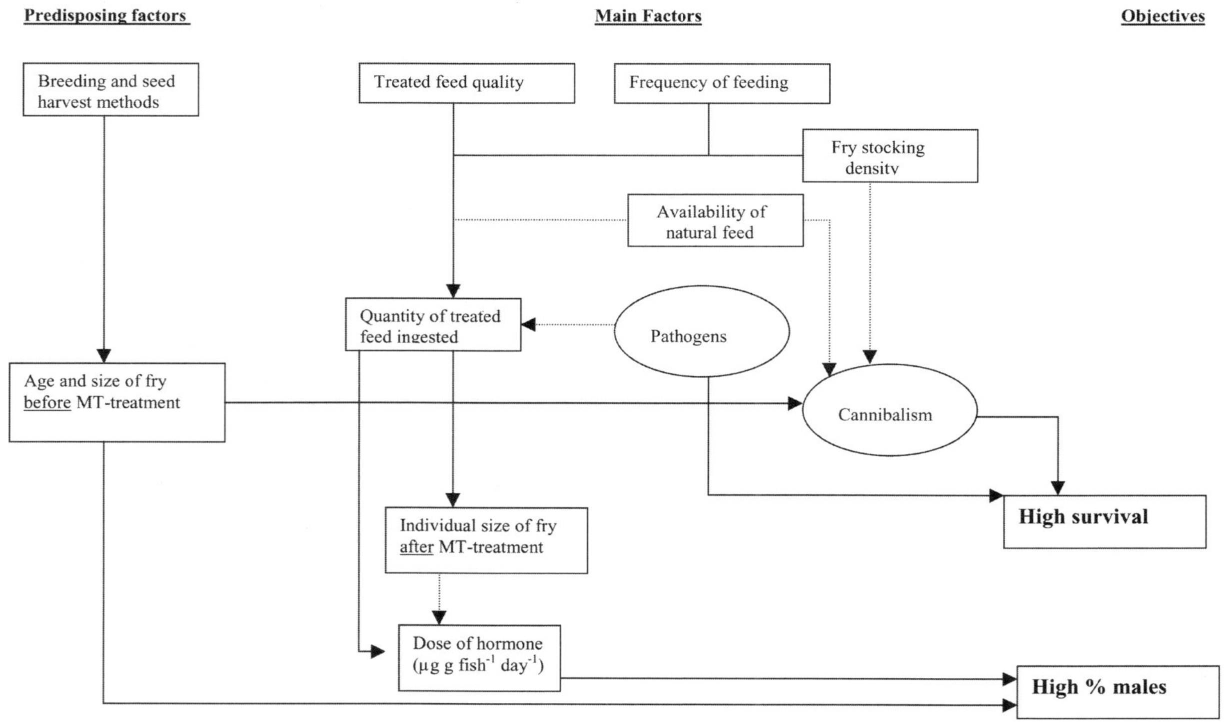

Fig. 8.2 Factors affecting the efficacy of hormonal sex reversal of tilapia (modified from Mair and Little, 1991). Solid arrow, major relationship; dotted arrow, minor relationship.

simple deepening of rice fields. Typically, demand for seed existed year-round, and family-based hatcheries, often specialized in the production of tilapia, have developed to supply the market. Such hatcheries tend to concentrate in certain locations, attracted by available year-round water, nutrients and infrastructure. This 'clustering' of enterprises allows a large quantity of seed to be assembled and marketed, even if individual hatchery output is low and unpredictable. Intermediaries provide a vital role in this respect and are critical to the effective dissemination of tilapia seed in many countries of the region (Yater and Smith, 1985; Little *et al.*, 1994). Although government agencies may have been involved in initial introductions of germ-plasm, farmers have developed the simple techniques used, and impressive levels of local knowledge have evolved.

Just as the simple nature of fry production systems may have been an important factor in the popularity of tilapias, it also may have contributed to deterioration in quality. Lack of control of breeding fish and harvest of their seed can result in poor quality seed. The inability of stocked fish to reach a minimum final size in a season (100–200 g), has been commonly linked to inbreeding and/or introgression with poor quality feral strains especially those derived from *O. mossambicus*. Such feral stocks are found in natural waters and irrigation systems of Asia and are practically difficult to exclude from pond-based hatcheries. But disappointing performance of tilapia can also have a non-genetic basis. Poor management of growout systems, especially over-stocking and lack of inputs, will result in poor individual growth, but older 'stunted' fry that breed shortly after stocking will exacerbate the problem greatly (see also Lorenzen, this volume). It has been demonstrated that the yields of mixed sex or MT-treated Nile tilapia of between 100 and 200 g individual size are similar provided that nutrient inputs are adequate and the stocked seed is a good strain and less than 2 months of age (AIT unpublished data). Market demand for larger fish is growing fast, however, and monosex stocks appear to have advantages in both semi-intensive and intensive systems. On balance, where they are available, the added expense of monosex fry is a marginal part of overall production costs and they are preferred for intensive systems because they lead to marketable fish more homogenous in shape and size.

Other factors affect the demand for tilapia seed. In areas of Asia in which aquaculture is well established but temperatures are seasonally cool, culture of carps tends to dominate and inhibit the development of tilapia production. Hatchery operators already producing carps have little incentive to produce tilapias as carp-only operations are more profitable in many areas (Little *et al.*, 1987). Many carps can be spawned early in the season, when tilapias have yet to begin conditioning for spawning, and their fingerlings also can be over-wintered to be sold and improve cash flow. Technologies such as over-wintering of monosex seed and broodfish using geothermal water and plastic tunnels for sheltering ponds, pioneered in Israel (e.g. Mires, 1995) to meet market needs and to allow tilapia and carp production to thrive together, are largely unexploited in Asia.

Box 8.1 Breeding tilapias in rice fields

Many pond-based tilapia hatcheries began from shallow, excavated rice fields, but special circumstances in Vietnam have led to the development of a system that retains rice in the field. Human population densities in the Red River delta in the north of the country are some of the highest in the world, and rice production is both a staple crop and considered strategically important by the government. The Government is therefore seeking to increase diversification and farmers' incomes without reductions in rice production, partly through the promotion of aquaculture. Nile tilapia has failed to become widely adopted in the region; this is partly because seed production methods have yet to develop, thus overcoming the restrictions to early seed production caused by seasonally cool temperatures. In one commune in Thai Binh Province, a unique technique is being assessed in which broodfish are conditioned in warm geothermal water-fed tanks and ponds by the commune before distribution to commune members and stocking into the spring irrigated rice fields. Little adaptation of the rice field is required or desirable, as the maintenance of rice yields is a priority. A simple trench at one end into which broodfish are stocked is all that is required, provided that water levels can be maintained. Growth of fry is rapid as broodfish and seed density is low and food plentiful. Fry and fingerlings are netted from the trench as they are observed and held in hapas. Advantages of the method include:

- maintenance of rice production while farmers diversify their farming system;
- no weeding of the rice crop is required because of the activity of the tilapias, reducing labour demand;
- low risk entry into value added aquaculture;
- production of large tilapia seed early in the season so that farmers can stock tilapia at the same time as carps;
- maintenance of equity; hatchery production frequently favours richer people but this technique allows anyone to participate;
- stimulation of other employment through increase in opportunities for trading of tilapia seed and supplies of feed and fertilizer; and
- role of the fish in integrated pest management in the rice field, reducing or eliminating the need for pesticide (Nguyen *et al.*, 1997).

Plate 8.1 Concrete blocks placed perpendicular to the wall of a circular plastered brick tank to provide space for high densities of male fish to make territories and reduce aggression.

Plate 8.2 *Oreochromis shiranus* mixed size fry and fingerlings harvested with stocked fish (large fish top, male; bottom, female). Courtesy Dr Malcolm Dickson.

Plate 8.3 Small-scale producers producing mixed sex and sized Nile tilapia fingerlings in Chacheonsao Province, Thailand.

Plate 8.4 Motorcycle trader in the Mekong Delta, Vietnam selling tilapias packed in plastic bags produced in sewage-fed ponds in Binh Chan District near Ho Chi Minh City.

Plates 8.5 and 8.6 Greenhouse-based hatcheries in Israel used for red and hybrid tilapia seed production.

In Israel, a pronounced cool season, a demand for a large marketable fish and the high cost of labour and water have combined to determine a very unique style of hatchery based on large earthen ponds. Recently, more intensive systems involving greenhouses and geothermal water together with smaller ponds with harvest sumps have become more common (Hulata, 1997).

The ability of tilapias to spawn in a range of artificial environments has given rise to aquarium, hapa and tank-based spawning systems. Although individual broodfish productivity can be high in aquaria, the relatively small numbers of broodfish that can be kept mean that such units only can serve niche, high-value markets, such as for pure strains and YY males. Scientists using low-cost, fine mesh hapas for holding fish or comparing their growth may have incidentally led to their use for breeding. Hapas allow land-less people access to community water bodies (Beveridge, 1984). Lack of a suitably resistant material for making hapas, or its cost, has however constrained adoption of hapas in many places. Tanks have been used in attempts to intensify output on smaller areas of land, where land or water is limited. Tanks may be managed intensively with water quality being maintained through filtration, aeration or water exchange despite high densities of fish. Alternatively, they may be managed like small, easily drained ponds. High initial capital costs compared to earthen ponds may be balanced by lower recurrent costs.

In situations where broodfish are maintained in more intensive systems (tanks and hapas) and the production of MT-treated fish is required, the development of artificial incubation and larval rearing systems becomes more attractive (Little, 1989; Macintosh and Little, 1995; Little *et al.*, 1997). This has occurred in areas with little prior hatchery development, and where the industry is based on an industrial, feedlot model (e.g. central Java). Such hatcheries also are rapidly developing in Thailand, where tilapia culture is well established, usually with semi-intensively cultured carps. Increasing domestic markets for larger fish and the expansion of production into more intensive cage systems and use of ex-shrimp ponds has fuelled demand for MT-treated seed (Little, 1998).

8.3 FACTORS AFFECTING REPRODUCTIVE SUCCESS AND HATCHERY EFFICIENCY

Environment

The natural breeding environment of tilapias is based around the lek, a breeding 'arena' in which males form and defend territories to attract, court and spawn with females. Vision has a dominant role in the spawning of tilapias; the coloration and behavioural display of spawning fish is often intense in the well-illuminated margins of water bodies that leks are typically situated. The territorial and hierarchic behaviour of males and both sexes respectively make successful spawning density-dependent. Densities of breeding fish in natural leks can be high; Fishelsen (1983) observed colonies of *O. aureus* males to frequently number over ten individuals in 30 m^{-2}, with a corresponding number of over 30 females. Post-spawning, mouthbrooding females seek solitude (Philippart and Ruwet, 1982), often in a reed bed, suggesting that avoidance of other fish is critical to the success of mouthbrooding.

Both egg-guarding and mouthbrooding species prefer a firm substrate that supports spawned eggs and facilitates their fertilization. Most *Oreochromis* species do not require gravel or sediment for males to dig a nest and to establish territories and court females (Costa Pierce, 1996). 'Nest-builders' adapt readily to artificial surfaces such as cement/fibreglass and fine-mesh net used in tank and hapa systems respectively. Reducing visual contact can control aggression between males, disturbance of spawning fish and, probably, hierarchies among females. Concrete blocks or shelters can be used to decrease visual contact between territorial males in clearwater tank systems (Plate 8.1). Artificial nests had no impact on spawning intensity in hapas suspended in fertile ponds, presumably as visual contact of broodfish was already reduced the phytoplankton densities (Bevis, 1994).

Water quality

Water quality, especially low dissolved oxygen and elevated levels of nitrogenous compounds, affects all stages of the reproductive cycle from broodfish production through to advanced nursing of fingerlings. Dissolved oxygen levels that optimize somatic growth also appear to enhance reproductive output. The diurnal cycle of dissolved oxygen and pH values typical of eutrophic ponds, in which values increase with photosynthesis during daylight hours and decline to minimum values before dawn, match with natural reproductive activity. Courtship of breeding fish appears to increase through the day for Nile tilapia breeders in tanks and hapas. Prolonged low early morning dissolved oxygen may constrain ovarian growth, courtship and the ability of female fish to mouthbrood eggs and larvae. Lowe-McConnell (1959) found lekking males in African swamps surfaced when dissolved oxygen concentrations were low, and air-gulping broodfish can also be observed in hatchery systems. Females can continue mouthbrooding under such conditions but clutch size is frequently reduced and seed quality poorer. Barlow (1991) raises the issue that mouthbrooding might be a pre-adaptation to breeding in periodically oxygen poor water or that such conditions might have stimulated its evolution. Female broodfish in hapas within a pond (Pond 1) with stable and high levels of phytoplankton produced significantly more seed than a pond with lower and more erratic levels (Pond 2) (Little *et al.*, 2000; Fig. 8.3). Suitable water quality parameters are given in Table 8.1 (see also Ross, this volume).

Temperature is a major constraint to hatchery production of tilapias in countries as diverse as Vietnam, Rwanda and Israel. Low water temperatures can result in reductions or delays in seed production both in the subtropics and at higher elevations in the tropics (Green *et al.*, 1997). Nile tilapia broodfish begin spawning at temperatures below 22°C in northern Vietnam but early

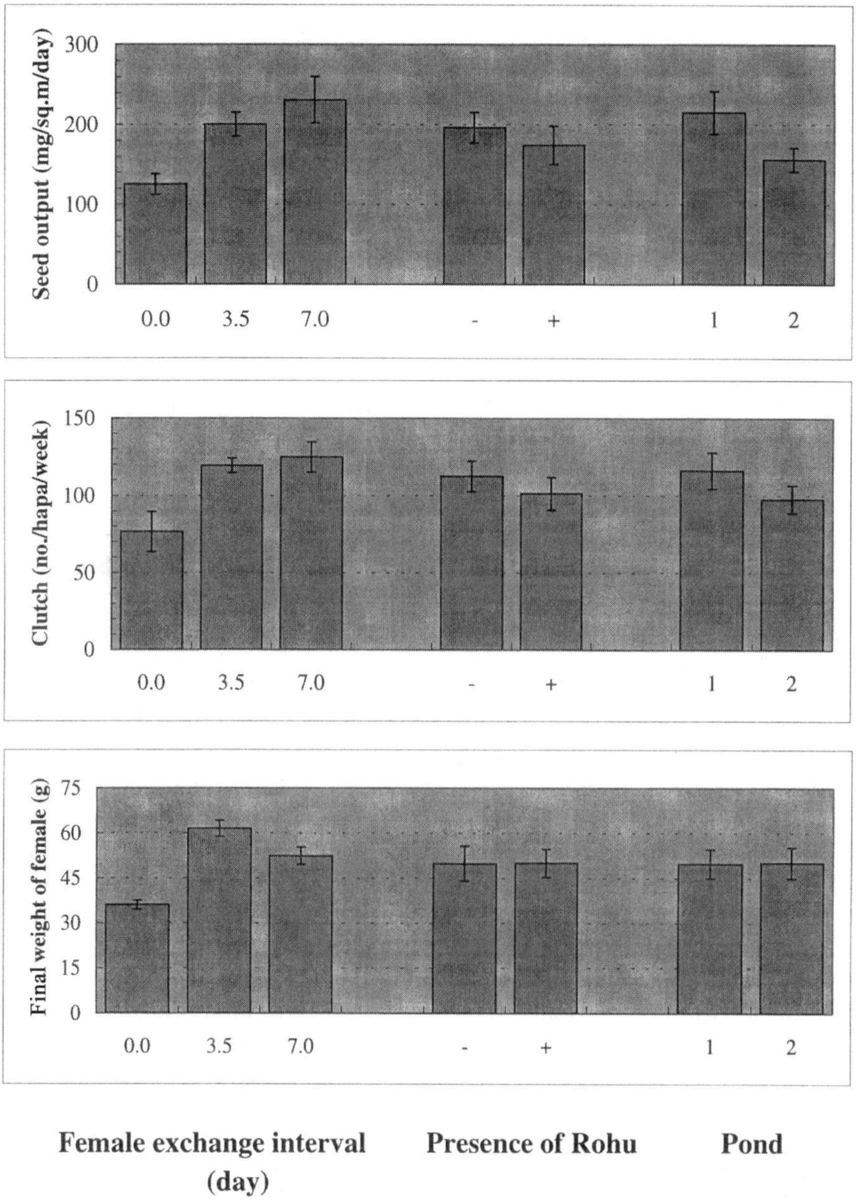

Fig. 8.3 Mean seed production, broodfish growth in female fish in hapas in two different ponds, with and without rohu (*Labeo rohita*) and maintained under different broodfish exchange regimes (modified after Little *et al.*, 2000).

Table 8.1 Water quality parameters for optimal tilapia seed production

Parameter	Time	Range
Temperature		28–32°C
Dissolved oxygen in fertilized ponds	06.00	>0.8 mg l^{-1}
	15.00	6–15 mg l^{-1}
pH		6–9
Chlorophyll a		100–300 mg m^{-3}
Total ammonia		<5 mg l^{-1}
Unionized ammonia		<0.1 mg l^{-1}

season eggs incubate poorly in artificial incubators and more fry can be produced by allowing females to mouthbrood naturally. Modifications in hatchery design to increase temperatures through passive daytime heating and improved conservation of warmer water overnight can improve hatchery performance. Wind breaks have been used in northern Vietnam to moderate the effects of cold northerly winds (Cao, 1998) and deep hapa-in-ponds for improving survival of seed and broodfish (Dan and Little, 2000). Innovative designs are also required in places such as Kuwait that experience extreme daytime/night-time variation in temperatures (K. Hopkins, personal communication). In contrast, high ambient temperatures (33–35°C) can reduce spawning intensity, egg quality and hatchery success of hapa-based Nile tilapia production in central Thailand during the hot season (Little *et al.*, 1997). Performance of similar stocks of fish in recirculated water-supplied tanks located under a roof, in which temperatures remained more moderate, remained stable during hot periods. Performance of hapas in deeper ponds can be improved with circulation of water and shading of the pond (Bhujel, 1999).

Salinity is a water quality parameter of importance in terms of both site and species selection. Saline tolerant species and strains tend to produce more seed at lower salinities than the levels at which they can be profitably raised as food fish. In practice most hatcheries producing tilapia seed under brackish water conditions are tank-based and this factor is considered at more length on p. 297.

Broodstock age, size, ratio and density

The working life span of tilapia broodfish varies widely with hatchery situation, and is related to the type of system, the cost of new fish, value of old fish, seasonality, etc. The management of same age, same size broodfish as 'sets' which are raised, managed and disposed of as discrete groups is good practice. It ensures that individual broodfish productivities and egg sizes are comparable. This can be important for ensuring best results during artificial incubation and larval rearing. Size and age of broodfish, stocking densities and sex ratios show wide variability among and within spawning systems. Larger, older broodfish are known to breed less frequently than smaller, younger fish, but the number

Plate 8.7 Hapa-in-pond production of Nile tilapia in northern Vietnam.

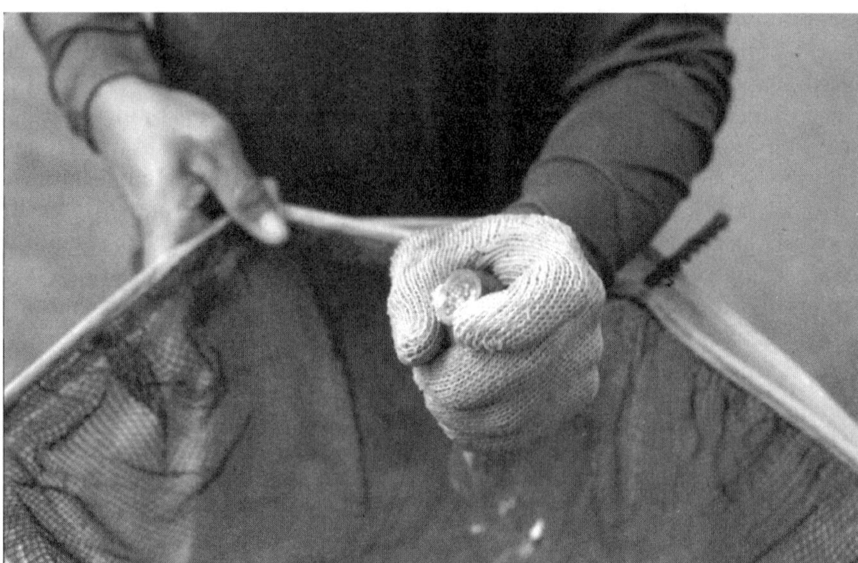

Plate 8.8 Eggs and yolk-sac fry can be harvested from the mouths of female spawned in hapas or tanks.

and size of eggs produced is larger (Rana, 1988). Also, more experienced, older fish incubate their eggs more successfully than younger fish (Lee, 1979; Siraj *et al.*, 1983). Larger fish are often preferred in pond systems, especially when fish can be conditioned for early synchronous spawning to meet a very seasonal demand for fry. Larger eggs produced by older broodfish hatch into larger larvae that may survive better during harvest and are also more tolerant of starvation (Rana and Macintosh, 1988). Larger, older fish can perform well in intensive systems, although frequency of spawning and relative clutch size decline with size and age (Rana, 1988; Little, 1989; Smith *et al.*, 1991). Poor handling efficiency of large fish is also a problem in hapa and tank systems in which broodfish are individually handled to harvest mouthbrooded seed. In practice, fish larger than 300 g are difficult to handle quickly and efficiently.

Sex ratio of broodfish can have both economic and technical impacts. Sub-optimal ratios, in which either males or females are unable to spawn, clearly waste resources and increase costs. In terms of maximizing genetic variability, a ratio of 1:1 is desirable but in practice subordinate fish may be excluded from spawning. Female to male broodfish sex ratios range between 5:1 in ponds and 10:1 (female:male ratio) in hapas and tanks. Fewer than five females for every male appears to be within the optimal range. In most situations, especially when synchronized spawning is required, less than three females per male is optimal. Although individual males can court and spawn with females successively, research shows that fertilization rates drop rapidly (Rana, 1988). Several studies have shown that male exchange improves fry production in Nile tilapia in hapas (Lovshin and Ibrahim, 1989; Amballi and Little, 1996).

Broodstock densities also vary considerably. Densities of less than 3 fish/m^2 for systems harvesting fry and 5–10 fish/m^2 for hapas and tanks in which eggs and yolk-sac fry are harvested from brooding females appear optimal. Optimal broodfish densities are also affected by the individual size and history of broodfish; larger fish require low stocking densities to spawn well, particularly if not acclimated to high densities.

High densities increase male pressure and reduce hierarchies that retard spawning in groups of tilapias until a critical point, when increased disturbance of spawning begins to reduce the success of courtship, fertilization and incubation. Spawning frequency and clutch size is known to decline with increased density (Little, 1989; Abrajano-Bongco, 1991).

Broodfish nutrition and feeding

Continuous broodfish growth during any period of seed production necessitates constant review of feeding requirements. As broodfish grow, inputs typically increase with consequent impacts on the system. Quantity and quality of feed and frequency of feeding can affect broodfish productivity both through direct nutrition and the impact on water quality. Some workers have attempted to ameliorate the negative effects of low early morning dissolved oxygen by reducing feed (Snow *et al.*, 1983), but this has negative impacts on broodfish nutrition over time. Low levels of natural food require the feeding of broodfish to support production, although in phytoplankton-rich systems broodfish can produce well on feeds of only 20% crude protein (Santiago *et al.*, 1985). In covered clear water tanks, *O. niloticus* broodfish fed complete pelleted feeds with a crude protein content of 30% have sustained high outputs of seed over periods of up to 200 days (Little, 1989). The overall performance (fecundity and frequency of spawning) of Nile tilapia broodfish is optimized at a dietary protein level of 25–30% in clear water systems (Wee and Nguyen, 1988; De Silva and Radampola, 1990).

Feeding a good quality ration (25–30% crude protein pellet) in both spawning and conditioning hapas results in fast fish growth and high seed output, but inevitably limits the reproductive life span of the broodstock as the fish quickly grow to a size difficult to handle efficiently (> 300 g).

Broodfish management – exchange and harvest interval

A major determinant to obtaining a consistent supply of seed over a long period of time is the maintenance of broodfish condition. Maintaining broodstock in good spawning condition is less important when the demand for seed is strongly seasonal and there is no economic value in repeated spawning through the year. The holding of single sex stocks at high density in hapas or tanks on a high plane of nutrition, or 'conditioning', also encourages synchrony of spawning when the fish subsequently are stocked at lower densities in spawning systems. Hypophysation of female broodfish did not improve synchrony of spawning above control levels (Srisakultiew and Wee, 1998) although short-term exposure to lower temperatures did improve spawning frequency. Conditioning of females for 10–15 days appears optimal, followed by a 5–7 day opportunity to spawn. A higher mean output of seed is maintained in both tank and hapa systems in which females are frequently exchanged rather than if fish are kept continuously in spawning systems (Fig. 8.4). More frequent disturbance and exchange of females, or longer or shorter periods for conditioning, tends to reduce seed yields and increase broodfish growth (Figs 8.3, 8.5). Selective exchange of females, in which spawned females and those with poor condition are removed from spawning units after harvest and replaced with 'ripe' fish selected from conditioning units, can further improve the productivity of broodfish and seed output (Fig. 8.5). Male exchange also improves seed production (Lovshin and Ibrahim, 1989; Little *et al.*, 1996) and is perhaps related to maintaining improving fertilization rather than spawning activity. A shorter period of conditioning (5 days) is adequate for males, reducing the numbers of fish required and support costs.

Harvest method and cannibalism

Frequent, complete harvest of seed improves the individual productivity of females and reduces the level of cannibalism that occurs between seed of

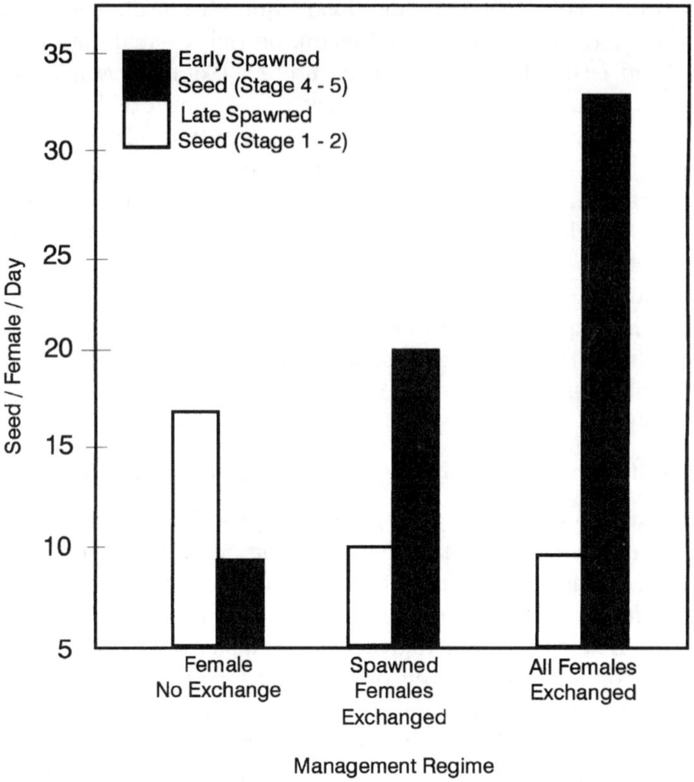

Fig. 8.4 Mean numbers of early spawned (stage 4 and 5) and late spawned (stage 1 and 2) seed harvested from individual female *Oreochromis niloticus* stocked in tanks under different management regimes over a 125-day experimental period (after Little *et al.*, 1993).

different sizes and by broodfish on fry. Fish aged between 10 and 30 days are particularly cannibalistic on newly hatched fry, and stocking fry of different age and sizes together appears to stimulate aggression (Pantastico *et al.*, 1988; Ellis *et al.*, 1993). Several measures may reduce losses from cannibalism. Sanitation of spawning and nursing units to remove uncaught fry and careful grading of fry removed intermittently from spawning systems may reduce losses. Using broodfish of the same age and size, which produce eggs and subsequently fry of similar size, should also reduce cannibalism.

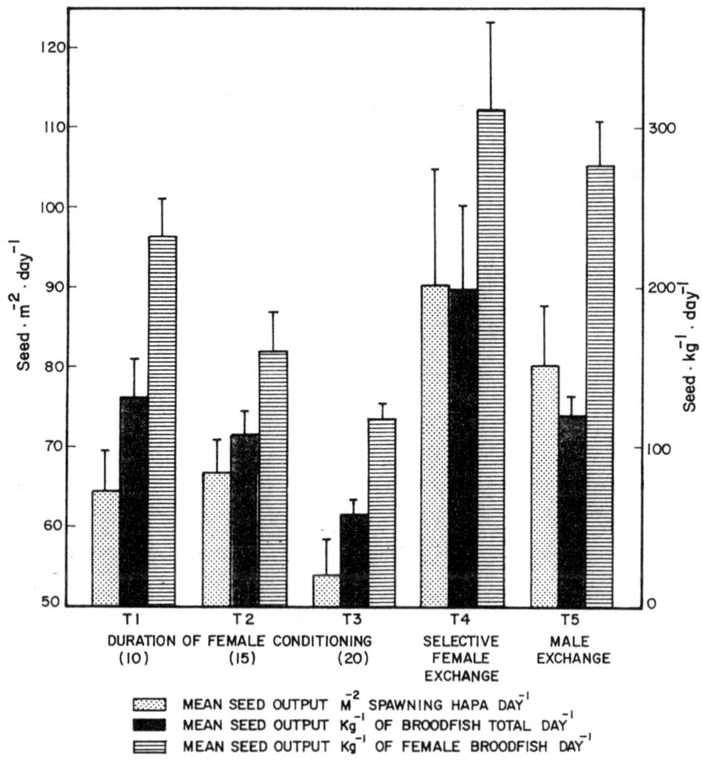

Fig. 8.5 Mean (±SE) values for performance of Nile tilapia (*Oreochromis niloticus*) broodfish spawned in large hapas after different periods of female conditioning (10 days, T1; 15 days, T2; and 20 days, T3), selective female conditioning (T3) and male conditioning (males and females conditioned for 10 days, T5) (after Little *et al.*, 1996).

8.4 HATCHERY SYSTEMS

Ponds

Spawning tilapia in earthen ponds is the oldest method reported and the only one recognized in early publications such as Bardach *et al.* (1972) and Hepher and Pruginin (1981). In its various forms, it is still the most widely used method of obtaining tilapia seed (Macintosh and Little, 1995). Tilapias spawn and produce fry easily in most pond environments, especially shallow ponds, and preventing them from spawning is more of a challenge (Bardach *et al.*, 1972). Synchronizing spawning to produce large numbers of same age fry has been a major constraint, however. Management and procedures for harvest of young depend on the scale of operation and the type of seed being produced. Guerrero (1987) describes small-scale, medium-scale and large-scale breeding/nursery ponds in the Philippines. Schooling fry are scooped daily from the surface of the ponds in the former two, while fingerlings most commonly are harvested by seines in the latter (Broussard *et al.*, 1983). Open ponds have been used for producing a range of tilapia seed types, namely mixed-sex pure species (e.g. Little *et al.*, 1987, 1994), monosex hybrids (Hepher and Pruginin, 1981), and even first feeding fry for sex reversal (Liao and Chen, 1983; Rothbard *et al.*, 1983; Verdegem and McGinty, 1989). Specific management adaptations may, however, be required to ensure large-scale production of seed of the latter types.

Pond management can be categorized into four main types depending on the size/age of the seed produced and the harvest method. Harvest of either early (eggs or yolk-sac fry) or late or mixed age seed (fry/fingerlings) can be either partial or complete. Partial harvest can be carried out on early seed by dip-netting spawning units or seining or trapping larger seed without draining the system. Complete harvest requires capture and checking or individual broodfish for early seed and draining of the pond to completely harvest fry and fingerlings. Most systems can be classified as those producing mixed sex and size seed (Table 8.2) or first feeding fry (Table 8.4). Factors that affect management include the need for fry of a quality suitable for hormonal sex reversal, availability of labour and the size of fry most in demand.

Mixed size and sex seed normally are produced by stocking of broodfish followed by partial, intermittent harvesting by seine net (Table 8.2). The production of fry and/or fingerlings and proportion of the harvest composed of different sized fish depend on a range of management factors. Clearly, there are species differences. Substrate spawning fish are known to have higher fecundities, producing more eggs on a single occasion than mouthbrooding *Oreochromis* (Noakes and Balon, 1982), and yet *T. rendalli* produced a relatively low output of fry compared to the Nile tilapia in similar systems in one trial (Table 8.3). Differences in seasonality of breeding can have a major impact on hatchery productivity. Temperature regimes can restrict hatchery output through limiting the duration of the breeding season or reducing spawning activity/ incubation success. In Mzuzu, Malawi, Brooks and Maluwa (1997) reported that *O. shiranus* could produce seed for 8 months of the year provided temperatures remained above 20°C, whereas production of *O. karongae* was limited to only 4 months. Harvest efficiency also affects hatchery productivity; the bamboo screens used in Malawi might be expected to have a lower harvest efficiency than well-managed seine nets. The efficient harvest of fry is a key factor in achieving high yields – more frequent harvesting to remove larger seed reduces cannibalism on younger fry and results in a higher proportion of the harvest being small and even-sized (Little, 1989). Attempting to produce larger individual seed, by extending the harvest interval generally results in much reduced yields. Harvest intervals of 7–60 days have been used, with shorter intervals producing the best results (Table 8.3). Clearly there is an inverse

Table 8.2 Management of broodfish in earthen ponds to produce mixed sex fry and fingerlings

Species/ strain	Mean pond size (m^2)	Period between draining (days)	Fertilizer inorganic	Fertilizer organic	Supple-mentary feed	Complete feed	Harvest method	Source
T. rendalli (Malawi)	200	309		++	++		Reed fence, 35 mm gap	Brummett, 1997
O. niloticus (Thailand)	1240	180–360		++	++	+	Seine net	Little *et al.*, 1994
O. aureus (USA)	390	194	+ (initial)	+ (initial)		+	Seine net, 3 mm mesh	Torrans and Hiott, 1990
O. niloticus (Philippines)	450	250	+	+			Nylon seine, 6 mm mesh	Broussard and Reyes, 1983
O. niloticus (Jamaica)	2500	150–240				++	nylon seine, 6 mm mesh	Popma *et al.*, 1984
O. niloticus (Vietnam)	1500	30–50		+ (sewage)	++		seine and drain	Little and Pham, 1996
O. karongae (Malawi)	300	180			++		Seined with 3 mm mesh net after 3 months and harvested after draining	Brooks and Maluwa, 1997
O. shiranus (Malawi)	300	90	++ weekly		++		Edge seining (2 mm mesh seine net), daily, weekly, monthly (8 mm mesh net) or after draining	Brooks and Maluwa, 1997
O. lidole (Malawi)	150			++ chicken manure	++		Seined after 3 months, drained after 6 months	Brooks and Maluwa, 1997

Table 8.3 Broodfish performance in earthen ponds to produce mixed sex fry and fingerlings

Species/ strain	Stocking density # m^{-2}	Stocking density g m^{-2}	Sex ratio F:M	Harvest interval (days)	Fry m^{-2} d^{-1}	Mean fry size (g)	Yield g m^{-2} d^{-1}	Female productivity Fry kg fem^{-1} $month^{-1}$	Source
T. rendalli (Malawi)			1:1	30	0.08	4.9			Brummett, 1997
				60	0.045	6.8			
				309	0.010	7.8			
O. niloticus (Thailand)	0.12–0.39	30–60	2005:1	After 45 and then every 10–15	2.5–5	1.0	2.44–4.88	1000–4000	Little *et al*., 1994
O. aureus (USA)	0.16	16.5		after 47	0.48	14.7	6.8	1285	Torrans and Hiott, 1990
	0.08	3.2	2:1	and then	0.38	17.1	6.4	2061	
	0.04	1.6		every 14	0.25	22.1	5.4	2694	
O. niloticus (Philippines)	0.16	29.8	3:1	After 60 and then every 28	0.29	4.3	1.2	360	Broussard and Reyes, 1983
O. niloticus (Jamaica)	0.75–1.0	90–120	3;1	After 35–50 and then every 7	1.8	0.9	1.66	655	Popma *et al*., 1984
O. niloticus (Vietnam)	1.0–2.0	140–279	2–3:1	After 21 then remove broodfish and then 10–30 for complete	2–3	0.3–1.0	1.3–1.96	approx 600	Little and Pham, 1996
O. karongae (Malawi)	0.047	16	1.8:1	84 and then at draining after a further 84	0.05–0.1	na	na	144	Brooks and Maluwa, 1997
O. shiranus (Malawi)	1.0	250	2:1	7	0.63	na	na	115	Brooks and Maluwa, 1997
				28	0.28			51	
				84	0.34			62	
O. lidole (Malawi)	0.007	13.3	2:3	84	0.022	3.0	0.073	117	Brooks and Maluwa, 1997

relationship between the final average size of seed and numbers of young harvested. Pond size and shape also affect harvest efficiency, and generally ponds of less than 1000 m^2, that can be managed and harvested more intensively, seem to optimize seed output across the range of species.

Female broodfish efficiency (fry kg female^{-1} month^{-1}) varies by a factor of 10 and can be explained partly by differences in stocking density (Table 8.3). Low stocking densities both improve broodfish efficiency and produce large seed quickly. Very high densities of Nile tilapia broodfish (e.g. up to 1.2 MT-treated *O. niloticus* ha^{-1} in Jamaica; Popma *et al.*, 1984) require the use of higher quality feeds but appear to produce lower yields than lower densities of broodfish maintained only through fertilization and supplementary feeding (e.g. Thailand; Little *et al.*, 1994).

The rotation of sets of broodfish from pond to pond, leaving fry to develop, can improve efficiency of broodfish use. Such a system demands that broodfish can be harvested efficiently by seine net in well-designed ponds. Regular complete draining of ponds after a certain temperature-dependent period of time increases the homogeneity of fry age and size. Practically, the period between harvests mainly reflects demand. In southern Vietnam ponds are drained and harvested more frequently to produce smaller fry when demand is high, but as demand declines the harvests are more infrequent and harvested fry size larger. A relatively high biomass of broodfish is stocked and the 'sets' are transferred after 3 weeks to a freshly prepared pond, fertilized with sewage, for renewed spawning. Broodfish are removed by careful seining, and fry nursed on for a further 10–30 days before complete harvest by draining. Rice bran is also given as a supplementary feed at rates of 1.5–3% body weight day^{-1} over the relatively short spawning season (150–200 days). Regular draining of earthen ponds has been advised for such systems and this is practised several times a season in southern Vietnam (Little and Pham, 1996). Mires (1982) observed that high fry densities inhibited production and linked the resumed spawning of broodfish removed to newly prepared ponds to changes in water quality.

Appropriate timing of draining and harvest of ponds to produce first feeding fry produces young and similar-aged fish suitable for sex reversal (Tables 8.4 and 8.5). In Israel and Central America, complete harvest is timed according to ambient temperature (Green and Teichert-Coddington, 1993) to ensure the maximum number of hormone treatable fry. In practice, pond harvest should occur between 195 and 220 cumulative degree-days after stocking. A major problem is the high mortality associated with harvesting such fry, but recently this has been improved by development of attached sex- reversal units, either small ponds or tanks. The procedures for collection of first feeding fry for sex reversal from pond-based hatcheries, described in detail by Rothbard *et al.* (1983), are efficient in terms of fry production per spawner, but require higher labour investment. Recently these procedures have been further improved. Since it was found that sex reversal can be efficiently performed in 'green water' ponds (Buddle, 1984; Berger and Rothbard, 1987), some hatcheries periodically

Table 8.4 Management of *Oreochromis* broodfish in earthen ponds to produce first feeding fry

Species/ strain	Site	Production /year (10^6)	Use of MT	System descriptors (m^2)	Duration	Production cycle (days)	Harvest strategy	Broodfish conditioning/ treatment	Source
O. niloticus × *O. aureus*	Israel Nir David	10	–	5000	April–August[a]	36	2	1 set for 2 spawning units	Hulata, 1997
O. aureus	Israel Nir David	4.5	–	300 plastic-lined	Jan–August[b]	15–12[d] 12–15[e]		1 set for 2 spawning units	Hulata, 1997
O. niloticus × *O. aureus*	Israel Sde Eliyahu	15	+	500–6000	April–August	24	1	Broodfish over-wintered in 100 m^2 concrete ponds (3MT/pond) with warm water (20°C)	Hulata, 1997
O. niloticus × *O. aureus* – Florida red	Israel Ein Hamifratz	15	+	4000	April–August	21–28	1		Hulata, 1997
O. niloticus	Ecuador/ Honduras		+	500–1000	year-round	14–23[f]	3	3–10 days holding mixed sex broodfish between cycles, fed 2–3% bodywt.d^{-1} high quality feed in cages (30 kg m^{-3})	Popma and Green, 1990
O. niloticus	Thailand (AIT-trial)	3.3[g]	+	1700	year-round	105	3	Harvest 6 × d^{-1}. Separate sex broodfish conditioned in hapas for 56 days, fed 3 × d^{-1} high quality feed to appetite	Little, 1989

Table 8.4 (cont)

Species/ strain	Site	Production n/year (10^6)	use of MT	System descriptors (m^2)	Duration	Production cycle (days)	Harvest strategy	Broodfish conditioning/ treatment	Source
O. niloticus	Philippines	2.0	–	300–500	Jan–July	30–40	3	Harvest of fry $2 \times d^{-1}$. Broodfish conditioned for 7 d prior to stocking, fed 1–2% bodywt.d^{-1}	Guerrero, 1986
O. shiranus	Malawi (trial)		–	200	Oct–Dec	84	3	Broodfish fed after stocking at 3% initial bodywt.d^{-1}	M'hango and Brummett, 1997
O. shiranus	Malawi (trial)		–	300	Nov–Feb	84	3	Broodfish fed at 4% bodywt.d^{-1} after stocking in fertilized ponts	M'hango and Brummett, 1997

Harvest strategy: 1, complete harvest of broodfish and fry/sex reversal in 350 m^3 concrete tanks in polytunnels/greenhouses; 2, removal of broodfish, nursing/sex reversal of fry in the same system; 3, partial harvest of fry by dip-netting/edge-seining, broodfish remain in the spawning system

[a]ambient water temperature; [b]warm, geothermal water; [c]duration until water completely drained; [d]upper spawning compartment; [e]lower nursery compartment; [f]T°C dependent; [g]extrapolated as 3 × 105 d cycles/year

Table 8.5 Performance of *Oreochromis* broodfish in earthen ponds to produce first feeding fry

Species/strain	Stocking density # m^{-2}	Stocking density g m^{-2}	Female, male size (g)	Sex ratio F:M	Fry m^{-2} d^{-1}	Female productivity Fry kg fem^{-1} month	Labour productivity Fry $manday^{-1}$	Source
O. niloticus × *O. aureus* (Nir David, Israel[1])	0.375	75	150–250	4:1	0.2-0.3[a]	2000–3000	15 000	Hulata, 1997
O. aureus (Nir David, Israel[1])	3.7	750	150–250	4:1	24[b]	1000	15000	Hulata, 1997
O. niloticus × *O. aureus* (Sde Eliyahu, Israel[1])	1.2	300	350–400, 1000	3–4:1	5[b]; 2.4[a]	600	20 000	Hulata, 1997
O. niloticus × *O. aureus* (Ein Hamifratz, Israel[1])	1.6	880	500+	4:1	28.5[b]	1000–1200	50 000	Hulata, 1997
O. niloticus (AIT, Thailand[2])		30–40	218[c]	1:1	6.2[b]	5077	30 000	Little, 1989
O. niloticus (Ecuador, Honduras[2])		100–200	na	1.5–2:1	1.6–8[b]	1518–1864	16 000–21 000	Popma and Green, 1990
O. niloticus (Philippines[1])	1.0	800	50–100	3:1	0.7	82.9		Guerrero, 1986
Oreochromis shiranus (Malawi[2])	1.25 1.25	56 550	43, 47.5 43, 47.5	1:1 3:1	0.8 0.7	860 512		M'hango and Brummett, 1997
Oreochromis shiranus (Malawi[2])	1.0	250	250, 250	2:1	1.2[b]	223.1		Brooks and Maluwa, 1997

[a]MT-treated fry, [b]swim-up fry, [c]females selected as smaller of two size classes from same-age broodfish
[1]commercial production; [2]experimental trial

seine out the broodstock (and transfer them to a fresh pond), leaving the first-feeding fry in the spawning pond for hormonal sex reversal. A recent improvement in Israeli commercial tilapia hatcheries originated from adaptation of a pond design described by Lovshin (1982; Fig. 8.6). The system consists of a plastic-lined pond with two compartments – a spawning pond and a fry collecting pond built at a lower level (see also Hulata, 1997). The latter is sometimes separated from the spawning pond with a fenced sluice gate that retains the spawners in the upper compartment. Alternatively, spawners and fry are drained together into the lower compartment, and separated there. This design and operation make batch harvesting of fry at short (10–14 days) intervals very easy and eliminates the need to handle the delicate fry. Hormonal treatment for sex reversal can start in the lower compartment after removal of spawners, and continue in a nursing pond to which the larger and stronger fingerlings are transferred in preparation for the next batch harvest of fry from the upper compartment. A problem with the method is that although fry mortality may be reduced, the numbers produced are likely to vary from batch to batch. This makes efficient utilization of space and the precise feeding of hormone-treated feed based on percentage body weight problematic.

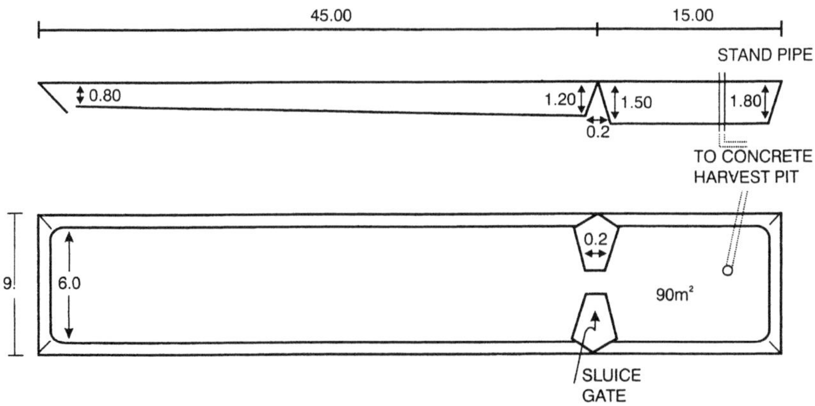

Fig. 8.6 Tilapia spawning/sex reversal ponds, Israel.

Stocking of broodfish at the beginning of warm temperatures after cooler weather is known also to synchronize the early spawning of the fish, which further improves efficiency. In Israel, where broodfish are over-wintered at high density in concrete ponds, spawning begins in April and lasts until August-September (Table 8.4).

A key management step to produce large quantities of fry is a high level of sanitation between crops, to remove larger fry that cannibalize newly spawned seed. This is done easily in the plastic-lined ponds described above. Conditioning of broodfish, especially in single sex groups between spawning events also improves results; good quality feed is eaten voraciously by broodfish after spawning and incubation, allowing rapid recovery.

In areas with high year-round temperatures, synchronization of brooders is less effective. The continuous harvest of first-feeding fry by edge seining, or dip-netting, earthen pond perimeters is common in a range of tilapia-producing countries from Taiwan and the Philippines to Brazil. A major advantage is the low labour requirement and modest equipment used. A problem is that as variable numbers of fry are produced daily it is difficult to ensure that early nursing (or hormone treatment) occurs at optimal density. Careful grading to remove larger, older fry is also required to achieve high rates of sex reversal (Popma and Green, 1990).

The impact of 'harvest intensity' has been demonstrated (Little, 1989) and appears important in both optimizing the number of fry caught and reducing the number of fingerlings that cannibalize fry and compete for food with the broodfish (Fig. 8.7). Commercial farms typically employ workers to remove fry constantly from the shallow margins of large ponds throughout the day. Frequent harvesting also suits family operations that can use household labour flexibly.

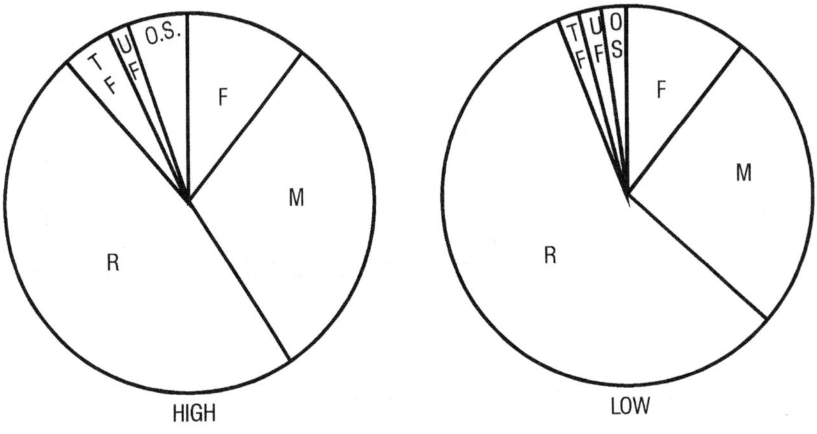

Fig. 8.7 Harvested net production (wet weight) from partial and final harvesting of earthen ponds (area 1740 m^2) by category at two levels of harvest intensity: High and low intensity harvest (pond perimeter dip-netted 6 and 3 × daily respectively). R, recruits; M, F, stocked males and females respectively; TF, UF, fry suitable and unsuitable for MT treatment; OS, other species.

Tanks

Tanks include a variety of systems used for the production of tilapia fry, of which the main distinguishing features include facility for rapid draining and easy harvest of broodfish and fry, and control of water quality. A range of sizes, shapes and construction materials have been classified as tanks, ranging from 1 m^2 experimental concrete units to large disused naval assault vessels (Table 8.6). Tanks have been used to produce seed in both fresh and brackish water environments, and for a range of species, especially *O. niloticus* in freshwater and *O. spilurus* and red tilapias in brackish water. A major advantage of using tanks in island and coastal areas with limited freshwater is their high seed output per unit of freshwater and area (Ernst *et al.*, 1991).

A key descriptor of tank systems is their management of water quality, through water exchange or some form of aeration or filtration (Fig. 8.8a). If managed as static water systems with intermittent or total water exchange at harvest, management may be similar to the earthen ponds described above. Broodfish stocking density and productivity tend to be low under such conditions.

The relatively high cost of tanks compared to hapas or ponds has stimulated research towards more intensive management. Tanks can be operated at high stocking densities, with intensive feeding and water quality control to yield high outputs of seed, but in practice productivity over long periods is often disappointing. Many of these problems can be related to the difficulties in maintaining the spawning intensity of broodfish at high density.

The size and design of a tank system has implications for the conditioning, courtship and spawning of broodfish. Post-spawning needs of the broodfish also must be considered both for the period of mouthbrooding and, after seed removal, for reconditioning. If tanks are managed to harvest free-swimming fry, then conditions suitable for their survival also have to be considered.

The large variation in tank size used by researchers and commercial operators has implications for comparing productivity (Tables 8.6 and 8.7). In large tank systems, it can be difficult to completely, and frequently, harvest seed, and to maintain water quality and broodfish condition. In contrast, broodfish in small systems are more likely to suffer from aggression caused by the more rigid hierarchies that are established among groups of fewer fish.

Tank design affects nest building, courtship, spawning and the incubation and release of fry by brooders. Spawning of Nile tilapia was found to be more intense in deep (1 m) compared to shallow (0.5 m) tanks (Nguyen, 1992). The dimensions and design of tanks can affect the density of males that can be involved in courtship. Balarin and Haller (1982) reinterpreted data from Uchida and King (1962) to show that male *O. mossambicus* required 0.3–0.4 m^2 $fish^{-1}$. Construction blocks have proved useful both in the PVC-lined ponds at Nir David farm in Israel and in intensive tanks at AIT to delineate areas between male nests and reduce aggression. Nguyen (1992) demonstrated a clear relationship between the intensity of spawning (seed kg $female^{-1}$ total) and the number of

Table 8.6 Management of *Oreochromis* broodfish in tanks

Species/strain (site)	System size (m^2)	Duration (days)	Water exchange (reuse/day)	Extra aeration in spawning tank	Water source salinity[d] (ppt)	Natural food-rich	Feeding practice	Water quality indicators	Harvest interval (days)	Type of seed	Source
O. mossambicus (USA)	4.46 wooden, assault boat		Continuous	Yes	Tap water	No	Trout feed, once daily (2% bodywt d^{-1})		Daily dip-netting	b	Uchida and King, 1962
O. urolepis honorum × *O. mossambicus* (Israel)	700 m^2 concrete tanks	112	Every 12–14 days	No	Overwintered in well water at 20°C	No	During over-wintering fed at 1% bodywt d^{-1} with 40% CP pellet		12–14	b	Hulata, 1997
O. niloticus (Thailand)	50		Complete – – 28 d – 15 d	No		Yes			28 15	c b	AIT, 1996
O. niloticus (Philippines)	100		To replace evaporation losses drain after – 40 days – 50–72 days – 21–30 days	No	Deep well	Yes	Fed in single sex cages at 4% body wt.d^{-1} for 7 d prior to stocking; in breeding tanks fed at 3% body wt.d^{-1}		Dip-netting $4.6 \times d^{-1}$	b	Guerrero and Guerrero, 1984, 1985[a] 1985[b]
O. niloticus (Philippines)	1		Daily to replace evap losses, complete change every 21 days	No	Laguna lake	Lake water, low NNP= <2 $gC/m^2/d$	– no feed – fed with 20% CP diet – fed with 40% CP diet	Early morning DO variable (0.5–6.6) and T°C (19–31°C)	Every 21 days	a, b	Santiago *et al.*, 1985
O. niloticus (Philippines)	4		Every 21 days	Yes	Laguna lake	na	Once daily with a commercial broiler starter (CP=21%) @5% bodywt d^{-1}	Variable mean monthly T°C 24–30°C, wet and dry seasons		a, b	Bautista *et al.*, 1988
Red tilapias[f] (Philippines)	4		Every 21 days	No	Lake water	na	Twice daily with a commercial feed (CP=42%)		Every 21 days	a, b	Eguia, 1996

Table 8.6 (cont)

Species/strain (site)	System size (m^2)	Duration (days)	Water exchange (reuse/day)	Extra aeration in spawning tank	Water source salinity[d] (ppt)	Natural food-rich	Feeding practice	Water quality indicators	Harvest interval (days)	Type of seed	Source
T. rendalli (Malawi)	5 m^3 concrete +/– gravel as spawning substrate		50% every 21 days	No	na	Secci disc reading 0.25–1.0 m, declined over 416 d period	Maize bran @ 5% bodywt d^{-1}	18–28°C tanks with substrate had higher DO levels at dawn than tanks without	Every 21 days	c	Costa-Pierce, 1991
O. aureus (USA)	7.3 plastic swimming pools		Continuous	No	Reservoir water	na	Trout feed @ 2–3% bodywt d^{-1}	sometimes9 dropped to <1 mg l^{-1}, feeding reduced	(a) dip-netted as observed or weekly (b) as (a) + 21 day draining (c) seining and egg removal from females every 10 –12 days	b a, b a	Snow *et al.*, 1983
O. spilurus (Kenya)	27[c]		Continuous 0.03 m^3 min^{-1}	No	Pumped ground water 9 ppt	Low <50 algal cells ml^{-1}	Separate sex initial conditioning of broodfish. In spawning tanks fed several times @1% body wt.d^{-1} poor quality feed (27% CP)	DO 0.3 mg l^{-1} at dawn to 9+ mg l^{-1} late afternoon. T°C 28–35°C	Drainage of external shallow channel to which fry tend to accumulate twice daily	b	Sampson, 1983

Table 8.6 (cont)

Species/strain (site)	System size (m^2)	Duration (days)	Water exchange (reuse/day)	Extra aeration in spawning tank	Water source salinity[d] (ppt)	Natural food-rich	Feeding practice	Water quality indicators	Harvest interval (days)	Type of seed	Source
O. spilurus (Kuwait)	2 × 2 × 0.5		Changed weekly	1 airstone $tank^{-1}$	3–4 ppt 38–41 ppt	Indoors	Twice daily @ 2% bodywt d^{-1}, 40% CP Pellet	20–30°C 21–31°C	Every 7 days	a, b	Al-Ahmad *et al.*, 1986, 1988
Taiwanese Red tilapia (Kuwait)	2 × 2 × 0.5 fibreglass tanks with rounded		Flushed daily to remove solids, weekly	yes	3–5 ppt	Indoors	Twice daily @ 3% bodywt d^{-1}, 40% CP pellet	20–30°C	Every 7 days	a, b	Ridha *et al.*, 1985
O. urolepis honorum × *O. mossambicus* (Bahamas)	7.6 × 4.5 × 1.0 plastic-lined tanks	89	Continuous 2 water exchnages (48 l min^{-1} $tank^{-1}$ through a bio-filter	yes	12 ppt	Enclosed in clear plastic greenhouse	Twice daily to 10 min satiation, floating pellet 32% CP	25–31°C	Every 8–16 days Every 15–16 days	b a,b	Watanabe *et al.*, 1992
		224			5 ppt (mixed (rain and ground water) 18 ppt (mixed (ground and (sea water)	Shaded to ensure T°C < 32°C	Ad libitum 1–3% bodywt d^{-1}	High water quality, transparency 22–77 cm	Fry harvested when nos of fry > 1000 $tank^{-1}$	a, b	Ernst *et al.*, 1991
O. niloticus (Thailand)	12.6, circular concrete spawning tanks (d=1.0), 1.8 circular conditioning tanks	155	Continuous 3.3 exchanges d^{-1} through bio-filter	no	Ground water	Covered by open roof, no significant plankton level	Fish in spawning and conditioning tanks fed to satiation 3 × daily with floating pellet (CP=30%) Feed intake 1.3–2.6% bodywt d^{-1}	28–30°C	Edge dip-netting > 6 × /d -complete harvest every 5 d -complete harvest every 10 d	b a a, b	Little *et al.*, 1993

[a]eggs and yolk-sac larvae, [b]first feeding fry, [c]fry (TL = 2–3 cm), [d]freshwater unless indicated, [e]spawning arena, comprising 3 concentric rings for spawning, incubation and fry harvest, [f]*O.niloticus* × *O. mossamibus*, *O. mossamibus* × *O. honorum* × *O.niloticus*

blocks in shallow water of high transparency in circular concrete tanks (3 m diameter). Baroiller *et al.* (1997) also demonstrated a trend towards improved seed production when a raceway environment was made more complex using shelters; the impact of shelter appeared to increase with the duration that fish were stocked in the spawning system.

Tank design also affects seed harvest efficiency. Tilapia seed can be partially harvested by dip-netting swim-up fry from the tank periphery (Guerrero and Guerrero, 1984, 1985b). Tank designs have been developed that ensure shallow perimeters to attract first-feeding fry and exclude broodfish (Balarin and Haller, 1982), but these may aggravate the problem of cannibalism of small fry by larger individuals if they are poorly managed (Sampson, 1984). Vertical walls and lack of cover in spawning tanks makes first-feeding fry vulnerable to predation by adults (Berrios-Hernandez and Snow, 1983; Little *et al.*, 1993). Capture of all broodfish after draining the tank allows complete removal of both free-swimming fry and, after checking of individual fish, any eggs and larvae in the mouths of females. Some operators use a combination of regular partial harvest with intermittent complete drainage and harvests when seed output declines below a certain level (Table 8.6; Snow *et al.*, 1983; Ernst *et al.*, 1991; Little *et al.*, 1993).

Many factors affect water quality in tank systems. The initial quality of the water available and its residence time in the spawning tank is important. Ground water is likely to be less variable in quality than surface waters such as Laguna Lake, Philippines (Santiago *et al.*, 1985; Table 8.6).

Filters are used often to maintain the quality of water in recirculated systems, but their efficiency depends on a wide range of design criteria (Muir, 1982). The residence time of the water can be affected by exchange rates in recirculating and flow-through systems, but also by the harvest strategy. Frequent harvests, which involve complete drainage of the system, can maintain water quality even in static water systems. Supplementary aeration in static water systems can maintain dissolved oxygen levels without high levels of water exchange. Low dissolved oxygen still can occur in tanks with initially high water quality and water exchange e.g. Baobab Farm in Kenya (Sampson, 1983). This is related usually to both poor system design and management.

Under brackish water conditions, availability of freshwater to reduce salinity is important for optimizing yields in saline-tolerant tilapias such as *O. spilurus* (Tables 8.6, 8.7; Al Ahmad *et al.*, 1984, 1988) and the Florida red tilapia (Ernst *et al.*, 1991; Watanabe *et al.*, 1997). Female productivity of the saline tolerant *O. spilurus* spawned in ground water (3–4 ppt) was twice that of fish in hyper-saline water (38–41 ppt) and similar differences were found in the Florida red tilapia over a narrower range of salinities (5 ppt versus 18 ppt).

Feeding practices can drastically affect water quality, especially in tank systems in which total water volume is limited and poor quality feeds are given. Poor quality feeds and lack of natural food were implicated in the poor productivity of *O. spilurus* broodfish in Baobab arena systems in Kenya (Table 8.7; Sampson, 1983). The amount and quality of natural feed can be as

A.

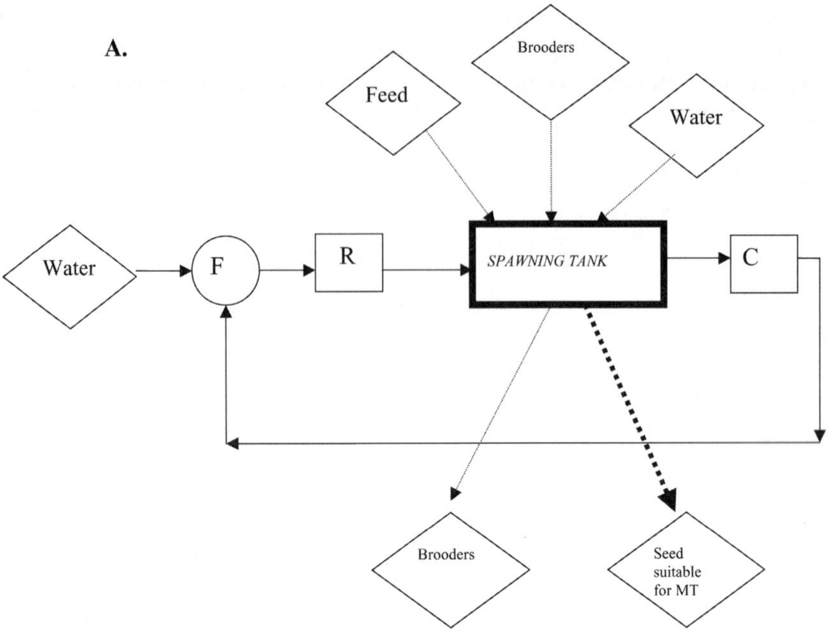

B.

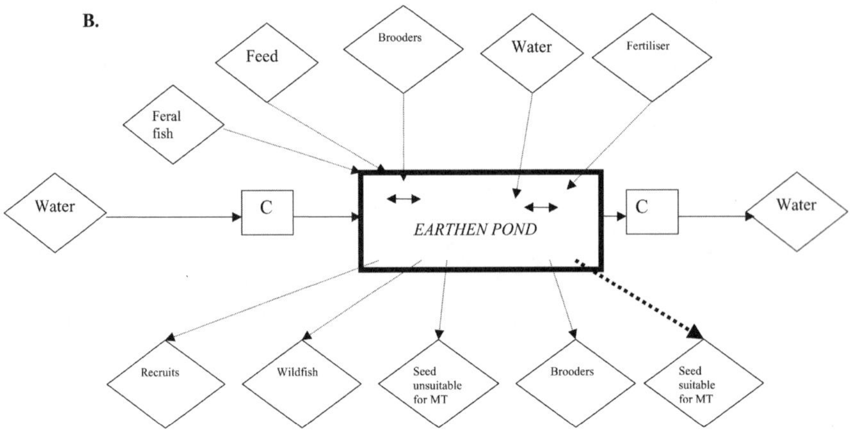

Fig. 8.8 (and opposite page) Process diagrams of *Oreochromis* hatchery systems: (A) tank within a recirculation system (B) earth pond and (C) hapa-in-pond (after Little, 1992).

C.

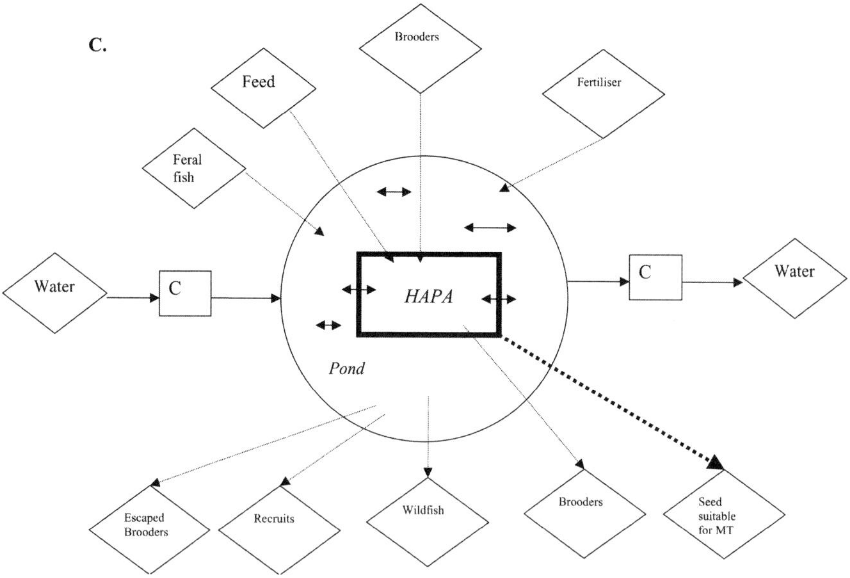

LEGEND (after EIFAC, 1986)

SPAWNING TANK — A rearing unit

F — A unit of process that changes water quality so as to make it significantly more or less suitable for rearing organisms within the system e.g Filter (F) or pond itself.

◇ — Input or output from the system

R — A process necessary to the system but has no significant direct effects on water quality e.g reservoir (R), channel or pipe ©

→ Movement of water

→ Other types of transfer

Table 8.7 Performance of tilapias in tanks

Species/strain	Stocking density # m^{-2}	Stocking density g m^{-2}	Female, male size (g)	Sex ratio F:M	Fry m^{-2} d^{-1}	Female productivity Fry kg fem^{-1} $month^{-1}$	Source
O. mossambicus (USA)	11.2	1350	113, 142	3	61[b]	1800	Uchida and King, 1962
							84 d study[d]
O. niloticus	1.2	420	250+, 450	2	*On* 7.2[c]	480	
					On × *Om* 3.6[c]		
O. niloticus							AIT data, 1996
O. mossambicus (red tilapia)	1.0	350	250+, 450	1	*On* 24[b]	2057	
(Thailand)					*On* × *Om*	1337	
					15.6[b,e]		
	3	300	95, 109	2	9.6[b]	1408	Guerrero and Guerrero, 1984
O. niloticus	3	340	108, 126	2	16.2[b]	2100	Guerrero, 1985
(Philippines)					9.0[b]	1166	50 d study, 72 d study
	1.3	400	183, 208	2[f]	5.7[b]	598	
	5.0	400	75, 91	2[f]	8.7[b]	912	Guerrero, 1985b
	1.9	400	219, 200	2[f]	7.9[b]	828	
	4.6	400	86.5, 85	2[f]	7.6[b]	797	
O. niloticus	9.3	1400	150, 146	2.75	nf [i]4.5–0.5[ab]	173–619	Santiago *et al.*, 1985
(Philippines)					20 36–56[ab]	734-2168	24 week trial
					40 59–75[ab]	1523–2278	
O. niloticus	6.3	600	88.8, 98.3	4	8.41[ab]	530	Bautista *et al.*, 1988
(Philippines)	6.0	500		7	6.96[ab]	419	12 month study
	5.5	500		10	5.18[ab]	328	
	5.0	400	81.1, 90.5	4	8.8[ab]	756	
	8.8	700		4	8.04[ab]	397	
	12.5	1400		4	4.11[ab]	142	

Table 8.7 (cont)

Species/strain	Stocking density # m^{-2}	Stocking density g m^{-2}	Female, male size (g)	Sex ratio F:M	Fry m^{-2} d^{-1}	Female productivity Fry kg fem^{-1} $month^{-1}$	Source
O. niloticus × *O. mossambicus*[j]	4	>100	18, 27	4	16.7–37.2[ab]	2165–4822	Eguia, 1996
O. mossambicus × *O. honorum*							12 month study
× *O. niloticus*[k]							
(Philippines)							
T. rendalli	1.8[g]	>100	40	2	0.9–4.2[b]	105–490	Costa Pierce, 1991. No
(Malawi)							difference between +/-
							substrate, 416 d study
O. aureus	1.6	300	206, 213	3	(i) 22.2b	337	Snow *et al.*, 1983,
					(ii) 17.1[b]	260	93 d study
					(iii) 23.1[ab]	351	
					(iv) 0.64[c]	9.7	
O. spilurus	13.9	240	400, 250	2.25	18.5[b]	215	Sampson, 1983[h]
(Kenya)							
		SW[l] *GW*	*SW* *GW*		*SW* *GW*	*SW* *GW*	Al Ahmad *et al.*, 1984, 1985
O. spilurus	10	890 710	76 46	3	72.1[ab] 96.2[ab]	3540 7804	215 d study
(Kuwait)						(2792) (5488)	
	3.3	1020 940	272 241		81.3[ab] 150	3347 7001	
						(3214) (6586)	
Taiwanese red tilapia	3.25	1100	308, 458	3.3	55.6[ab]	2158	Ridha *et al.*,1985
(Kuwait)							105 d study
O. urolepis honorum			145, 207	4	LS 77.7[ab]	2434 (1650)	Ernst *et al.*, 1991
× *O. mossambicus*	8.2	1300			HS 47.4[ab]	1355 (940)	232 d study
(Bahamas)							

Table 8.7 (cont)

Species/strain	Stocking density # m^{-2}	Stocking density g m^{-2}	Female, male size (g)	Sex ratio F:M	Fry m^{-2} d^{-1}	Female productivity Fry kg fem^{-1} $month^{-1}$	Source
O. urolepis honorum	7.1	3300	na	3	Yr I 52.3[ab]	1358	Smith *et al.*, 1991,
× *O. mossambicus*							140 d study
(Bahamas)	5.9	1000			Yr II 36.9[ab]	398	
O. urolepis honorum					3.3[b]	106	Watanabe *et al.*, 1992,
× *O. mossambicus*	7.0	1000	na	3			89 d study
(Bahamas)					91.7[ab]	2820	
O. niloticus	8	530	58.6, 75.3	1	302[a]	12130	Little, 1989
(Thailand)						(5053)	201 d study

[a]egg and yolk-sac larvae; [b]yolk-sac absorbed fry; [c]fry (TL; 2–3 cm); [d]optimal treatment of a range of densities; [e]broodfish exchanged every 15 days
[f]by weight; [g]assuming 1 m depth; [h]based on observation of commercial system; [i]no feeding (nf), 20% CP diet, 40% CP diet; [j]Thai strain of red tilapia; [k]3 local strains of red tilapia;
[l]SW seawater, GW ground water; [m]data in parentheses based on mean female biomass, other data based on initial biomass

important in maintaining water quality in tanks exposed to sunlight as it is in pond and hapa systems. High quality supplementary feed (40% crude protein) improved fry yields considerably over fish fed poorer quality feeds (20% crude protein) or unfed fish in tanks filled with water of low primary productivity (Table 8.7; Santiago *et al.*, 1985).

Hapas

The hapa environment can have the benefits of intensive management of broodfish and seed harvest without the high costs typically associated with construction of tanks. Hapas can be suspended in either individually owned ponds, in which case the land, construction and maintenance costs have to be considered, or in community water bodies.

The advantage of hapas, however, depends on the control of fouling. Blockage of the fine mesh netting used in hapas can prevent water exchange, leading quickly to deterioration in water quality inside the hapa. The degree of fouling appears to be related to a complex of factors including densities of fish inside and outside the hapa, the level of management, particularly feeding, and level of fertilization. The nature of fouling also may vary between the accumulation of inorganic particles to the active growth of a variety of epiphytic organisms. Unfortunately, the levels of fouling, or methods used in its control, rarely are reported in accounts of hapa-based production (Table 8.8). The characteristic deterioration in water quality that occurs as a result of fouling probably explains much of the high variability in seed yields observed in hapas. A variety of practical fouling control measures have been tried (Kingnate, 1995), but hapa exchange every 5–10 days is most effective.

Hapas are used in a variety of ways, suspended as floating structures in deeper water bodies or attached to poles erected in the sediments of shallow ponds or the perimeters of water bodies. The management of broodfish in hapas is affected by density in two respects; both the density of fish in individual hapas and the overall density of fish (and hapas) in the water body need to be considered (Fig. 8.8c). If the latter is expressed as the proportion of total water surface area covered by hapas, it can be observed that a wide range of 'densities' is apparent across the research and commercial data presented in Table 8.8. The overall density of fish affects both the availability of natural food and water quality inside any single hapa. The numbers of feral and/or escaped fish outside the hapas is therefore also an important factor in ensuring that the carrying capacity of the overall system is not exceeded.

The control of water quality may be more difficult in large community water bodies than in individually owned fishponds. Seasonal variability in water quality such as occurs in Laguna de Bay, Philippines appears to account for much of the monthly variation in fry yields recorded by Santiago *et al.* (1985) and Bautista *et al.* (1988) (Table 8.9). The effects of wind, turbidity and variable water level are more controllable for hapas suspended in smaller water bodies.

Table 8.8 Management of tilapia broodfish in hapas

Species/ strain (site)	Hapa size (m) suspended to d [m]	Water body [% covered by nets]	Fish outside hapas/fouling control	Natural food level indicators	Feeding practice	Water quality indicators	Harvest interval (days)	Type of seed harvested	Source
O. niloticus (Philippines)	10 × 2 × 1.5 –single –double, inner large mesh side panels (30 mm) with fine mesh base	200 ha of cages /8000 ha pens in 900 km^2 Laguna lake [11%]		0.45–8.31 gC/m^{-2} d^{-1}	– rice bran (12% CP)@ 2.5% bodywt d^{-1} –pellet (20% CP) @1.5% bodywt d^{-1}	T°C 20–33.5°C, DO 6.7–13.2 mg l^{-1}	Daily dipnetting	b, c	Guerrero and Garcia, 1983
O. niloticus (USA)	2.74 × 1.22 × [0.5] 1.66 mm mesh	Hapas in concrete tank [50%]			Pellet (35%CP) fed @3% initial body wt d^1		1–18 after crowding	a, b	Hughes and Baerends, 1983
O. niloticus (Puerto Rica)	2 × 1.05 × [0.7] 1.6 mm mesh	700 m^2 covered with hapas [<5%]	150 × 74 g male *O. niloticus* stocked outside	Mean Secchi disk 35 cm	32% CP diet, @ 3% bodywt d^{-1}	min.DO 3.9 mg l^{-1}	2 4 7 10	q a a, b a, b	Verdegem and McGinty, 1987
O. niloticus (Philippines)	1.5 × 1.5 × 1.5	200 h of cages/ 800 ha pens in 900 km^2 Laguna lane [11%]		Typhoon reduced seed yield low DO at 07.30 (0.50–2.9 mg l^{-1}) T°C	– no feed – 20% CP pellet – 40% CP pellet @1% bodywt d^{-1}		21		Santiago *et al.*, 1985
O. niloticus (Philippines)	2 × 2 × 1.5 [1 m] 0.25 mm mesh	Placed 2.5 × 2.5 m apart 200 ha of cages/ 800 ha pens in 900 km^2 Laguna lake [11%]		algal blooms (May–June) correlated with low DO and poor yields	Commercial broiler starter (21% CP)@5% bodywt d^{-1} once daily	High T°C in May (>31°C) and low T°C (24–26°C) Nov–Jan believed to depress yields	21	a,b	Bautista *et al.*, 1988

Table 8.8 (cont)

Species/ strain (site)	Hapa size (m) suspended to d [m]	Water body [% covered by nets]	Fish outside hapas/fouling control	Natural food level indicators	Feeding practice	Water quality indicators	Harvest interval (days)	Type of seed harvested	Source
O. niloticus (Philippines)	Spawning hapas; 1.96 × 0.67 × 1.5 [0.7] conditioning hapas; 1.96 × 2.5 × 1.5 [0.7]	21 m^2 in 500 m^2 earthen pond [4%]	Fish eradicated before experiment. Hapas cleaned *in situ* after every harvest	No fertilization, benthic algae removed after every harvest	Pelleted feed made from broiler starter and fish meal; fed @10% bodywt (adjusted every 21 d) d^{-1} twice daily		21	a, b	Batao, 1988
O. niloticus (Philippines)	1.0 × 1.0 × 1.0	60 m^2 in 500 m^2 earthen pond [12%]			Pelleted feed ad libitum twice daily	Remained within optimal range (Rana, 1988)	4 or after release by female	a, b	Mair *et al.*, 1993
O. niloticus (USA)	Spawning and conditioning hapas 2.7 × 1.2 × 0.8 1.6 mm mesh, mesh cover	10 m^2 in 20 m^2 concrete tank [50%]		No water exchange, added to replace evaporation losses	Sinking trout pellet (38% CP) in spawning and conditioning hapas @2 and 2.5% body wt d^{-1} respectively	DO levels 0.7 –10 mg l^{-1}; T°C 24–31°C	21	a, b	Lovshin and Ibrahim, 1988
O. niloticus (Thailand)	Spawning hapas; 5 × 8 × 0.9 [0.6] Conditioning hapas; 5.4 × 0.9 [0.6]	620 m^2 hapa area in 1740 m^2 earthen pond [36%]	Wild tilapias eradicated before experiment and 20 (80 g each) snake-head stocked. All hapas were exchanged every 10 days	fertilization @ 4.7 and 1.5 kg N and P ha^{-1} d^{-1}, respectively/ twice weekly to maintain eutrophic conditions	30% CP pellet, twice daily *ad libitum* in both spawning and conditioning hapas	Early morning DO was >1 mg l^{-1} for most of the trial, 3 pm values 4–13 m l^{-1}. Phytoplankton biomass increased during the experiment. T°C 28–35°C	Every 5,	a	Abrajano-Bonco, 1991

Table 8.8 (cont)

Species/ strain (site)	Hapa size (m) suspended to d [m]	Water body [% covered by nets]	Fish outside hapas/fouling control	Natural food level indicators	Feeding practice	Water quality indicators	Harvest interval (days)	Type of seed harvested	Source
	Spawning hapas, 5 × 8 × 0.9 [0.6] conditioning hapas; 3 × 1.8 × 0.9 [0.6]	280 m^2 in 1740 m^2 earthen pond [16%]			30% CP pellet, twice daily *ad libitum* in both spawning and conditioning hapas		5	a	Little *et al.*, 1991
	Spawning hapas; 5 × 8 × 0.9 [0.6] conditioning hapas; 1.25 × 4 × 0.9 [0.6]	330 m^2 in 2304 m^2 earthen pond [14%]	Hapas were exchanged every 10 d	Fertilized @ 4 kg N and 2 kg P ha^{-1} d^{-1} as urea and TSP respectively					Ungsethaphan, 1995
O. aureus (USA)	2.7 × 1.2 × 1.2 [0.4-0.8] 1.6 mm mesh	65 m^2 in 1000 m^2 earthen pond [6.5%]			32% CP catfish [pellet @25 body wt.d^{-1} fed twice daily		7–13	a, b	Behrends *et al.*, 1993
O. shiranus (Malawi)	3 × 3 × 1 1.6 mm mesh size	27 m^2 in 200 m^2 earthen pond [13.5%]			chicken layer mash @5% body wt d^{-1} three times daily	No regular harvest of fry if daily T°C below 19°C (min)–24°C (max)	14 after stocking and then 21	a, b	Maluwa and Costa Pierce, 1993

[a]eggs and yolk-sac larvae, [b]first feeding fry, [c]fry (TL = 2–3 cm)
CP = crude protein; TSP = triple superphosphate

Controlled fertilization can counteract turbidity, and well-designed and positioned ponds can moderate changes in water level and reduce the effects of wind, respectively. When hapas have been suspended in easily drained concrete-lined ponds or tanks in research trials, inorganic turbidity and external fish biomass can be controlled even more easily. Other factors that negatively impact on seed production such as theft and physical damage from floating debris may also be less problematic in smaller individually-owned systems.

A variety of broodfish management strategies have been described, but most commercial operators maintain the same set in the spawning hapa continuously. It has been shown in both tanks (see above) and hapas (Little *et al.*, 1996) that frequent exchange and reconditioning of broodfish improves intensity of spawning, but the number of broodfish and hapas required increases considerably. Broodfish exchange also affects feeding strategy since a decision to feed only in the conditioning hapas can reduce fouling in the spawning hapa, albeit at a lower level of seed production. A feeding level of 1% body weight day^{-1}, given once a day only to fish in conditioning hapas, was technically and economically optimal for Nile tilapia brooders in Thailand. Females were completely exchanged after seed harvest every 5 days and conditioned for a 10-day period. (Ungsethaphan, 1995). In less intensive systems, in which first-feeding fry were removed every 14 days from a hapa in a fertilized pond, no feeding produced higher seed yields (>20 fry m^{-2} hapa d^{-1}) than feeding the broodstock during the spawning period (AIT, 1996). Fry yields were related to water quality, which remained high where fish were not fed a supplement and where fish intensively grazed fouling organisms off the hapa walls (AIT, 1996).

A variety of feeds and feeding methods have been used for tilapia broodfish in hapas (Table 8.8; see also Jauncey, this volume). These range from rice bran, to mixes with commercial concentrates fed as a dry mash to pelleted fish feeds. The level of natural feed and fish density affects the impact of feed quality on seed production.

The use of double hapa systems, in which broodfish are held within a larger mesh net, allowing first feeding fry access to an outer fine mesh hapa net, or directly to the larger tank or pond, has been advocated. Use of an outer fine mesh hapa is claimed to reduce disturbance of broodstock and cannibalism of fry and to be particularly suitable for a lacustrine environment, but performance has been disappointing (Guerrero and Garcia, 1983; Costa Pierce and Hadikusumah 1995) and costs are higher. The low yields attained may reflect problems specific to the particular hapa environment used, however (see above). Argue and Phelps (1995) found that allowing fry to escape from spawning hapas to concrete tanks in which the hapas were located did not improve yields, but fry were larger at the end of the production cycle, potentially allowing more frequent, shorter cycles. Labour to harvest the fry, and presumably, other costs inherent in draining the system were greater, however.

Table 8.9 Performance of tilapia broodfish in hapas

Species/strain (site)	Stocking density # m^{-2}	Stocking density g m^{-2}	Female, male size (g)	Sex ratio F:M	Fry m^{-2} d^{-1}	Female productivity fry kg fem.$^{-1}$ month^{-1}	Source
O. niloticus (Philippines)	single net 4	250	50–103, 75–118	3	2.9–7.6	391–1419	Guerrero and Garcia, 1983;
O. niloticus (USA)	5	270–780	46[b]–185[c], 58	2	44.7–73.4	2520[c]–4480[b]	Hughes and Behrends, 1983; 70 d trial
				3	19.8	840[d]	
	10	850–970		2	32.8	1120[d]	
				3	48.7	1120[d]	
O. niloticus (Puerto Rico)	5.7	420	74[e]	1			Verdegem and McGinty, 1987; 35 d trial
O. niloticus (Philippines)	6.2	600	96, 102	6	0[f] 4.1	191	
					20[g] 4.4	191	
					40[h] 11.1	409	Santiago *et al*., 1985; 2 sequential 168 d trials
	7.1	1100	147, 149	3	0[f] 7.7	289	
					20[g] 20.4	608	
					40[g] 26.4	674	
	5.5–6.5	490–570	88.8, 98.3	4	31.0	1955[i]	
				7	25.5	1531[i]	
				10	27.8	1750[i]	Bautista *et al*., 1988; 365 d trial
	5.0	400		4	28.8	2482[i]	
	8.8	700	81.1, 90.5		22.0	1085[i]	
	12.5	1000			27.3	941[i]	
					27.2–30.1[j]	1069–1920[j]	Batao, 1988;
	3	370	121, 212	3	41.6[k]	1371[k]	105 d study
	2		98	1	7.2	2035	Mair *et al*., 1993;
			121		15.1	3519	90 d study

Table 8.9 (cont)

Species/strain (site)	Stocking density # m^{-2}	Stocking density g m^{-2}	Female, male size (g)	Sex ratio F:M	Fry m^{-2} d^{-1}	Female productivity fry kg $fem.^{-1}$ $month^{-1}$	Source
O. niloticus	6	290	89, 138	3	34.7[l]	4161	Lovshin and Ibrahim, 1988;
(USA)					37.5[l]	2094	105 d study
					43.3[m]	2459	
	0.6	50	95, 85	1	27.8	6210	Abrajano-Bongco, 1991;
O. niloticus	1.25	110			65.0	6225	76 d study
(Thailand)	2.5	230			118.6	5980	
	5	450			182.0	4727	
	10	900			107.0	1407	
		244–249[a]	107, 139[u]		71.4–53.3[r]	6530–3293[r]	Little *et al.*, 1989, 1996;
	2	247[u]	103, 145[u]	1	91.8[s]	8957[s]	106 d study
		224[u]	103, 120[u]		79.5[t]	7627[t]	
	5	360	59, 84	1	64.0[n]–77.9[o]	3384[n]–4185[o]	Ungsethaphan, 1995;
					79.9[n]–82.1[o]	3851[n]–4058[o]	75 d study
					91.4[n]–77.4[o]	4086[n]–3422[o]	
O. aureus	3.7	1190	225, 415	1	78-82[p]	5111–5601[p]	Behrends *et al.*, 1993;
(USA)					48–60[q]	3329–4035[q]	50 d trial
O. shiranus	0.6	30	46, 43	2	0.8	612	
(Malawi)	1.0	50			1.2	566	Maluwa and Costa-Pierce,
	1.5	80			1.3	503	1993; 413 d study

[a]based on area of outer hapa, [b]year I female, [c]year II females, [d]mixed year I and II females (1:1), [e]mean of both sexes [f,g,h]0, 20 and 40% crude protein diet respectively, [i]based on initial stocking weights, [j]no exchange, female only or male only exchange, [k]male and female exchange, [l]no exchange or female only exchange, [m]male and female exchange, [n]fed once daily in conditioning hapas only, [o]fed twice daily in conditioning hapas only, [p]no pre-maxilla removal or only male, [q]pre-maxilla removed of female only or both male and female, [r]female exchange, 10–20 days conditioning, no male exchange, [s]selective female exchange, no male exchange, [t]female exchange, male exchange (10 days conditioning), [u]based on mean of initial and final individual weights

8.5 NURSING OF TILAPIA

In pond production, natural incubation and early nursing of fry by the female is followed by a prolonged period of female protection, generally ensuring high survival of the large, but slow-moving first-feeding fry. First-feeding fry, produced by artificial incubation of eggs and yolk-sac fry harvested from hapas or tanks, when stocked directly into earthen ponds, usually survive poorly. A series of trials at AIT found that survival was normally under 30% (unpublished AIT data). Post-hatchery fingerling rearing has been reviewed by Green *et al.* (1997) for freshwater and by Watanabe *et al.* (1997) for saltwater environments.

Hormonal sex reversal in hapa-in-pond systems produces fish of less than 0.2 g after 21 days, a size that often does not meet market requirements. Sex reversed fish can be nursed further in nylon hapas at densities of between 100 and 2500 fry m^{-2} to reach marketable size.

Advanced nursing in hapas

A common problem in commercial hatcheries is the difficulty in relating production of seed to demand. Advanced nursing of MT-treated fry in hapas at high density can be used to attain optimal market size, improve survival and to 'store' fish until demand is highest. Ensuring that this 'stunting' process does not negatively impact subsequent performance was a key part of a trial to develop optimal holding strategies by Wangpen (1996). He compared a range of holding regimes using an on-farm feed (fishmeal:fine rice bran, 1:2) for post-sex reversed fry raised at 2160 fish m^{-2} in small hapas over a 3-month period, adjusting feeding levels monthly. A feeding level of 3% for the first month, 2% for the second month and 1% for the third month gave optimal survival, size range and economic returns overall and had no adverse effect on subsequent survival and growth.

8.6 SYSTEMS COMPARISON

There is wide variation in the efficiency of seed production both within and between pond, hapa and tank-based systems, and this is partly related to how systems are designed and operated. For instance, tanks and hapas may be managed in a similar way to small pond units, and well-designed earthen ponds can be managed as intensively as many hapa and tank systems.

The intensity of management explains the large differences in efficiency reported among systems. The type (age/size) of seed produced is a major determinant of efficiency. A comparison of major inputs, outputs and processes in earthen ponds, hapas and tanks managed to produce seed that is hormone treatable (i.e. first feeding or eggs/yolk-sac fry) can identify strengths and weaknesses of each system. In tanks supplied with recirculated, bio-filtered water, frequent seed harvest ensures that the total system net yield is made up only of seed suitable for hormone treatment and broodfish (Fig. 8.8a). Unharvested fry, often as recruits, typically make up the largest part of the net yield in earthen ponds (Figs 8.7, 8.8b). Smaller fish, unsuitable for sex reversal, but still saleable as mixed sex seed, are also important. Hormone treatable fry can form a high proportion of net yields harvested from hapas but if the larger water body is considered, growth of wild spawned fry, wild fish, etc., is often important (Fig. 8.8c). The water body itself acts as the 'filter' to maintain water quality in hapa systems, and water quality in hapas suspended in larger, less manageable water bodies will usually be harder to control.

The process of tilapia seed production can be divided into seven stages, which can occur in separate systems, completely within the same system, or an intermediate number of steps (Fig. 8.9). The flexibility of tilapia with regard to culture environments has resulted in a range of combinations for managing broodfish (stage 1–3). Critical factors include the period and environment for conditioning between spawning and the inter-harvest interval (IHI) of spawning systems.

A comparison of the tabulated data (Tables 8.2–8.9) indicates that there is a good deal of overlap in the productivity of both system (seed m^{-2} day^{-1}) and broodfish (seed kg $fish^{-1}$ $month^{-1}$) among pond, tank and hapa systems. Given the quite different investment options, this raises important questions about optimal management methods.

Clearly more frequent and complete harvest of eggs and yolk-sac fry, rather than first feeding or larger fry, improves productivity of hatcheries. System design and scale affects how often broodfish can be handled cost-effectively and seed harvested. Early harvest of seed requires more complex management and necessitates artificial incubation and larval rearing.

The management of water quality is critical to good seed production in any system and is related to stocking density, feed and fertilization practices. Good water quality can be maintained in simple earthen ponds using low stocking densities of broodfish and relatively little feed and fertilizer. Most producers using ponds and hapas-in-ponds also find it necessary to regularly remove sediments, particularly if organic fertilizers such as sewage or pig manure are used, to maintain high water quality and a firm substrate. Higher broodfish stocking densities in tanks and hapas necessitate greater attention to water quality or high seed yields are not sustained.

Temperature is a major constraint to hatchery production of tilapias in subtropical countries that can restrict both seasonal duration and the productivity of hatchery output. Although some species and strains are known to be more cold tolerant than others (e.g. *O. aureus*) optimal seed production still occurs across a relatively narrow temperature range. Even minor changes in temperature regime moderated by simple environmental changes can have important impacts on breeding intensity and seed survival.

The frequent exchange of broodfish conditioned as separate stocks can

Site	Mixed sex or MT fry	Broodfish			Egg and Larvae		Fry	Fingerling	Fry/fingerling	Source
		1	**2**	**3**	**4**	**5**	**6**	**7**	**8**	
		Production	**Conditioning**	**Spawning**	**Incubation**	**Early rearing**	**Early nursing Hormone treatment**	**Advanced nursing**	**Marketing**	
Thailand Southern Vietnam	Mixed sex Mixed size	EARTH POND	EARTH POND						HAPA	Little *et al.*, 1994, Little and Pham, 1996
Honduras, Ecuador	MT	EARTH POND	EARTH POND	EARTH POND			EARTH POND			Popma and Green, 1984
Israel, Nir David		EARTH POND	EARTH POND	EARTH POND				EARTH POND		Hulata, 1997
Thailand,	MT	EARTH POND	HAPA EARTH POND	HAPA	INCUBATORS	TRAYS	HAPA EARTH POND	HAPA EARTH POND	CONCRETE TANKS	Little, 1989
Thailand, DOF	Mixed sex	EARTH POND	EARTH POND	CONCRETE TANK						AIT data
Taiwan, Brazil	MT	EARTH POND	EARTH POND	EARTH POND			CONCRETE TANKS		CONCRETE TANKS	Liao and Chen, 1983
Northern Vietnam	Mixed sex	CAGE EARTH POND	CONCRETE TANK	RICE FIELDS					HAPA	Nguyen *et al.*, 1997
Bahamas	MT	POND/ CAGE *SEAWATER*	WOOD/LINER TANK *BRACKISH WATER*	WOOD/ LINER TANK *BRACKISH WATER*	INCUBATORS *BRACKISH WATER*	TROUGHS *BRACKISH WATER*	PLASTIC TANKS *BRACKISH WATER*	TANK/ CAGE SEA*WATER*		Watanabe *et al.*, 1997

Fig. 8.9 Stages in tilapia hatchery systems.

increase productivity of pond, hapa and tank systems. Broodfish can be conditioned at high density and when space is limiting, this option becomes more attractive. Despite the advantages, broodfish conditioning may be limited by other practical considerations. The costs of purchase and broodstock maintenance increase by a function of 2–3. The labour and management costs of handling large numbers of broodfish regularly are also considerable.

High densities of broodfish can result in intense spawning activity if their condition and water quality remain high, but sub-optimal seed production is likely if prolonged mouthbrooding occurs under such conditions. The stress of being unable to escape other fish is indicated in mouthbrooding females by body coloration and behaviour, and, frequently, the small and poor quality broods of fry produced. Early and complete harvest by seed robbing is therefore advisable in high-density systems.

Broodfish efficiency in ponds (fry kg^{-1} $female^{-1}$ $month^{-1}$), despite incomplete harvest of seed, often compares favourably with efficiencies in hapas and tanks, probably because mouthbrooding females can escape from the courtship and spawning area, and fry can be released in protective shallows. Social hierarchies, which inhibit spawning of subordinate fish (Turner, 1986), also appear to be related to stocking density, transparency of water and broodfish management. Exchange of broodfish appears to disrupt the dominance of fish in spawning groups, leading to greater participation of subordinates (Little *et al.*, 1996). Spawning frequencies of groups of *O. niloticus* females were normally distributed in fish spawned in hapas in plankton-rich water with low transparency, but highly skewed in groups within clear-water tanks suggesting stronger hierarchies in the latter environment (Fig. 8.10; Little, 1989).

8.7 SUMMARY

Four major factors combine to affect overall seed production through their impact on the condition of broodfish, the intensity and success of spawning and the survival and harvest of seed.

The first is the area of the spawning system and stocking density of breeding fish. Tilapias are territorial and hierarchical and reproductive behaviour, especially courtship and spawning, is density-dependent. The area available to breeding fish determines the intensity of spawning and the success of natural incubation. The second is water quality, particularly dissolved oxygen, a chronic deficit of which negatively affects both the intensity and success of reproduction. Hatchery systems maintaining dissolved oxygen levels within the optimal range sustain higher levels of seed production. Temperature and salinity are key parameters that affect species, season and site of seed production. Third, the intensity and efficiency of harvest of both broodfish and seed greatly affects the level of seed production possible. Complete harvest of seed is desirable to both maximize output directly and to prevent seed remaining in the system to

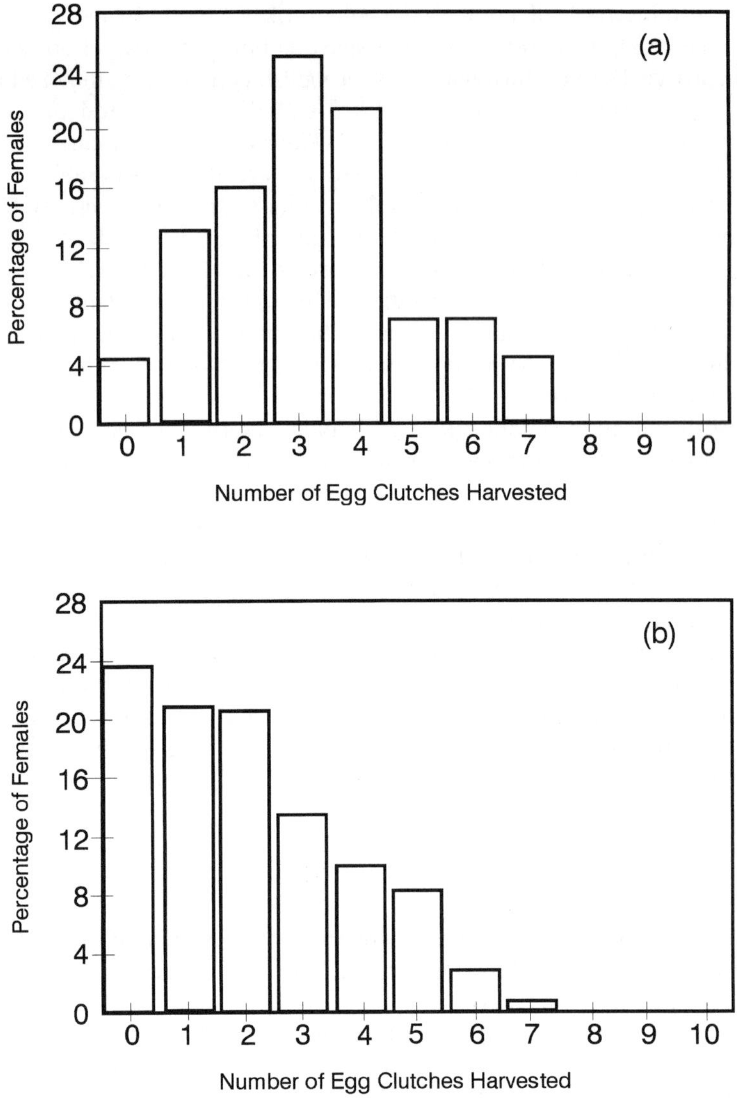

Fig. 8.10 Frequency of seed harvest from tagged individual females in (a) hapas-in-ponds (n = 69) and (b) tanks within a recirculation system over period of 116 days (n = 224) (after Little, 1989).

compete with, or predate, on fry. Fourth, a dependence on natural feed that allow tilapias, which are to varying degrees phytophagous, to grow and reach breeding condition by feeding on natural foods occurring in spawning systems alone. However, breeding tilapias are often fed supplements, or if natural food is lacking as in clear water tank systems, fed complete diets.

The ease with which tilapia reproduce has assisted their spread without any major novel hatchery development being required. Indeed, an ability to reproduce in growout systems has helped tilapias spread in situations without hatcheries and well developed seed distribution networks. Ease of reproduction by mixed sex stocked fish can also contribute positively to overall economic returns by reducing costs and increasing the diversity of produce. The deterioration in tilapia quality, particularly the problem of early stunting, that is frequently reported as a major constraint to freely-breeding tilapia may be related to farmers' management of such stocks.

Hatchery systems for tilapias have evolved through farmer trial and error, and it is only recently that commercial systems based on research results have been developed. The approaches to sex control are very dependent on the hatchery strategy employed, but well-managed ponds, hapas and tanks can all produce high quality monosex fry. Importantly, earthen ponds can match the productivity and efficiency of tanks and hapa systems. If investment capital is limiting, hapas and tanks should be used for conditioning broodfish, early fry nursing and marketing rather than spawning. Tanks and hapas stocked with high densities of broodfish have higher management requirements than extensive earthen ponds. The harvest of eggs and yolk-sac fry also increases investment and management costs but can result in significantly higher productivities. Use of artificially incubated fry of known age also appears to improve the consistency of sex reversal after hormone treatment.

Ensuring consistent broodstock condition, enhancing the capacity to harvest seed completely and frequently, and maintaining high water quality are the key factors to optimize seed production in any hatchery system.

ACKNOWLEDGEMENTS

The authors would like to thank Leonard Lovshin, Graham Mair and Eric Hallerman for reviewing the manuscript.

REFERENCES

Abrajano-Bongco, A. (1991) *Effect of Stocking Density on the Reproductive Performance of Nile Tilapia (*Oreochromis niloticus *L) Spawned in Net Hapas*, Unpublished MSc thesis, Asian Institute of Technology, Bangkok, Thailand.

AIT (1986) *Buffalo/fish and Duck/Fish Integrated Systems for Small-Scale Farmers at the Family Level.* AIT Research Report No. 198, Asian Institute of Technology, Bangkok, Thailand.

AIT (1996) *ODA Support for Aquaculture at the Asian Institute of technology (December 1992–May 1995) and Aquaculture Project Support Project (June-November 1995). Final Report,* Asian Institute of Technology, Bangkok, Thailand.

Al-Ahmad, T.A., Ridha, M. and Al-Ahmed, A.A. (1988) Reproductive performance of the tilapia *Oreochromis spilurus* in seawater and brackish groundwater. *Aquaculture* **73**, 323–332.

Amballi, A.J.D and Little, D.C. (1996) Studies on the effect of manipulating hapa size on broodfish conditioning of *Oreochromis niloticus* in fertilized earthen ponds, in *Proceedings of The Third International Symposium on Tilapias in Aquaculture* (ICLARM Conference Proceedings Vol. 41) (eds R.S.V. Pullin, J. Lazard, M. Legendre, J.B. Amon Kothias and D. Pauly), ICLARM, Manila, Philippines, pp. 223–228.

Argue, B.J. and Phelps, R.P. (1995) Hapa-based systems for producing *Oreochromis niloticus* fry suitable for hormone sex reversal. *J. Appl. Aquacult.* **5**, 21–27.

Balarin, J.D and Haller, R.D. (1982) The intensive culture of tilapia in tanks, raceways and cages, in *Recent Advances in Aquaculture* (eds J.F. Muir and R.J. Roberts), Croom Helm, London, pp. 265–356.

Bardach, J.E., Ryther J.H. and McLarney W.O. (1972) *Aquaculture, the Farming and Husbandry of Freshwater and Marine Organisms,* Wiley, New York.

Barlow, G.W. (1991) Mating systems among cichlid fishes, in *Cichlid Fishes. Behaviour, Ecology and Evolution* (ed. M.H.A. Keenleyside), Chapman and Hall, London, pp. 173-190.

Barman, B.K., Little, D.C. and Edwards, P. (1998) Impacts of tilapia on poor farmers in Northwest Bangladesh. Paper presented at the Asian Fisheries Forum, Chiang Mai, Thailand 11–14th November 1998.

Baroiller, J.F., Desprez, D., Carteret, Y. *et al.* (1997) Influence of environment and social factors on the reproductive efficiency in three tilapia species, *Oreochromis niloticus, O. aureus* and the red tilapia (red Florida strain), in *Proceedings from the Fourth International Symposium on Tilapias in Aquaculture* 1 (ed. K. Fitzsimmons), Northeast Regional Agricultural Engineering Service, Ithaca, New York, pp. 238–252.

Bautista, A.M., Carlos, M.H. and San Antonio, A.I. (1988) Hatchery production of *Oreochromis niloticus* L. at different sex ratios and stocking densities. *Aquaculture* **73**, 85–95.

Batao, M.N. (1988) *Broodstock Exchange Technique for Maximum Production of* Oreochromis niloticus *Egg and Fry in Hapas,* Unpublished MSc thesis, Central Luzon State University, Muñoz, Philippines.

Behrends, L.L., Kingsley, J.B. and Price, A.H. II (1993) Hatchery production of blue tilapia, *Oreochromis aureus* (Steindachner), in small suspended hapa nets. *Aquacult. Fish. Manage.* **24**, 237–243.

Berger, A. and Rothbard, S. (1987) Androgen induced sex reversal of red tilapia fry stocked in cages within ponds. *Bamidgeh* **39**(2), 49–57.

Berrios Hernanadez, J.M. and Snow, J.R. (1983) Comparison of methods for reducing fry losses to cannibalism in *Tilapia* production. *Prog. Fish Cult.* **45**(2), 116–118.

Beveridge, M.C.M. (1984) Tilapia hatcheries-lake or land based? *ICLARM Newsl.* **7**(1), 10–11.

Bevis, R. (1994) *The Effect of Artificial Nests on Reproductive Performance in the Nile Tilapia* Oreochromis niloticus *L. Spawned in Net Hapas,* Unpublished MSc thesis, Asian Institute of Technology, Bangkok, Thailand.

Bhujel, R.C. (1999) *Management of Nile Tilapia (*Oreochromis niloticus*) Broodfish for Commercial Seed Production in Thailand,* Unpublished PhD thesis, Asian Institute of Technology, Bangkok, Thailand.

Brooks, A.C. and Maluwa, A.O. (1997a) The development of fish farming technologies, in *Fish Farming in Malawi: a Case Study of the Central and Northern Regions Fish Farming Project* (eds M.W. Dickson and A.C.Brooks), Stirling Aquaculture, University of Stirling, Scotland.

Brooks, A.C. and Maluwa, A.O. (1997b) On-station trials, in *Fish farming in Malawi: a Case Study of the Central and Northern Regions Fish Farming Project, Technical Supplement* (ed. M.W. Dickson) Stirling Aquaculture, University of Stirling, Scotland.

Broussard, M.C., Reyes, R. and Raguindin, F. (1983) Evaluation of hatchery management schemes for large scale production of *Oreochromis niloticus* fingerlings in Central Luzon, Philippines, in *Proceedings of the International Symposium on Tilapia in Aquaculture* (eds L. Fishelson and Z. Yaron), Tel Aviv University, Tel Aviv, Israel, pp. 414–424.

Buddle, C.R. (1984) Androgen-induced sex inversion of *Oreochromis* (Trewavas) hybrid fry stocked into cages standing in an earthen pond. *Aquaculture* **40**, 233–239.

Brummett, R.E. (1997). Production of *Tilapia rendalli* fingerlings under Malawian smallholder Conditions, in *Proceedings from the Fourth International Symposium on Tilapias in Aquaculture* 1 (ed. K. Fitzsimmons), Northeast Regional Agricultural Engineering Service, Ithaca, New York, pp. 273–286.

Cao, T.B. (1998) *Development of Pond Culture of Thai Nile Tilapia (*Oreochromis niloticus *L.) and its Marketability in Hanoi, Vietnam*, Unpublished PhD thesis, Asian Institute of Technology, Bangkok, Thailand.

Capili, J.B. (1995) *Growth and Sex Determination in the Nile tilapia,* Oreochromis niloticus *(L),* Unpublished PhD thesis, University of Wales, Swansea, Wales, UK.

Costa Pierce, B.A. (1996) Effects of substrate and water quality on seasonal fry production by *Tilapia rendalli* in tanks, in *Proceedings of The Third International Symposium on Tilapias in Aquaculture* (ICLARM Conference Proceedings, Vol. 41) (eds R.S.V. Pullin, J. Lazard, M. Legendre, J.B. Amon Kothias and D. Pauly), ICLARM, Manila, pp. 280–289.

Costa Pierce, B.A. and Hadrikusumah, H. (1995) Production management of double net tilapia *Oreochromis* spp. hatcheries in a eutrophic tropical reservoir. *J. World Aquacult. Soc.* **26**, 453–459.

Dan, N.C. and Little, D.C. (2000) The culture performance of monosex and mixed sex new season and over-wintered fry in two strains of Nile tilapia (*Oreochromis niloticus*) in northern Vietnam. *Aquaculture* **184**(3–4), 221–231.

De Silva, S.S. and Radampola, K. (1990) Effect of dietary protein level on the reproductive performance of *Oreochromis niloticus*, in *Proceedings of the Second Asian Fisheries Forum* (eds R. Hirano and I. Hanyu), Asian Fisheries Society, Manila, Philippines, pp. 559–563.

Dickson, M.W. and Brooks, A.C. (1997) The development of fish farming technologies, in *Fish Farming in Malawi: A Case Study of the Central and Northern Regions Fish Farming Project*, Stirling Aquaculture, University of Stirling, Scotland.

Edwards, P., Kaewpaitoon, K., Little, D.C. and Siripandh, N. (1994) An assessment of the role of buffalo manure for pond culture of tilapia. II. *Aquaculture* **126**, 97–106.

Eguia, M.R.R. (1996) Reproductive performance of four red tilapia strains in different seed production systems. *Israeli J. Aquacult.-Bamidgeh* **48**(1), 10–18.

EIFAC (1986) *Report of the Working Group on Terminology, Format and Units of Measurement as Related to Flow Through and Recirculation Systems*. EIFAC Tech. Paper, 49, FAO, Rome.

Ellis, S.C., Watanabe, W.O. and Head, W.D. (1993) Effect of initial age variation on production of Florida red tilapia fry under intensive, brackishwater tank culture. *Aquacult. Fish. Manage.* **24**, 465–471.

Engle, C. (1997) Optimal resource allocation by fish farmers in Rwanda. *J. Appl. Aquacult.* **7**, 1–18.

Ernst, D.H., Watanabe, W.O. and Ellingson, L.J.(1991) Commercial-scale production of Florida red tilapia seed in low-and Brackish-salinity tanks. *J. World Aquacult. Soc.* **22**, 36–44.

Fishelsen, L. (1983) Social behaviour of adult tilapia fish in nature and captivity (Israel), in *Proceedings of the International Symposium on Tilapia in Aquaculture* (eds L. Fishelson and Z. Yaron), Tel Aviv University, Tel Aviv, Israel, pp. 48–58.

Fryer, G. and Iles T.D. (1972) *The Cichlid Fishes of the Great Lakes of Africa*, Oliver & Boyd, Edinburgh.

Green, B.W. and Teichert-Coddington, D.R. (1993) Production of *Oreochromis niloticus* fry for hormonal sex reversal in relation to water temperature. *J. Appl. Ichthyol.* **9**, 230–236.

Green B.W. and Teichert-Coddington, D.R. (1994) Growth of control and androgen-treated Nile tilapia during treatment, nursery and grow-out phases in tropical fish ponds. *Aquacult. Fish. Manage.* **25**, 613–621.

Green, B.W., Verrica, K.L. and Fitzpatrick, M.S. (1997) Fry and fingerling production, in *Dynamics of Pond Aquaculture* (eds H.S Egna and C.E Boyd), CRC Press, Boca Raton, pp. 215–243.

Guerrero, R.D. III (1982) Control of tilapia reproduction, in *The Biology and Culture of Tilapias (*eds R.S.V. Pullin and R.H. Lowe-McConnell), ICLARM, Manila, Philippines, pp. 309–316.

Guerrero, R.D. III (1986) Production of Nile tilapia fry and fingerlings in earthen ponds at Pila, Laguna, Philippines, in *Proceedings of The First Asian Fisheries Forum* (eds J.L. Maclean, L.B. Dizon and L.V. Hosillos), Asian Fisheries Society, Manila, Philippines, pp. 49–52.

Guerrero, R.D. III (1987) *Farming Tilapia in the Philippines*, National Book Store Publishers, Manila, Philippines.

Guerrero, R.D. III and Garcia, A.M. (1983) Studies on the fry production of *Saratherodon niloticus* in a lake-based hatchery, in *Proceedings of the International Symposium on Tilapia in Aquaculture* (eds L. Fishelson and Z. Yaron), Tel Aviv University, Tel Aviv, Israel, pp. 388–393.

Guerrero, R.D. III and Guerrero, L.A. (1984) Commercial fry production of *Tilapia nilotica* using concrete tanks in the Philippines. Paper presented at the Aquaculture Symposium, Auburn University, Auburn, Alabama, USA.

Guerrero, R.D. III and Guerrero, L.A. (1985a) Effect of breeder size on fry production of Nile tilapia in concrete pools. *Trans. Nat. Acad. Sci. and Tech. (Phils.)* **7**, 63–66.

Guerrero, R.D. III and Guerrero, L.A. (1985b) Further observations on the fry production of *Oreochromis niloticus* in concrete tanks. *Aquaculture* **47**, 257–261.

Gupta, M.V., Ahmed, M., Bimboa, M.A.P. and Lightfoot, C. (1992) *Socioeconomic Impact and Farmers' Assessment of Nile tilapia (*Oreochromis niloticus*) Culture in Bangladesh.* ICLARM Tech. Rep., 35, ICLARM, Manila, Philippines.

Hanson, B.J., Moehl, J.F., Veverica, K.L, Rwangano, F and Van Speybroeck, M. (1988) Pond culture of tilapia in Rwanda, a high altitude equatorial African country, in *Proceedings of The Second International Symposium on Tilapias in Aquaculture* (*ICLARM Conf. Proc.*, Vol. 15) (eds R.S.V. Pullin, T. Bhukaswan, K, Tonguthai, and J.L. Maclean), ICLARM, Manila, Philippines, pp. 553–559.

Hepher, B. and Pruginin, Y. (1981) *Commercial Fish Farming*, Wiley, New York.

Hughes, D.G. and Behrends, L.L. (1983) Mass production of *Tilapia nilotica* seed in suspended net enclosures, in *Proceedings of the International Symposium on Tilapia in Aquaculture* (eds L. Fishelson and Z. Yaron), Tel Aviv University, Tel Aviv, Israel, pp. 394–401.

Hulata, G. (1997) Large-scale tilapia fry production in Israel. *Israeli J. Aquacult. – Bamidgeh* **49**, 174–179.

Kingnate, D. (1995) *Productivity of, and Fouling Control in, Hapa-Based Seed Production of the Thai Red Tilapia (*Oreochromis niloticus × Oreochromis mossambicus), Unpublished MSc thesis, Asian Institute of Technology, Bangkok, Thailand.

Lee, J.C. (1979) *Reproduction and Hybridization of Three Cichlid Fishes,* Tilapia aurea *(Steindachner),* T. hornorum *(Trewavas) and* T. nilotica *(Linneus) in Aquaria and Plastic Pools*, Unpublished PhD thesis, Auburn University, Alabama.

Liao, I.C and Chen, T.P. (1983) Status and prospects of tilapia culture in Taiwan, in *Proceedings of the International Symposium on Tilapia in Aquaculture* (eds L. Fishelson and Z. Yaron), Tel Aviv University, Tel Aviv, Israel, pp. 588–598.

Little, D.C. (1989) *An evaluation of strategies for production of Nile tilapia (*Oreochromis niloticus*) fry suitable for hormonal treatment*, Unpublished PhD thesis, University of Stirling, Scotland.

Little, D.C. (1992) Efficiency indicators in intensive *Oreochromis* seed production systems, in *Proceedings of The Third Asian Fisheries Forum* (eds L.M Chou Chou, A.D. Munro, T.J. Lam, T.W. Chen, L.K.K. Cheong, J.K. Ding, K.K. Hooi, H.W. Khoo, V.P.E, Phang, K.F. Shim and C.H. Tan), Asian Fisheries Society, Manila, Philippines, pp. 61–64.

Little, D.C. (1998) Options in the development of the 'aquatic chicken'. *Fish Farmer* (July/August), 23–24.

Little, D.C. and Pham, T.A. (1996) *Overview of Freshwater Fish Seed Production and Distribution in Vietnam*, AIT Aquaculture Outreach Working Paper NV6, AIT, Bangkok, Thailand.

Little, D.C., Skladany, M. and Rode, R. (1987) Small-scale hatcheries in N.E. Thailand. *Aquacult. Fish. Manage.* **18**, 15–3.

Little, D.C., Innes Taylor, N.L., Turongruang, D. and Komolmarl, S. (1991) Large seed for small-scale aquaculture. *Aquabyte* **4**(2), 2–3.

Little, D.C., Macintosh, D.J. and Edwards, P. (1993) Improving spawning synchrony in the Nile Tilapia, *Oreochromis niloticus* (L). *Aquacult. Fish. Manage.* **24**, 319–325.

Little, D.C., Sikawa, D. and Juntana, J. (1994) Commercial production and marketing of Nile tilapia (*Oreochromis niloticus)* fry in Chonburi and Chachoengsao Provinces, Thailand. *NAGA ICLARM Q.* **17**(2), 14–17.

Little, D.C., Macintosh, D.J. and Edwards, P. (1996) Selective broodfish exchange of *Oreochromis niloticus* in large breeding hapas suspended in earthen ponds, in *Proceedings of The Third International Symposium on Tilapias in Aquaculture* (ICLARM Conference Proceedings, Vol. 41) (eds R.S.V. Pullin, J. Lazard, M. Legendre, J.B. Amon Kothias and D. Pauly), ICLARM, Manila, Philippines, pp. 92–99.

Little, D.C., Bhujel, R. and Turner, W.A. (1997) Commercialization of a hatchery process to produce MT-treated Nile tilapia in Thailand, in *Proceedings of the IV Symposium on Aquaculture in Central America: Focusing on Shrimp and Tilapias, 22–24th April 1997, Tegucigalpa, Honduras* (eds D.E. Alston, B.W. Green and H.C. Clifford), National Association of Honduran Aquaculturists and the Latin American Chapter of the World Aquaculture Society, Boca Raton, Florida, pp. 108–118.

Little, D.C., Coward, K., Bhujel, R.C., Pham, T.A. and Bromage, N.R. (2000) Effect of broodfish exchange strategy on the spawning performance and sex steroid levels of *Oreochromis niloticus* broodfish in hapas. *Aquaculture* **186**(1–2), 77–88.

Lovshin, L.L. (1982) Tilapia hybridization, in *The Biology and Culture of Tilapias* (ICLARM Conference Proceedings, Vol. 7) (eds R.S.V. Pullin and R.H. Lowe Mc Connell), ICLARM, Manila, Philippines, pp. 279–308.

Lovshin, L.L. and Ibrahim, H.H. (1989) Effects of broodstock exchange on *Oreochromis niloticus* egg and fry production in net enclosures, in *Proceedings of The Second International Symposium on Tilapias in Aquaculture* (ICLARM Conference Proceedings Vol. 15) (eds R.S.V. Pullin, T. Bhukaswan, K, Tonguthai and J.L. Maclean), ICLARM, Manila, Philippines, pp. 231–236.

Lovshin, L.L. and Pretto, R. (1993) A strategy for the use of tilapias in rural Latin America: the Panamanian integrated approach, in *Proceedings of the International Symposium on Tilapia in Aquaculture* (eds L. Fishelson and Z. Yaron), Tel Aviv University, Tel Aviv, Israel, pp. 494–505.

Lowe McConnell, R.H. (1959) Breeding behaviour patterns, and ecological differences between *Tilapia* species and their significance for evolution within the genus *Tilapia* (Pisces: Cichlidae). *Proc. Zool. Soc. Lond.* **132**, 30 p.

Macaranas, J.M., Taniguchi, N., Pante, M.J.R., Capili, J.B. and Pullin, R.S.V. (1986) Electrophoretic evidence for extensive hybrid gene introgression into commercial (*Oreochromis niloticus* L.) stocks in the Philippines. *Aquacult. Fish. Manage.* **17**, 249–258.

Macintosh, D.J. and Little, D.C. (1995) Broodstock management and fry production of the Nile tilapia (*Oreochromis niloticus*), in *Broodfish Management and Egg and Larval Quality* (eds N.R. Bromage and R.J. Roberts), Blackwell Science, Oxford, pp. 277–320.

Mair, G.C. (1997) *Genetic Manipulations for Improved Tilapia-Technology Adaptation and Development II (R 6070A). Final report, April 1, 1994 to March 31, 1997*, University of Wales, Swansea, Central Luzon State University, Asian Institute of Technology, National Aquaculture Genetics Research Institute.

Mair, G.C. and Little, D.C. (1991) Population control in farmed tilapias. *NAGA ICLARM Q.*, **14**(3), 8–13.

Mair, G.C. and Van Dam, A. (1996) The effect of sex ratio at stocking on growth and recruitment in Nile tilapia (*Oreochromis niloticus*) ponds, in *Proceedings of The Third International Symposium on Tilapias in Aquaculture* (ICLARM Conference Proceedings, Vol. 41) (eds R.S.V. Pullin, J. Lazard, M. Legendre, J.B. Amon Kothias and D. Pauly), ICLARM, Manila, Philippines, pp. 100–107.

Mair, G.C., Estabillo, C.C., Sevilleja, R.C. and Recometa, R.D. (1993) Small-scale fry production systems for Nile tilapia, *Oreochromis niloticus* L. *Aquacult. Fish. Manage.* **24**, 229–235.

Mair, G.C., Abucay, J.S., Skibinski, D.O.F., Abella, T.A. and Beardmore, J.A. (1997) Genetic manipulation of sex ratio for the large scale production of all-male tilapia *Oreochromis niloticus* L. *Can. J. Fish. Aquat. Sci.* **54**, 396–404.

Maluwa, A.O. and Costa Pierce, B.A. (1993) Effect of broodstock density on *Oreochromis shiranus* fry production in hapas. *J. Appl. Aquaculture* **2**(2), 63–76.

McAndrew, B.J. (1993) Applications of biochemical genetics to the management of wild and cultured aquatic organisms, in *Proceedings of the AADCP Workshop on Genetics in Aquaculture and Fisheries Management* (eds D. Penman, N. Roongratri, B. McAndrew), AADCP, Bangkok, Thailand.

McAndrew, B.J. and Majumdar, K.C. (1983) Tilapia stock identification using electrophoretic markers. *Aquaculture* **30**, 249–261.

Meyer, D.E., Torres, D. and Aceituno, C. (1997) Comparscion del Cultivo de Tilapia con dos Technologias diferentes. *Proceedings of the IV Symposium on Aquaculture in Central America: Focusing on Shrimp and Tilapias, 22–24th April 1997, Tegucigalpa, Honduras* (eds D.E. Alston, B.W. Green and H.C. Clifford), National Association of Honduran Aquaculturists and the Latin American Chapter of the World Aquaculture Society, Boca Raton, Florida, pp. 228–229.

Mires, D. (1982) A study of the problems of mass production of hybrid tilapia fry, in *The Biology and Culture of Tilapias* (eds R.S.V. Pullin and R.H. Lowe-McConnell), ICLARM, Manila, Philippines, pp. 317–330.

Mires, D. (1995) The tilapias, in *Production of Aquatic Animals – Fishes* (eds C.E. Nash and A.J. Novotny), Elsevier, Amsterdam, pp. 133–152.

M'hango, T. and Brummett, R.E. (1997) Fry production of Shire River tilapia *Oreochromis shiranus* at two broodstock ratios. *J. World Aquacult. Soc.* **28**(2), 188–192.

Msiska, O.V. and Costa Pierce, B.A. (1997) Factors influencing the spawning success of *Oreochromis karongae* (Trewavas) in ponds. *Aquacult. Res.* **28**, 87–99.

Muir, J.F. (1982) Recirculated water systems in aquaculture, in *Recent Advances in Aquaculture* (eds J.F. Muir and R.J. Roberts), Croom Helm, London, pp. 265–356.

Nguyen, T.T. (1992) The effects of nest, water level and broodstock size on Nile tilapia (*Oreochromis niloticus*) seed production. Special Study, Asian Institute of Technology, Bangkok, Thailand.

Nguyen, C.D, Tran, M.T and Dinh, V.T. (1997) Tilapia breeding in rice fields in Vietnam. *NAGA ICLARM Q.* **20**(1), 23–25.

Noakes, D.L.G. and Balon, E.K. (1982) Life histories of tilapias: An evolutionary perspective, in *The Biology and Culture of Tilapias* (eds R.S.V. Pullin and R.H. Lowe-McConnell), ICLARM, Manila, Philippines, pp. 61–82.

Pantastico, J.B., Dangilan, M.M.A. and Egui, R.V. (1988) Cannibalism among different sizes of tilapia (*Oreochromis niloticus*) fry/fingerlings and the effect of natural food, in *Proceedings of the Second International Symposium on Tilapia in Aquaculture* (ICLARM Conference Proceedings Vol. 15) (eds R.S.V. Pullin, T. Bhukaswan, K. Tonguthai and J.L. Maclean), ICLARM, Manila, Philippines, pp. 465–468.

Phelps, R.P. and Cerezo, G. (1992) The effect of confinement in hapas on sex reversal and growth of *Oreochromis niloticus. J. Appl. Aquaculture* **1**(4), 73–81.

Philippart, J.C.L. and Ruwet, J.C.L. (1982) Ecology and distribution of tilapias, in *The Biology and Culture of Tilapias* (ICLARM Conference Proceedings, Vol. 7) (eds R.S.V. Pullin and R.H. Lowe McConnell), ICLARM, Manila, Philippines, pp. 15–60.

Popma, T.J. and Green, B.W. (1990) *Sex Reversal of Tilapia in Earthen Ponds. Aquacultural production manual,* Research and Development Series No. 35, International Center for Aquaculture, Auburn.

Popma, T.J., Ross, F.E., Nerrie, B.L. and Bowman, J.R. (1984*) The Development of Commercial Farming of Tilapia in Jamaica 1979–1983,* Research and Development Series No. 31, International Center for Aquaculture, Auburn.

Rana, K.J. (1988) *Parental Influences on Egg Quality, Fry Production and Fry Performance in* Oreochromis niloticus *(L) and* O. mossambicus *(Peters),* Unpublished PhD thesis, University of Stirling, Stirling.

Rana, K.J and Macintosh, D.J. (1988) A comparison of the quality of hatchery-reared *Oreochromis niloticus* and *O. mossambicus* fry, in *Proceedings of The Second International Symposium on Tilapias in Aquaculture* (ICLARM Conference Proceedings, Vol. 15) (eds R.S.V. Pullin, T. Bhukaswan, K, Tonguthai and J.L. Maclean), ICLARM, Manila, Philippines, pp. 497–502.

Ridha, M., Al-Ahmad, T.A. and Al-Ahmed, A.A. (1985) *Tilapia Culture in Kuwait: Spawning Experiements, 1984.* Technical Report, Kuwait Institute for Scientific Research, Safat, Kuwait.

Rothbard, S., Solnik, E., Shabbath, S. *et al.* (1983) The technology of mass production of hormonally sex-inversed all-male tilapias, in *Proceedings of the International Symposium on Tilapia in Aquaculture* (eds L. Fishelson and Z. Yaron), Tel Aviv University, Tel Aviv, Israel, pp. 425–434.

Sampson, D.R.T. (1983) *An Evaluation of the "Arena Method" of Tilapia Fry Production in Kenya (with Observations of Fingerling Grow-On and the Potential for Producing All Male Tilapia),* Unpublished MSc thesis, University of Stirling, Stirling, Scotland.

Santiago, C.B., Aldaba, M.B., Abuan, E.F. and Laron, M.A. (1985) The effects of artificial diets on fry production and growth of *Oreochromis niloticus* breeders. *Aquaculture* **47**, 193–203.

Siraj, S.S., Smitherman, R.O. and Casitllo-Gallusser, S. (1983) Reproductive traits for three year classes of *Tilapia nilotica* and maternal effects of their progeny, in *Proceedings of the International Symposium on Tilapia in Aquaculture* (eds L. Fishelson and Z. Yaron), Tel Aviv University, Tel Aviv, Israel, pp. 210–218.

Smith, S.J., Watanabe, W.O., Chan, J.R., Ernst, D.H., Wicklund, R.I. and Olla, B.L. (1991) Hatchery production of Florida red tilapia seed in brackishwater tanks: the influence of broodstock age. *Aquacult. Fish. Manage.* **22**, 141–147.

Snow, J.R., Berrios Hernandez, J.M. and Ye, H.Y. (1983) A modular system for producing tilapia seed using simple facilities, in *Proceedings of the International Symposium on Tilapia in Aquaculture* (eds L. Fishelson and Z. Yaron), Tel Aviv University, Tel Aviv, Israel, pp. 402–413.

Srisakultiew, P. and Wee, K.L. (1988) Synchronous spawning of the Nile tilapia through hypophysation and temperature manipulation, in *Proceedings of the Second International Symposium on Tilapia in Aquaculture* (ICLARM Conference Proceedings, Vol. 15) (eds R.S.V. Pullin, T. Bhukaswan, K. Tonguthai and J.C. Maclean), ICLARM, Manila, Philippines, pp. 275–284.

Torrans, L. and Hiott, A. (1990) Effects of brood-stock density on production of bait-or forage-sized blue tilapias. *Prog. Fish-Cult.* **52**, 9–14.

Trewavas, E. (1983) *Tilapiine Fishes of the Genera* Sarotherodon, Oreochromis *and* Danakilia, British Museum (Natural History), London.

Turner, G.F. (1986) Territory dynamics and cost of reproduction in a captive population of the colonial nesting mouthbrooder *Oreochromis mossambicus* (Peters). *J. Fish Biol.* **29**, 573–587.

Uchida, R.N. and King, J.E. (1962) Tank culture of *Tilapia. Fish. Bull.* **62**, 21–52.

Ungsethaphan, T. (1995) *An On-Farm Trial to Investigate Feeding Strategies for Nile Tilapia (*Oreochromis niloticus*) Broodfish*, Unpublished MSc thesis, Asian Institute of Technology, Bangkok, Thailand.

Verdegem, M.C. and McGinty, A.S. (1987) Effects of frequency of egg and fry removal on spawning by *Tilapia nilotica* in hapas. *Prog. Fish-Cult.* **49**, 129–131.

Wangpen, P. (1996) *Nursing Strategies for MT Sex Reversed Tilapia Fry* (Oreochromis niloticus *L.*), Unpublished MSc thesis, Asian Institute of Technology, Bangkok, Thailand.

Watanabe, W.O., Smith, S.J., Wicklund, R.I. and Olla, B.L. (1992) Hatchery production of Florida red tilapia seed in brackishwater tanks under natural-mouthbrooding and clutch-removal methods. *Aquaculture* **102**, 77–88.

Watanabe, W.O., Wicklund, R.I., Olla, B.L. and Head, W.D. (1997) Saltwater culture of the Florida red tilapia and other saline tolerant tilapias: A review, in *Tilapia Aquaculture in the Americas* 1 (eds B.A. Costa Pierce and J.E. Rakocy), World Aquaculture Society, Baton Rouge, Florida, pp. 54–141.

Wee, K.L. and Nguyen, A.T. (1988) Effects of dietary protein level on growth and reproduction in Nile tilapia *(Oreochromis niloticus)*, in *Proceedings of The Second International Symposium on Tilapias in Aquaculture* (ICLARM Conference Proceedings, Vol. 15) (eds R.S.V. Pullin, T. Bhukaswan, K, Tonguthai, and J.L. Maclean), ICLARM, Manila, Philippines, pp. 401–410.

Yang Yi, (1992) *Investigation into the Immersion of Known-Age* Oreochromis niloticus *Juveniles for Hormonal Sex Reversal*, Unpublished MSc thesis, Asian Institute of Technology, Bangkok, Thailand.

Yater, L.R. and Smith, I.R. (1985) Economics of private tilapia hatcheries in Laguna and Rizal Provinces, Philippines, in *Philippine Tilapia Economics* (ICLARM Conference Proceedings, Vol. 15) (eds I.R. Smith, E.B. Torres and E.O. Tan), ICLARM, Manila, Philippines, pp. 15–32.

Chapter nine

Nutritional requirements

K. Jauncey

9.1 INTRODUCTION

This chapter considers all known nutrient requirements of tilapias and is intended to complement Chapter 3, which considers feeding mechanisms, diet and trophic efficiencies in tilapias. It should be noted that the current chapter does not attempt to extend requirement data into the development of practical feeds or feeding strategies and that this topic is amply covered elsewhere (De Silva and Anderson, 1995; Jauncey, 1998).

Consideration is given to requirements for proteins, lipids, carbohydrates, vitamins and minerals as well as to the particularly important relationships between dietary protein and energy. Caution is advised in the interpretation of the data presented since it has been provided from a number of different experimental studies, using varying approaches and in different laboratories.

'Older' data are most suspect as less was understood about nutrient requirements, interactions and the effects of experimental design at the time that they were derived. However, even recent experimental work may not be without its deficiencies. The chapter concludes with tables summarizing nutrient requirement data for tilapias. It must be remembered throughout that tilapia nutrition is still very much an inexact and evolving science.

9.2 PROTEIN REQUIREMENTS

Dietary protein is always considered to be the most important nutrient component of complete formulated fish feeds and, for this reason, is considered first when reviewing the nutrient requirements of fish. Dietary protein require-

M.C.M. Beveridge and B.J. McAndrew (eds), Tilapias: Biology and Exploitation, 327–375.

ments of fish for maximum growth appear, as a percentage of the diet, to be much higher than those of farmed terrestrial animals. This is principally an artefact of the way in which requirements are expressed, attributable to the relatively lower energy requirements of poikilotherms. The protein requirements of fish, per unit weight gain or per unit protein gain, are much the same as those of terrestrial monogastric animals such as poultry. Indeed, the values for net protein retention for fish and terrestrial monogastric animals are in the region of 20–50% for both groups of vertebrate (NRC, 1993).

The level of dietary protein which results in maximum growth (the so called 'optimum' for growth) is influenced by a number of factors including:

- the energy content of the diet,
- the physiological state of the animal (age, weight, maturity),
- environmental factors (water temperature, salinity, dissolved oxygen),
- protein 'quality' (levels and availability of essential amino acids),
- food intake.

Each of these factors is considered in the following text but some are much more significant than others. In order of 'effect' the most significant factors are probably ranked as age/size, energy content of diet, protein quality and food intake.

Consideration of protein requirements is generally divided into two parts, namely the quantitative requirements (the absolute levels in the diet) and the qualitative requirements (essential amino acid levels and availability).

Quantitative dietary protein requirements of tilapias

Dietary protein is principally required for the functions of:

- Repair and replacement of worn or wasted tissue, protein 'lost' in such processes being referred to as the endogenous protein loss. The protein intake required to cover these losses is known as the maintenance requirement.
- Provision of an energy substrate (catabolism).
- Synthesis of new body protein (anabolism, growth).

In general, whenever nutritionists refer to the optimum level of any dietary nutrient they are referring to the level which results in maximum growth and this is usually expressed as a proportion (%) of the diet. There are many reservations about this approach, in particular this narrow definition of optimum needs considerable qualification when referring to practical feeds as opposed to theoretical or experimental ones. A number of studies have reported dietary protein optima for tilapias and these are presented in Table 9.1. Most of this work involved dose-response experiments (where the response was growth) of some 6–12 weeks duration.

Fig. 9.1 presents data from Jauncey (1982) and Ross (1982, unpublished) for two size categories of *Oreochromis mossambicus*. Typically, at 0% protein in the diet growth is negative (the endogenous losses being met by catabolism of body tissues) and the fish loses weight. As the protein content increases it reaches a point at which there is no change in fish body weight (zero growth, this is the maintenance requirement for protein). Above this growth increases to a maximum (the optimum for growth) after which it either continues as a plateau or declines slightly (possibly due to energy used to metabolize excess absorbed amino acids) (Jauncey, 1982). In addition, it can be seen that for fish of 0.5–10.0 g a protein level of 24% still produces 80% of the maximum growth rate obtained (at 40% protein) whilst, for this size of fish, even a protein level of 16% produces 64% of the maximum growth rate.

Jackson and Capper (1982) showed that for 1.5–6.0 g *Oreochromis mossambicus* 20% protein resulted in 80% of the growth response recorded with a 40% protein feed. These results suggest that for tilapias the optimum dietary protein level for growth is unlikely to correspond to the optimum economic dietary protein level in terms of feed cost per unit weight gain. Dietary protein level may also be used to manipulate growth rates (as may energy or feed intake) and thus stagger production or hold fish at market size until the selling price is optimized.

Table 9.1 shows the range of dietary protein optima experimentally determined for tilapia. It can be seen that there are reasonably large variations even for fish of the same species and size. Most of these apparent discrepancies are likely to be attributable to limitations in experimental design. The most pronounced effects are likely to be due to differences in feeding husbandry as it is well established that restricting food intake (as in most experiments to date) will elevate the percentage protein in the diet giving maximum growth. In many of the fry studies reported underfeeding would appear to be a problem and authors frequently fail to report survival data. The most important factor affecting dietary protein requirements in tilapia would appear to be age/size. The optimum dietary protein requirements of tilapia can be broadly generalized as shown in Table 9.2. This assumes the absence of any other feed, moderate levels of food and energy intake and a good quality protein.

Protein requirements are likely to be affected by environmental factors, such as temperature and salinity, but there is as yet little or no experimental evidence for this in tilapias. De Silva and Perera (1984, 1985) investigated the effects of salinities of 0, 5, 10 and 15 ppt on the dietary protein requirements of *Oreochromis niloticus* fry and showed that for diets of 30% protein fish grew best at 10 ppt whilst for higher protein diets fish grew best in fresh water. Salinity appeared to have little effect on the level of dietary protein producing maximum growth.

In addition, protein requirements are likely to be affected by reproductive

Table 9.1 The protein requirements of tilapias (for maximum growth and expressed as a percentage of the diet)

Species	Requirement	Size (g)	Protein source	Gross energy[1] $kJ.g^{-1}$	Comments	Reference
Mixed	34%	1–5 g			Review data	De Silva *et al.* (1989)
O. mossambicus	50%	First feeding	Fishmeal		No other data	Jackson (pers. com.)
O. mossambicus	40%	0.5–1.0	Fishmeal	14.77	Fed $6\%.d^{-1}$, FCR 1.5	Jauncey (1982)
O. mossambicus	30–35%	6-30	Fishmeal		Fed $3\%.d^{-1}$, FCR 1.6–1.8	Ross (1982)
O. niloticus	35	0.013–0.26	Fishmeal		Fed $15\%.d^{-1}$, FCR 1.8	Santiago *et al.* (1982)
O. niloticus	45	0.012	Fishmeal (white)	16.7	Satiation FCR 1.1	El-Sayed and Teshima (1992)
O. niloticus	28–30	0.024	Fishmeal	19.22	Fed $6\%.d^{-1}$	De Silva and Perrera (1985)
O. niloticus	40	1.0	Casein	13.86		Kesamaru *et al.* (1982)
O. niloticus	40	0.7-3.0	Casein		Fed $6\%.d^{-1}$, FCR 2.00	Toledo *et al.* (1983)
O. niloticus	30	3.5–10.0	Casein	20.482	Ad lib. FCR 0.85	Wang *et al.* (1985)
O. niloticus	25	9–17	Casein		Ad lib. FCR 0.80	Wang *et al.* (1985)
O. niloticus	40	0.8	Fishmeal	19.10	Fed $10–5\%.d^{-1}$ FCR 1.89,	Siddiqui *et al.* (1988)
O. niloticus	30	40.0	Fishmeal	18.83	Fed $3\%.d^{-1}$ FCR 1.72	Siddiqui *et al.* (1988)
O. niloticus	35	0.56	Casein[2] Gelatine	18.42	Fed $4\%.d^{-1}$ FCR 1.0	Teshima *et al.* (1985)
O. niloticus	19–29				Fed $3\%.d^{-1}$	Wannigama *et al.* (1985)

Table 9.1 (cont)

Species	Requirement	Size (g)	Protein source	Gross energy[1] $kJ.g^{-1}$	Comments	Reference
O. niloticus × aureus	30				Fed 2–2.5%.d^{-1}	Viola and Zohar (1984)
O. niloticus × aureus	24	3–8	Fishmeal (white)	17.39	Fed 3%.d^{-1} FCR 1.42	Shiau and Huang (1989)
O. aureus	56	Fry to 2.5	Casein/ albumin		Fed 20%.d^{-1}, FCR 2.5	Winfree and Stickney (1981)
O. aureus	36	0.3–0.5	Soy or fishmeal	17.522	Levels above 36% not studied	Davis and Stickney (1978)
O. aureus	30	1.0-3.0	Casein		Fed 6%.d^{-1}, FCR 2.3	Toledo *et al.* (1983)
O. aureus	34%	2.5–7.5	Casein/ albumin		Fed 10%.d^{-1}, FCR 2.8	Winfree and Stickney (1981)
T. zillii	35	1.3–3.5	Casein	15.222	Fed 5%.d^{-1}	Mazid *et al.* (1979)
T. zillii	35–40	1.6	Casein	14.632	Fed 4%.d^{-1}	Teshima *et al.* (1978)

[1]Gross energy calculated where possible using the figures of Steffens (1989) of 23.9 $kJ.g^{-1}$ for protein, 39.8 $kJ.g^{-1}$ for lipid and 17.6 $kJ.g^{-1}$ for carbohydrate
[2]Author's own data

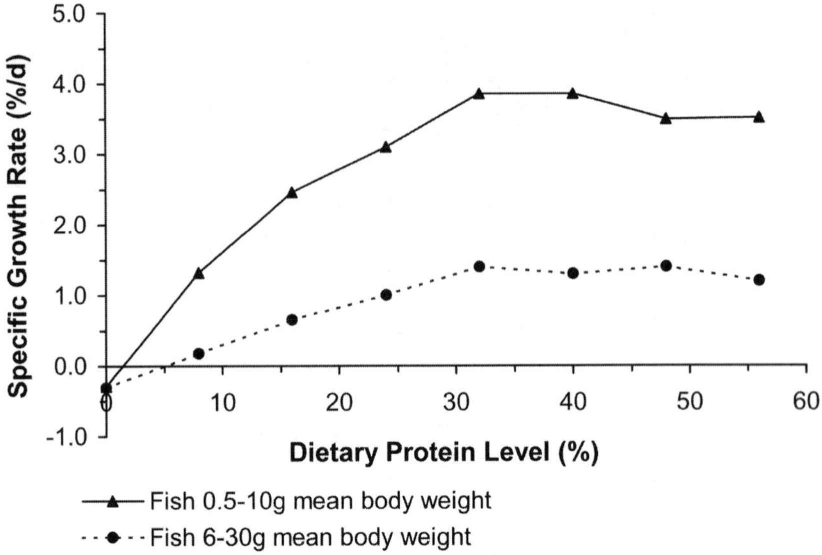

Fig. 9.1 The variation in specific growth rate (SGR) with dietary protein level for two size classes of *O. mossambicus.*

Table 9.2 Approximate dietary protein requirements of tilapia

Approximate wet body weight (g)	Optimum dietary protein content (%)
Fry to 0.5	30–56, recommend 40–45
0.5 to 10.0	30–40, recommend 30–35
10.0 to 30.0	recommend 25–30
30.0 to market size	recommend 25–30

state, especially in females. Broodstock females might be expected to require up to 5% more protein than equivalent sized non-reproducing or male fish but this is largely hypothetical. Cisse (1988) showed that a high protein (50%) diet did not affect spawning frequency or fecundity in tilapias but did promote better growth of males. Chang *et al.* (1988) compared 24% protein pellets and 44% protein eel feed as broodstock tilapia rations and concluded that the latter promoted better broodstock growth and fry prodution. Gunasekera *et al.* (1996, 1997) examined the effects of three isoenergetic diets containing 10, 20 or 35% protein on reproductive performance in *O. niloticus*. Inclusion rates of 10% crude protein impaired reproduction and the 35% diets resulted in eggs with a higher protein content.

The level of dietary protein producing maximum growth is also affected by protein quality and dietary energy level (see later in this chapter).

Luquet (1989) has attempted to re-evaluate the data on tilapia protein requirements with a view to calculating the protein requirement per kilogram of body weight per day and per kilogram of liveweight gain. Per kilogram of body weight per day the protein requirements range from 9.2 to 68 g with the majority of values ranging from 10 to 20 g. Surprisingly, there is no clear relationship between fish size and this requirement. The requirement per kilogram of liveweight gain is more consistent, 264–697 g, with most values in the range 400–600 g.

Qualitative dietary protein requirements

Proteins are complex structures composed of amino acids of which approximately 25 commonly occur in the types of feed proteins likely to be encountered by tilapia. These can be sub-divided into two groups:

- Essential amino acids (EAA) that cannot be synthesized by tilapia, are constituents of essential proteins and must be supplied pre-formed in the diet.
- Non-essential amino acids (NEAA) which can be synthesized by tilapia given suitable starting materials. NEAA are made by inter-conversion of other amino acids as this group of compounds cannot be synthesized *de novo* by animals. NEAA are only non-essential in the dietary context, they still perform many essential metabolic functions.

Qualitative essential amino acid (EAA) requirements

The commonest experimental approach to the identification of essential amino acids has been to use purified amino acid test diets in dose/response experiments. A more novel approach has been use of injected C^{14} labelled glucose with subsequent examination of carcass amino acids for presence of the label.

All fish for which such studies have been performed exhibit the same 10 EAA, namely arginine, histidine, isoleucine, leucine, lysine, methionine, phenylalanine, threonine, tryptophan and valine. The qualitative EAA have been determined for *Tilapia zillii* (Mazid *et al.*, 1978) and it is likely that all fish have the same qualitative EAA requirements.

Quantitative essential amino acid (EAA) requirements

Not only must the essential amino acids be present in the dietary protein, they must also be present in the correct relative proportions in order that *in vivo* protein synthesis can occur. The relative proportions of EAA in a dietary protein are referred to as its amino acid profile. If the EAA profile of the dietary protein matches that of the requirements (largely equivalent to the tissue EAA profile) then that, in general terms, defines a good quality protein. Of course the fact that EAA may be chemically measurable in the dietary protein does not necessarily mean that they will be biologically available and this should also be taken into account when considering protein quality.

Jackson and Capper (1982) determined the quantitative requirements for arginine, lysine and methionine for *O. mossambicus* using EAA deficient diets supplemented with specific purified crystalline amino acids. Their results, along with other relevant data, are shown in Table 9.3.

Jauncey *et al.* (1984) determined the complete EAA requirements of *O. mossambicus* using the daily deposition method (after Ogino, 1980) and the carcass amino acids profiling technique (after Tacon and Cowey, 1985; see Table 9.3). Santiago (1986) defined the quantitative EAA requirements of *O. niloticus* using a basal casein/gelatine diet supplemented with crystalline amino acids. She also compared muscle and egg EAA profiles with these requirements and found a good correlation with the former. Interestingly, Mambrini and Kaushik (1994) recommended that total NEAA should be $\leqslant 60\%$ of the total protein in *O. niloticus* feeds when alanine, glutamic acid or glycine were fed singly and $\leqslant 50\%$ when all three were used in combination.

The requirement for the EAA methionine is affected by the level of the NEAA cystine which appears able to substitute (spare) part of the methionine requirement. The same effect is noted with the EAA phenylalanine and the NEAA tyrosine. In general it is believed that of the total requirement (as determined in the absence of the NEAA), not less than 60% should be EAA and up to 40% can be NEAA.

Several of the EAA are of particular significance when referring to dietary protein sources. Lysine tends to occur at low levels except in feedstuffs of animal origin. In addition, lysine has two amino groups both of which must be free and reactive for it to have any biological value. The second ε amino group is very reactive and frequently becomes 'bound' during processing so that chemical measurements of lysine should be of available rather than total lysine. As an example of the importance of lysine Viola *et al.* (1992) found, with common carp, that 0.5% dietary lysine supplement was nutritionally equivalent, and economically superior, to 5% crude protein in the feed.

Methionine is relatively easily oxidized, especially during processing, to form the sulphoxide or sulphone. The biological values of these have yet to be established in fish. Methionine also tends to be in short supply in plant proteins but it is spared by the NEAA cystine. Phenylalanine levels in plant proteins also tend to be low but this EAA is relatively stable and is spared by the NEAA tyrosine.

EAA deficiency in the diet leads to poor protein utilization, growth and food conversion efficiency as well as loss of appetite. In some species of fish tryptophan deficiency results in scoliosis (dorso-ventral curvature of the spine)

Table 9.3 The quantitative essential amino acid requirements of freshwater fish[1]

	Trout[2]	Trout[3]	Carp[4]	Carp[3]	Catfish[5]	Tilapia[6]	Tilapia[7]	Tilapia[8]
Arginine	3.3–5.9	3.5	3.3	3.8	4.3	4.2	2.8	0.9
Histidine		1.6	2.1	1.4	1.5	1.7	1.1	0.9
Isoleucine		2.4	2.5	2.3	2.6	3.1	2.0	0.97
Leucine	9.1	4.4	3.3	4.1	3.5	3.4	3.4	1.14
Lysine	3.7–6.1	5.3	5.7	5.3	5.1	5.1	3.8	1.17
Methionine	2.2–2.9	1.8	2.1	1.6	2.3	2.7	1.0	0.6
Phenylalanine	2.0[9]	3.1	3.4	2.9	5.0	3.8	2.5	0.7
Threonine		3.4	3.9	3.3	2.0	3.8	2.9	0.73
Tryptophan	0.5–1.4	0.5	0.8	0.6	0.5	1.00	0.4	0.53
Valine		3.1	3.6	2.9	3.0	2.80	2.20	0.77

[1]All values as % of dietary protein
[2]Experimentally determined data for *Oncorhynchus mykiss* from the review of Cho and Cowey (1991)
[3]Estimated for *Cyprinus carpio* and *O .mykiss* from tissue deposition (Ogino, 1980a)
[4]Experimentally determined data for *Cyprinus carpio* from review of Tacon (1987)
[5]Experimentally determined data for *Ictalurus punctatus* from review of Wilson (1991)
[6]Experimentally determined data for *Oreochromis niloticus* from Santiago (1986)
[7]Estimated for *O. mossambicus* from carcass composition and tissue deposition (Jauncey *et al.*, 1984)
[8]Estimated for *O .mossambicus* from carcass deposition and tissue deposition (Gaber, 1994)
[9]From Kim (1993), total phenylalanine + tyrosine is 4.3% of protein
The EAA requirements of *Catla catla* fry have also been determined experimentally (Ravi and Deveraj, 1991):. arginine, 4.80%; histidine, 2.45%; isoleucine, 2.35%; leucine, 3.70%; lysine, 6.23%; methionine (no cystine), 3.55%; phenylalanine (with 1% tyrosine), 3.70%; threonine, 4.95%; tryptophan, 0.95%; valine, 3.55%
In addition to the above, Odum and Ejike (1991) report arginine and lysine requirements of 1.48% and 1.76% respectively for *O. niloticus* as a percentage of the diet

and both tryptophan and methionine deficiency may result in cataracts, although none of these conditions have been recorded for tilapias (Tacon, 1985).

Essential amino acid supplementation

It is possible to improve the quality of EAA deficient diets by supplementation with a purified form of the EAA in deficit. Methionine and lysine are quite commonly used as feed supplements in terrestrial animal feeds to improve protein quality and they are readily commercially available. EAA supplementation of fish feeds could be used to raise amino acids levels to those optimal for the target species.

Jackson and Capper (1982) assert that free EAA are well utilized by *O. mossambicus* although Jauncey *et al.* (1984) reported considerable difficulty in using purified amino acid test diets with this species. Viola *et al.* (1993) found no improvement in performance of pond-fed tilapia hybrids (*O. niloticus* × *O. aureus*) when 0.5% lysine hydrochloride was added to the feed. Teshima and Kanzawa (1988) and Teshima *et al.* (1990) found no improvement in performance of *O. niloticus* fed a soybean diet supplemented with crystalline methionine unless in the form of a methionine-enriched soybean plastein. Obviously the question of EAA supplementation of tilapia feeds requires further study and careful economic analysis to determine whether any improvements in growth justify the additional feed cost. Amino acid supplements are also unlikely to be available in many tilapia producing areas.

Yamada *et al.* (1982) report that *O. niloticus* absorbed free amino acids more quickly (plasma peak values 2 hours after the meal) than amino acids from a casein based diet (plasma peak values 4 hours after the meal) and that plasma amino acid concentrations were sustained in tilapia fed protein (24 hours to reach fasting levels) compared to those fed free amino acids (4–8 hours to reach fasting levels). This suggests that amino acid supplementation may not work as expected because all the amino acids required for protein synthesis must be present at the same time.

Protein digestibility

A further aspect of protein quality that must be addressed is digestibility. This also applies to other nutrients, although in feed formulation protein and energy digestibility are of the greatest significance. Digestible protein refers to the proportion (%) of the ingested protein that is hydrolysed in the gastro-intestinal tract and subsequently absorbed. In practical terms it is usually measured as the difference between the protein ingested and the protein voided and because it does not allow for any endogenous losses of protein in the faeces it is more appropriately referred to as the 'apparent digestibility'.

Most of the studies referred to in this section on protein and EAA requirements employed very good quality protein sources where the digestibility is high. Table

9.4 gives some values for protein digestibility in tilapias derived from a variety of sources. In general, animal proteins are more digestible that plant ones but it must be remembered that the types of processing to which these materials have been subjected will have had an enormous effect on the values. Any digestibility data can only be indicative because no two batches of materials are identical and because of great differences in the methodology used to determine digestibility by different authors. It is practically impossible to obtain absolute values for digestibility because collection of true faeces in the aquatic environment is so difficult.

Table 9.4 Apparent crude protein digestibilities of selected feedstuffs in tilapias

Product	Apparent crude protein digestibility (%)	Reference
Alfalfa meal	66	Popma (1982) from NRC (1993)
Azolla	75	Lorico-Querijero and Chiu (1989)
Brewers' grain	62	Luquet (1989)
Brewers' grain	63	Hanley (1987)
Casein	97.2	Lorico-Querijero and Chiu (1989)
Copra meal	56	Lorico-Querijero and Chiu (1989)
Copra meal	99.1	Kamarudin *et al.* (1989)
Corn (maize) grain	84	Popma (1982) from NRC (1993)
Corn (maize) ground	83	Hanley (1987)
Corn (maize) cooked	79	Popma (1982) from NRC (1993)
Corn gluten meal	97	Lorico-Querijero and Chiu (1989)
Corn meal	89.4	Kamarudin *et al.* (1989)
Cottonseed meal	31	Luquet (1989)
Fish meal	72	Luquet (1989)
Fish meal	86	Hanley (1987)
Fish meal	92	Lorico-Querijero and Chiu (1989)
Fish (menhaden) meal	85	Popma (1982) from NRC (1993)
Fish meal	99.5	Kamarudin *et al.* (1989)
Groundnut meal	79	Luquet (1989)
Poultry offal meal	74	Hanley (1987)
Rice bran	93	Lorico-Querijero and Chiu (1989)
Rice bran	99.9	Kamarudin *et al.* (1989)
Shrimp meal	87	Lorico-Querijero and Chiu (1989)
Shrimp meal	99.4	Kamarudin *et al.* (1989)
Soybean meal	91	Hanley (1987)
Soybean meal	93	Lorico-Querijero and Chiu (1989)
Soybean meal	94	Popma (1982) from NRC (1993)
Soybean meal	91.6	Kamarudin *et al.* (1989)
Wheat bran	20	Luquet (1989)
Wheat grain	90	Popma (1982) from NRC (1993)
Wheat middlings	75	Hanley (1987)

9.3 LIPID

Lipids are an extremely diverse group of compounds that function as important sources of metabolic energy, as essential components of animal cell membranes, as carriers for fat soluble vitamins, as precursors of prostaglandins and steroids as well as having physical properties affecting buoyancy and acting as mechanical cushions/supports for various organs. Of the various types of lipid it is the simple glycerol-based fats and oils that are of most interest in terms of tilapia nutrition.

Dietary lipid has two principle functions:

- as a source of metabolic energy,
- as a source of essential fatty acids (EFA).

These are dealt with separately in the following sections and additional information on the lipid requirements of warm water fish, including tilapias, can be found in the excellent review of Stickney and Hardy (1989).

Lipid as an energy source

One gram of lipid contains almost twice as much total energy as either one gram of carbohydrate or one gram of protein. The high energy density of lipid has prompted numerous studies of the 'protein sparing' capacity of dietary lipid in finfish feeds. In addition the natural diets of many fish, particularly carnivores, are very high in lipid. There is, however, relatively little information on this aspect of the nutrition of tilapias. Most currently available commercial tilapia feeds contain 10% or less lipid, not for nutritional reasons but because higher levels can cause difficulty with conventional feed pelleting techniques.

The first published data on the effect of varying dietary lipid levels on tilapias appear to be those of Winfree and Stickney (1981) who fed *Oreochromis aureus* with diets varying in lipid level from 2 to 8.6% and with protein:energy ratios (P:E) ranging from 19.6 to 30.9 (mg protein kJ^{-1}). It was concluded that for fish up to 2.5 g the optimum P:E ratio was 29.9 (mg protein kJ^{-1}) (56% protein, 5.2% lipid, 19.2 kJ g^{-1}) decreasing to 25.8 (mg protein kJ^{-1}) (34% protein, 4.4% lipid, 13.4 kJ g^{-1}) for fish up to 7.5 g.

Jauncey and Ross (unpublished data) have shown that it is possible to reduce the protein content of diets for 10 to 40 g *O. aureus* × *O. niloticus* hybrids from 40 to 30% with little reduction in growth performance, and with improved protein utilization, provided that diets contained 12% lipid. This corresponds to reducing the P:E ratio from 23.9 to 19.9 mg protein kJ^{-1}. Dietary lipid levels in excess of 12% depressed growth and caused significant accumulation of carcass lipid.

Dietary lipid will not be totally available as an energy source as it will not be completely digested and absorbed. Wang *et al.* (1985b) reported that in diets for

O. niloticus corn oil was >90% digestible at 2–15% dietary inclusion levels, suggesting a digestible energy (DE) of 37.6 kJ g^{-1}. In this study the DE of the diet, as measured experimentally, was very similar to that predicted using DE values of 18.8 kJ g^{-1} for protein (casein), 16.7 kJ g^{-1} for carbohydrate (dextrin and starch) and 37.6 kJ g^{-1} for lipid (corn oil). It was also concluded that the optimum P:E ratio was 16–17 mg protein kJ^{-1} (30% protein, 8–12% lipid, 18.8 to 19.6 kJ g^{-1}) for fish of 6 g.

There have been a number of studies on the efficacy of various sources of dietary lipid for tilapias. Takeuchi *et al.* (1983) fed corn oil, soybean oil, pollack liver oil and beef tallow to *O. niloticus*. Corn oil and soybean oil were found to be superior to pollack oil while the performance of fish fed beef tallow oil was very poor. Stickney and McGeachin (1984a) confirm the poor performance of beef tallow in *O. aureus*. In general tilapia do not seem to be able to utilize high levels of dietary lipid as effectively as salmonids or carp (Jauncey, 1979). Viola and Amidan (1980) reported that the addition of 5% lipid (poultry oil) to diets of *O. aureus* × *O. niloticus* hybrids led to significant accumulation of visceral lipid (much higher than in carp under the same conditions).

Stickney and Wurts (1986) examined the growth responses of *O. aureus* fed menhaden and catfish oils. Menhaden oil at 10% (the highest level tested) gave the best response. The authors speculate that 18:3n-3 levels should not exceed 1% in tilapia feeds but that this effect only applies to that specific fatty acid and not to the additive effects of total n-3 levels.

In contrast to the previous studies, De Silva *et al.* (1991), working with 'red' *O. mossambicus* × *O. niloticus*, found up to 18% lipid (range 6–24%) spared protein and a 30% protein, 18% lipid diet was recommended. However, the authors fail to state the source of lipid that they used (it is merely referred to as oil) so that the effects of high linolenic acid levels cannot be related to other studies. Chou and Shiau (1996) examined the effect of five isoproteic and isoenergetic diets containing 0–20% lipid on *O. niloticus* × *O. aureus* hybrids and report an optimal lipid level of 12%. Fitzsimmons *et al.* (1997) fed isoenergetic (11.9 MJ kg^{-1}), isoproteic (30%) feeds containing 3, 6 and 8% lipid to 68–250 g *O. mossambicus*' × *O. aureus* hybrids. There were no detectable effects of varying the energy source between lipid and carbohydrate on fish performance, feed performance or carcass lipid.

To maximize protein utilization, lipid levels between 6 and 12% would appear optimal in tilapia feeds. It is tentatively suggested that dietary lipid levels decrease from 8 to 12% in diets for tilapia up to 25 g to 6 to 8% in diets for larger fish. In ponds fed supplementary feeds Hanley (1991) could find no differences between feeds containing 5, 9 or 12% lipid fed to *O. niloticus* and concluded that protein rather than energy was of greater significance in such feeds.

Lipids as a source of essential fatty acids (EFA)

Just as there are certain amino acids indispensable for growth, that cannot be synthesized by fish and must be supplied pre-formed in the diet, so there are fatty acids that fall into this 'essential' category. These are termed essential fatty acids (EFA). With the exception of the land snail (*Cepaea nemoralis*) animals are incapable of *de novo* synthesis of fatty acids with double bonds in the n-6 (linoleic) or n-3 (linolenic) series (Tacon, 1987). These fatty acids must therefore be supplied pre-formed in the diet.

For land animals fatty acids of the n-6 series have been found to have the highest EFA activity and the polyunsaturated fatty acids (PUFA) of carcass lipids of terrestrial animals are rich in n-6 fatty acids. In contrast the PUFA in carcasses of fish tend to be of the n-3 series (although freshwater fish are richer in n-6 PUFA than marine species).

Most fish have demonstrated a requirement for EFA of the n-3 series; however, studies by Kanazawa *et al.* (1980) showed an EFA requirement for 1% n-6 fatty acid and no n-3 requirement in the diet of *T. zillii*. While the EFA requirements of tilapias under a range of environmental conditions have yet to be determined, current findings are summarized in Table 9.5.

Table 9.5 The essential fatty acid requirements of tilapias

Species	Requirement	Reference
Tilapia zillii	1% 18:2n-6 or 1% 20:4n-6	Kanazawa *et al.* (1980)
Oreochromis niloticus	0.5–1% 18:2n-6 or 1% 20:4n-6	Teshima *et al.* (1982)
O. niloticus	0.5% 18:2n-6	Takeuchi *et al.* (1983)
Oreochromis aureus	18:2n-6 or 18:3n-3 $\leqslant$1%	Stickney and McGeachin (1983)

EFA deficiency results in reduced growth, inappetance, fin erosion (especially the caudal fin), fatty livers and erratic and pronounced responses to shock (Tacon, 1985).

Stickney and Wurts (1986) suggest that dietary levels of >1% linolenic acid, although not those of longer chain n-3 fatty acids, depress growth of *O. aureus*. In contrast, it is also suggested that levels of linoleic acid $\leqslant$2.5% of the diet are acceptable (Stickney and McGeachin, 1983).

Olsen *et al.* (1990) have shown that *O. niloticus* can chain elongate and desaturate both linoleic and linolenic acids but that conversion is suppressed by dietary longer chain fatty acids of these series. Santiago and Reyes (1993) found that n-6 fatty acids enhanced spawning success and fry production in *O. niloticus* and that n-3 fatty acids increased weight gain but reduced reproductive performance.

Rancidity (oxidation) of dietary lipid

The requirement for PUFA in tilapia feeds means that these must be adequately protected from oxidation. PUFA are subject to oxidation (rancidity) of the double bonds they contain on exposure to air, heat, light, moisture, etc. The products of oxidation of PUFA (peroxides, hydroperoxides, aldehydes (alkanals) and ketones (alkanones)) are themselves toxic and react with other essential dietary nutrients reducing their biological availability. They also impart an 'off-flavour' to both feeds and to fish that consume them.

A whole variety of pathological signs of lipid rancidity has been reported in tilapias (Soliman *et al.*, 1984). These include haemorrhage around the snout and the bases of the fins, lordosis, exophthalmia (popeye), oedema (abdominal swelling), cataract, darkening of the liver, steatitis of all abdominal fat and mortality.

Other, non-triglyceride, dietary lipids

There has been much speculation about the possible requirements of farmed species (some fish and crustacea) for phospholipids. These are the second largest component of body lipid after triglycerides and there is some evidence that they are beneficial in the diets of penaeid shrimp and some marine fish – especially at the larval stages (Tacon, 1987). There is no suggestion of any dietary benefit in tilapias. The only other dietary lipid group of interest is the sterols and there is no dietary requirement or benefit to be derived from these in fish. In contrast crustacea cannot synthesize sterols *de novo* and must have them supplied preformed in the diet (Tacon, 1987).

9.4 CARBOHYDRATE

Carbohydrates are the third most abundant group of organic compounds in the animal body (Tacon, 1987) and in contrast to the proteins and lipids there are no absolute dietary requirements for carbohydrate. Fish, including tilapias, can obtain the glucose they require by synthesis (gluco-neogenesis) from amino acid and fatty acid precursors. Overall carbohydrate metabolism in fish is considered in an excellent review by Cowey and Walton (1989) and the carbohydrate utilization of tilapias by Shiau (1997).

Carbohydrates as a source of energy

Although carbohydrates are 'non-essential' dietary nutrients for tilapias they should still be included in the feed for a number of reasons:

- they are a cheap source of dietary energy (in terms of cost per kJ),
- they can, to some extent, 'spare' dietary protein for growth,

- they improve the pelletability and water stability of feeds,
- they are essential as 'bulking' agents to make formulations add up to 100%.

Despite the fact that carbohydrates supply less energy per gram than either dietary protein or lipid they usually represent the cheapest form of dietary energy in terms of cost per kJ. Cereals contain 60% or more carbohydrate (mostly in the form of starches) and oilseed meals 40% or more and these are often among the cheapest materials for tilapia feeds.

Although glycogen (the storage form of carbohydrate) is the major energy source (through glycolysis) in fish white muscle during peak activity, fish liver and muscle tissues store little glycogen (less than 1% of wet tissue) (Cowey and Sargent, 1979). Fish do not mobilize liver glycogen stores during starvation. In fact it has been frequently stated that carbohydrate metabolism in fish resembles that of diabetic terrestrial animals. In general terms this means that the value of dietary carbohydrate as an energy source in fish feeds is limited. For carnivorous fish species dietary levels of digestible carbohydrates must be restricted to prevent pathological effects.

Relatively few data exist on the nutritional value of dietary carbohydrates for tilapias. Initially, recommendations for carbohydrate levels in tilapia feeds were based on those for channel catfish on the grounds that there was, at least superficially, a resemblance between their nutritional requirements. For channel catfish Page and Andrews (1973) found 25% corn starch to be equivalent to 12% lipid as a dietary energy source with the starch being 63% digestible and the lipid 75% digestible (digestible energies of 10.9 kJ g^{-1} and 28.4 kJ g^{-1} for starch and lipid respectively). It was thus assumed that 2.25 g starch was equivalent in energy terms to 1 g lipid.

Cowey and Sargent (1979) have suggested a maximum of 25% digestible carbohydrate (equivalent to 40% corn starch if 63% digestible) for most finfish species. Cisneros *et al.* (1984) varied the dextrin levels (15–60%) in a diet with a fixed protein level (30% casein) and found that an inclusion level of 30% gave the best performance. Wang *et al.* (1985b) reported that crude starch was well utilized by *O. niloticus* at levels between 30 and 70% of the diet. Teshima *et al.* (1985) found that levels of 30 or 40% digestible carbohydrate were well utilized by *O. niloticus*. Anderson *et al.* (1984) also reported that a variety of carbohydrates were well utilized by this species at levels up to 40% in the diet. Interestingly, Tung and Shiau (1991) reported that feeding frequency affected the utilization of dietary carbohydrates by *O. niloticus* x *O. aureus* hybrids. As feeding frequency increased from 2–6 times per day, so did carbohydrate utilization – especially of glucose although this was still much lower than for fish fed either starch or dextrin. Tung and Shiau (1992, 1993) also demonstrated in *O. niloticus* × *O. aureus* hybrids that larger fish utilized carbohydrate better than smaller ones.

Shiau and Peng (1993) followed up earlier studies on protein-sparing by dietary carbohydrate in *O. niloticus* × *O. aureus* hybrids, investigating three protein levels, three carbohydrate levels and three carbohydrate sources. The study seems merely to confirm those of others that carbohydrate does spare protein in tilapias, that glucose is inferior to starch and dextrin as a carbohydrate source and that even the highest carbohydrate level tested (41%) was well utilized.

Shiau and Lin (1993), Shiau and Chen (1993) and Shiau and Shy (1998) suggest that dietary chromium as Cr_2O_3 can improve glucose utilization in tilapias. Initially a level of 2 mg kg^{-1} chromium was recommended but the later work established a requirement of 204 mg kg^{-1} Cr_2O_3, equivalent to 139.6 mg Cr kg^{-1}.

Shimeno *et al.* (1993) investigated at the effects of carbohydrate:lipid ratios on intermediary metabolism in *O. niloticus*. They suggest that as dietary carbohydrate levels increase, and dietary carbohydrate:lipid levels also increase, it accelerates glycolysis and lipogenesis and decelerates gluconeogenesis and amino acid degradation in the liver. This would seem to support the protein-sparing action of dietary carbohydrates and suggest that they are valuable sources of energy in this species.

Chuang and Shiau (1993) and Shiau and Chuang (1995) studied the effect of five carbohydrates on intestinal disaccharidase activity in *O. niloticus* × *O. aureus* hybrids. Disaccharidase activity was unaffected by carbohydrate source and performance was starch fed > disaccharide fed > glucose fed.

Carbohydrate digestibility

The carbohydrate content of feedstuffs can be broadly divided into digestible carbohydrate and fibre. The digestible portion is usually assumed to be principally starches with an average digestibility of 60% (giving a digestible energy (DE) of 10.5 kJ g^{-1}). The indigestible portion, fibre, is composed of complex polysaccharides which, in the case of plant material, are principally celluloses.

Fish do not produce endogenous enzymes capable of hydrolysing fibre and thus are unlikely to utilize it as an energy source to any significant extent. However, cellulase activity has been detected in the gut of a number of fish species as a result of the activity of intestinal microflora (bacteria) (Stickney and Shumway, 1974). To all intents and purposes this activity is likely to be very slight and fibre can thus be assumed to be indigestible in fish.

High levels of dietary fibre have been shown to reduce growth in other finfish species (Leary and Lovell, 1975) and are not recommended. Low levels of fibre, however, may improve the pelletability of feeds. For tilapias dietary fibre levels >5% reduce food utilization and digestibility and levels >10% reduce protein utilization (Anderson *et al.*, 1984). In low protein diets high fibre levels reduce food intake (Wang *et al.*, 1985a). Dioundick and Stom (1990) tested a range (0–10%) of fibre levels in feeds of *O. mossambicus* and found best feed performance and growth with diets containing 2.5–5% fibre. Shiau and Kwok (1989) showed that a range of complex polysaccharides (cellulose, carboxymethyl cellulose, agar, carageenan, guar gum) all resulted in depressed tilapia growth. Teshima *et al.* (1987) reported that growth and protein utilization decreased with increasing cellulose level in *O. niloticus*. Dyundik *et al.* (1991) detected cellulase activity in the digestive tract of young *O. mossambicus* and recommended a dietary concentration of 2.5–5.0% cellulose.

Shiau *et al.* (1988) reported that dietary cellulose levels >2% depressed growth in tilapia and Yong *et al.* (1989) recommended <4.5% cellulose as higher levels resulted in reduced feed intake in *O. niloticus*. Shiau *et al.* (1989) showed that dextrin absorption from the intestine of tilapia was lowered in the presence of fibre.

Carbohydrate digestibility in tilapias varies with the nature of the carbohydrate, its level in the feed, the dietary fibre level and possibly other as yet undetermined factors. Barash *et al.* (1984) report, for *O. niloticus* × *T. honorum* hybrids, that at 85% of the diet potato starch was 50–56% digestible whilst wheat flour starch at 35% of the diet was 75–79% digestible. Wang *et al.* (1985a) report that crude starch digestibility in *O. niloticus* is 86–97% with dietary levels from 30 to 70%. These authors also report a slight depression of starch digestibility with high dietary fibre levels (14.5%). Nagase (1964) reports that dietary carbohydrate levels are positively correlated with digestive amylase (the principal carbohydrase) activity in *O. mossambicus*. Degani and Revach (1991) conclude that tilapias digest carbohydrate better than African catfish (*Clarias* sp.).

9.5 PROTEIN AND ENERGY RELATIONSHIPS

There are three classes of dietary compounds which can provide substrates for energy production:

- protein – total energy content 23.6 kJ g^{-1} (5.64 kcal g^{-1})
- lipid – total energy content 39.5 kJ g^{-1} (9.44 kcal g^{-1})
- carbohydrate – total energy content 17.2 kJ g^{-1} (4.11 kcal g^{-1})

These figures are averages for a range of types of each class of compounds (Jobling, 1983).

One reason proposed for the high dietary protein requirements of fish, compared to terrestrial animals, is that fish preferentially catabolize protein as a source of energy rather than depositing it as tissue. In terms of cost per kilojoule of energy supplied, protein is the most expensive energy source. It must therefore be a goal in tilapia nutrition to maximize the use of protein for growth (anabolism), thus minimizing its use for energy (catabolism), by supplying adequate amounts of alternative dietary energy sources. In fact, recent re-evaluation of this concept shows that fish do not really have higher dietary protein requirements than terrestrial monogastric animals in terms of the absolute quantity of protein required to produce unit weight gain. The apparent difference arises because fish are poikilotherms and have much lower requirements for energy and this increases the ratio between protein and energy in fish feeds. Thus, the high dietary protein requirements of fish are an artefact of their expression as a proportion of the diet.

Reduction of the dietary protein requirements for maximum growth, by increasing the level of dietary non-protein energy, is termed 'protein-sparing'. The relationship between the protein and energy levels in a feed is usually expressed as the protein:energy ratio (P:E) (mg protein kJ^{-1} energy). Occasionally the reverse is used, energy:protein ratio (E:P) (kJ mg^{-1} protein).

Table 9.6 summarizes some of the relevant information with respect to optimizing the protein:energy ratio in tilapia feeds.

According to Wang *et al.* (1985a) starch at 30–70% in tilapia feeds is 86–97% digestible and according to Anderson *et al.* (1984) and Teshima *et al.* (1985) tilapia feeds should contain up to 40% digestible carbohydrate. If we assume that 'natural' (i.e. inherent in the feedstuffs) starch is around 75% digestible and that the carbohydrate portion of practical feeds (excluding fibre) is equivalent to starch then this gives us a maximum of 53% carbohydrate in the feed.

Lipid, as mentioned earlier, is limited to 12% of the feed or less. Based on this and the P:E ratios given above it is possible to estimate optimal protein:lipid: carbohydrate levels in the feed, except that it is also necessary to consider the levels of digestible energy in the feed. El-Sayed and Garling (1988) examined carbohydrate to lipid ratios in *T. zillii* feeds and concluded that as long as EFA requirements were met these energy sources can be substituted for one another based on their physiological fuel values at a ratio of 2.25:1 (carbohydrate to lipid). Teshima *et al.* (1985) recommend 30–40% protein, 12–15% lipids and 30–40% digestible carbohydrate for *O. niloticus*. Li *et al.* (1991) concluded that 18–22% of dietary protein in tilapia feeds can be spared by supplementation with lipids and carbohydrates.

Meyer-Burgdorff *et al.* (1989) studied energy metabolism in *O. niloticus* and concluded that increasing the feeding rate, and thus energy intake, resulted in a decline in the availability of gross energy. They concluded that tilapia should not be fed more than 400 kJ ME kg^{-1} d^{-1}. Kaushik *et al.* (1995) determined protein and energy needs in juvenile *O. niloticus*. Protein intake for maximum carcass nitrogen gain was 12 g kg^{-1} d^{-1} and maintenance was 2 g kg^{-1} d^{-1}. Optimal protein:energy ratio was 18 mg kJ^{-1}.

Anderson *et al.* (1991) investigated the measurement and prediction of digestible energy values of a range of feedstuffs for *O. niloticus*. The authors reported that the level of a component in test diets for digestibility determination had little or no effect on DE values. However, they also reported that soybean DE was significantly higher after a 15 week adaptation period and that 'typical' 28 day digestibility studies might underestimate DE in this species. The DE values for 16 feedstuffs were determined and are presented in Table 9.7. In general the DE values for plant material were higher than reported for either trout or

Table 9.6 Protein:energy ratios for tilapia

Species	Size (g)	P:E (mgP.kJ^{-1}	Other feed data	Reference
O. aureus	2.5	29.9	56% CP, 5.2% EE, 19.2kJ.g^{-1}	Winfree and Stickney (1981)
O. aureus	7.5	25.8	34% CP, 4.4% EE, 13.4 kJ.g^{-1}	Winfree and Stickney (1981)
O. mossambicus	10–40	19.9	30% CP, 12% EE, 18.9 kJ.g^{-1}	Jauncey and Ross (1982) unpub.
O. niloticus	6	16–17	30% CP, 8–12% EE, 19.6 kJ.g^{-1}	Wang *et al.*(1985b)
O. niloticus	14	17.9	30% CP, 16.7 kJ.g^{-1}	Yong *et al.* (1989)
T. zillii	1	21	32% CP, 5% EE, 15.1 kJ.g^{-1}	Akand (1987)
S. melanotheron	7.5	17	30% CP	Cisse (1996)
O. niloticus × *O. aureus*	1.6	25.0	24% CP, 8.9% EE, 10.8 kJ.g^{-1}	Shiau and Huang (1990)
O. niloticus × *O. aureus*	1.6	16.2	21% CP, 15% EE, 14.6 kJ.g^{-1}	Shiau and Huang (1990)
O. niloticus × *O. aureus*	0.16	26.5	40% CP. 15.1 kJ.g^{-1}	Santiago and Lovell (1991)
O. niloticus	12 mg	26.3	45% CP, 7% EE, 16.7 kJ.g^{-1}	El-Sayed and Teshima (1992)

CP = crude protein; EE = ether extract, all energy contents on the basis of gross energy because of lack of digestibility data, protein is also as total, rather than digestible, crude protein channel catfish whilst those for animal materials were lower. These authors also found that DE could be predicted from chemical analyses with a reasonable degree of certainty, obviating the need for costly and technically difficult direct measurements. The predictive equations that they recommended are:

- DE (MJ kg^{-1} dry matter) = 16.88 – 0.021 NDF (neutral detergent fibre) (r^2 0.87)

  or

- DE (MJ kg^{-1} dry matter) = 0.034 CP + 0.022 AC – 3.5 (r^2 0.87)

(where CP is crude protein and AC is available carbohydrate measured as glucose released from starch by amyloglucosidase).

Table 9.7 Digestible energy contents of a range of feedstuffs for *O. niloticus*

Feedstuff	DE (MJ kg^{-1} dry matter)
Cassava[1]	13.4
Copra cake[1]	6.5
Copra[2]	15.6
Fish meal[1]	16.2
Fish meal[2]	13.9
Groundnut cake[1]	17.9
Maize[1]	13.1
Maize[2] (corn)	14.5
Meat and bone meal[1]	9.1
Palm kernel meal[1]	1.7
Poultry by-product meal[1]	10.0
Rapeseed meal[1]	10.6
Rice bran[1]	5.0
Rice bran[2]	17.9
Shrimp meal[2]	10.4
Soybean meal[1]	14.4
Soybean meal[2]	14.6
Sorghum[1]	12.4
Sunflower seed meal[1]	3.6
Wheat[1]	13.3
Wheat bran[1]	6.1
Wheat middlings[1]	10.4

[1]Anderson *et al.* (1991)
[2]Kamarudin *et al.* (1989) (recalculated)

There is a trend towards calculating all nutrient requirements relative to the digestible energy intake of fish feeds, especially for carnivores such as trout and salmon. For example, if the optimal protein:energy ratio is known the 'nutrient density' of the feed can be changed by varying the concentrations of these two nutrients whilst retaining their correct relative proportions. The consequence of this, assuming that fish feed (as seems likely) to a set energy intake, is that as the nutrient density increases the feeding rate would decrease so that intakes of protein and energy remained constant.

Why is this of any significance? Where the environmental impact of aquaculture is a problem this approach has obvious benefits. The impact, in terms of nutrient loading, of aquaculture is largely due to uneaten or unretained food and nutrients – particularly suspended solids, nitrogen and phosphorus outputs. In high nutrient density feeds, protein:energy ratios are optimized thus reducing nitrogen output as ammonia. In addition, feeding rates are lower and more controlled leading to less food wastage and such feeds utilize high quality ingredients of high digestibility, thus reducing faecal losses. Finally, it is possible to manipulate such feeds to reduce the levels of total phosphorus and increase its availability, reducing phosphorus loading.

With respect to tilapias the approach is likely to be more of academic than practical interest as tilapias are rarely cultured in areas where such environmental considerations are a serious problem. In addition, tilapias would generally seem poorly adapted to high nutrient density feeds (they are not obligatory carnivores) and economic considerations are, as yet, unlikely to justify the inevitably high cost feeds. An exception might be in the use of intensive feeds for tank based hatcheries where water quality/treatment becomes a significant economic factor. The same would apply to farms practising intensive tank-based culture of tilapias in either open flow or recirculated water systems.

9.6 VITAMINS

Vitamins are dietary nutrients that are essential at relatively low dietary inclusion levels and are incapable of supplying dietary energy. The vitamin requirements of finfish have been reviewed in some detail by Halver (1979) and Tacon (1987, 1991).

Vitamins are an extremely variable group of organic compounds that are essential for normal metabolism. Vitamins are generally 'essential' in the same sense as EAA or EFA in that they either cannot be synthesized by a particular species or they are not synthesized fast enough to meet requirements. Approximately 15 vitamins have been identified and they form a group of chemically unrelated organic compounds required in trace amounts (Table 9.8).

Vitamins are generally subdivided into water soluble (sometimes subdivided into B vitamins and others, or possibly into macrovitamins, on the basis of the

Table 9.8 The vitamins

Water soluble vitamins	Fat soluble vitamins
B-complex vitamins	
Vitamin B_1 (thiamine)	Vitamin A (retinol)
Vitamin B_2 (riboflavin)	Vitamin D (D_3 – cholecalciferol)
Vitamin B_6 (pyridoxine)	Vitamin E (tocopherol)
Pantothenic acid	Vitamin K (K_3 menadione)
Nicotinic acid (niacin)	
Biotin	
Folic acid	
Vitamin B_{12}	
Macrovitamins	
Inositol (myo-inositol)	
Choline	
Vitamin C (ascorbic acid)	

levels required) and fat soluble groups. Water soluble vitamins are not stored to any significant extent in animal tissues and turnover is rapid. Fat soluble vitamins, in contrast, are absorbed along with dietary fat and are stored in the fatty tissues of animals. Because of this difference, deficiency of water soluble vitamins is more common (a regular dietary intake is required to avoid deficiency) whereas toxicity of these compounds is rare (as they cannot be stored). The situation with fat soluble vitamins is quite the reverse, deficiency is rare (because they are stored) and toxicity, if fed to excess, is more likely.

Water soluble vitamins

Vitamin B_1 (thiamine)

Thiamine is essential in the metabolism of carbohydrates as the coenzyme co-carboxylase (thiamine pyrophosphate) which is required for the oxidative decarboxylation of pyruvic acid and α-ketoglutaric acid, as well as being an activator of the enzyme transketolase. The thiamine requirements of tilapia have not been reported although dietary thiamine did reduce lead intoxication in *Tilapia zillii* (Ghazaly, 1991).

Vitamin B_2 (riboflavin)

Riboflavin plays an essential role in energy metabolism in the form of flavin mononucleotide (FMN) and flavin adenine dinucleotide (FAD), coenzymes associated with the energy yielding breakdown of pyruvic acid, fatty acids and amino acids. The riboflavin requirements of *O. aureus* fingerlings (0.7 g) have been investigated (Soliman and Wilson, 1992a). Fish fed a riboflavin free diet exhibited lethargy, fin erosion, anorexia, loss of colour, short body dwarfism and cataracts. Based on feed efficiency data and levels of hepatic d-amino acid oxidase these authors recommend a level of 6 mg kg^{-1}diet. For 'red' *O. mossambicus* × *O. niloticus* fingerlings Lim *et al.* (1993a,b) reported a requirement of about 5 mg kg^{-1} diet.

Vitamin B_6 (pyridoxine)

Pyridoxine plays an essential role in protein metabolism as the coenzyme pyridoxal phosphate which is required for the non-oxidative degradation of amino acids including transamination, deamination, decarboxylation and sulphhydration. Pyridoxal phosphate is required for the metabolic breakdown of tryptophan to nicotinic acid and the synthesis of haemoglobin, acetyl co-A and mRNA and for the release of glycogen from the muscle and liver. The pyridoxine requirements of tilapia have been reported by Lim *et al.* (1995) as 3 mg kg^{-1} for red tilapia (*O. mossambicus* × *O. niloticus*) in seawater. Previously Oyetano *et al.* (1985) had suggested a requirement of 0.5–1.17 mg 100 g^{-1} in *O. mosssambicus*.

Pantothenic acid

Pantothenic acid plays an essential role in protein, lipid and carbohydrate metabolism as a component of acetyl co-A which is necessary for acetylation reactions and the subsequent oxidation and release of energy from the major food nutrients. It is also involved in the synthesis of fatty acids, cholesterol, steroid hormones, phospholipids, haemoglobin, etc. Soliman and Wilson (1992b) have reported the pantothenic acid requirements of *O. aureus*. Deficient fish showed poor growth, haemorrhage, sluggishness, high mortality, anaemia and hyperplasia of gill lamellae. The requirement was ascertained to be 10 mg calcium d-pantothenate kg^{-1} diet, duplicating, almost exactly, the findings of an earlier study on the same species (Roem *et al.*, 1991). In an earlier study (Roem *et al.*, 1990) no pantothenic acid requirement could be demonstrated, possibly because requirements were met by browsing on attached microflora in the experimental tanks.

Nicotinic acid (niacin)

Nicotinic acid plays an essential role in the metabolism of carbohydrates, lipids and proteins as a component of the coenzymes NAD and NADP. NADP is essential for the release of energy from the major food nutrients and for the synthesis of cholesterol. NAD is essential for the synthesis of fatty acids. Nicotonic acid requirements have not been reported for tilapia.

Biotin

Biotin plays an important role in the metabolism of carbohydrates, lipids and proteins as a coenzyme for reactions involving carboxylation and decarboxylation and, as such, is essential for the synthesis of fatty acids and microsomal protein and for the catabolism of certain amino acids. Biotin is also believed to be important in cellular defence mechanisms as an activator of lysozyme. The biotin requirements of tilapia have not been reported.

Folic acid

Folic acid plays an important role in protein metabolism as a component of the coenzyme tetrahydrofolic acid which is required for the transfer of one carbon units in the synthesis of haemoglobin, glycine, methionine, choline, thymine and purines and in the metabolism of phenylalanine, tyrosine and histidine. Requirements of tilapia have not been reported.

Vitamin B_{12} (cyanocobalamin)

Cyanocobalamin plays an essential role in the formation of red blood cells and the maintenance of nerve tissue as the coenzyme cobamide. Cobamide is involved in the metabolism of single carbon units, for example in the synthesis of nucleic acids, the recycling of tetrahydrofolic acid, the maintenance of glutathione activity, the conversion of methyl malonyl co-A to succinyl co-A and in the methylation of homocysteine to methionine. Lovell and Limsuwan (1982) reported that during a 16 week growth trial fingerling *O. niloticus* were unable to demonstrate a dietary requirement for vitamin B_{12}. The intestinal microflora of this species was assessed to be capable of synthesizing at least 11.2 ng g^{-1} fish body weight day^{-1}, thus meeting requirements. The results of Shiau and Lung (1993) confirm this. As the study was conducted with small fish in aquaria it is likely that larger fish in ponds would be in an even better position with respect to synthesis of B_{12}. The work of Sugita *et al.* (1989, 1990, 1991) supports this view.

Inositol (myo-inositol)

Myo-inositol is an essential component of the inositol containing phospholipids and thus is structurally important in skeletal, brain and heart tissues. It is believed to be important in the growth of liver and bone marrow cells, in liver lipid (cholesterol) transport and in the synthesis of RNA. The inositol requirements of tilapia have not been reported.

Choline

Choline is an essential component of acetylcholine and the phospholipids lecithin and sphingomyelin, playing a vital role in both cell structure and neurotransmission. Choline is also a methyl donor and plays an important role in lipid transport. Roem *et al.* (1990) were unable to detect a requirement for choline in *O. aureus* and the authors postulated that tilapia were able to substitute methionine for choline. In practice many feeds contain marginal levels of methionine so that supplementation with choline to 'spare' methionine may be appropriate.

Vitamin C (ascorbic acid)

Ascorbic acid plays an essential role in maintaining the integrity of collagenous connective tissues, blood vessels, bone tissue and wound tissue (largely through the hydroxylation of proline to hydroxyproline but also by hydroxylation of tryptophan, lysine and phenylalanine). It also acts as a strong reducing agent required for the conversion of folic acid to tetrahydrofolic acid, of tryptophan to serotonin and the synthesis of steroid hormones. Jauncey *et al.* (1985) and Soliman *et al.* (1986ab, 1987, 1994) have conducted a detailed investigation of vitamin C requirements of tilapias, encompassing quantitative requirements, the ability (or otherwise) to synthesize the vitamin *de novo*, the utilization and stability of various chemical forms, its role in wound healing and its effect on hatchability and fry performance.

Many animals are capable of synthesizing ascorbic acid; those that cannot lack the specific enzyme L-gulonolactone oxidase. There is some contention about whether or not some fishes can synthesize ascorbic acid and it would seem safer to assume that tilapias in general cannot until or unless new evidence to the contrary is produced. The quantitative requirement of tilapia for vitamin C was found to be 1250 mg kg^{-1} diet for both *O. niloticus* and *O. mossambicus*, equivalent to a level of 420 mg kg^{-1} body weight (Soliman *et al.*, 1994). Free L-ascorbic acid is very unstable with large losses occurring during processing, storage and leaching of diets.

Al Amoudi *et al.* (1992) investigated the ascorbic acid requirements of fingerling *O. spilurus* and found that 75 mg kg^{-1} prevented signs of deficiency but that vertebral collagen was reduced in fish fed less than 150 mg kg^{-1}. Oyetano (1985) reported a requirement of 40–50 mg kg^{-1} for *O. mossambicus* fry and Stickney *et al.* (1984) a requirement of 50 mg kg^{-1} in *O. aureus* fingerlings. Shiau and Jan (1992) found the requirements of 1 g *O. niloticus* × *O. aureus* to be 79 mg kg^{-1}.

Soliman *et al.* (1986a,b) evaluated five forms of vitamin C (L-ascorbic acid, sodium salt of L-ascorbic acid, glyceride coated L-ascorbic acid, L-ascorbic acid-2-sulphate and ascorbyl palmitate) for their biological availability to tilapias and their chemical stability. All forms were found to be equally effective, on an equimolar basis, as dietary sources of the vitamin for *O. niloticus*. The glyceride coated form and the 2-sulphate were very stable whereas during pelleting alone up to 70% of the free vitamin was lost. A more recently advocated ester is ascorbyl phosphate (mono or poly) where losses during processing and storage are at most around 20%. Shiau and Hsu (1995) compared L-ascorbic acid (AA), L-ascorbyl-2-sulphate (AS) and L-ascorbyl-2-monophosphate (A2MP) in feeds for 1.5 g *O. niloticus* × *O. aureus*. Requirements were assessed as 41–48 mg AS kg^{-1} (≡19–23 mg AA kg^{-1}), 37–42 mg A2MP kg^{-1} (≡17–20 mg AA kg^{-1}). Abdelghany (1996) compared equimolar additions of AA, AS and L-ascorbyl-2-polyphosphate (A2PP) in *O. niloticus*. A2PP at 50 mg kg^{-1} was best, closely followed by AS and both were much better than AA. Kittakoop and Piamsak (1995) demonstrated that *O. niloticus* possessed the acid phosphatase activity appropriate to hydrolysis of ascorbyl phosphate esters.

The effects of ascorbic acid deficiency on *O. niloticus* included poor growth, anorexia, surface haemorrhage, tail erosion, exophthalmia and spinal deformity; deficiency signs appeared after 6 weeks on an ascorbic acid free diet (Soliman *et al.*, 1986a).

Fat soluble vitamins

Vitamin A (retinol)

Vitamin A plays an essential role as a component of visual pigments and in the secretion of epithelial mucus. It is believed to be involved in the release of proteolytic enzymes from lysosomes. The vitamin A requirements of tilapia have not been reported.

Vitamin D_3 (cholecalciferol)

Vitamin D_3 is important in calcium and phosphorus metabolism as the hormone 1,25-dihydroxycholecalciferol which is believed to affect calcium absorption, the conversion of organic phosphorus to inorganic phosphorus, the resorption of phosphate and amino acids from kidney tubules, the maintenance of blood calcium levels and the deposition and oxidation of citrate in bone. Wendelar-Bonga *et al.* (1983) and O'Connel and Gatlin (1992, 1994) have investigated the effects of dietary vitamin D_3 on bone formation in *O. aureus*. Interestingly, the latter authors found that vitamin D_3 did not affect the ability of the species to utilize dietary calcium. Shiau and Hwang (1993) established a quantitative requirement for D_3 in *O. niloticus* × *O. aureus* fry (0.78 g) of 375 IU kg^{-1}. Sunita-Rao and Raghuramulu (1996) demonstrated that *O. mossambicus* does not have the ability to synthesize vitamin D_3 from cholesterol or acetate through non-photomediated pathways.

Vitamin E (tocopherol)

Tocopherol plays an essential role as a fat-soluble anti-oxidant within animal bodies, where it protects reactive compounds (e.g. PUFA, vitamins A and C) from oxidative damage by acting as a free radical acceptor. It has been suggested that it may also be involved in cellular respiration and biosynthesis of both DNA and coenzyme-Q. Satoh *et al.* (1987) studied the vitamin E requirements of *O. niloticus*. With a dietary lipid level of 5% (supplied as 1% n-3 in the diet) no deficiency signs, other than lowered HSI (hepatosomatic index), were found after feeding a diet devoid of vitamin E for 28 weeks. On the basis of HSI a dietary level of 50 mg kg^{-1} of the diet was recommended. In a second experiment with a diet containing 10–15% lipid (fish oil fatty acid methyl esters) fish fed a vitamin E free diet showed inappetance and lower growth after 3 weeks. A level of 500 mg kg^{-1} was recommended to overcome this. In practical tilapia diets levels of lipid are likely to be quite low and thus lipid will be low in n-3 fatty acids. Roem *et al.* (1990b) further investigated the tocopherol requirements of *O. aureus* and proposed a requirement of 3–4 mg corn oil kg^{-1} diet.

Vitamin K_3 (menadione)

Vitamin K is essential for the maintenance of normal blood clotting by facilitating the production or release of plasma proteins associated with coagulation including prothrombin, proconvertin, plasma thromboplastin and the 'Stuart-Prower' factor. Vitamin K may also play a part in electron transport and oxidative phosphorylation. The vitamin K requirements of tilapia have not been reported.

General comments on vitamin requirements, deficiency signs and supplements

By balancing the dietary feedstuffs used Dickson (1987) showed that a vitamin supplement was probably unnecessary in feeds for intensively farmed tilapia in Zambia. Viola (1989) showed that with tilapia densities of 100 fish m^{-2} and yields of 20 t ha^{-1} no beneficial effects of a vitamin premix could be detected in ponds, cages or concrete tanks. There are many other reports of the 'non-essentiality' of supplemented dietary vitamins for tilapia (Pantastico and Baldia, 1979; Guerrero, 1980; Campbell, 1985; Wannigama *et al.*, 1985; Wee and Ng, 1986). Recently Roem *et al.* (1990a) reported that *O. aureus* fingerlings could satisfy their requirements for pantothenic acid, choline and possibly other (but not all) vitamins by browsing on the walls of tanks housed in a laboratory recirculation system and fed purified diets. However, there still remain many unanswered questions concerning the vitamin requirements of tilapias and it is exceedingly difficult to make general recommendations as to the levels required in supplements.

Table 9.9 summarizes the vitamin requirements of tilapias, the likely deficiency signs and recommended levels for inclusion in diets. It must be remembered that this is based on little experimental data in the case of tilapia, and represents the 'worst' case assuming no contribution from natural food, endogenous vitamins in the feedstuffs selected or from microbial biosynthesis in tilapia intestine. It is possible that all these 'other' vitamin sources may combine to produce a situation where no supplementary vitamins are, in fact, required. NRC (1993) contains tables of the likely vitamin content of a range of feedstuffs for finfish species.

9.7 MINERALS

The mineral nutrition of fish has been reviewed a number of times in recent years and the present review draws heavily on the publications by Tacon (1987) and Davis and Gatlin (1991). Except for organically bound carbon, hydrogen, oxygen and nitrogen there are about 20 or so inorganic mineral elements that are considered essential for animal life and thus, presumably, for tilapias (Table 9.10).

Of the total of 23 minerals 13 are required in the diets of most terrestrial animals and only nine of these have been demonstrated to be essential in the diets of fish (Davis and Gatlin, 1991). It may seem obvious but it perhaps should be noted that minerals are rarely effective in pure form, they are generally presented as salts which can be solubilized in fish guts. The general functions of essential minerals can be summarized as follows (Tacon, 1987):

- minerals are essential components of hard skeletal structures such as bones and teeth,
- minerals play a key role in osmoregulation,
- mineral are essential structural components of soft tissues,
- minerals are essential for nerve impulse transmission and muscle contraction,
- minerals play vital roles in the body acid-base equilibrium and thus regulate the pH of both blood and other body fluids,
- minerals serve as essential components of many enzymes, vitamins, hormones and respiratory pigments as well as being co-factors in metabolism, catalysts and enzyme activators.

With the exception of osmoregulation, the functions of minerals in fish appear to be broadly similar to those of terrestrial animals. However, terrestrial animals are largely restricted to dietary sources of minerals whereas fish, and other aquatic animals, may be able to use minerals dissolved in the water to meet some or all of their physiological requirements.

Table 9.9 The vitamin requirements (mg kg^{-1}) of tilapia and associated deficiency signs

Vitamin/Species	Requirement	Deficiency signs	Reference
Thiamine			
O. mossambicus × *O. urolepsis honorum*	2.5	Anorexia, light coloration, nervous disorders, poor feed efficiency and growth, low haematocrit	Lim and Leamaster (1991)
Riboflavin			
O. aureus	6	Lethargy, fin erosion, anorexia, loss of colour, short body dwarfism, cataracts	Soliman and Wilson (1992)
O. mossambicus × *O. niloticus*	5	Anorexia, reduced growth, nervousness, short body dwarfism, cataracts	Lim *et al.* (1993a)
O. mossambicus × *O. urolepsis honorum*	5	Anorexia, reduced growth, light coloration, nervous symptoms, mortality, dwarfism, cataract	Lim *et al.* (1993b)
Pyridoxine			
Channel catfish	3	Anorexia, nervous disorders, erratic swimming, opercule extension, tetany, blue-green coloration of the dorsal surface	Andrews and Murai (1979)
O. mossambicus × *O. niloticus*	3	Anorexia, muscle spasm, convulsions, lethargy, caudal fin erosion, mouth lesions	Lim *et al.* (1995)
Pantothenic acid			
O. aureus	10	Exudated gills, fusion of gill filaments and lamellae, fin erosion	Roem *et al.* (1991)
O. aureus	10	Poor growth, haemorrhage, sluggishness, high mortality, anaemia and hyperplasia of gill lamellae	Soliman and Wilson (1992)
O. mossambicus × *O. urolepsis honorum*	NR		Roem *et al.* (1990)
Nicotinic acid			
Channel catfish	14	Haemorrhage and lesions of skin/fin, deformed jaws, anaemia, exopthalmia, high mortality	Dupree (1966), Andrews and Murai (1978)

Table 9.9 (cont)

Vitamin/species	Requirement	Deficiency signs	Reference
Biotin			
Channel catfish	$\leqslant 1$	Depigmentation, anaemia, anorexia, reduced growth, hypersensitivity	Robinson and Lovell (1978), Lovell and Buston (1984)
Folic acid			
Channel catfish	0.5–1.0	Anorexia, increased mortality, lethargy, reduced growth, low haematocrit	Dupree (1966), Duncan and Lovell (1991)
Cyanocobalamin			
O. niloticus	NR		Limsuwan and Lovell (1981)
Inositol			
Channel catfish	NR		Burtle (1981)
Choline			
O. aureus	NR		Roem *et al.* (1990)
Ascorbic acid			
O. niloticus	1250 420 in diet)	Scoliosis, lordosis, reduced growth, poor wound repair, haemorrhage, exopthalmia, anaemia	Soliman *et al.* (1986, 1986a, 1994)
O. spilurus	75–150	Haemorrhage and jaw deformity	Al-Amoudi *et al.* (1992)
O. mossambicus	40–50	Soft opercular bones	Oyetano (1993)
O. aureus	50	Mild scoliosis, occasional haemorrhages, gill abnormalities	Stickney *et al.* (1984)
O. niloticus × *O. aureus*	79	Poor growth, poor FCR, poor PER, low haematocrit, mortality	Shiau and Jan (1992)
Vitamin A			
Channel catfish	1000–2000 (IU kg^{-1})	Depigmentation, opaque and protruding eyes, oedema, atrophy, kidney haemorrhage, increased mortality	Dupree (1966)

Table 9.9 (cont)

Vitamin/species	Requirement	Deficiency signs	Reference
Vitamin D			
Channel catfish	NR – 1000 IU kg^{-1})	Reduced growth	Andrews *et al.* (1980), Lovell and Li (1978), Launer *et al.* (1978), Brown (1988)
O.niloticus × *O. aureus*	374.8 IU kg^{-1}	Poor growth and FCR, low haemoglobin, reduced HSI and plasma alkaline phosphatse	Shiau and Hwang (1993)
Vitamin E			
O. niloticus	50–100	Anorexia, reduced growth, poor feed efficiency, skin and fin haemorrhage, muscle degeneration, impaired erythrocyte production, ceroid deposition in liver and spleen, lack of skin colour, increased mortality	Satoh *et al.* (1987)
O. aureus	10–25		Roem *et al.* (1990a)
Vitamin K			
Lake trout	0.5–1.0	Increased clotting time, anaemia, haemorrhagic gills, eyes, and vascular tissue	Poston (1964)

Note: where no data yet exists for tilapia the 'nearest' species investigated has been used
NR = not required

Table 9.10 The essential mineral elements

Macroelements		Trace/microelements	
Principal cations	Principal anions		
Calcium (Ca)	Phosphorus (P)	Iron (Fe)	Fluorine (F)
Magnesium (Mg)	Chlorine (Cl)	Zinc (Zn)	Vanadium (V)
Sodium (Na)	Sulphur (S)	Manganese (Mn)	Chromium (Cr)
Potassium (K)		Copper (Cu)	Molybdenum (Mo)
		Iodine (I)	Selenium (Se)
		Cobalt (Co)	Tin (Sn)
		Nickel (Ni)	Silicon (Si)

After Tacon (1987), in addition Davis and Gatlin (1991) list aluminium (Al) and Arsenic (As)

Calcium

Calcium is an important component of bone and cartilage as well as being essential for blood clotting. It activates several key enzymes including pancreatic lipase, acid phosphatase, cholinesterase, ATPases and succinic dehydrogenase. Through enzyme activation calcium stimulates muscle contraction and regulates nerve impulse transmission (through control over acetylcholine production). Calcium, with phospholipids, regulates cell membrane permeability and is believed to be essential for B_{12} uptake from the gut.

Robinson *et al.* (1987) investigated the dietary calcium requirements of *O. aureus* reared in calcium free water. It appeared the 0.70% calcium was required for good growth. However, it should be noted that calcium requirements may be easily met by the culture water, as demonstrated in an elegant radiotracer study with *O. mossambicus* by Flik *et al.* (1985), and that almost any combination of feed ingredients would result in sufficient dietary calcium, so that supplementation is unlikely to be necessary in any practical tilapia farming context. McCormick *et al.* (1992) showed that tilapia skin can transport calcium against the ionic gradient and Yarzhombed and Bekina (1987) that calcium is absorbed from water passing over the gills. Takagi and Yamada (1992) concluded that dietary calcium alone would not preclude loss of bone calcium if the water was low in this mineral. Hwang *et al.* (1996) showed that *O. mossambicus* eggs and larvae modulated calcium uptake from high and low water calcium levels in order to maintain normal body calcium levels.

Phosphorus

Phosphorus is an essential component of both hard skeletal materials (bone and teeth) and cartilage. It is also an essential component of phospholipids, nucleic acids, phospho-proteins, high energy esters (ATP), hexose phosphates, creatine phosphate and several key enzymes. Phosphorus plays a key role in energy and cell metabolism and inorganic phosphates serve as buffers regulating body fluid pH.

Unlike calcium, and although soluble phosphorus salts can be absorbed through the skin, gills and fins, the concentration of phosphorus in most waters is low so that dietary sources are important. The availability (digestibility) of dietary phosphorus from different sources and in different chemical forms varies between species. Many plant sources contain phosphorus bound to phytic acid which reduces bioavailability.

Because of its greater dietary significance phosphorus has received more attention from researchers than calcium. Robinson *et al.* (1987) reported a requirement, in *O. aureus*, of 0.50% for normal bone mineralization and no pathological signs of phosphorus deficiency were recorded. Watanabe *et al.* (1980) recorded a requirement of less than 0.9% in *O. niloticus* again using bone mineralization as the index. The authors also found the availability of phosphorus in white fishmeal to be 65%. In larger fish Viola *et al.* (1986) report a requirement of 0.7–1.0% depending upon dietary ingredients and fish size. Haylor *et al.* (1988) estimated the available phosphorus requirements for *O. niloticus* as 0.46% of the diet where 70% of the phosphorus was available.

There has been much discussion of the need to optimize the calcium to phosphorus ratio in fish feeds, as is commonly practised for terrestrial animals. This work has been inconclusive although high phosphorus levels may be detrimental where dietary, and culture water, levels of calcium are low.

Magnesium

Magnesium is an essential component of bone and cartilage and is the activator of a large range of enzymes including kinases, mutases, muscle ATPases, cholinesterase, alkaline phosphatase, enolase, isocitric dehydrogenase, arginase, deoxyribonuclease and glutaminase. Through these enzyme functions magnesium stimulates muscle and nerve irritability (contraction), it is involved in intracellular pH regulation and plays an important role in carbohydrate, lipid and protein metabolism.

Magnesium is readily absorbed from the culture water through the skin, gills and fins as well as from the gut (Van der Velden *et al.*, 1991a). In addition, much magnesium in plant materials is bound to phytic acid which may reduce bioavailability. Magnesium has received quite a lot of attention in tilapia mineral nutrition research. Van der Velden *et al.* (1991a) reported aspects of magnesium distribution in the tissues of *O. mossambicus*. Dabrowska *et al.* (1989a,b) investigated the availability of magnesium compounds to tilapia (*O. niloticus*) and subsequently the interaction between dietary protein and magnesium levels in this species. Of the three sources tested the acetate performed best and the authors concluded that with 0.5–0.7 g kg^{-1} magnesium in the diet neither the sulphate nor the oxide could assure maximum growth. In their second study these authors demonstrated a requirement of 0.59–0.77 g kg^{-1} magnesium. Reigh *et al.* (1991) reported the optimum magnesium content of diets for *O. aureus* to be 0.5–0.65 g kg^{-1}. Van der Velden *et al.* (1991b, 1992) also looked at growth rates and tissue magnesium levels in large (70–300 g) tilapia and found that although the growth rate decreased on low magnesium diets the tissue levels did not and they concluded that large tilapias derive a substantial proportion of their magnesium requirement from the culture water. Bijvelds *et al.* (1996, 1997) examined the uptake, distribution and excretion of magnesium in *O. mossambicus*. They report that internal distribution of magnesium changes when the level is low in both food and water. Under these conditions, magnesium is lost from bones and scales and muscle levels are maintained. In addition, under low magensium conditions, magnesium accumulation is almost equal to intake suggesting highly efficient uptake.

Sodium, potassium and chlorine

Sodium, potassium and chlorine occur almost entirely in body fluids and soft tissues with sodium and chlorine principally in the body fluids and potassium within cells. They serve vital roles in controlling osmoregulation (water metabolism) and pH. Sodium is the main monovalent ion of the extracellular fluids constituting about 98% of the ions in the blood stream. Sodium additionally has roles in muscle irritability and carbohydrate absorption. Potassium is the major cation of the intracellular fluid where it regulates osmotic pressure and pH. It also has an effect on stimulating muscle irritability and is required for glycogen and protein synthesis and glucose breakdown. Chlorine is the main monovalent anion of extracellular fluids where it regulates osmotic pressure and pH. It also has roles in the transport of oxygen and carbon dioxide in the blood as well as in stomach pH.

There are no studies on these elements in tilapia nutrition, however the reviews of Tacon (1987) and Davis and Gatlin (1991) suggest that in most species dietary sources are dispensable and requirements can be met from the culture water. The exception may be potassium with requirements from 0.21 to 1.2% quoted.

Sulphur

Sulphur is an essential component of several key amino acids (methionine and cystine), vitamins (thiamine and biotin) and insulin. As the sulphate it is a component of heparin, chondroitin, fibrinogen and taurine. It is also, as the free sulphydryl group, an activator of enzyme systems such as coenzyme A and glutathione. It is believed to be involved in the detoxification of aromatic compounds. There are no reports on specific sulphur requirements for any fish species.

Iron

Iron is an essential component of the respiratory pigments haemoglobin and myoglobin (and thus oxygen and electron transport) and is also an essential component of various enzyme systems including cytochromes, catalases, peroxidases, xanthine and aldehyde oxidases and succinic dehydrogenase.

Iron is readily absorbed through the gut and integument (skin, gills and fins) and availability is depressed by high phosphate, calcium, phytic acid, copper or zinc. In general, inorganic iron is more available than organic, especially as the Fe^{2+} rather than Fe^{3+} ion. There are no data on the iron requirements of tilapias but channel catfish appear to require $\leqslant 30$ mg kg^{-1} (Gatlin and Wilson, 1986).

Zinc

Zinc is an essential component of more than 80 metalloenzymes including carbonic anhydrase (carbon dioxide transport in the blood and acid secretion in the stomach), glutamic dehydrogenase, alkaline phosphatase, pyridine nucleotide dehydrogenase, alcohol dehydrogenase, superoxide dismutase, pancreatic carboxypeptidase and tryptophan desmolase. Zinc is also an enzyme cofactor playing a role in the metabolism of protein, lipid and carbohydrate being particularly active in RNA and protein synthesis and metabolism. Zinc may play a role in hormones such as insulin, glucagon, corticotrophin, follicle-stimulating hormone (FSH) and luteinizing hormone (LH) and in wound healing.

Zinc is readily absorbed from the gut or via the integument and absorption is reduced by phytic acid and high levels of calcium, phosphorus or copper. While water borne zinc is toxic to tilapias (Hilmy *et al.*, 1987) dietary zinc is unlikely to be (Tacon, 1987). McClain and Gatlin (1988) reported a requirement for 20 mg kg^{-1} in *O. aureus* and found that phytic acid reduced zinc bioavailability in this species so that phytic acid containing diets should contain higher levels of supplementation. Eid and Ghoneim (1994) suggested a requirement of 30 mg kg^{-1} in *O. niloticus.*

Manganese

Manganese functions as an enzyme activator for enzymes which mediate phosphate group transfer (phosphate transferases, phosphate dehydrogenases) especially in the citric acid cycle. It is a component of pyruvate carboxylase and is essential for bone formation, erythrocyte regeneration, carbohydrate metabolism and the reproductive cycle.

Manganese is readily absorbed from the gut or via the integument and availability is reduced in the presence of phytic acid or high levels of dietary calcium. Ishac and Dollar (1967) found that *O. mossambicus* required manganese in both the feed and culture water to meet a daily requirement of 1.7 mg kg^{-1} live fish per day.

Copper

Copper is an essential component of numerous oxidation-reduction enzymes (cytochrome oxidase, uricase, tyrosinase, superoxide dismutase, amine oxidase, lysyl oxidase, caeruloplasmin). Through caeruloplasmin (ferroxidase) copper is involved in haemoglobin synthesis and erythrocyte production and maintenance. It is believed to be important in melanin synthesis (skin pigmentation), in bone and connective tissue formation and in the integrity of the myelin sheath of nerve fibres.

Copper is readily absorbed from the gut and via the skin, gills and fins and availability is reduced in the presence of phytic acid or high levels of zinc, iron, molybdenum, cadmium, inorganic sulphates and calcium carbonate. There have been no studies on copper in tilapias but channel catfish require around 5 mg kg^{-1} (Gatlin and Wilson, 1986).

Cobalt

Cobalt is an essential component of vitamin B_{12} (cyanocobalamin) and, as such, is essential for erythrocyte formation and the maintenance of nerve tissue. Cobalt may have unconfirmed enzyme activation functions. Cobalt is readily absorbed from the gut and surrounding water and availability is reduced by high iodine intakes.

There are no reports of cobalt requirements in tilapias or in any other fish, but there is some suggestion that dietary cobalt supplementation allows B_{12} synthesis by gut flora thus reducing, or eliminating, a dietary need for this vitamin. Cobalt has been shown to improve glucose utilization in tilapia (Shiau and Lin, 1993).

Iodine

Iodine is an integral component of the thyroid hormones thyroxine and tri-iodothyronine and is essential for regulating metabolic rate in all body processes. It is readily absorbed from the gut or water and availability is reduced by high levels of cobalt. There are no reports on this mineral in tilapia nutrition but Chinook salmon (*Oncorhynchus tschawytscha*) require 0.6–1.1 mg kg^{-1} (Woodall and LaRoche, 1964).

Selenium

Selenium is an essential component of the enzyme glutathione peroxidase and as such (together with tocopherols – especially vitamin E) protects cellular tissues and membranes from oxidative damage. Selenium may also participate in the biosynthesis of coenzyme Q and in the absorption and retention of vitamin E. Selenium is readily absorbed from both the diet and surrounding water. There are no studies on tilapia but it is suggested that selenium requirements may be influenced by the levels of dietary PUFA. In channel catfish a requirement of 0.1–0.25 mg kg^{-1} has been shown (Gatlin and Wilson, 1984).

Chromium

Trivalent chromium is an integral component of the glucose tolerance factor and is a cofactor for insulin. It may also play an important role in cholesterol and amino acid metabolism and there have been recent suggestions, in terrestrial animals, that it is a stress reduction factor (possibly via effects on insulin) (Mowat, 1994). It is readily absorbed from the gut and culture water. Shiau and Lin (1993) found that glucose utilization in tilapias was improved when diets were supplemented with chromium at 2 mg kg^{-1}. Shiau and Chen (1993) report that 2 mg kg^{-1} Cr_2O_3 was more effective in this respect than the same level as $CrCl_3.6H_2O$ or Na_2CrOH_2O. Shiau and Shy (1998) established a quantitative requirement for 204 mg Cr_2O_3 kg^{-1} ($\equiv$139.6 mg Cr kg^{-1}).

General comments on mineral requirements

Because aquatic animals are able to absorb minerals from both the diet and the culture water, and because of the complexities of osmoregulation, there is relatively little data on the mineral requirements of fish. In a related study Al-Amoudi (1987) reported that adding 10% sodium chloride to the diets of various *Oreochromis* species prior to sea water transfer enhanced the survival rates of fish post-transfer. In saltwater it is obvious that most of the mineral requirements are likely to be met from either drinking sea water or absorption across the integument. Even in fresh water a substantial proportion of the requirements for many minerals will be met this way. Despite this, and the intrinsic levels of minerals in the feed ingredients, most feeds contain supplemental minerals in the form of a premix. This is largely to protect against possible occurrences of mineral deficiency caused by reduced bioavailability (phytic acid) or antagonism. In the latter category high dietary calcium levels have been shown to cause phosphorus, zinc, iron and manganese deficiencies. Table 9.11 summarizes the levels of minerals recommended for inclusion in tilapia feeds.

Mineral toxicity

Some potentially accumulative toxic dietary elements occasionally give rise to concern. Certain raw materials are potentially very high in some of elements, principally heavy metals, depending upon the methods of preparation. Table 9.12 summarizes the reports in the scientific literature in this respect although none of these is specific to tilapias.

Table 9.11 Recommended dietary mineral levels for tilapias

Mineral	Species	Requirement	Reference
Calcium	*O. aureus*	7 g kg^{-1}	Robinson *et al.* (1987)
Phosphorus	*O. niloticus*	<9 g kg^{-1}	Watanabe *et al.* (1980)
	O. niloticus	4.6 g kg^{-1}	Haylor *et al.* (1988)
	O. aureus	7–10 g kg^{-1}	Viola *et al.* (1986)
	O. mossambicus	4.5 g kg^{-1}	Robinson *et al.* (1987)
Magnesium	*O. niloticus*	0.6–0.8 g kg^{-1}	Dabrowska *et al.* (1986a,b)
	O. niloticus	0.5–0.65 g kg^{-1}	Reigh *et al.* (1991)
Zinc	*O. aureus*	20 mg kg^{-1}	McClain and Gatlin (1988)
	O. niloticus	30 mg kg^{-1}	Eid and Ghoneim (1994)
Chromium	*O. niloticus* × *O. aureus*	2 mg kg^{-1}	Shiau and Lim (1993), Shiau and Chen (1993)
	O. niloticus × *O. aureus*	139.6 mg kg^{-1}	Shiau and Shy (1998)

Table 9.12 Potential toxicity of dietary minerals in fish (after Tacon, 1987)

Element	Species	Toxicity
Zinc	Common carp	Reduced growth above 300 mg kg^{-1}
Copper	Channel catfish	Reduced growth, feed efficiency and haematocrit above 15 mg kg^{-1} but recent trials refute this at up to 500 mg kg^{-1}
Selenium	Rainbow trout and channel catfish	Reduced growth and feed efficiency, high mortality, nephrocalcinosis above 13 mg kg^{-1}
Cadmium	Rainbow trout and common carp	Scoliosis, hyperactivity
Lead	Rainbow trout	Scoliosis, lordosis, black tail, anaemia
Chromium	Rainbow trout	Reduced growth and feed efficiency

CONCLUSIONS

It can be seen that although much progress has been made in the field of tilapia nutrition over the last 20 years or so that there is still much to learn. Most importantly, the interpretation of this laboratory derived data in practical tilapia culture systems needs to be addressed. Our understanding of nutrient interactions is still weak with most studies taking the 'one nutrient variable' at a time approach. Finally, much more attention needs to be paid to nutrient source/ chemical form, bioavailability and the effects of processing.

REFERENCES

Abdelghany, A.E. (1996) Growth response of Nile tilapia *Oreochromis niloticus* to dietary L-ascorbic acid, L-ascorbyl-2-sulfate and L-ascorbyl-2-polyphosphate. *J. World Aquaculture Society* **27**(4), 449–455.

Akand. A.M. (1987) Availability of carbohydrate and lipid as energy source in tilapia, *Tilapia zillii. Bangladesh J. Zool.* **15**(1), 65–69.

Al-Amoudi, M.M. (1987) The effect of high salt diet on the direct transfer of *Oreochromis mossambicus, O. spilurus* and *O. aureus/O. niloticus* hybrids to sea water. *Aquaculture* **64**(4), 333–338.

Al-Amoudi, M.M., El-Nakkadi, A.M.N. and El-Nourman, B.M. (1992) Evaluation of optimum dietary requirement for vitamin C for the growth of *Oreochromis spilurus* fingerlings in water from the Red Sea. *Aquaculture* **105**, 165–173.

Anderson, J., Jackson, A.J., Matty, A.J. and Capper, B.S. (1984) Effects of dietary carbohydrate and fibre on the tilapia *Oreochromis niloticus* (Linn.). *Aquaculture* **37**, 303–314.

Anderson, J., Capper, B.S. and Bromage, N.R. (1991) Measurable and prediction of digestible energy values in feedstuffs for the herbivorous fish tilapia (*Oreochromis niloticus* Linn.). *Brit. J. Nutr.* **66**, 37–48.

Andrews, J.W. and Murai, T (1978) Dietary niacin requirements for channel catfish. *J. Nutr.* **108**, 1508–1511.

Andrews, J.W. and Murai, T (1979) Pyridoxine requirements of channel catfish. *J. Nutr.* **109**, 533–537.

Andrews, J.W., Murai, T. and Page, J.W. (1980) Effects of cholecalciferol and ergocalciferol on catfish. *Aquaculture* **19**, 49–54.

Barash, H., Neumark, H. and Heffer, E. (1984) Diet absorption determination in tilapia using magnesium ferrite as a reference and Michel wound clips for increasing the amount of feces available. *Nutr. Rep. Int.* **29**(3), 527–532.

Bijvelds, M.J.C., Flik, G., Kolar, Z.I. and Wendelaar-Bonga, S.E. (1996) Uptake, distribution and excretion of magnesium in *Oreochromis mossambicus*: Dependance on magnesium in diet and water. *Fish Physiol. Biochem.* **14**(4), 287–298.

Bijvelds, M.J.C., Flik, G. and Wendelaar-Bonga, S.E. (1997) Mineral balance in *Oreochromis mossambicus*: Dependence on magnesium in diet and water. *Fish Physiol. Biochem.* **16**(4), 323–331.

Brown, P.B. (1988) *Vitamin D Requirements of Juvenile Channel Catfish Reared in Calcium Free Water*, PhD thesis, Texas A&M University, College Station, Texas.

Burtle, G.J. (1981) *Essentiality of Dietary Inositol for Channel Catfish*, PhD dissertation, Auburn University, Auburn, Alabama.

Campbell, D. (1985) Large scale cage farming of *Sarotherodon niloticus*. *Aquaculture* **48**(1), 57–69.

Chang, S.-L., Huang, C.-M. and Liao, I.-C. (1988) The effects of various feeds on seed production by Taiwanese red tilapia, in *Second International Symposium on Tilapia in Aquaculture, Bagkok, Thailand, March 16–20 1987* (eds R.S.V. Pullin, T. Bhukaswan, K. Tonguthai and J.L. Maclean), ICLARM Conference Proceedings No. 15, ICLARM, Manila, Philippines, pp 319–322.

Cho, C.Y. and Cowey, C.B. (1991) Rainbow Trout *Oncorhynchus mykiss*, in *Handbook of Nutrient Requirements of Finfish* (ed. R.P. Wilson), CRC Press, Florida, pp. 131–144.

Chou, B.S. and Shiau, S.Y. (1996) Optimal dietary lipid level for growth of juvenile hybrid tilapia, *Oreochromis niloticus* × *Oreochromis aureus*. *Aquaculture* **143**(2), 185–195.

Chuang, J.C. and Shiau, S.Y. (1993) Intestinal disaccharidase activity, plasma glucose level, body composition and growth of tilapia *Oreochromis niloticus* × *O. aureus* fed different carbohydrates, in *From Discovery to Commercialisation* (eds M. Carillo, L. Dahle, J. Morales, P. Sorgeloos, N. Svennig and J. Wyban), EAS publications No. 19, Oostende, Belgium, p. 213.

Cisneros, J.A., Toledo, J. and Oritz, E. (1984) Requerimientos nutricionales en alevines *de Oreochromis aureus* (antes Tilapia) II – Relacion Proteina-Carbohideato. *Rev. Lat. Acui. Lima-Peru* **19**, 29–36.

Cisse, A. (1988) Effects of varying protein levels on spawning frequency and growth of *Saortherodon melanotheron*, in *Second International Symposium on Tilapia in Aquaculture, Bagkok, Thailand, March 16–20 1987* (eds R.S.V. Pullin, T. Bhukaswan, K. Tonguthai and J.L. Maclean), ICLARM Conference Proceedings No. 15, ICLARM, Manila, Philippines, pp. 329–333.

Cisse, A. (1996) Effects of varying protein-energy levels on food consumption, growth and body composition of *Sarotherdon melanotheron* (Rueppel 1852), in *Proceedings of the Third International Symposium on Tilapia in Aquaculture* (eds R.S.V. Pullin, J. Lazard, M. Legendre, J.B. Amon-Kottias and D. Pauly), ICLARM No. 41, Makati-City, Philippines, pp. 193–197.

Cowey, C.B. and Sargent, J.R. (1979) Nutrition, in *Fish Physiology: Bioenergetics and Growth*, Vol. VIII (eds W.S. Hoar, D.R. Randall and J.R. Brett), Academic Press, New York, pp. 1–69.

Cowey, C.B. and Walton, M.J. (1989) Intermediary metabolism, in *Fish Nutrition* (ed. J.E. Halver), 2nd edition, Academic Press, London, pp. 259–329.

Dabrowska, H., Guenther, K.-.D. and Meyer-Burgdorff, K. (1989a) Availability of various magnesium compounds to tilapia (*Oreochromis niloticus*). *Aquaculture* **76**(3–4), 269–276.

Dabrowska, H., Meyer-Burgdorff, K. and Guenther, K.-.D. (1989b) Interaction between dietary protein and magnesium level in tilapia (*Oreochromis niloticus*). *Aquaculture* **76**(3–4), 277–291.

Davis, D.A. and Gatlin, D.M. III (1991) Dietary mineral requirements of fish and shrimp, in *Proceedings of the Aquaculture Feed Processing and Nutrition Workshop, Thailand and Indonesia, September 19–25* (eds D.N. Akiyama and R.K.H. Tan), Singapore American Soybean Assoc., pp. 49–67.

Davis, A.T. and Stickney, R.R. (1978) Growth responses of *Tilapia aurea* to dietary protein quality and quantity. *Trans. Am. Fish. Soc.* **107**(3), 479–483.

Degani, G. and Revach, A. (1991) Digestive capabilities of three commensal fish species: Carp *Cyprinus carpio* L., tilapia *Oreochromis aureus* × *O. niloticus* and African catfish *Clarias gariepinus* Burchell 1822. *Aquaculture and Fisheries Management* **22**, 397–403.

De Silva, S.S. and Anderson, T. (1995) *Fish Nutrition in Aquaculture*, Chapman and Hall, London.

De Silva, S.S. and Perera, M.K. (1984) Digestibility in *Sarotherodon niloticus* fry: Effect of dietary protein level and salinity with further observations on variability in daily digestibility. *Aquaculture* **38**(4), 293–306.

De Silva, S.S. and Perera, M.K. (1985) Effects of dietary protein level on growth, food conversion, and protein use in young *Tilapia nilotica* at four salinities. *Trans. Am. Fish. Soc.* **114**, 584–589.

De Silva, S.S., Gunasekera, R.M. and Atapattu, D. (1989) The dietary protein requirements of tilapia and an evaluation of least-cost dietary protein levels. *Aquaculture* **80**, 271–284.

De Silva S.S., Gunasekera, R.M. and Shim, K.F. (1991) Interactions of varying dietary protein and lipid levels in young red tilapia: Evidence of protein sparing. *Aquaculture* **95**(3–4), 305–318.

Dickson, M.W. (1987) The supply of vitamins in feed for intensive tilapia farming in Zambia. *Aquacult. Fish. Manage.* **18**(2), 221–230.

Dioundick, O.B. and Stom, D.I. (1990) Effects of dietary alpha -cellulose levels on the juvenile tilapia, *Oreochromis mossambicus* (Peters). *Aquaculture* **91**(3–4), 311–315.

Dupree, H.K. (1966) Vitamins essential for growth of channel catfish *Ictalurus punctatus.* Technical Paper U.S. Bureau of Sport Fisheries and Wildlife, No. 7.

Duncan, P.L. and Lovell, R.T. (1991) Effect of folic acid on growth, survival and hematology in channel catfish (*Ictalurus punctatus*), 22nd Annual Conference, World Aquaculture Society, San Juan, Puerto Rico, June 16–20.

Dyundik, O.B., Gil, T.A., Salovarova, V.P. and Shul'gina, O.L. (1991) The ability of *Oreochromis mossambicus* (Peters) to assimilate cellulose from feeds. *Biol. Nauki.* **2**, 51–56.

Eid, A.E. and Ghoneim, S.I. (1994) Dietary zinc requirement of fingerling *Oreochromis niloticus. Aquaculture* **119**, 259–264.

El-Sayed, A.-.F.M. and Garling, D.L., Jr. (1988) Carbohydrate-to-lipid ratios in diets for *Tilapia zillii* fingerlings. *Aquaculture* **73**(1–4), 157–163.

El-Sayed. A-F. and Teshima, S. (1992) Protein and energy requirements of Nile Tilapia, *Oreochromis niloticus*, fry. *Aquaculture* **103**, 55–63.

Flik, G., Fenwick, J.C., Kolar, Z., Mayer-Gostan, N. and Wendelaar-Bonga, S.E. (1985) Whole-body calcium flux rates in cichlid teleost fish *Oreochromis mossambicus* adapted to freshwater. *Am. J. Physiol.* **249**, 432–437.

Gaber, M.M.A. (1994) Amino acid requirements and composition of *Tilapia mossambica (Oreochromis mossambicus)*, *Ann. Agric. Sci. Moshtohor.* **32**(3), 1445–1460.

Gatlin, D.M. and Wilson, R.P. (1984) Dietary selenium requirement of fingerling channel catfish. *J. Nutr.* **114**, 627–633.

Gatlin, D.M. and Wilson, R.P. (1986) Dietary copper requirement of fingerling channel catfish. *Aquaculture* **54**, 277–285.

Ghazaly, K.S. (1991) Influences of thiamin on lead intoxication, lead deposition in tissues and lead hematological responses of *Tilapia zillii. Comp. Biochem. Physiol.* **100C**(3), 417–421.

Guerrero, R.D. (1980) Studies on the feeding of *Tilapia nilotica* in floating cages. *Aquaculture* **20**, 169–175.

Gunasekera, R.M., Shim, K.F. and Lam, T.J. (1996) Effect of dietary protein level on spawning performance and amino acid composition of eggs of Nile tilapia, *Oreochromis niloticus. Aquaculture* **146**(1–2), 121–134.

Gunasekera, R.M., Shim, K.F. and Lam, T.J. (1997) Influence of dietary protein content on the distribution of amino acids in oocytes, serum and muscle of Nile tilapia, *Oreochromis niloticus. Aquaculture* **152**(1–4), 205–221.

Halver, J.E. (1979) Vitamin requirements of finfish, in *Finfish Nutrition and Fishfeed Technology*, Vol. 1 (eds J.E. Halver and K. Tiews), Heenemann Verlag, Berlin, pp. 45–58.

Hanley, F. (1987) The digestibility of foodstuffs and the effects of feeding selectivity on digestibility determinations in tilapia, *Oreochromis niloticus* (L). *Aquaculture* **66**(2), 163–179.

Hanley, F. (1991) Effects of feeding supplementary diets containing varying levels of lipid on growth, food conversion, and body composition of Nile tilapia, *Oreochromis niloticus* (L.). *Aquaculture* **93**(4), 323–334.

Haylor, G.S., Beveridge, M.C.M. and Jauncey, K. (1988) Phosphorus nutrition of juvenile *Oreochromis niloticus*, in *Second International Symposium on Tilapia in Aquaculture, Bagkok, Thailand, March 16–20 1987* (eds R.S.V. Pullin, T. Bhukaswan, K. Tonguthai and J.L. Maclean), ICLARM Conference Proceedings No. 15, ICLARM, Manila, Philippines, pp. 341–345.

Hilmy, A.M., El-Domiaty, N.A., Daabees, A.Y. and Latife, H.A.A. (1987) Some physiological and biochemical indices of zinc toxicity in two freshwater fishes, *Clarias lazera* and *Tilapia zillii*. *Comp. Biochem. Physiol.* **87C**(2), 297–301.

Hwang, P.P., Tung, Y.C. and Chang, M.H. (1996) Effect of environmental calcium levels on calcium uptake in tilapia larvae *Oreochromis mossambicus*. *Fish Physiol. Biochem.* **15**(5), 363–370.

Ishac, M.M. and Dollar, A.M. (1968) Studies on manganese uptake in *Tilapia mossambica* and *Salmo gairdneri*. I – growth and survival of *Tilapia mossambica* in response to manganese. *Hydrobiologia* **31**, 572–584.

Jackson, A.J. and Capper, B.S. (1982) Investigations into the requirements of the tilapia *Sarotherodon mossambicus* for dietary methionine, lysine and arginine in semi-synthetic diets. *Aquaculture 29*, 289–297.

Jauncey, K. (1979) *Growth and Nutrition of Carp in Heated Effluents*, PhD thesis, University of Aston in Birmingham.

Jauncey, K. (1982) The effects of varying dietary protein level on the growth, food conversion, protein utilization and body composition of juvenile tilapias (*Sarotherodon mossambicus*). *Aquaculture* **27**(1), 43–54.

Jauncey, K. (1998) *Tilapia Feeds and Feeding*, Pisces Press Ltd., Bridge of Allan, Scotland.

Jauncey, K., Tacon, A.G.J. and Jackson, J. (1984) The quantitative essential amino acid requirements of *Oreochromis (=Sarotherodon) mossambicus*, in *International Symposium On Tilapia In Aquaculture, Nazareth, Israel 8–13 May 1983* (eds L. Fishelson and Z. Yaron), Proceedings, Tel Aviv University, Tel Aviv, Israel, pp. 328–337.

Jauncey, K., Soliman, A. and Roberts, R.J. (1985) Ascorbic acid requirements in relation to wound healing in the cultured tilapia *Oreochromis niloticus* (Trewavas). *Aquacult. Fish. Manage.* **16**(2), 139–149.

Jobling, M. (1983) A short review and critique of the methodology used in fish growth and nutrition studies. *J. Fish Biol.* **23**, 685–703.

Kamarudin, M.S., Kaliapan, K.M. and Siraj, S.S. (1989) The digestibility of several feedstuffs in red tilapia, in *Fish Nutrition Research In Asia* (ed. S.S. De Silva), Proceedings Of The Third Asian Fish Nutrition Network Meeting, Spec. Publ. Asian Fish. Soc., pp. 118–122.

Kanazawa, A., Teshima, S., Sakamoto, M. and Awal, A. (1980) Requirements of *Tilapia zillii* for essential fatty acids. *Bull. Jap. Soc. Sci. Fish.* **46**(11), 1353–1356.

Kaushik, S.J., Doudet, T., Medale, F., Aguirre, P. and Blanc, D. (1995) Protein and energy needs for maintenance and growth of Nile tilapia (*Oreochromis niloticus*). *J. Appl. Ichthyol. Z. Angew. Ichthyol.* **11**(3–4), 290–296.

Kesamaru, K., Okumura, N., Takeda, H. and Kuroki, A. (1982) Studies on the nutrition of *Tilapia nilotica* III. Protein requirement of fingerling *Tilapia nilotica*. *Bulletin of the Faculty of Agriculture, Miyazaki University* **29**, 285–291.

Kittakoop, P. and Piamsak, M. (1995) Acid phosphatase in black tiger prawn *Penaeus monodon* and red tilapia *Oreochromis nilotica*: An expected enzyme in animals for utilisation of phosphorylated derivatives of ascorbic acid. *ARRI Newsletter* **2**(1), 15–17.

Launer, C.A., Tiemeier, O.W. and Deyoe, C.W. (1978) Effects of dietary addition of vitamins C and D_3 on growth and calcium and phosphorus content of pond cultured channel catfish. *Prog. Fish Cult.* **40**, 16–20.

Leary, D.F. and Lovell, R.T. (1975) Value of fibre in production type diets for channel catfish. *Trans. Am. Fish. Soc.* **104**(2), 328–332.

Li, Z., Wu, L., Jun, Y. and He, X. (1991) The nutritional value of commercial feed ingredients for Nile tilapia (*Oreochromis niloticus* L.) in China, in *Fish Nutrition Research in Asia* (ed. S.S. De Silva), Proceedings of the Fourth Asian Fish Nutrition Workshop, pp. 101–106.

Lim, C. and Leamaster, B. (1991) Thiamine requirement of red hybrid tilapia grown in seawater. *J. World. Aquacult. Soc.* **22**, 36A.

Lim, C., Leamaster, B. and Brock, J.A. (1993a) Riboflavin requirement of red hybrid tilapia grown in seawater. *J. World Aquaculture Society* **24**(4), 451–458.

Lim, C., Leamaster, B. and Brock, J.A. (1993b) Riboflavin requirement of red hybrid tilapia grown in seawater, in *Fish Nutrition in Practice* (eds S.J. Kaushik and P. Luquet), Institut National de la Recherche Agronomique No. 61, Paris, France, pp. 743–752.

Lim, C., Leamaster, B. and Brock, J.A. (1995) Pyridoxine requirement of fingerling red hybrid tilapia grown in seawater. *J. Applied Aquaculture* **5**(2), 49–60.

Limsuwan, T. and Lovell, R.T. (1981) Intestinal synthesis and absorption of vitamin B_{12} in channel catfish. *J. Nutr.* **111**, 133–140.

Lorico-Querijero, B. and Chiu, Y.N. (1989) Protein digestibility studies on *Oreochromis niloticus* using chromic oxide indicator. *Asian Fish. Sci.* **2**(2), 177–191.

Lovell, R.T. and Buston, J.C. (1984) Biotin supplementation of practical diets for channel catfish. *J. Nutr.* **114**, 1092–1096.

Lovell, R.T. and Li, Y.P. (1978) Essentiality of vitamin D in diets of channel catfish (*Ictalurus punctatus*). *Trans. Am. Fish. Soc.* **107**, 809–811.

Lovell, R.T. and Limsuwan, T. (1982) Intestinal synthesis and dietary nonessentiality of vitamin B_{12} for *Tilapia nilotica*. *Trans. Am. Fish. Soc.* **111**(4), 485–490.

Luquet, P. (1989) Practical considerations on the protein nutrition and feeding of tilapia. *Aquat. Living Resour.* **2**(2), 99–104.

Mambrini, M. and Kaushik, S.J. (1994) Partial replacement of dietary protein with dispensable amino acids in diets of Nile tilapia, *Oreochromis niloticus*. *Comp. Biochem. Physiol.* **109A**, 469–477.

Mazid, M.A., Tanaka, Y., Katayama, T., Simpson, K.L. and Chichester, C.O. (1978) Metabolism of amino acids in aquatic animals – III. Indispensable amino acids for *Tilapia zillii*. *Bull. Jap. Soc. Sci. Fish.* **44**(7), 739–742.

Mazid, M.A., Tanaka, Y., Katayama, T., Asdur Rahman, M., Simpson, K.L. and Chichester, C.O. (1979) Growth response of *Tilapia zillii* fingerlings fed isocaloric diets with variable protein levels. *Aquaculture* **18**, 115–122.

McClain, W.R. and Gatlin D.M. I.I.I. (1988) Dietary zinc requirement of *Oreochromis aureus* and effects of dietary calcium and phytate on zinc bioavailability. *J. World Aquacult. Soc.* **19**(3), 103–108.

McCormick, S.D., Hasegawa, S. and Hirano, T. (1992) Calcium uptake in the skin of a freshwater teleost. *Proc. Natl. Acad. Sci. USA* **89**(8), 3635–3638.

Meyer-Burgdorff, K.-H., Osman, M.F. and Guenther, K.D. (1989) Energy metabolism in *Oreochromis niloticus*. *Aquaculture* **79**(1–4), 283–291.

Mowat, D.N. (1994) Organic chromium, a new nutrient for stressed animals, in *Proceedings of Feed Ingredients Asia '94, Raffles City, Singapore, April 1994*, Turret Group plc., pp. 166–173.

Nagase, G. (1964) Contributions to the physiology of digestion in *Tilapia mossambica* Peters: digestive enzymes and the effects of diets on their activity. *Zeitschrift fur vergleichende Physiologie* **49**, 270–284.

NRC (1993) (National Research Council), *Nutrient Requirements of Fish*, National Academy Press, Washington D.C.

O'Connell, J.P. and Gatlin, D.M. III (1992) Interaction of dietary calcium and cholecalciferol on growth and mineral composition of the blue tilapia *Oreochromis aureus*, in *Aquaculture '92: Growing Towards the 21st Century*, World Aquaculture Society, Baton Rouge, Louisiana, USA, pp. 173–174.

O'Connell, J.P. and Gatlin, D.M. III (1994) Effects of dietary calcium and vitamin D on weight gain and mineral composition of blue tilapia *Oreochromis aureus* in low calcium water. *Aquaculture* **125**, 107–117.

Odum, O. and Ejike, C. (1991) Aspects of amino acid utilization in the cichlid *Oreochromis niloticus* L. *Acta Hydrobiol* **33**(3–4), 345–352.

Ogino, C. (1980) Requirements of carp and rainbow trout for essential amino acids. *Bull. Jap. Soc. Sci. Fisheries* **46**, 171–174.

Olsen, R.E., Henderson, R.J. and McAndrew, B.J. (1990) The conversion of linoleic acid and linolenic acid to longer chain polyunsaturated fatty acids by *Tilapia (Oreochromis) nilotica in vivo*. *Fish Physiol. Biochem.* **8**(3), 261–270.

Oyetano, A.S. (1985) Dietary ascorbic acid requirement of tilapia *Sarotherodon mossambicus* Peters. *Biol. Afr.* **2**(2), 50–57.

Oyetano, A.S., Thorburn, C.C., Matty, A.J. and Jackson, A. (1985) Pyridoxine and survival of tilapia (*Sarotherodon mossambicus* Peters), in *Proceedings of the 4th Annual Conference of the Fisheries Society of Nigeria, Port Harcourt, Nigeria 26th–29th November 1985* (eds E.O. Ita, T.O. Ajayi, B. Ezenwa, A.A. Olaniawo, R.E.K. Udolisa and P.A. Taggert), pp. 223–230.

Page, J.W. and Andrews, J.W. (1973) Interactions of dietary levels of protein and energy on channel catfish *Ictalurus punctatus*. *J. Nutrition* **103**, 1339–1346.

Pantastico, J.B. and Baldia, J.P. (1980) Ipil-ipil leaf meal as supplemental feed for *T. nilotica* in cages. *Fish. Res. J. Philipp.* **5**(2), 63–68.

Poston, H.A. (1964) Effect of dietary vitamin K and sulphaguanidine on blood coagulation time, microhematocrit and growth of immature brook trout. *Prog. Fish Cult.* **26**, 59–64.

Ravi, J. and Deveraj, K.V. (1991) Quantitative essential amino acids requirements of catla *Catla catla* (Hamilton), *Aquaculture* **96**, 281–291.

Reigh, R.C., Robinson, E.H. and Brown, P.B. (1991) Effects of dietary magnesium on growth and tissue magnesium content of blue tilapia, *Oreochromis aureus*. *J. World. Aquacult. Soc.* **22**(3), 192–200.

Robinson, E.H. and Lovell, R.T. (1978) Essentiality of biotin for channel catfish (*Ictalurus punctatus*) fed lipid and lipid free diets. *J. Nutr.* **108**, 1600–1605.

Robinson, E.H., LaBomascus, D., Brown, P.B. and Linton, T.L. (1987) Dietary calcium and phosphorus requirements of *Oreochromis aureus* reared in calcium-free water. *Aquaculture* **64**(4), 267–276.

Roem, A.J., Stickney, R.R. and Kohler, C.C. (1990a) Vitamin requirements of blue tilapias in a recirculating water system. *Prog. Fish-cult.* **52**(1), 15–18.

Roem, A.J., Kohler, C.C. and Stickney, R.R. (1990b) Vitamin E requirements of the blue tilapia *Oreochromis aureus* (Steindachner) in relation to dietary lipid level. *Aquaculture* **87**, 155–164.

Roem, A.J., Stickney, R.R. and Kohler, C.C. (1991) Dietary pantothenic acid requirements of the blue Tilapia. *Prog. Fish. Cult.* **53**, 216–219.

Santiago, C.B. (1986) Amino acid requirements of Nile tilapia. *Diss. Abst. Int. Part B Sci. and Eng.* **46**(12), 141 pp.

Santiago, C.B. and Reyes, O.S. (1993) Effects of dietary lipid source on reproductive performance and tissue lipid levels of Nile tilapia *Oreochromis niloticus* (Linnaeus) broodstock. *J. Applied Ichthyology* **9**, 33–40.

Santiago, C.B., Banes-Aldaba, M. and Laron, M.A. (1982) Dietary crude protein requirement of *Tilapia nilotica* fry. *Kalikasan* **11**(2–3), 255–265.

Satoh, S., Takeuchi, T. and Watanabe, T. (1987) Requirements of tilapia for alpha-tocopherol. *Bull. Jap. Soc. Sci. Fish.* **53**(1), 119–124.

Shiau, S.Y. (1997) Utilisation of carbohydrates in warmwater fish with partciular reference to tilapia, *Oreochromis niloticus* × *O. aureus*. *Aquaculture* **151**(1–4), 79–96.

Shiau, S.Y. and Chen, M.J. (1993) Carbohydrate utilisation by tilapia *Oreochromis niloticus* × *O. aureus* as influenced by different types of chromium supplementation, in *From Discovery to Commercialisation* (eds M. Carillo, L. Dahle, J. Morales, P. Sorgeloos, N. Svennig and J. Wyban), EAS publications No. 19, Oostende, Belgium, p. 266.

Shiau, S.Y. and Chuang, J.C. (1995) Utilisation of disaccharides by juvenile tilapia *Oreochromis niloticus* × *O. aureus*. *Aquaculture* **133**(3–4), 249–256.

Shiau, S. Y. and Hsu, T. S. (1995) L-ascorbyl–2-sulfate has equal antiscorbutic activity as L-ascorbyl-2-monophosphate for tilapia *Oreochromis niloticus* × *O. aureus*. *Aquaculture* **133**(2), 147–157.

Shiau, S.Y. and Huang, S.L. (1989) Optimal dietary protein level for hybrid tilapia (*Oreochromis niloticus* × *O. aureus*) reared in seawater. *Aquaculture* **81**, 119–127.

Shiau, S.Y. and Hwang, J.Y. (1993) Vitamin D requirements of juvenile hybrid tilapia *Oreochromis niloticus* × *O. aureus*. *Bull. Jap. Soc. Sci. Fish.* **59**(3), 553–558.

Shiau, S.Y. and Jan, F.L. (1992) Dietary ascorbic acid requirement of juvenile tilapia *Oreochromis niloticus* × *O. aureus*. *Bull. Jap. Soc. Sci. Fish.* **58**(4), 671–675.

Shiau, S.Y. and Kwok, C.C. (1989) Effects of agar, carrageenan, guar gum and carboxymethylcellulose on tilapia growth. *J. World Aquaculture Soc.* **20**(2), 60.

Shiau, S.Y. and Lin, S.F. (1993) Effect of supplemental dietary chromium and vanadium on the utilisation of different carbohydrates in tilapia *Oreochromis niloticus* × *O. aureus*. *Aquaculture* **110**(3–4), 321–330.

Shiau, S.Y. and Lung, C.Q. (1993) No dietary B_{12} required for juvenile tilapia *Oreochromis niloticus* × *O. aureus*. *Comp. Biochem. Physiol.* **105A**, 147–150.

Shiau, S.Y., Peng, C.Y. (1993) Protein-sparing effect by carbohydrates in diets for tilapia, *Oreochromis niloticus* × *O. aureus*. *Aquaculture* **117**(3–4), 327–334.

Shiau, S.Y. and Shy, S.M. (1998) Dietary chromic oxide inclusion level required to maximise glucose utilisation in hybrid tilapia *Oreochromis niloticus* × *O. aureus*. *Aquaculture* **161**(1–4), 355–362.

Shiau, S.Y., Yu, H.L., Hwa, S., Chen, S.Y. and Hsu, S.I. (1988) The influence of carboxymethylcellulose on growth, digestion, gastric emptying time and body composition of tilapia. *Aquaculture* **70**, 345–354.

Shiau, S.Y., Kwok, C.C., Chen, C.J., Hong, H.T. and Hsieh, H.B. (1989) Effects of dietary fibre on the intestinal absorbption of dextrin, blood sugar level and growth of tilapia *Oreochromis niloticus* × *O. aureus*. *J. Fish Biol.* **34**(6), 929–935.

Shimeno, S., Ming, D.C. and Takeda, M. (1993) Metabolic response to dietary carbohydrate to lipid ratios in *Oreochromis niloticus*. *Bull. Jap. Soc. Sci. Fish.* **59**(5), 827–833.

Siddiqui, A.Q., Howlader, M.S. and Adam, A.A. (1988) Effects of dietary protein levels on growth, feed conversion and protein utilization in fry and young Nile tilapia, *Oreochromis niloticus*. *Aquaculture* **70**(1–2), 63–73.

Soliman, A.K. and Wilson, R.P. (1992a) Water soluble vitamin requirements of Tilapia. 2. Riboflavin requirement of blue tilapia *Oreochromis aureus*. *Aquaculture* **104**(3–4), 309–314.

Soliman, A.K. and Wilson, R.P. (1992b) Water-soluble vitamin requirements of tilapia. 1. pantothenic acid requirements of blue tilapia *Oreochromis aureus*. *Aquaculture* **104**, 121–126.

Soliman, A.K., Roberts, R.J. and Jauncey, K. (1984) The pathological effects of feeding rancid lipid in diets for *Oreochromis niloticus* (Trewavas), in *International Symposium on Tilapia in Aquaculture* (eds L. Fishelson and Z. Yaron) Tel Aviv University Press, Tel Aviv, Israel, pp. 193–199.

Soliman, A.K., Jauncey, K. and Roberts, R.J. (1986a) The effect of dietary ascorbic acid supplementation on hatchability, survival rate and fry performance in *Oreochromis mossambicus* (Peters). *Aquaculture* **59**(3–4), 197–208.

Soliman, A.K., Jauncey, K. and Roberts, R.J. (1986b) The effect of varying forms of dietary ascorbic acid on the nutrition of juvenile tilapias (*Oreochromis niloticus*). *Aquaculture* **52**(1), 1–10.

Soliman, A.K., Jauncey, K. and Roberts, R.J. (1987) Stability of L-ascorbic acid (vitamin C) and its forms in fish feeds during processing, storage and leaching. *Aquaculture* **60**, 73–83.

Soliman, A.K., Jauncey, K. and Roberts, R.J. (1994) Water-soluble vitamin requirements of tilapia: Ascorbic acid (vitamin C) requirement of Nile tilapia, *Oreochromis niloticus* (L.). *Aquacult. Fish. Management* **25**(3), 269–278.

Stickney, R.R. and Hardy, R.W. (1989) Lipid requirements of some warmwater species. *Aquaculture* **79**(1–4), 145–156.

Stickney, R.R. and McGeachin, R.B. (1983) Responses of *Tilapia aurea* to semi-purified diets of differing fatty acid composition, in *International Symposium on Tilapia in Aquaculture* (eds L. Fishelson and Z. Yaron) Tel Aviv University Press, Tel Aviv, Israel,(eds L. Fishelson and Z. Yaron), pp. 346–355.

Stickney, R.R. and McGeachin, R.B. (1984) Growth, food conversion and survival of fingerling *Tilapia aurea* fed differing levels of beef tallow. *Prog. Fish Cult.* **46**, 102–105.

Stickney, R.R. and Shumway, S.E. (1974) Occurrence of cellulase activity in the stomach of fishes. *J Fish Biol.* **6**, 779–790.

Stickney, R.R. and Wurts, W.A. (1986) Growth response of blue tilapias to selected levels of dietary menhaden and catfish oils. *Prog. Fish-cult.* **48**(2), 107–109.

Stickney, R.R., McGeachin, R.B., Lewis, D.H., Marks, J., Riggs, A., Sis, R.F., Robinson, E.H. and Wurts, W. (1984) Response of *Tilapia aurea* to dietary vitamin C. *J. World Mariculture Society* **14**, 179–185.

Sugita, H., Miyajima, C. and Deguchi, Y. (1989) Vitamin B_{12} producing bacteria in the intestine of freshwater fishes, in *Program of the 1st International Marine Biotechnology Conference, IMBC*, p. 60.

Sugita, H., Miyajima, C. and Deguchi, Y. (1990) The vitamin B_{12} producing ability of intestinal bacteria isolated from tilapia and channel catfish. *Bull. Jap. Soc. Sci. Fish.* **56**(4), 701.

Sugita, H., Miyajima, C. and Deguchi, Y. (1991) The vitamin B_{12}-producing ability of the intestinal microflora of freshwater fish. *Aquaculture* **92**(2–3), 267–276.

Sunita-Rao, D. and Raghuramulu, N. (1996) Lack of vitamin D_3 synthesis in Tilapia mossambica from cholesterol and acetate. *Comp. Biochem. Physiol.* **114A**(1), 21–25.

Tacon, A.G.J. (1985) Nutritional Fish Pathology. UNDP/FAO, ADCP/REP/85, FAO, Rome, 22 pp.

Tacon, A.G.J. (1987) The Nutrition and Feeding of Farmed Fish and Shrimp – A Training Manual. Part 1. The Essential Nutrients, 117 p (Field Document 2/E); Part 2. Nutrient Sources and Composition, 129 p (Field Document 5); Part 3. Feeding Methods, 208 p (Field Document 7/E), FAO Support to the Regional Aquaculture Activities For Latin America and the Caribbean Project GCP/ RLA/ 075/ ITA. Brasilia, Brazil.

Tacon, A.G.J. (1991) Vitamin nutrition in shrimp and fish, in *Proceedings of the Aquaculture Feed Processing and Nutrition Workshop* (eds D.M. Akiyama and R.K.H. Tan), American Soybean Association, Singapore, pp. 10–41.

Tacon, A.G.J. and Cowey, C.B. (1985) Protein and amino acid requirements, in *Fish Energetics: New Perspectives* (eds P. Tytler and P. Calow), Croom Helm Ltd, London, pp. 155–183.

Takagi, Y. and Yamada, J. (1992) Effects of calcium deprivation on the metabolism of acellular bone in tilapia *Oreochromis niloticus. Comp. Biochem. Physiol.* **102A**(3), 481–485.

Takeuchi, T., Satoh, S. and Watanabe, T. (1983) Dietary lipids suitable for the practical feed of *Tilapia nilotica. Bull. Jap. Soc. Sci. Fish.* **49**, 1361–1365.

Teshima, S. and Kanazawa, A. (1988) Nutritive value of methionine-enriched soybean plastein for *Oreochromis niloticus* fry, in *Second International Symposium on Tilapia in Aquaculture, Bagkok, Thailand, March 16–20 1987* (eds R.S.V. Pullin, T. Bhukaswan, K. Tonguthai and J.L. Maclean), ICLARM Conference Proceedings No.15, ICLARM, Manila, Philippines, pp. 393–399.

Teshima, S.I., Gonzalez, G.M.O. and Kanazawa, A. (1978) Nutritional requirements of Tilapia: utilization of dietary protein by *Tilapia zillii. Mem. Fac. Fish., Kagoshima Univ.* **27**(1), 49–57.

Teshima, S., Kanazawa, A. and Koshio, S. (1987) Effect of feeding rate, fish size and dietary protein and cellulose levels on the growth of *Tilapia nilotica. Mem. Fac. Fish. Kagoshima Univ.* **36**, 7–15.

Teshima, S.I., Kanazawa, A. and Sakamoto, M. (1982) Essential fatty acids of *Tilapia nilotica. Mem. Fac. Fish. Kagoshima Univ.* **31**, 201–204.

Teshima, S., Kanazawa, A. and Uchiyama, Y. (1985) Optimum protein levels in caesein-gelatin diets for *Tilapia nilotica. Mem. Fac. Fish. Kagoshima Univ./Kagoshimadai Suisangakubu Kiyo.* **34**(1), 45–52.

Teshima, S., Kanazawa, A. and Koshio, S. (1990) Effects of methionine enriched plastein supplemented to soybean-protein based diets on common carp *Cyprinus carpio* and tilapia *Oreochromis niloticus*, in *Proceedings of the Second Asian Fisheries Forum. Tokyo, Japan, 17–22 April 1989* (eds R. Hirano and I. Hanyu), Asian Fisheries Forum, Tokyo, pp. 279–282.

Toledo, J., Cisneros, J.A. and Ortiz, E. (1983) Niveles de adicion de alimento segun peso y contenido de proteina cruda en la dieta de de alevines de *Oreochromis aureus* (*Tilapia nilotica*). *Rev. Latinoam. Acuicult.* **17**, 38–42.

Tung, P.H. and Shiau, S.Y. (1991) Effects of meal frequency on growth performance of hybrid tilapia, *Oreochromis niloticus* × *O. aureus*, fed different carbohydrate diets. *Aquaculture* **92**(4), 343–350.

Tung, P.H. and Shiau, S.Y. (1992) The influence of body size on the utilisation of different carbohydrates in tilapia, *Oreochromis niloticus* × *O. aureus*, in *Aquaculture-92: Growing towards the 21st Century.* Aquaculture '92, Orlando, Florida, 21–25 May 1992, World Aquaculture Society, pp. 219–220.

Tung, P.H. and Shiau, S.Y. (1993) Carbohydrate utilisation versus body size in tilapia, *Oreochromis niloticus* × *O. aureus. Comp. Biochem. Physiol.* **104A**(3), 585–588.

Van der Velden, J.A., Spanings, F.A.T., Flik, G., Zegers, C., Kolar, Z.I. and Wendelaar Bonga, S.E. (1991a) Growth rate and tissue magnesium concentration in adult freshwater tilapia, *Oreochromis mossambicus* (Peters), fed diets differing in magnesium content. *J. Fish Biol.* **39**(1), 83–91.

Van der Velden, J.A., Kolar, Z.I. and Flik, G. (1991b) Intake of magnesium from water by freshwater tilapia fed on a low-Mg diet. *Comp. Biochem. Physiol.* **99A**, 103–105.

Van der Velden, J.A., Flik, G. and Wendelaar-Bonga, S.E. (1992) Prolactin cell activity and ion regulation in tilapia *Oreochromis mossambicus* (Peters): Effects of a low magnesium diet. *J. Fish Biol.* **40**, 875–885.

Viola, S. (1989) Production of commercial feeds for warm water fish, in *Fish Culture in Warm Water Systems: Problems and Trends* (eds M. Shilo and S. Sarig), CRC Press, Boca Raton, Florida, pp. 143–162.

Viola, S. and Amidan, G. (1980) Observations on the accumilation of fat in carp and *Sarotherodon (Tilapia)* fed oil-coated pellets. *Bamidgeh* **32**(2), 33–40.

Viola, S. and Zohar, G. (1984) Nutrition studies with market size hybrids of tilapia (*Oreochromis*) in intensive culture. 3. Protein levels and sources. *Bamidgeh* **36**(1), 3–15.

Viola, S., Zohar, G. and Arieli, Y. (1986) Phosphorus requirements and its availability from different sources for intensive pond culture species in Israel. Part 1. Tilapia. *Bamidgeh* **38**(1), 3–12.

Viola, S., Lahav, E. and Angeoni, H. (1992) Reduction of feed protein levels and of nitrogenous excretions by lysine supplementation in intensive carp culture. *Aquatic Living Resources* **5**, 277–285.

Viola, S., Gur, N., Rimsky, Y., Angeoni, H. and Bahat, M. (1993) Lysine supplementation of tilapia feeds. *Fisheries and Fishbreeding in Israel* **26**, 55–61.

Wang, K.-W., Takeuchi, T. and Watanabe, T. (1985a) Effect of dietary protein levels on growth of *Tilapia nilotica. Bull. Jap. Soc. Sci. Fish.* **51**(1), 133–140.

Wang, K.-W., Takeuchi, T. and Watanabe, T. (1985b) Optimum protein and digestible energy levels in diets for *Tilapia nilotica. Bull. Jap. Soc. Sci. Fish.* **51**(1), 141–146.

Wannigama, N.D., Weerakoon, D.E.M. and Muthukumarana, G. (1985) Cage culture of *S. niloticus* in Sri Lanka: Effect of stocking density and dietary crude protein levels on growth, in *Finfish Nutrition in Asia – Methodological Approaches to Research and Developments* (eds C.Y. Cho, C.B. Cowey and T. Watanabe), Vol. IDRC-233e, IDRC, pp. 113–117.

Watanabe, T., Takeuchi, T., Murakami, A. and Ogino, C. (1980) The availability to *Tilapia nilotica* of phosphorus in white fish meal. *Bull. Jap. Soc. Sci. Fish.* **46**(7), 897–899.

Wee, K.L. and Ng, L.T. (1986) Use of cassava as an energy source in a pelleted feed for the tilapia, *Oreochromis niloticus* L. *Aquacult. Fish. Manage.* **17**(2), 129–138.

Wendelar-Bonga, S.E.P., Lammers, P.I and van der Meij, J.C.A. (1983) Effects of a,25 and 24,25-dihydroxyvitamin D_3 on bone formation in the cichlid teleost *Sarotherodon mossambicus. Cell and Tissue Research* **228**, 117–126.

Wilson, R.P. (1991) *Handbook of Nutrient Requirements of Finfish*, CRC Press, Florida.

Winfree, R.A. and Stickney, R.R. (1981) Effects of dietary protein and energy on growth, feed conversion efficiency and body composition of *Tilapia aurea. J. Nutr.* **111**, 1001–1012.

Woodall, A.N. and LaRoche, G. (1964) Nutrition of salmonid fishes. II. Iodide requirements of chinook salmon. *J. Nutr.* **84**, 475–482.

Yamada, S., Tanaka, Y., Katayama, T., Sameshima, M. and Simpson, K.L. (1982) Plasma amino acid changes in *Tilapia nilotica* fed a casein and a corresponding free amino acid diet. *Bull. Jap. Soc. Sci. Fish.* **48**(12), 1783–1787.

Yarzhombed, A.A. and Bekina, E.N. (1987) Efficiency of extraction of soluble matter from water by fish. *J. Ichthyology* **27**, 22–29.

Yong, Wen.-Yue., Takeuchi, T. and Watanabe, T. (1989) Relationship between digestible energy contents and optimum energy to protein ratio in *Oreochromis niloticus* diet. *Bull. Jap. Soc. Sci. Fish.* **55**(5), 869–873.

Chapter ten

Semi-intensive pond aquaculture

P. Edwards, C. K. Lin and A. Yakupitiyage

10.1 INTRODUCTION

Semi-intensive production of tilapia in ponds using fertilizers and supplementary feeds is a means to produce low-cost fish which contributes to national food security in many developing countries. Culture of herbivorous and omnivorous tilapias feeding low down on the food chain provides produce that is potentially available to a wide range of consumers as a staple food. This feature led to tilapia being referred to as an 'aquatic chicken' (Maclean, 1984). Indeed, global production and value of farmed tilapias increased at a recorded average rate of 11.4 and 15.5% between 1986 and 1995, respectively (FAO, 1997). The rate of increase has probably been higher because of considerable underestimation of farmed tilapia production, particularly in Asia which dominates production, because of the difficulty of collecting statistics from diverse and widely scattered farms in developing countries. Using hatchery data as a proxy for output may be useful for carps and other species produced in conventional hatcheries but as tilapia can breed in food fish production systems this method cannot be used (Little and Hulata, this volume). Tilapias are currently farmed commercially or experimentally in at least 75 countries (Pullin *et al.*, 1994).

Over the past decade a low-tech/high-tech 'divide' has developed in tilapia culture similar to that which occurred with poultry (Pullin *et al.*, 1994). At the same time that small-scale farmers in developing countries in Asia are introducing tilapias into traditional carp polycultures, entrepreneurs are exploring ways to farm tilapia as an international commodity, usually by using relatively high-cost intensive culture using complete feeds. This is to satisfy a

M.C.M. Beveridge and B.J. McAndrew (eds), Tilapias: Biology and Exploitation, 377–403.

rising demand for high quality 'white fish' products in the developed world, especially in Europe and North America. However, the major factor inhibiting the purchase of more tilapia in North America is price; tilapia is overpriced compared to chicken and most other forms of animal protein using current, largely intensive methods of aquaculture (Smiley, 1995).

The case is made in this chapter that a semi-intensive mode of pond culture can be used effectively to satisfy both the subsistence needs and growing desire of small-scale farmers to intensify production to generate income and to lower the cost of production of industrial scale farming. The current status of production and regional evolution of culture of tilapias is compared with carps, the major group of inland herbivorous/omnivorous finfish. Culture systems of tilapias in three major resource zones (inland, rural areas; wastewater-fed culture in periurban areas; and in coastal areas) are discussed. The scientific basis of pond fertilization and supplementary feeding strategies are reviewed, followed by consideration of environmental and long term sustainability of semi-intensive pond culture of tilapias.

10.2 COMPARISON OF TILAPIAS AND CARPS

Global and regional statistics

Carnivorous and non-carnivorous fish

FAO statistics do not distinguish between culture systems and degree of intensification so the distinction between carnivorous and non-carnivorous species was considered to be the most suitable criterion to distinguish finfish that feed low in the food chain which are significantly cheaper to produce than carnivores (Csavas, 1993). Most carnivores are cultured in intensive monoculture. Although tilapias are increasingly being cultured in intensive systems for luxury markets, the bulk is undoubtedly harvested from semi-intensive pond culture. Csavas considered only the Asia and Pacific Region, but it produces about 90% of global aquaculture. Of 7 million t of finfish produced in 1990, only 7% were carnivorous species (Csavas, 1993). Carps from inland waters comprised 86% of the 93% of non-carnivorous species, 7% were milkfish from mainly coastal waters, about 2% were mullets and other species. Tilapias constituted only 5% of the non-carnivorous fish production.

Tilapias and carps

Although tilapias comprised only 5% of non-carnivorous fish production, global tilapia production is increasing rapidly: production of tilapias almost tripled between 1986 and 1995 from 225 693 to 658 178 t, respectively (FAO, 1997). The bulk of production is in Asia and the Pacific, with 84% the total global tilapia production of 658 000 t. Africa, despite being the home of tilapias, produced only 31 000 t or 5.4% (and 62.3% of the total for continental Africa came from Egypt), with 59 000 t or 8.9% coming from the Americas. The remaining 9500 t, 1.4% of the global total, came from the Mediterranean and the Middle East.

The growing importance of tilapias in Asia is masked by China, and also by India, because of their huge and increasing production of carps. Although carps dominate inland finfish production in China (8.3 million t of carps reported in 1995 compared to 0.7 million t of tilapias), the annual growth rate of tilapia production in China from 1984–95 was double that of carps (29.7% and 14.9%, respectively). Chinese tilapia production was almost half (47.8%) of global production in 1995, five times that of the next largest producer, the Philippines (FAO, 1997).

Evolution of tilapia aquaculture

Despite Nile tilapia (*Oreochromis niloticus*) being depicted in a pond on a bas-relief from an Egyptian tomb dating back over 4000 years (Hickling, 1963), inland aquaculture in Africa began to develop only this century. Widespread attempts were made to culture tilapia in ponds in their native Africa from 1940–60 during the colonial era but it has rarely been sustainable (Balarin, 1988). The relatively recent, successful development of tilapia culture began when Mozambique tilapia (*O. mossambicus*) mysteriously appeared in Asia in Java in 1939. It was widely distributed in Asia in the 1940s–1950s, followed by the more desirable, for both culture and consumer, Nile tilapia in the 1960s–1980s, which currently dominates global production (FAO, 1997). Tilapias have steadily increased in cultured production, particularly in East and Southeast Asia, and are beginning to do so in South Asia.

It was in Taiwan that tilapias first began to gain popularity at the expense of traditionally cultured Chinese carps (Chen, 1990); nearly 20 years ago tilapias had already become dominant in the traditional Chinese carp polyculture which then was mainly integrated with either ducks or pigs (Chen and Li, 1980). Subsequently there has been an increase in intensive monoculture of tilapias in concrete ponds and cages with fish fed formulated feeds as people began to reject relatively small fish raised in manured ponds with an increase in living standards (Chen, 1990). A similar trend is now underway in the People's Republic of China where the proportion of tilapias has increased in the traditional Chinese polyculture, and intensive monoculture of tilapias in expanding (Li, 1990).

In tropical Southeast Asia where tilapias have greater regional production than carps, semi-intensive pond culture systems of tilapias are diverse (Edwards *et al.*,1997). As in East Asia tilapias were usually first introduced into polyculture with, and have gradually replaced, Chinese carps and common carp in many countries, particularly in Malaysia, Singapore and Thailand even though there are large numbers of Overseas Chinese. Tilapias are commonly cultured also with exotic Indian major carps (mrigal and rohu) which are more popular than Chinese carps, and with other native carps, particularly silver barb or tawes, which retain widespread popularity. There is an increasing tendency in mainland Southeast Asia to culture tilapias in monoculture but this has always been the predominant mode of culture in the Philippines without a tradition of carp culture (Guerrero, 1996).

Culture of tilapias has developed the slowest in the Indian subcontinent, despite introduction of *O. mossambicus* in the early 1950s and the more desirable *O. niloticus* into Bangladesh in 1974 (Rahman, 1992) and India in the early 1980s (Jhingran, 1992). India has started to record finfish species statistics, although not tilapias (FAO, 1997), despite being cultured. According to Jhingran (1992) "there is no other species of fish that has generated so much controversy in India as tilapia" due largely to concern about adverse impacts on Indian major carps which usually fetch a far higher price than tilapias. Nevertheless, tilapias are widely cultured in wastewater-fed ponds in West Bengal and all male culture has been taken up by the commercial sector as consumer acceptance is growing. Jhingran (1992) points out that irrespective of whether there is an adverse impact on the endemic fauna or not, tilapias are widespread in India and practically impossible to eradicate. He therefore recommends making the best use of tilapias which are likely to considerably increase inland fisheries production. A recent rapid survey of fish markets in Calcutta revealed the popularity of tilapia as well as the traditional Indian major carps (Morrice *et al.*, 1998). In Bangladesh, culture of tilapias has been promoted in small, seasonal roadside ditches for poor marginal farmers (Gupta *et al.*, 1992; Rahman, 1992). They retain water for only 3–5 months annually, too short a time to culture carps which are marketed at a relatively large size, but long enough for tilapias to be cultured profitably by low-income households.

Tilapias were introduced into Latin America, which also lacks a tradition of pond culture, in the 1950s, but production is expanding rapidly in some countries (FAO, 1997), mainly because of good access to North American markets (D.C. Little, personal communication).

10.3 CULTURE SYSTEMS

Tilapias are cultured in diverse systems in inland rural areas, in periurban areas in wastewater-fed ponds, and in coastal areas in brackishwater ponds.

Rural inland systems

Integrated agriculture/aquaculture systems

Most semi-intensive pond culture of tilapia (usually in polyculture with carps) in rural areas is associated with crop and/or livestock in integrated agricultural/ aquaculture systems (IAAS), particularly in East and Southeast Asia (Edwards *et al.*, 1986). Crop residues such as brans and oil cakes, vegetation and manures (human as well as livestock depending on cultural acceptance) are used as pond inputs. Annual extrapolated fish yields can attain or exceed 5 t/ha in fertile areas with high levels of inputs and good stock management (large fingerlings, multiple harvests) such as West Java, Indonesia and the Red River delta, Vietnam (Edwards *et al.*, 1997).

Although Asia has the widest range of semi-intensive pond systems stocked with tilapia, they are far less widespread and productive than is commonly believed. Previously abundant wild fish formerly supported smaller human populations; but in most areas changing circumstances now provide a stimulus for the development of aquaculture. Limited indigenous technical knowledge of aquaculture and the poor resource base of most small-scale farms constrain aquaculture also in Asia, as well as in Africa and Latin America.

Where fish ponds have been developed in many parts of Asia, small-scale farmer production usually is <1 t/ha/yr due to the resource poor nature of most farms in the tropics (Edwards *et al.*, 1997). Poultry usually scavenge which normally precludes manure collection although a novel technique has been researched and extended to farmers in which scavenging poultry are partially penned over a pond at night and given supplementary feed (Little and Satapornvanit, 1997). Large ruminant manure (from buffalo or cattle used for draught) was shown to be a poor fertilizer due to low nutrient content and tannin in the manure which stained pond water brown (Edwards *et al.*, 1994a). *O. niloticus*, the major cultured species in Asia, is unable to consume most terrestrial and aquatic vegetation; grass carp which consumes vegetation and supports growth of planktivorous tilapias through copious faeces production is not commonly cultured except in China and north Vietnam.

Such systems have developed especially slowly in Africa where tilapias are indigenous for a variety of technical, social and economic reasons (Harrison *et al.*, 1994) despite impressive technical demonstrations of their productive potential over three decades ago (Hickling, 1963). More recently, however, methods have been developed in Malawi to culture indigenous tilapias (*Oreochromis shiranus* and *Tilapia rendalli*) using on-farm resources such as firewood ash and maize bran, and forages such as napier grass (Jamu and Costa-Pierce, 1995). IAAS using on-farm inputs have been shown to be very attractive to Malawi's smallholder farmers as they do not depend on costly and often unavailable pond inputs. In Latin America, small-scale IAAS involving chickens have been promoted in Panama (Lovshin *et al.*, 1986) and Guatemala (Popma *et al.*, 1995) although in Guatemala most farmers involved in a development project managed low-quality nutrient input, non-integrated systems.

IAAS are almost considered to be synonymous with integrated feedlot-livestock/fish farms. These have been developed in East and Southeast Asia in particular over the past few decades by wealthier livestock farmers as they depend on off-farm formulated feed. There have been numerous attempts to promote scaled-down, feedlot livestock/fish systems for small-scale farmers but they usually fail, e.g. in Thailand 30 egg-laying ducks confined above a 200 m^2 pond produced an extrapolated yield of almost 9 t/ha/yr of *O. niloticus* but farmers were unable to run the system when project support was withdrawn due to problems of input supply and marketing for the duck subsystem (Edwards, 1983).

Use of off-farm inputs

The poor quality and/or short supply of on-farm inputs for semi-intensive tilapia culture suggests that off-farm inputs are also required for the pond to generate income from sale of surplus fish rather than contributing only towards subsistence (Edwards *et al.*, 1996). Farming households initially become interested in aquaculture for subsistence to compensate for declining availability of wild fish, at least in areas where fish are consumed traditionally. However, once confidence is gained with what for most farmers is novel technology, interest changes to income generation with the rapid expansion of the market economy in even remote areas. Production of tilapia on large-scale commercial farms is almost totally dependent on off-farm feeds although in tilapia pond culture integrated with feedlot livestock this invariably is indirectly through the livestock subsystem.

A case can be made for increased use of inorganic fertilizers which usually are the cheapest form of nitrogen (N) and phosphorus (P) (Edwards, 1993). Inorganic fertilization has been effectively used in pond tilapia culture in Bangladesh (Hussain *et al.*, 1989; Gupta *et al.*, 1992), the Philippines (Guerrero, 1996) and Thailand (Edwards *et al.*, 1996). However, to determine the most effective way to use inorganic fertilizer in an IAAS, research on nutrient cycling in crop and livestock as well as fish subsystems is required to answer a range of questions. Should fertilizer be applied to the fishpond or to crops? If it is applied to crops should the crops be consumed by humans, livestock or fish? And should the resulting human or livestock manure be used to fertilize the pond or crops (Edwards, 1993)?

Fertilization with waste from intensive fish culture

Effluents from intensive fish culture are of major concern as a source of pollutants to natural waters but they are a potential source of fertilizer that can be reused in an integrated system with semi-intensive culture of tilapia (Fig. 10.1). An initial series of experiments was conducted on integration of hybrid catfish (*Clarias macrocephlus* × *C. gariepinus*) and Nile tilapia, the former cultured in cages with pelleted feed and the latter in ponds in which the cages were installed (Lin *et al.*, 1989; Lin and Diana, 1995). The integrated system yielded

144 kg/m^3 catfish and 0.35 kg/m^2 tilapia in a 4-month trial, during which the wastes from caged catfish were estimated to contain 30.9 kg N and 11.6 kg P, giving a fertilization rate of 5.1 kg N and 2.0 kg P/ha/day with a N:P ratio of 2.6:1. The extrapolated tilapia standing crop of 3.4 t/ha is comparable to ponds fertilized with chicken manure or chemical fertilizers.

A similar approach was subsequently used to produce large Nile tilapia in cages fed with pelleted feed and fingerlings nursed in ponds fertilized with wastes from caged fish (Fig. 10.1) (Yang *et al.*, 1996). The optimal standing crop of caged fish was 24 kg/m^3 with a mean daily weight gain of 4.6 g/fish; the growth rate of fingerlings in the pond was linearly correlated to feed consumption (biomass) of caged fish. This fish-fish integrated system may have relevance for small-scale fish farmers who can use their small ponds for the dual purposes of nursing and grow-out/fattening of tilapia to achieve maximum biomass production and economic return.

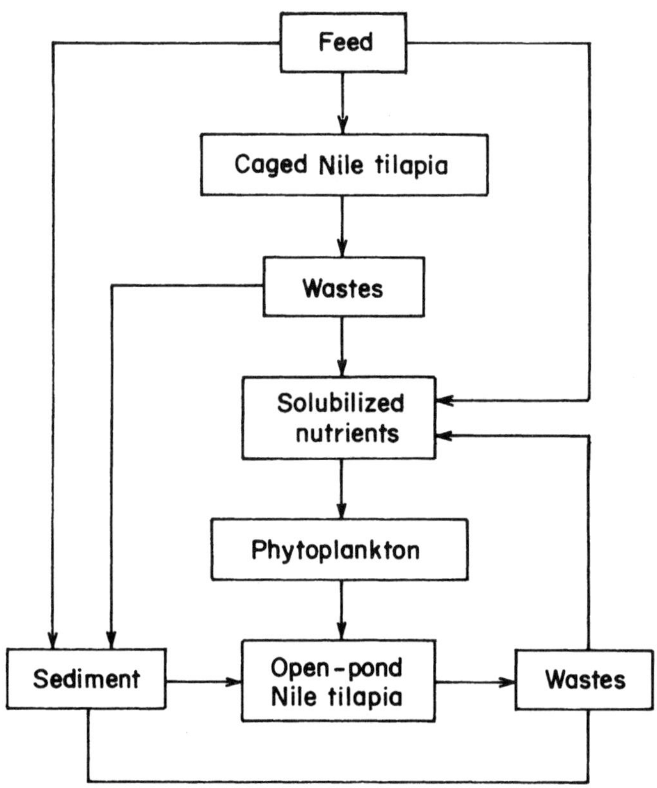

Fig. 10.1 Fingerlings of Nile tilapia may be raised in open-pond culture on wastes produced from grow-out of pellet-fed tilapia cultured in cages installed in the same pond.

Wastewater reuse

Overview

Wastewater reuse in aquaculture is receiving increasing scientific attention because it treats waste inexpensively, provided that land is available for pond construction at reasonable cost, recycles nutrients to produce fish, and human health can be safeguarded (Edwards and Pullin, 1990; Edwards 1992). In peri-urban areas in certain cities in China, India and Vietnam where there is water-borne transport of excreta or sewage, wastewater-fed fishponds may occupy large areas. There are also over 50 ha of 'sewage-fed' ponds and lakes in Dhaka, Bangladesh from which tilapias are harvested (Rahman, 1992).

Tilapia culture in wastewater

Tilapias are cultured in wastewater-fed ponds in China with Chinese carps and in India and Vietnam with Chinese and Indian major carps. A constraint is the poor quality strains of tilapia (invariably *O. mossambicus* or hybrids with *O. niloticus*) which do not attain an attractive marketable size due to prolific breeding. The recent introduction of the Chitralada and GIFT strains of *O. niloticus* into the Hanoi wastewater-fed system has generated considerable farmer interest because of its relatively rapid growth rate and ability to readily reach a target size of 150 g which doubles the farm gate price compared to that of 60–100 g fish.

Research has been carried out to assess the feasibility of culturing tilapia in the treated effluents of the San Juan stabilization ponds in Lima, Peru to produce food and employment as well as to improve the efficiency of water use in a desert environment (Cavallini, 1996). Sixty-gram *O. niloticus* stocked at 2 fish/m^2 attained 250 g in 4 months. The growth rate of the fish in tertiary stabilization ponds was 30.8 kg/ha/day without supplementary feed, with a carrying capacity of 4.4 t/ha. A market study carried out in Lima revealed complete acceptance of wastewater-fed tilapia even though consumers were aware of the origin of the fish. It was estimated that in the tropics a commercial 18 ha farm may produce 127 t of tilapia annually (7 t/ha) and cover the entire treatment cost for a 100 l/s conventional wastewater treatment plant with primary and secondary treatment located in an arid area with land of little to no commercial value. The corresponding figures for a subtropical area are a 32 ha farm producing 106 t annually (3.3 t/ha).

Considerably increased yields of microbiologically safe tilapia from waste-water reuse should be possible without treating wastewater to the degree in the case study from Lima reported above. A design has been proposed which simultaneously optimizes wastewater treatment and fish production in a practical and economic way based on the Calcutta system (Mara *et al.*, 1993). The system would pretreat the wastewater minimally in a 1-day anaerobic pond followed by a 5-day facultative pond before effluent is discharged to fishponds.

The microbiological quality of fishpond water was calculated to comply with the WHO recommendation of ⩾1000 faecal coliforms/100 ml, due to a sufficiently long retention time in the anaerobic (primary treatment) and facultative (secondary treatment) ponds. The area used for wastewater treatment is only 6% of the total pond area for wastewater treatment and aquaculture compared to 73% for the Lima system where fish were cultured in tertiary ponds following more extensive treatment in primary and secondary stabilization ponds. Furthermore, it was estimated that a tilapia yield of about 13 t/ha/yr may be feasible in a wastewater-fed pond in the tropics, assuming the fishponds have a carrying capacity of 5 t/ha and are stocked and harvested three times annually.

Tilapia for high-protein animal feed

Most wastewater-fed aquaculture systems produce fish for direct human consumption but a case has been made that indirect reuse to produce animal feed has potential (Edwards, 1990). Lengthening the food chain so that wastewater is used to produce tilapia for animal feed which is subsequently fed to fish, shrimp or livestock for human food in a separate system (Fig. 10.2) may lead to the acceptance of wastewater reuse in societies in which it is currently

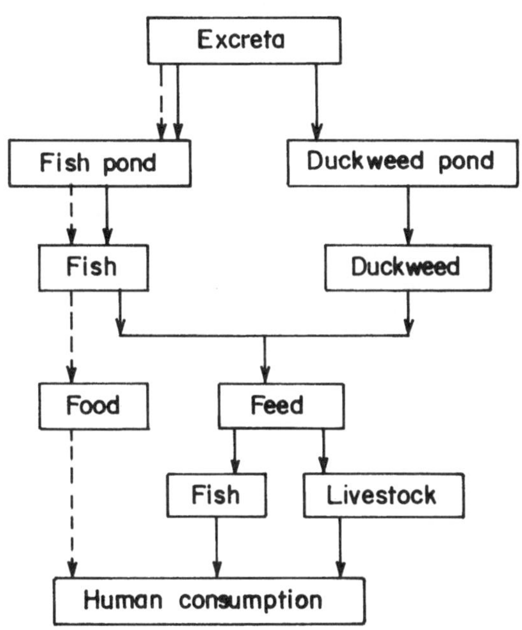

Fig. 10.2 Major excreta-reuse strategies in aquaculture. Excreta may be used directly to produce tilapia for human consumption (broken line) or indirectly through the production of animal feed for fish or livestock (solid line). High-protein feed may be either fish or duckweed, and it may be used to feed either fish or livestock. Source: Edwards (1990).

unacceptable to directly consume such tilapia. Furthermore, wastewater-fed ponds stocked with tilapia could provide an alternative source of fish meal for pelleted feed for intensive aquaculture and feedlot livestock. The concept was inspired by a report that tilapia (probably *O. mossambicus*) used to be cultured in old milkfish ponds in Taiwan. Fish larger than 40 g from frequent seining of ponds could be sold for human consumption but smaller fish were sold as feed for livestock. Maximum yields reached 6.5–7.8 t/ha/yr. Such a practice is commonly carried out informally by farmers who feed small amounts of small fish to pigs and ducks in many countries.

Research has been conducted on the biomass production of a freely breeding population of *O. niloticus* in experimental septage-fed ponds (Edwards, 1988). The size of harvested fish is relatively unimportant if they are intended for animal feed rather than direct human consumption. Net extrapolated yields of 5–6 t/ha/yr were obtain on draining 200 m^2 ponds at the end of 5–7 month experiments, with the mean maximum standing stock in five experiments being only 2.8 t/ha. However, a sustainable and higher extrapolated net average yield of 6.8 t/ha/yr resulted from seining more commercially realistic 1740 m^2 ponds, which was three times the mean standing stock of 2–4 t/ha at the end of the experiment. It is noteworthy that this yield obtained experimentally was similar to the yield reported above from Taiwan in a commercially operated system.

Research has also been conducted on the production of duckweed in septage-fed ponds as a high-protein feed for *O. niloticus* (Edwards *et al.*, 1990). An experiment compared direct use of septage to fertilize ponds stocked with tilapia with indirect use in which duckweed was cultivated in septage-fed ponds and subsequently harvested to feed tilapia *ad libitum* in adjacent ponds. The mean extrapolated net yield of tilapia fed with duckweed was 6.7 t/ha/yr, almost double that of 3.7 t/ha/yr from fishponds directly fertilized with septage. However, the area required for indirect septage reuse was about three times greater than that for direct reuse because duckweed culture needed almost twice the area for fish culture. The feed conversion ratio of duckweed (duckweed dry weight:fish fresh weight) was 3.7, but was 61 on a fresh weight basis. Nevertheless, indirect reuse was more efficient in terms of septage reuse at 1.7 kg fish/m^3 septage compared to only 0.9 kg fish/m^3 septage for direct reuse. The considerably greater land use requirement for indirect reuse may well constrain the feasibility of the system where fish ponds can be constructed on unshaded land. However, duckweed grows better in partial shade than in full light, particularly during the hot season in the tropics when high light intensity and temperature constrain growth. Duckweed may thus be cultivated in derelict ditches as occur in Bangladesh, and in ditches constructed in the conversion of rice fields to fruit orchards, and not compete with fish culture for space for pond construction.

Brackishwater pond culture

The Mozambique tilapia (*O. mossambicus*) has long been a pest in brackishwater milkfish ponds in Indonesia and the Philippines (Guerrero, 1996). It is now known that several tilapia species are euryhaline and can grow in brackishwater through proper acclimation (Suresh and Lin, 1993). A variety of red tilapia has been successfully cultured in brackishwater in the Caribbean (Watanabe, 1991). More recently, a genetic breeding programme has created a new hybrid tilapia with high salinity tolerance, fast growth and, a uniform red skin colour (Lahav and Ra'anan, 1997).

There has been an increasing desire to culture tilapia in brackishwater ponds, particularly in Southeast Asia due to the large number of ponds available, either from a failure of, or a desire to diversify, shrimp culture e.g. in Thailand and Vietnam shrimp culture is now commonly reduced to one crop a year, leaving an empty pond for 6 months. Recently brackishwater tilapia, both Nile tilapia and a red variety, raised intensively on pelleted feed have become a popular commodity in domestic supermarkets in Thailand. Tilapia culture has also attracted shrimp farmers as a potential means to utilize phytoplankton in both shrimp ponds and their effluents (Akiyama and Anggawati, 1998). Recently it has also been reported that hybrids of female *O. niloticus* and male *O. mossambicus* are cultured in fertilized brackishwater ponds in the Philippines and fed supplementary feed (Guerrero, 1996).

Compared to the voluminous literature on semi-intensive culture of tilapia in freshwater ponds, there are limited data for brackishwater. Further research is required on fertilization of brackishwater, and the species composition and nutritional value of phytoplankton to tilapia, to optimize semi-intensive culture.

10.4 POND FERTILIZATION

Rationale

Various chemical elements are required to stimulate phytoplankton productivity in ponds. In practice only nitrogen (N), phosphorus (P) and carbon (C) need to be considered as inputs in pond fertilization. Other elements required in trace amounts are contributed either from the natural environment (air, soil, source water) or from fertilizer and supplementary feed inputs. Unlike field crops, fertilization practices for fish ponds do not consider potassium (K). The need for K and other micro nutrients for pond fertilization has rarely been investigated experimentally, except for the earlier work of Hickling (1962) who reported that K fertilization was ineffective in pond production. Most pond fertilization experiments have emphasized only N and P inputs.

Under optimal growth conditions, the average nutrient composition of phytoplankton biomass is approximately 45–50% C, 8–10% N and 1% P by weight, giving a typical C:N:P ratio of about 50:10:1 (Goldman, 1980) which gives a general idea of the relative amounts of the three nutrients required. Although most pond fertilization experiments have emphasized only N and P inputs, phytoplankton require considerable amounts of C in heavily fertilized fish ponds. For fertilized temperate ponds the total alkalinity (derived from atmospheric CO_2, dissolution of carbonaceous material and respiration of organisms) should be >20 mg/l (Boyd, 1974). A higher concentration is probably required for highly productive ponds in the tropics. Organically fertilized ponds are probably seldom C limited because of bacterial breakdown of organic matter and release of CO_2 but for inorganically fertilized ponds with low alkalinity, liming is recommended (Lin *et al.*, 1997a).

Pond fertilization has a long and confusing history concerning types and amounts of nutrients required and the relative merits of organic and inorganic fertilizers which was reviewed by Colman and Edwards (1987). It is now appreciated that significant amounts of N and P are required to optimize tilapia production through pond fertilization. The need for N fertilization of fish ponds was even initially questioned as several studies failed to demonstrate significant benefits to fish production with applications of N. Most of the earlier studies, in which fish yields were relatively low (<1500 kg/ha), probably failed to show significant benefits of N fertilization in fish ponds due to either a relatively low stocking density (1000–2000 fish/ha) or stocking non-microphagous filter feeding species. Concern has now turned to adverse effects of excessive N loading rates, particularly from chemical fertilizers such as urea. A N loading rate higher than 4 kg N/ha/day, the current maximum recommended rate, led to a significant decrease in net yield of Nile tilapia which was correlated with total ammonia concentrations in pond water of >1.5 mg/l (AIT, unpublished data). High daytime pH $>$ 9 leads to high un-ionized ammonia concentrations which are toxic to fish. The potential impacts of sustained use of high inputs of inorganic fertilizers remain unclear.

Available dissolved P in water is curtailed by its rapid reaction with cations such as Fe, Ca, Mg, and Al, forming biologically inactive precipitates. Phosphorus supply to the water column is also removed by pond mud which strongly adsorbs P (Boyd and Musig, 1981). The mud adsorption capacity of P is linearly correlated to the clay content of the sediments (Shrestha and Lin, 1996a,b). Thus, the amount of P required for pond fertilization is influenced by the type of bottom soils and their % P saturation factors, which are often related to the fertilization history of the ponds (Knud-Hansen, 1992). A P fertilization rate of 1 kg/ha/day is recommended but additional P, 2 kg P/ha/day may be required to maintain N limitation in ponds in acid sulphate soils, which are common in the tropics, due to the soil's sequestering action (Knud-Hansen *et al.*, 1991).

Livestock manures as fertilizers

Fish yields reported from manured ponds range from 7 to 36 kg/ha/day (2555–13 140 kg/ha/yr) (Wohlfarth and Hulata, 1987). Almost all the higher yields (> 30 kg/ha/day) were polyculture of tilapia and Chinese carps in Israel, integrated with feedlot cattle, ducks and chickens. The yields of tilapia monoculture from manured ponds in tropical developing countries have usually been lower, ranging from 8.6 to 19.2 kg/ha/day (Hopkins and Cruz, 1982; Diana *et al.*, 1991; Green *et al.*, 1994; Knud-Hansen and Lin, 1996).

An extrapolated net fish yield of 10.1 t/ha/yr of tilapia monoculture integrated with feedlot ducks has been reported in Thailand (AIT, 1986). However, much of the production probably resulted from formulated feed falling into the pond from inefficient feeding of ducks. Assuming that 10% of feed fed to ducks was 'waste', the pond received an estimated 8.6 t/ha annually of formulated feed in addition to duck manure. The higher yields reported in other integrated feedlot livestock and fish studies may be similarly attributed in part to spilled formulated livestock feed being consumed by fish. Furthermore, earlier studies stocked freely breeding rather than monosex fish; this appears to have further increased yield in longer term experiments because of precocious breeding leading to increased stocking densities.

High fish production in manured ponds is mainly due to protein-rich phytoplankton for filter feeding fish such as Nile tilapia. The phytoplankton also provide dissolved oxygen to support the respiration of aerobic organisms in the pond, including the fish (Colman and Edwards, 1987; Knud-Hansen *et al.*, 1993). Animal manures may also enhance fish pond production through detritus as well as being the elemental nutrient source (Wolhlfarth and Schroeder, 1979; Coleman and Edwards, 1987). However, Nile tilapia is primarily a column feeder and benefited little from detritus in ponds when fish standing crop was low and phytoplankton were abundant (Xu, 1989). Further research is required on the formation and nutritional value of detritus for tilapia in manured ponds.

Livestock manures usually have a N:P ratio well below the optimal for phytoplankton growth but this can be compensated for by increasing the manuring rate. However, heavy fertilization with manure causes severe oxygen depletion which adversely affects yield, particularly if the manure has a high C:N ratio (Edwards *et al.*, 1994a). As grazing buffalo manure (C:N ratio, 26) had a lower N and P content of only 1.4% N and 0.2%P on a dry matter (DM) basis compared to 4.4% N and 1.1% P for feedlot duck manure (C:N ratio, 10) (AIT, 1986), it was loaded into ponds at rates up to 300 kg DM/ha/day to achieve a desirable nutrient loading rate of 4 kg N/ha/day comparable to duck manured ponds. In duck manured ponds the DM loading rate was < 100 kg/ha/day and green water with mean phytoplankton densities of about 35 mg/l led to high fish yields (AIT, 1986). In contrast, buffalo manured ponds had adverse dissolved oxygen regimes due to the respiratory demand of bacterial loadings caused by high buffalo manure loadings and corresponding low fish fields. An additional inhibitory factor was the presence of tannins in buffalo manure which stained pond water brown; this reduced light penetration and inhibited phytoplankton growth (usually < 15 mg/l or half that of duck manured ponds) (Shevgoor *et al.*, 1994). Consequently, it is recommended that buffalo manuring rates should not exceed 50 kg/ha/day (Edwards *et al.*, 1994b) and small-scale farmers in the poor northeast region of Thailand have been recommended to supplement buffalo manure with urea, and poultry manure when it can be collected, to substitute for ruminant manure (Edwards *et al.*, 1996). Inorganic P fertilizers were not readily available but buffalo manure had a low N:P ratio and thus excess P relative to N; farmers generally also fed fish with rice bran which contained significant amounts of P.

A global fertilization research programme

Our knowledge of fertilized ponds has been considerably increased through the Pond Dynamics/Aquaculture Collaborative Research Support Program (PD/A CRSP) which includes US universities and collaborating institutions in several tropical host countries. Experimentation has sought, over the past decade, to elucidate the underlying principles of pond fertilization and to progressively intensify production using Nile tilapia as a single test species (Szyper *et al.*, 1995; Lin *et al.*, 1997a).

Simple addition of inorganic P was assessed in stage 1 according to the traditional limnological expectation that biological production in freshwater would be limited by P but this led to low fish yields of only about 0.8 t/ha/yr.

Stage 2 moved to manuring ponds with aged chicken manure (collected periodically from beneath feedlot birds and sold in sacs), a readily available, standard organic matter input. Nitrogen deficiency became apparent when ponds were fertilized with less than 500 kg chicken manure/ha/week (71 kg/ha/day) (McNabb *et al.*, 1990; Diana *et al.*, 1991). Although this chicken manuring rate led to extrapolated net yields of approximately 5 t/ha/yr (about 2 t/ha in the standard 150 day experimental period), symptoms of excessive organic matter loading rate occurred.

To strike a balance between organic overload and N deficiency, ponds were fertilized at a reduced organic manure rate which was supplemented with urea and triple superphosphate (TSP) in stage 3. It was concluded that the most appropriate fertilization strategy is to apply chicken manure weekly at 200–250 kg DM/ha (29–36 kg DM/ha/day) and to supplement it with urea and TSP to attain a total N rate of 28 kg N/ha (4 kg N/ha/day) and a total P rate of 7 kg P/ha (1 kg P/ha/day). For new ponds and ponds on acidic soils it is recommended that P be increased to 10–14 kg/ha/week (1.4–2 kg/ha/day) (Lin *et al.*, 1997a). Fish yields increased markedly to 23–30 kg/ha/day when the stocking density of fish was increased from 1 to 3 fish/m^2 over the 150 day experiments, i.e. a net harvest of 3.4–4.5 t/ha or an extrapolated net yield of 8–11 t/ha/yr.

Stage 4 experimentation has involved sole use of inorganic fertilizers as elemental nutrients are cheaper than in the form of bulky chicken manure if the latter have to be purchased and transported (if feasible it is better to use fresh chicken manure from poultry directly integrated with fish as fewer nutrients are lost and fresh spilled feed also falls into the pond). Providing that pond water alkalinity is adequate, similar yields to stage 4 have been obtained (Lin *et al.*, 1997a).

Yields in the various fertilization regimes outlined above may be increased by multiple stocking and harvesting strategies rather than the single initial stocking and single final harvesting of fish (Knud-Hansen and Lin, 1996). Physical variables such as pond depth and pond area had generally small effects on fish yield (Szyper *et al.*, 1991).

10.5 SUPPLEMENTARY FEEDING

Rationale

There are a number of economic, social and technical issues which determine the feasibility and effectiveness of supplementary feeding. Optimal pond fertilization strategies stocked with tilapia described in the previous section produced fish of only about 200–250 g in 5 months. Beyond this weight growth slowed considerably, presumably because larger fish were unable to obtain sufficient nutrition from plankton alone to sustain a high growth rate. Larger sized fish which can readily be grown on formulated feed fetch a higher market price in many countries, particularly if they are for the export market (Szyper *et al.*, 1995). However, tilapia market prices are often low, especially domestic markets in developing countries. Thus, there is a need to explore ways to reduce the cost of supplementary feeding. Two strategies are outlined below: delayed feeding with commercial formulated feed; and supplementary feeding based upon nutritional complementarity of natural food in fertilized ponds.

Small-scale farmers may also hesitate to continue to fertilize their ponds if they do not respond to initial attempts to 'green' the water because of highly turbid water. As small ponds are often multipurpose, providing also a source of domestic water, farmers may not wish to fertilize the water and instead use only supplementary feed. An assessment of such a strategy is also outlined below.

Delayed feeding with formulated feed

Initial experiments determined that a daily supplementary feed ration (commercial floating pellets, 30% crude protein) of 25% of satiation was inadequate to maintain growth in an optimally inorganically fertilized pond but that feeding 50% of satiation level produced growth as high as fish fed a 100% satiation ration (Fig. 10.3) (Diana *et al.*, 1994).

Subsequent experiments investigated timing of supplementary feeding to 50%

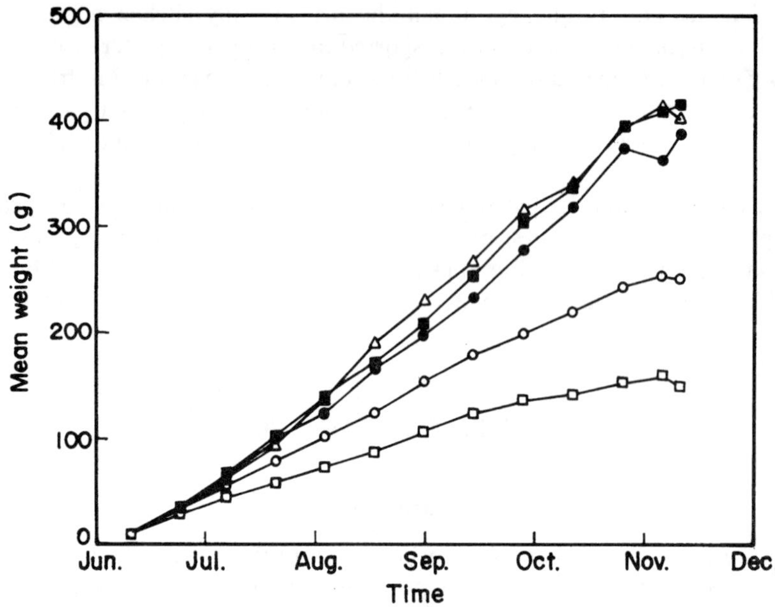

Fig. 10.3 The effect of supplementary feeding with pelleted feed on growth of Nile tilapia in fertilized ponds. All ponds were fertilized at the same optimal rate with inorganic fertilizers. In addition, five treatments received supplementary feed with decreasing percentages of satiation ration. □, Fertilization only, remaining treatments fertilization plus supplementary feed: ■, 100% satiation ration; △, 75% satiation ration; ●, 50% satiation ration; ○, 25% satiation ration. Source: Diana *et al.* (1994).

satiation level in optimally fertilized ponds (Diana *et al.*, 1996). Fish were stocked at an average size of 15 g at $3/m^2$. Treatments varied with size of fish (50, 100, 150, 200 and 250 g) and thus timing of first supplementary feeding at 50% satiation rate. Fish growth at 1.17 g/day was similar for all treatments when pond were only fertilized and was also similar at 3.1 g/day in all treatments after fish began to receive supplementary feed (Fig. 10.4). Each treatment was run until the fish reached a target size of 500 g, which ranged from 236–328 days. Feed conversion ratios averaged 1.0, indicating substantial use of natural feed. The most economical treatment for Thai conditions was to grow fish to 100–150 g for about 3 months with fertilizer alone, prior to commencement of supplementary feeding. Net harvests for the six ponds at which feeding commenced when fish had grown to 100 or 150 g averaged 14.0 t/ha, or an annual average extrapolated net yield of 21.0 t/ha. This is approximately double the extrapolated annual net yield of 8–11 t/ha with fertilization alone. Nor was there an unacceptable deterioration of water quality with supplementary feeding in the static water ponds.

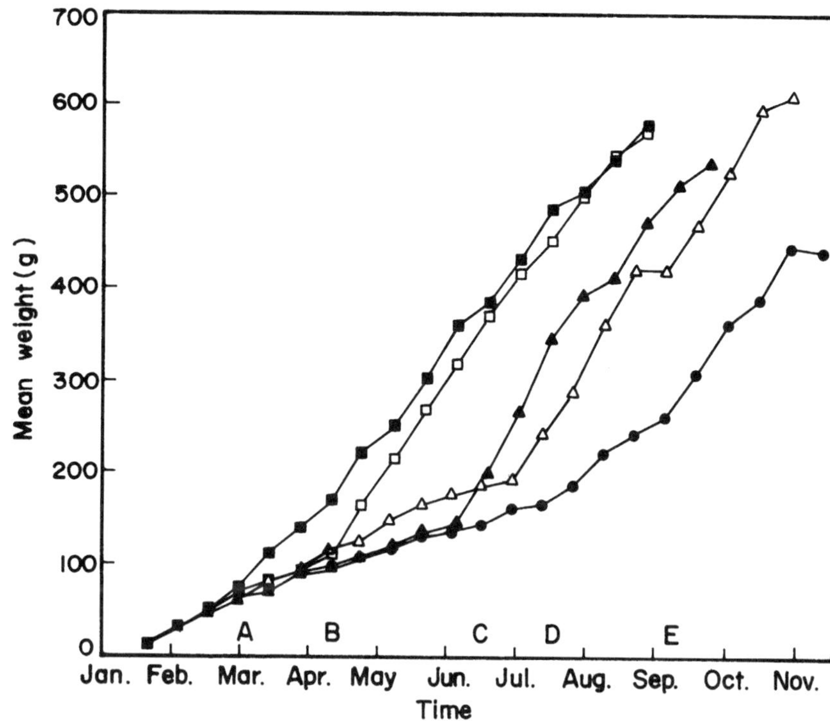

Fig. 10.4 The effect of timing of supplementary feeding on growth of Nile tilapia with pelleted feed at 50% satiation level in fertilized ponds. All ponds were fertilized at the same optimal rate with inorganic fertilizers. Treatments were first feeding at: ■, 50 g; □, 100 g; ▲ 150 g; △, 200 g; and ●, 250 g. The letters on the x axis indicates the dates when supplementary feeding was initiated for each treatment. Source: Diana *et al.* (1996).

Nutritional complementarity of supplementary feeding

Ideally, supplementary feed added to a fertilized pond should complement limiting nutrients in natural food. Hepher (1988) introduced the concept of critical standing crop (CSC), the biomass of fish in the pond at which grow slows due to the development of a nutritional deficit. Tacon and De Silva (1997) broadened the concept to include both standing crops of fish and natural food. To develop a practical method of determining the nutritional demand of the fish in the pond, the quality of different feeds fed to fish was related to the growth and body composition of Nile tilapia (Yakupitiyage, unpublished data).

The experiment assessed how tilapia in fertilized ponds responded to feed containing energy only, energy + protein, energy + protein + minerals, and energy + protein + minerals + vitamins compared to fish grown in a fertilization regime only (Table 10.1). The experiment consisted of five treatments, all fertilized with the control receiving only fertilizer and no supplement. Fish of approximately 25 g initial weight were stocked at $4/m^2$ in the five treatment in the 200 m^2 fertilized ponds.

All nutrients except vitamins were limiting in the fertilized green water ponds (Table 10.1). Fish fed the nutritionally complete diet as a supplementary feed increased in biomass up to 146 kg in the 200 m^2 fertilized pond after 4 months. When vitamin, vitamin + mineral, and vitamin + mineral + protein were deleted from formulated feed, and when supplementary feed was not given, the harvests were reduced to 135, 114, 70, and 50 kg, respectively. Only added vitamin did not have a significant effect on growth of tilapia.

Thus, the following classification of semi-intensive fish culture is proposed based on the above results:

- stage 1: provision of enhanced natural food only i.e. fertilized pond fish culture,
- stage 2: provision of natural food + substrate for energy metabolism e.g. a fertilized pond and use of rice bran as a supplementary feed,
- stage 3: provision of natural food + energy + some building blocks e.g. soybean cake and noodle-waste fed ponds,
- stage 4: provision of natural food +energy + building blocks +catalysers e.g. use of formulated feed containing oil cakes, rice bran, di-calcium phosphate (DCP) and salt.

Pond cultured tilapia have been fed a wide range of feedstuffs including rice bran, aquatic macrophytes, terrestrial macrophytes, a dry feed-mix containing agricultural by-products, and commercial feeds (De Silva, 1993; Yakupitiyage, 1993; Tacon and De Silva, 1997). Feed mixtures are further processed into wet dough or to dry pellet to minimize dispersion losses. Simple feeds such as brans and macrophytes are used in semi-intensive stages 1–3. Dry pelleted feeds are usually used from semi-intensive stage 4 to intensive tilapia culture.

Addition of an energy supplement has been mainly used in carp culture. Benthic feeders appear to mainly benefit from an energy supplement rather than filter feeders (Yakupitiyage, 1993), possibly because the relative scarcity of benthic organisms leads to a pronounced protein sparing effect of the energy supplement. As phytoplankton are abundant in well-fertilized ponds, the protein sparing action of an energy supplement may not be important to plankton feeding fish. Szumeic (1969) and Gurzeda (1969) recommended that approximately one unit of ingested natural feed should be supplemented with three units of energy supplement e.g. barley by weight for common carp. Viola *et al.* (1988) reported that at least 2% of lipid supplement was necessary in animal protein free formulated feed for tilapia. However, oil had no significant positive effect on tilapia growth in 'green water' (Hung, 1989). Commonly used energy

Table 10.1 Growth response of fish fed supplementary feed having varied nutrient composition (Yakupitiyage, unpublished data)

Parameter	Control[1]	Energy supplement[2]	Energy + protein supplement[3]	Energy + protein + P supplement[4]	Energy + protein + P + vitamin supplement[5]
Mean initial weight (g)	24.5 $\pm$2.1[a]	25.5$\pm$4.0[a]	23.6 $\pm$2.8[a]	25.4 $\pm$1.9[a]	26.0 $\pm$1.3[a]
Mean final weight (g)	86.0 $\pm$25.6[d]	124.2$\pm$6.5[c]	189.7 $\pm$13.3[b]	199.5 $\pm$9.9[b]	227.1 $\pm$8.8[a]
Daily weight gain (g)	0.56$\pm$0.24[d]	0.9$\pm$0.08[c]	1.51$\pm$0.15[b]	1.58$\pm$0.07[ab]	1.83$\pm$0.07[a]
Survival (%)	73.4 $\pm$11.1[a]	71.9$\pm$8.4[a]	76.4 $\pm$5.9[a]	85.3 $\pm$1.0[a]	83.8 $\pm$2.7[a]
Gross yield/pond	50.1 $\pm$10.0[d]	70.2$\pm$4.3[c]	113.8 $\pm$5.5[b]	135.7 $\pm$7.3[a]	146.4 $\pm$4.8[a]
Gross yield (kg/ha/day)	22.8 $\pm$4.5[d]	31.9$\pm$2.0[c]	51.7 $\pm$2.5[b]	61.7 $\pm$3.4[a]	66.6 $\pm$2.2[a]
Net yield (kg/ha/day)	13.9 $\pm$4.2[d]	22.7$\pm$0.8[c]	43.1 $\pm$2.7[b]	52.5 $\pm$2.7[a]	57.1 $\pm$1.7[a]

Numbers after $\pm$ indicate standard deviation. The same letter indicates that treatments are not significantly different

[1]4 kg urea N + 2 kg TSP.P ha/day

[2]pelleted cassava starch + lipid

[3]fish meal + soy bean meal mixture

[4] DCP (di-calcium phosphate) mixture

[5]vitamin mixture

supplements are various agricultural by-products such as rice bran, broken rice and maize. If the particle size is small, special preparation of these ingredients is not necessary. However, cooked carbohydrate, e.g. cassava tubers, has certain advantages over uncooked feed due to its gel-forming property, such as increased water stability and digestibility.

It has been demonstrated that plant protein can be used as the main protein source for tilapias without addition of an animal protein component (Viola *et al.*, 1988; Hung, 1989. However, the presence of anti-nutritional factors such as mimosine in *Leucaena* and trypsin inhibitors in soybean should be taken into consideration during the selection of ingredients for feed preparation.

Terrestrial based live feed, such as earthworms, maggots, termites and snails can also be used depending upon their availability or can be produced on-farm using various organic wastes (New *et al.*, 1993).

Viola *et al.* (1988) demonstrated that the addition of 2% di-calcium phosphate is necessary for both carps and tilapia to use soybean based diets efficiently. Hung (1989) also showed a positive growth effect of addition of DCP in a well fertilized culture system. As Ca and P are supplied together it is reasonable to conclude that either Ca or P, or both should be added to supplementary feeds but Dato-Cajegas and Yakupitiyage (1996) found that addition of minerals other than P did not enhance pond cultured tilapia yield.

Supplementary feeding of fish in turbid ponds

Farmers often experience turbid water in their ponds that does not respond to inorganic fertilization as 'green water' does not develop. To develop guidelines for farmers in such a situation, a comparison was made experimentally of the yields fish farmers might obtain if they were to use the same amount of money they would spend on inorganic fertilizers to buy three supplementary feeds (commercial pelleted feed, rice bran and dried cassava chips) using Thai prices in 1994 (Yakupitiyage, unpublished data).

Turbid 200 m^2 ponds were fertilized at a daily rate of 4 kg N and 2 kg P/ha using urea and TSP. An equal amount of cash was used to purchase rice bran, minced cassava chips, and commercial pelleted feed to feed fish in three more treatments.

The yields of fish fed pelleted feed, natural feed from fertilization, rice bran, and cassava chips were 25, 21, 17 and 11 kg/200 m^2 pond, respectively. The survival rate of fish fed cassava chips (70%) was relatively lower that fish grown in other treatments (78–83%). Since pelleted feed was fed to fish and residual feed acts as an indirect fertilizer, an extremely low apparent food conversion ratio (FCR) of 0.2 for fish fed with pelleted feed was observed. Fish fed rice bran and cassava ships also gave relatively low FCR of 0.7 and 1.5, respectively.

In turbid water ponds, farmers can be recommended to use equal costed amounts of inorganic fertilizers or commercial pelleted feed. Rice bran, which is widely used by small-scale farmers, would have produced only about three-quarters of the yield and cassava chips only about half. Turbidity is a complex factor and is often seasonal, being more prevalent in the rainy season, especially in older ponds with unstable dykes due to run-off. Inorganic fertilizers may also be 'washed out' if ponds overflow or function as flow-through systems in the rainy season.

10.6 ENVIRONMENTAL CONCERNS

Nutrient aspects of semi-intensive pond aquaculture

In recent years, aquaculture wastes have been identified as a major source of pollution that deteriorates water quality and accelerates eutrophication in many aquatic ecosystems, particularly from 'open' water cage and raceway systems (Pillay, 1992; Beveridge and Phillips, 1993). However, semi-intensive pond systems are essentially 'closed' or static water, with little to no exchange of water (and nutrients) to the outside environment except when the pond is drained. A preliminary nutrient budget for a pond fertilized with bagged chicken manure and urea and TSP (Edwards, 1993) based on CRSP data showed that 83.1% of the total N and 86% of the total P added in fertilizers was not utilized by fish (15.1% N and 8.1% P) or remained in the water (1.8% N and 5.9% P) but disappeared within the system (Table 10.2). Similar observations have been made subsequently for ponds receiving fertilizers and supplementary feed (Lin *et al.*, 1996). Most of these nutrients were probably deposited in pond mud (Acosta-Nassar *et al.*, 1994) rather than being released to the environment.

Table 10.2 Preliminary nutrient budget for a semi–intensive pond fertilized with aged chicken manure supplemented with urea and triple superphosphate and stocked with Nile tilapia (*Oreochromis niloticus*). The 280 m^2 pond gave a net fish yield of 96.3 kg over the 146 day experiment

Nutrient	Added in fertilizer (kg)	Utilized by fish (kg)[1]	Remaining in water (kg)	Disappeared within the system
N	16.352 (100)	2.466 (15.1)	0.298 (1.8)	13.588 (83.1)
P	4.088 (100)	0.328 (8.1)	0.242 (5.9)	3.518 (86.0)

Figures in parentheses are percentages of nutrients. N, total nitrogen; P, total phosphorus. Source: Edwards (1993)

[1]Assumptions: fish 16% crude protein and protein 16% N; fish 0.34% P, both on fresh weight basis

Much lower losses to pond mud (70% N and 35–40% P) with a correspondingly higher loss in pond effluents were reported from tilapia ponds in Honduras (Green and Boyd, 1995). As pond water contains a relatively small proportion of the total nutrients added, the increased effluent nutrient load was due to disturbance and suspension of pond sediments during fish seining and pond draining. It has been recommended that 70–80% of the pond water be drained before seining the fish to reduce turbidity and nutrient loss, and the remaining water be kept in a retention pond (Lin *et al.*, 1997b).

The optimal fertilization rate recommended for tilapia culture is 4 kg N and 1 kg P/ha/day which is equivalent to 1460 kg N and 895 kg P_2O_5 /ha/yr or a total of 2355 fertilizer nutrients. This rate is over five times higher than the highest fertilization rate in Asia for arable and permanently cropped land, 425 kg/ha for the Republic of Korea (ESCAP, 1991). A priority research area is nutrient budgets of semi-intensive fish ponds because of the large nutrient inputs with potential to cause eutrophication. However, as most of the nutrients added to the pond are sequestered in pond mud, strategies can be developed to recycle the nutrients in agriculture and contribute to agricultural diversification also.

Water use

Adequate water is a pre-requisite for aquaculture. Water consumption for fish ponds depends on pond depth, seepage rate, weather conditions and management. Tilapia culture in a pond of depth of 1.0–1.5 m requires 10 000 to 15 000 m^3 of water to fill a 1 ha pond. Tropical ponds normally lose water at a rate of 1–2 cm/day (100–200 m^3/ha) through seepage and evaporation (Hepher and Pruginin, 1982) giving an estimated annual water consumption range from 40 000 to 80 000 m^3/ha/yr. For an average tilapia production of 5000–8000 kg ha/yr in a semi-intensive pond (Lin *et al.*, 1997a), it would require 5 to 16 m^3 of water to produce 1 kg of fish, which is extremely inefficient water use. More efficient water use would result from increased fish yield through intensification of the culture system. The annual extrapolated production of tilapia may be increased to 20 000 kg/ha with partial supplementary feeding with pelleted feed which reduces water consumption to 2–4 m^3/kg fish. For tilapia production in fertilized ponds, fish yield and water quality were not affected significantly by pond depths of 0.6, 1.0 and 1.5 m (Szyper *et al.*, 1991). Shallower ponds would require less water to fill, but would have no effect on the major loss through seepage and evaporation.

REFERENCES

Acosta-Nassar, M.V., Morell, J.M. and Corredor, J.E. (1994) The nitrogen budget of a tropical semi-intensive freshwater fish culture pond. *Journal of the World Aquaculture Society* **25**(2), 261–270.

Akiyama, D.M. and Anggawati, A.M. (1998). Growing tilapia with shrimp increased shrimp production, tended to improve pond conditions. *Aquaculture Asia* **3**(2), 18–19.

AIT (1986) *Buffalo/Fish and Duck/Fish Integrated Systems for Small-Scale Farmers at the Family Level.* AIT Research Rpt, 198, Asian Institute of Technology, Bangkok.

Balarin, J.D. (1988) Development planning for tilapia farming in Africa, in *The Second International Symposium on Tilapia in Aquaculture* (eds R.S.V. Pullin, T. Bhukaswan, K. Tonguthai and J.L. Maclean), ICLARM Conference Proceedings 15, Department of Fisheries, Bangkok and ICLARM, Manila, pp. 531–538.

Beveridge, M.C.M. and Phillips, M.J. (1993) Environmental impact of tropical inland aquaculture, in *Environment and Aquaculture in Developing Countries* (eds R.S.V. Pullin, H. Rothenthal and J. L. Maclean), ICLARM Conference Proceedings 31, pp. 213–236.

Boyd, C.E. (1974) *Lime Requirements of Alabama Fish Ponds*, Alabama Agricultural Experiment Station, Bull. 459, Auburn University, Alabama.

Boyd, C.E. and Musig, Y. (1981) Orthophosphate uptake by phytoplankton and sediment. *Aquaculture 22*, 165–173.

Cavallini, J.M. (1996) *Aquaculture Using Treated Effluents from the San Juan Stabilization Ponds, Lima, Peru*, CEPIS, Lima.

Chen, L.C. (1990) *Aquaculture in Taiwan*, Fishing News Books, London.

Chen, T.P. and Li, Y.P. (1980) Integrated agriculture-aquaculture studies in Taiwan, in *Integrated Agriculture-Aquaculture Farming Systems* (eds R.S.V. Pullin and Z.H. Shehadeh), ICLARM Conference Proceedings 4. ICLARM, Manila, pp. 239–241.

Coleman, J.A. and Edwards, P. (1987) Feeding pathways and environmental constraints in waste-fed aquaculture: balance and optimization, in *Detritus and Microbial Ecology in Aquaculture* (eds D.J.W. Moriarty and R.S.V. Pullin), ICLARM Conference Proceedings 31, ICLARM, Manila, pp. 240–281.

Csavas, I. (1993) Aquaculture development and environmental issues in the developing countries of Asia, in *Environment and Aquaculture in Developing Countries* (eds R.S.V. Pullin, H. Rosenthal and J.L. Maclean), ICLARM Conference Proceedings 31, ICLARM, Manila, pp. 74–101.

Dato-Cajegas, C.R.S. and Yakupitiyage, A. (1996) The need for dietary mineral supplementation for Nile tilapia, *Oreochromis niloticus*, cultured in a semi-intensive system. *Aquaculture* **144**, 227–237.

De Silva, S.S. (1993) Supplementary feeding in semi-intensive aquaculture systems, in *Farm-Made Aquafeeds* (eds M.B. New, A.G.J. Tacon and I. Csavas), Proceedings of the FAO/AADCP Regional Expert Consultation on Farm-Made Aquafeeds, 14–18 December 1992, Bangkok, Thailand. FAO-RAPA/AADCP, Bangkok, pp. 24–60.

Diana, J.S., Lin, C.K. and Schneeberger, P.J. (1991) Relationship among nutrient inputs, water nutrient concentrations, primary production and yield of *Oreochromis niloticus* in ponds. *Aquaculture* **92**, 323–341.

Diana, J.S., Lin, C.K. and Jaiyen, K. (1994) Supplemental feeding of tilapia in fertilized ponds. *Journal of the World Aquaculture Society* **25**(4), 497–506.

Diana, J.S., Lin, C.K. and Yang Yi. (1996) Timing of supplemental feeding for tilapia production. *Journal of the World Aquaculture Society* **27**(4), 410–419.

Edwards, P. (1983) The future potential of integrated farming in Asia. *Proceedings V World Conference on Animal Production* **1**, 273–281.

Edwards, P. (1988) Tilapia raised on septage as high protein animal feed, in *the Second International Symposium on Tilapia in Aquaculture* (eds R.S.V. Pullin, T. Bhukaswan, K. Tonguthai and J.L. Maclean), ICLARM Conference Proceedings 15, Department of Fisheries, Bangkok, and ICLARM, Manila, pp. 7–13.

Edwards, P. (1992) *Reuse of Human Wastes in Aquaculture: A Technical Review*, Water and Sanitation Report No. 2, World Bank, Washington D.C.

Edwards, P. (1993) Environmental issues in integrated agriculture-aquaculture and wastewater-fed fish culture systems, in *Environment and Aquaculture in Developing Countries* (eds R.S.V. Pullin, H. Rosenthal and J.L. Maclean), ICLARM Conference Proceedings 31, ICLARM, Manila, pp. 139–170.

Edwards, P. and Pullin, R.S.V. (1990) *Wastewater-Fed Aquaculture*, Proceedings of the International Seminar on Wastewater Reclamation and Reuse for Aquaculture, Calcutta, India, 6–9 December, 1988, Environmental Sanitation Information Center, Asian Institute of Technology, Bangkok.

Edwards, P., Pullin, R.S.V. and Gartner, J.A. (1986) *Research and Education for the Development of Integrated Crop-Livestock-Fish Farming Systems in the Tropics*, ICLARM Studies and Reviews 16, ICLARM, Manila.

Edwards, P., Pracharaprakiti, C. and Yomjinda, M. (1990) Direct and indirect reuse of septage for culture of Nile tilapia *Oreochromis niloticus*, in *The Second Asian Fisheries Forum* (eds R. Hirano and I. Hanyu), Asian Fisheries Society, Manila, pp. 165–168.

Edwards, P., Pacharaprakiti, C. and Yomjinda, M. (1994a) An assessment of the role of buffalo manure for pond culture of tilapia, 1. On-station experiment. *Aquaculture* **126**, 83–95.

Edwards, P., Kaewpaitoon, K., Little, D.C. and Siripandh, N. (1994b) An assessment of the role of buffalo manure for pond culture of tilapia. II. Field trial. *Aquaculture* **126**, 97–106.

Edwards, P., Demaine, H., Innes-Taylor, N. and Turongruang, D. (1996) Sustainable aquaculture for small-scale farmers: need for a balanced model. *Outlook on Agriculture* **25**(1), 19–26.

Edwards, P., Little, D.C. and Yakupitiyage, A. (1997) A comparison of traditional and modified inland artisanal aquaculture systems. *Aquaculture Research* **28**, 777–788.

ESCAP (1991) Comparative economic indicators of the fertilizer sector 1989/90. *Agro-Chemicals News in Brief*, Special Issue, November 1991, FADINAP/ARSAP, Agriculture and Rural Development Division, ESCAP, Bangkok.

FAO (1997) *Aquaculture Production Statistics 1986–1995*, FAO Fisheries Circular No. 815, Revision 9. FAO, Rome.

Goldman, J.C. (1980) Physiological processes, nutrient availability and concept of relative growth rate in marine phytoplankton ecology, in *Primary Productivity in the Sea* (ed. P.G. Falkowski), Plenum Press, New York, pp. 179–194.

Green, B.W. and Boyd, C.E. (1995) Chemical budgets for organically fertilized ponds in the dry tropics. *Journal of the World Aquaculture Society* **26**, 284–296.

Green, B.W., Teichert-Coddington, D.R. and Hanson, T.R. (1994) *Development of Semi-intensive Aquaculture Technologies in Honduras: Summary of Freshwater Aquacultural Research Conducted from 1983 to 1992*, Research and Development Ser. 39. Auburn University, Alabama.

Guerrero, R.D. III. (1996) Aquaculture in the Philippines. *World Aquaculture* **27**(1), 7–13.

Gupta, M.V., Ahmed, M., Bimbao, M.A.P. and Lightfoot, C. (1992) *Socio-economic Impact and Farmers' Assessment of Nile Tilapia (*Oreochromis niloticus*) Culture in Bangladesh*, ICLARM Tech. Rep. 35, ICLARM, Manila.

Gurzeda, A. (1969) *Relationship Between Natural and Artificial Food in Carp Nutrition.* EIFAC Tech. Paper 9, pp. 118–130.

Harrison, E., Stewart, J.A., Stirrat, R.L. and Muir, J. (1994) *Fish Farming in Africa – What's the Catch?* Institute of Aquaculture, University of Stirling, Stirling, University of Sussex, Brighton, and Overseas Development Administration, London.

Hepher, B. (1972) *Supplementary Feeding in Fish Culture*, Proceedings of the 9th International Congress of Nutrition, Mexico 1972, Vol 3, pp. 183–198.

Hepher, B. (1988) *Nutrition of Pond Fishes*, Cambridge University Press, Cambridge.

Hickling, C.F. (1962) *Fish Culture*, Faber and Faber, London.

Hickling, C.F. (1963) The cultivation of tilapia. *Scientific American* **208**, 143–152.

Hopkins, K.D. and Cruz, E.M. (1982) *The ICLARM-CLSU Integrated Animal- Fish Farming Project: Final Report,* ICLARM Tech. Rep 5, ICLARM, Manila.

Hung, L.T. (1989) *Evaluation of Soybean Meal as a Supplementary Feed for Red Tilapia,* MSc thesis, Asian Institute of Technology, Bangkok.

Jamu, D.M. and Costa-Pierce, B.A. (1995) Culture of indigenous species using on–farm resources: a rational approach to the development and evolution of aquaculture in rural Africa. *World Aquaculture* **26**(2), 45–49.

Jhingran, A.G. (1992) Performance of tilapia in Indian waters and its possible impact on the native ichthyofauna, in *Paper Contributed to the Workshop on Tilapia in Capture and Culture-enhanced Fisheries in the Indo-Pacific Fishery Commission Countries* (ed. E.A. Balayut), 27–29 June 1991, Bogor, Indonesia. FAO Fisheries Report No. 458, Supplement, FAO, Rome, pp. 143–161.

Knud-Hanson, C. F. (1992) Pond history as a source of error in fish culture experiments: a quantitative assessment using covariate analysis. *Aquaculture* **105**, 21–36.

Knud-Hansen, C.F. and Batterson, T.R. (1994) Effect of fertilization frequency on the production of Nile tilapia (*Oreochromis niloticus*). *Aquaculture* **123**, 271–280.

Knud-Hansen, C.F. and Lin, C.K. (1996) Strategies for stocking Nile tilapia (*Oreochromis niloticus*) in fertilized ponds, in *The Third International Symposium on Tilapia in Aquaculture* (eds R.S.V. Pullin, J. Lazard, M. Legendre, J.B.A. Kothias and D. Pauly), ICLARM Conference Proceedings 41, ICLARM, Manila, pp. 70–76.

Knud-Hansen, C.F., McNabb, C.D. and Batterson, T.R. (1991) Application of limnology for efficient nutrient utilization in tropical pond aquaculture. *Verh. Internat. Verein. Limnol.* **24**, 2541–2543.

Knud-Hanson, C.F., Batterson, T.R. and McNabb, C.D. (1993) The role of chicken manure in the production of Nile tilapia, *Oreochromis nolotícus* (L.). *Aquaculture and Fisheries Management* **24**, 483–493.

Lahav, E. and Ra'anan, Z. (1997) Salinity tolerance of genetically produced tilapia (*Oreochromis*) hybrids. *Bamidgeh* **49**(3), 160–165.

Li, S.F. (1990) Recent advances in freshwater aquaculture in China, in *Aquaculture in Asia* (ed M.M. Joseph), Asian Fisheries Society, Indian Branch, Mangalore, pp. 141–161.

Lin, C.K. and Diana, J.S. (1995) Co-culture of Nile tilapia (*Oreochromis niloticus*) and hybrid catfish (*Clarias mocrocephalus* × *C. gariepinus*) in ponds. *Aquatic Living Resources* **8**, 449–454.

Lin, C.K., Jaiyen, K. and Muthuwan, V. (1989) Intergration of intensive and semi–intensive aquaculture: concept and example. *Thai Fisheries Gazette* **42**, 425–430.

Lin, C.K., Yang, Y. and Diana, J.S. (1996) The effect of pond management strategies on nutrient budgets: Thailand, in Pond Dynamics/Aquaculture CRSP 14th Ann. Rpt., Oregon State University, Covallis, Oregon, pp. 19–24.

Lin, C.K., Teichert-Coddington, D.R. Green, B.W. and Veverica, K. (1997a) Fertilization regimes, in *Dynamics of Pond Aquaculture* (eds H.S. Egna and C.E. Boyd), CRC Press, New York, pp. 73–108.

Lin, C.K., Shrestha, M.K., Thakur, D.P. and Diana, J.S. (1997b) Management to minimize the environmental impacts of pond draining, in *Pond Dynamics/Aquaculture CRSP 15th Ann. Rpt.*, Oregon State University, Covallis, Oregon, pp.132–139.

Little, D. and Satapornvanit, K. (1997) Poultry and fish production: a framework for their integration in Asia. Second FAO Electronic Conference on Tropical Feeds, Livestock Feed Resources within Integrated Farming Systems (9 September 1996 28 February 1997), 21 p.
(http://www.fao.org.org/waicent/faoinfo/agricult/aga/agap/conf96.htm/little2.htm)

Lovshin, L.L., Schwartz, N.B., de Castillo, V.G., Engle, C.R. and Hatch, U.L. (1986) *Cooperatively Managed Panamanian Rural Fish Ponds: the Integrated Approach.* International Center for Aquaculture Research and Development Series No. 33, Alabama Agricultural Experiment Station, Auburn University, Alabama.

Maclean, J.L. (1994) Tilapia – the aquatic chicken. *ICLARM Newsletter* **7**(1), 17.

Mara, D.D., Edwards, P., Clark, D. and Mills, S.W. (1993) A rational approach to the design of wastewater-fed fishponds. *Water Research* **27**(12), 1797–1799.

McNabb, C.D., Batterson, T.R., Premo, B.J., Knud-Hansen, C.F., Eidman, H.M., Lin, C.K., Jaiyen, K., Hanson, J.E. and Chuenpagdee, R. (1990) Managing fertilizers for fish yield in tropical ponds in Asia, in *The Proceedings of the Second Asian Fisheries Forum* (eds R. Hirano and I. Hanyu), The Asian Fisheries Society, Manila, pp. 169–172.

Morrice, C., Chowdhury, N.I. and Little, D.C. (1998) Fish markets of Calcutta. *Aquaculture Asia* **3**(2), 12–14.

New, M.B., Tacon, A.G.J. and Csavas, I. (eds) (1993) *Farm–made Aquafeeds*, Proceedings of the FAO/AADCP Regional Expert Consultation on Farm-made Aquafeeds, 14–18 December 1992, Bangkok, Thailand. FAO-RAPA/AADCP, Bangkok.

Pillay, T.V.R. (1992) *Aquaculture and the Environment*, Fishing News Books, Oxford.

Popma, T.J., Phelps, R.P., Castillo, S., Hatch, L.U., Hanson, T.L. and Duncan, B. (1995) Family-scale fish farming in Guatemala: an example of sustainable aquacultural development through national and international collaboration, *Proceedings of the PACON Conference on Sustainable Aquaculture 95*, Pacific Congress on Marine Science and Technology, Pacon International, June 11–14, Honolulu, Hawaii.

Pullin, R.S.V., Bimbao, M.A.P. and Bimbao, G.B. (1994) World outlook for tilapia farming. *Paper presented at the First International Symposium on Aquaculture*, 9–11 June Boca del Rio, Vera Cruz, Mexico.

Rahman, A.K.A. (1992) Tilapia in Bangladesh, in *Paper Contributed to the Workshop on Tilapia in Capture and Culture-enhanced Fisheries in the Indo-Pacific Fishery Commission Countries* (ed E.A. Balayut), 27–29 June 1991, Bogor, Indonesia. FAO Fisheries Report No. 458, Supplement, FAO, Rome, pp. 139–142.

Schroeder, G.L., Wohlfarth, G.W., Alkon, A., Halevy, A. and Krueger, H. (1990) The dominance of algal–based food webs in fish ponds receiving chemical fertilizers plus organic manures. *Aquaculture* **86**, 219–229.

Shevgoor, L., Knud-Hansen, C.F. and Edwards, P. (1994) An assessment of the role of buffalo manure for pond culture of tilapia. III. Limiting factors. *Aquaculture* **126**, 107–118.

Shrestha, M. and Lin, C.K. (1996a) Phosphorus fertilization strategy in fish ponds based on phosphorus saturation level. *Aquaculture* **142**, 207–219.

Shrestha, M. and Lin, C.K. (1996b) Determination of phosphorus saturation level in relation to clay content in formulated pond muds. *Aquacultural Engineering* **15**(6), 441–459.

Smiley, F.F. (1995) Tilapia: the fish of the decade, in *Aquaculture Towards the 21st Century* (eds K.P.P. Nambiar and T. Singh), INFOFISH, Kuala Lumpur, pp. 117–121.

Suresh, R.V. and Lin, C.K. (1992) Tilapia culture in saline waters: a review. *Aquaculture* **106**, 201–226.

Szumeic, J. (1969) Relationship between natural and artificial food in carp feeding, in *New Developments in Carp and Trout Nutrition*, Papers submitted to a symposium held in conjunction with the 5th session of EIFAC, Rome, May 1968, EIFAC Tech. Paper 9, FAO, Rome, pp. 140–160.

Szyper, J.P., Hopkins, K.D. and Lin, C.K. (1991) Production of Oreochromis niloticus (L.) and ecosystem dynamics in manured ponds of three depths. *Aquaculture and Fisheries Management* **22**, 169–180.

Szyper, J.P., Lin, C.K., Little, D., Setboonsarng, S., Yakupitiyage, A., Edwards, P. and Demaine, H. (1995) Techniques for efficient and sustainable mass production of tilapia in Thailand, in *Proceedings of the PACON Conference on Sustainable Aquaculture 95*, Pacific Congress on Marine Science and Technology, Pacon International, Honolulu.

Tacon, A.G.J. and De Silva, S.S. (1997) Feed preparation and feed management strategies within semi-intensive fish farming systems in the tropics. *Aquaculture* **151**, 379–404.

Watanabe, W.O. (1991) Saltwater culture of tilapia in the Caribbean. *World Aquaculture* **22**, 49–54.

Wohlfarth, G. and Hulata, G. (1987) Use of manures in aquaculture, in *Detritus and Microbial Ecology in Aquaculture* (eds D.J.W. Moriarty and R.S.V. Pullin), ICLARM Conference Proceedings 14, ICLARM, Manila, pp. 353–367.

Wohlfarth, G.W. and Schroeder, G.L. (1979) Use of manure in fish farming – a review. *Agriculture Wastes* **1**(4), 279–299.

Viola, S., Rappaport, O. and Zohar, G. (1988) Animal protein free feeds for hybrid tilapia (*O.niloticus* × *O. aureus*) in intensive culture. *Aquaculture* **75**, 115–125.

Xu, Hui. (1989) *A Comparison of Tilapia* (Oreochromis niloticus) *Production in Pond, Cage and Pen Culture*, AIT thesis, Asian Institute of Technology, Bangkok.

Yakupitiyage, A. (1989) *Quantitative and Qualitative Aspects of Energy Acquisition of the Cichlid Fish* Oreochromis niloticus, PhD thesis, Institute of Aquaculture, University of Stirling, Stirling.

Yakupitiyage, A. (1993) On-farm feed preparation and feeding strategies for carps and tilapias, in *Farm-Made Aquafeeds* (eds M.B. New, A.G.J. Tacon and I. Csavas), Proceedings of the FAO/AADCP Regional Expert Consultation on Farm-Made Aquafeeds, 14–18 December 1992, Bangkok, FAO-RAPA/AADCP, Bangkok, pp.87–100.

Yang, Y., Lin, C.K. and Diana, J.S. (1996) Influence of Nile tilapia (*Oreochromis niloticus*) stocking density in cages on their growth and yield in cages and in ponds containing the cages. *Aquaculture* **146**, 205–215.

Chapter eleven

Production in intensive and recycle systems

J. Muir, J. van Rijn and J. Hargreaves

11.1 INTRODUCTION

The production of tilapia is most widely associated with tropical or sub-tropical semi-intensive pond systems, and it is towards these systems, described elsewhere in this text, that significant research activity and development interest has focused. As for other warm-water species, advantages of semi-intensive pond culture are widely recognized for simple artisanal production and rural food supply. Semi-intensive and intensive cage culture can also be viable in a range of circumstances, from small to large scale (Beveridge, 1996; Costa-Pierce, 1995), and can contribute usefully to rural livelihoods. However, production has also moved along a parallel direction, involving heavily stocked ponds, tanks or raceways, normally as commercial enterprises in better resourced locations, supplying more developed markets. Such systems are characterized by their substantial dependence on water exchange and external feeding, and are therefore designated as intensive; in many cases, given the extremely high stocking densities employed, and the very controlled management systems, the term super-intensive may be more appropriate. Production ranges of 100–500 t ha^{-1} yr^{-1} are typical.

Many intensive systems have been based on the copious replenishment of flowing water from streams and rivers, or pumped from surface, ground or coastal waters. However, interest has also increased in reuse or recycle systems, in which water is maintained within the holding unit, or treated and returned. Such systems can be installed in existing units to provide more stable and secure water supplies, or to increase production from existing resources. They also

M.C.M. Beveridge and B.J. McAndrew (eds), Tilapias: Biology and Exploitation, 405–445.

permit production in otherwise non-viable sites – e.g. where temperature, salinity or other parameters are outside acceptable ranges – or near urban centres where land and water access or costs might be prohibitive for conventional forms of production, but where improved market access could confer sufficient advantage.

The potential of intensive tilapia production has been realized slowly, as the species has not had the immediate appeal of the high value candidates for intensive aquaculture, such as salmon, trout, turbot and eel. However, a number of pioneering producers had recognized the possibilities for production in warm waters, using a relatively robust species with undemanding feed requirements. In the 1970s, prospects of replacing channel catfish with tilapia were explored in the USA, particularly in geothermal systems (Lauenstein, 1978; Ray, 1978) and in warmer waters of the southern states, by intensive and recycle tank producers facing increased problems of competing with pond catfish production in major markets (Muir, 1975). Work in Mombasa, Kenya at the Baobab farm (Balarin and Haller, 1982) also did much to establish basic operating parameters linking behaviour, stocking density, feeding patterns and growth; this was complemented by research in the University of Stirling on nutrition and physiology (Jauncey and Ross, 1982; Al Amoudi, 1987). A further key development was the intensive tilapia system developed by the Tate and Lyle sugar company in Zambia, an associated development linked with the University of Liege (Melard and Philippart, 1980), and the establishment of a commercial intensive tilapia unit using simple elliptical raceways fed with heated water from the nuclear power plant at Tihange, Belgium. At the time of writing, this system is still in production, producing some 250 t per year.

The conditions for viability in intensive systems are more restricted than those for semi-intensive production. Although land area and system volume requirements are lower, significant costs are typically incurred in the continuous supply and/or re-use of water and the provision of complete feeds. In addition, the intensive concentration of energy and nutrients results in definable and increasing costs of waste treatment and disposal, and hence increasing incentives for developing self-contained recycle systems. This chapter seeks to explain the significance and potential of tilapia in intensive and recycle production systems, describes factors that contribute to opportunities and constraints, and provides examples of current developments in this field. Finally, a brief conspectus is provided of the longer-term directions that might be expected.

11.2 THE SUITABILITY OF TILAPIA FOR INTENSIVE PRODUCTION

Various species of tilapia, particularly *Oreochromis niloticus* and *O. mossambicus*, and their associated hybrids, including the various 'red' tilapias, have been reared intensively, and it is useful to consider those attributes favouring their use, and to identify means by which systems might be established or operated to meet husbandry requirements. Tilapias have occupied habitats differing widely in physical and chemical parameters (Balarin and Hatton, 1979; Philippart and Ruwet, 1982). Temperature is a key property determining their distribution, and within extremes of 6°C to 40°C for natural populations, the optimum for most cultivated species varies from 24°C to 32°C. Traditionally, therefore, tilapia have been cultured in tropical and subtropical regions, and most intensive systems are located accordingly. However, by using heated, power-plant effluents or heated and recycled culture units, they can also be viable in temperate regions.

Tilapia can reproduce and grow under a wide range of water quality conditions typical of natural and enriched waters. Hybridization has also been effective in exploiting the tolerance range among different tilapia species (Pullin, 1988). A trait of particular interest for intensive culture is tolerance of organic-rich waters, for which the genera *Sarotherodon* and *Oreochromis*, characterized by larger numbers of gill rakers, are particularly well suited. Table 11.1 outlines some key relationships and their implications for intensive culture, though values defining sub-optimal conditions (e.g. effect concentrations, such as the EC_{90} for oxygen, at which 90% of optimal condition growth rate is attained) are more difficult to determine. However, once primary environmental conditions can be specified, the system can be defined and species chosen. The introduction of new species might be an issue where intensive or particularly recycle production provides opportunities to locate in non-indigenous areas. However in cold-water, arid and/or more highly saline environments there is a natural biological barrier which can usually provide the necessary safeguards. Though intensive systems may be designed and operated to reduce problems of mixed-sex culture, all-male hybrids offer advantages as with other tilapia production systems. Indeed, well-defined intensive systems may offer a more specific template for targeting selection attributes.

Intensive systems are generally characterized by high degrees of water replenishment and hence relatively high velocities and moderately active conditions, relatively high volumetric stocking density (ρ_v, e.g. kg m^{-3}), and high application rates of external feed. A complex of factors interact in determining the husbandry environment, including volume and configuration of the holding unit, water quality and turnover rate, behaviour and stock distribution, individual location, swimming speed and energy expenditure, feed intake, metabolic output, individual growth, size distribution, dominance and mass growth. Fig. 11.1 outlines key relationships, which are developed further. As also discussed later, the situation is even more complex for water reuse, where a range of additional interactions may occur between the 'stock sub-system' and the 'water quality management sub-system'.

Within this complex of relationships, basic energetic, growth and nutritional requirements of the stock must be defined and provided for (Balarin and Haller, 1982; Chervinski, 1982; Rakocy, 1989). Key issues are summarized in Table 11.2. By their nature, these systems are concentrated in energetic and other

Table 11.1 Water quality parameters relevant for intensive tilapia culture

Factor	Implications for intensive culture	Critical/ideal values
Oxygen	Water saturation levels, scope for growth, extent water supplies can be used/need for aeration or oxygenation	$<2/>6$ mg l^{-1}
Salinity	Gas saturation, energetics, spawning and early rearing	$<40/<10$ ppt
Temperature	Metabolic rates, growth, respiration, extreme limits	10–35/28–32°C
TAN (total ammonia nitrogen)	Metabolite, toxicity of unionized ammonia (UIA) fraction in higher pH water, extent to which water can be reused without ammonia removal	$>15/<3$ mg l^{-1} TAN $>1/<0.2$ mg l^{-1} UIA
Suspended solids	Respiratory occlusion, disease risks, reduced metabolism	$>200/<20$ mg l^{-1}
BOD	Organic oxygen uptake, additional water/O_2 demand	$>200/<5$ mg l^{-1}
CO_2	Metabolite, builds up in reused low pH water, toxicity and possible kidney function impairment	$>20/<3$ mg l^{-1}
NO_2	Possible buildup from partial TAN oxidation and NO_3 reduction; interferes with O_2 uptake in blood haemoglobin; reduced effect in higher chloride levels	$>5/<1$ mg l^{-1}
NO_3	Buildup from complete TAN oxidation, limited toxicity, usually diluted out	$>500/<20$ mg l^{-1}
H_2S, methane	May outgas from anaerobic breakdown of collected wastes; can depress growth, H_2S may be toxic	Usually <1 mg l^{-1}
Heavy metals	Usually external or from leaching system components; more toxic at low pH, usually sublethal effects; smaller fish more sensitive	Usually <0.1 mg l^{-1}

Notes: based on typical design standards

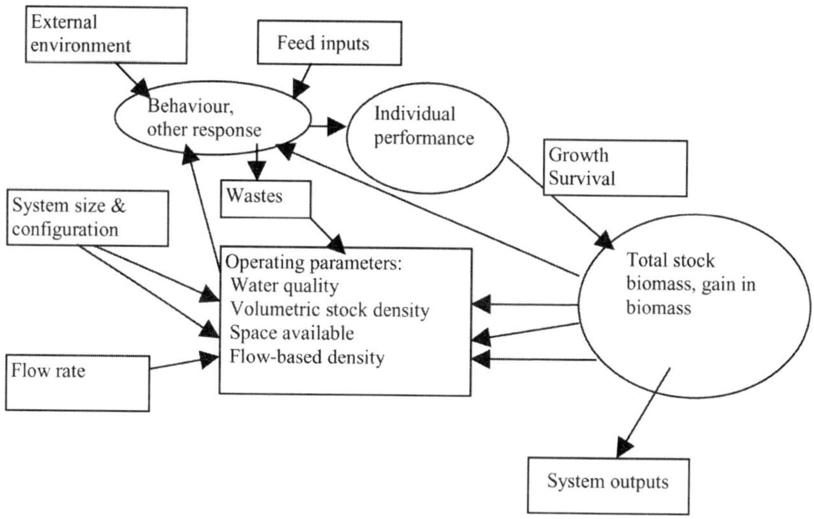

Fig. 11.1 Relationship between stock, system, biomass and output in intensive systems.

Table 11.2 Key physiological aspects of intensive tilapia production

Feature	Implications for intensive culture
Behaviour	Acceptance of crowding, aggression, dominance, opportunities to feed, size distribution
Swimming speed	Ability to maintain position; avoid stress; energy intake and body condition; effects on territoriality; flow rates, circulation and water exchange
Respiration	O_2 and CO_2 balances, scope for growth, activity, response to stress; relationship with feed input
Energy requirements	Feed composition, growth, relationship with swimming speed and other activity; metabolic outputs
Nutritional requirements	Composition, size, presentation, density, health, growth
– Protein	nitrogen metabolism, water exchange, muscle growth
– Carbohydrate	energy, CHO metabolism, water exchange, growth
– Lipids	energy, velocities, flesh quality, growth, wastes growth,
– Key elements	disease resistance
Growth	Production regimes, size distribution, market outputs
Metabolic effects	Patterns/with T, feed, etc; water exchange, discharges

resource terms, and their potential as sources of environmental impact must be considered. In particular, intensive feeding of a large contained biomass may result in significant output of waste nutrients, and where diseases require control, the potential discharge of therapeutants. These issues are encountered in other aquaculture sectors, and are not unique to intensive tilapia production (see e.g. Baird *et al.*, 1996) nor indeed are they absent from other forms of tilapia culture (see previous chapters). However, the higher potential for production control in intensive systems may confer advantages if efficient management and husbandry techniques are applied. The practical effectiveness of this is discussed later.

Table 11.3 provides a summary of the comparative features of intensive and recycle systems with semi-intensive ponds or cages, and intensive cages (see also Chapter 10).

11.3 SYSTEM DESIGN AND OPERATION – THEORETICAL AND PRACTICAL ASPECTS

Introduction

Intensive systems are expensive in capital and operating cost, and need to be designed carefully and operated efficiently to be viable. Through a combination of theoretical and empirical approaches, a range of design criteria has been developed, essentially linking elements described in Fig. 11.1. A balance has to be sought between inter-related but not always coincident aims:

- to maximize the output of stock, and hence the biomass held and its growth rate,
- to provide effective physico-chemical and behavioural environments for stocks,
- to create efficient and reliable operating conditions,
- to minimize wastes, environmental impacts, and stock losses.

System design is primarily dictated by the water quality requirements of the species chosen, with a wide range of water exchange rates (Table 11.4), from typical high flow rate/low-retention time conditions to extremely long retention times. In either case, the system can be replenished with clean or recirculated water. With recirculation, high flow rate systems may consume no more water than low water exchange systems with long retention times. However, in the former, water is cleaned in recirculation, while in the latter it will usually have high organic matter content (fish faeces, uneaten feed), a rich bacterial fauna and, in moderate sunlight levels, a rich phytoplankton community.

Table 11.3 Comparative features of intensive and other tilapia systems

Factor	Semi-intensive pond	Semi-intensive to intensive cage	Intensive pond/ tank/raceway	Recycle system
Development				
Location	Open, extensive level land adjacent to water supplies Impermeable, easily formed soils	Protected water body with adequate flushing	Limited open land with excellent water supplies	Small and well serviced land area; market access
Construction	Simple; varies with site condition	Usually artisanal, simple plastic/timber plus nets	Specialist build, GRP concrete, steel, pipes etc.	Highly specialized; as intensive system
Capital cost	Usually moderate	Usually low-moderate; can be more costly	Med-high – depends on service level, materials	Highest; depends on degree of water reuse
Services	Access, harvest nets, possibly feeders, aeration	Boat access, feeders, harvest facilities	Power, access, feeders, pipes/pumps, controls	As intensive system, plus water treatment
Operation				
Stocking	Fry/fingerling direct or in nets, often seasonally	Direct into suitable nets, often seasonally	Direct or in nets, usually periodic, controlled	Direct or in nets, usually periodic, controlled
Feeding	Natural productivity plus fertilization, some feeds	Natural productivity plus feeds, varying levels	Complete feeds, usually highly controlled	Complete feeds, highly controlled
Disease	Variable; limited treatment possible, may reinfect via water column, sediment	Variable; bag treatment reinfection via water column, sediment	May be serious; flush, bath treatments, good control potential	May be serious; flush, baths, poss. reinfection through filters, etc.
Maintenance/ Husbandry	Pond walls, sediment removal, periodic water exchange, some grading	Cage frame, net changes grading, feed systems	Cleaning, water supply, grading, feed systems	Cleaning, water supply, grading, filters, waste removal, feed systems
Other inputs	Aeration for peak loadings Algal control, storage	Boat, anchors, lines, land access, storage	Pumps, pipes, valves, water treatment, storage	Pumps, pipes, valves, water treatment, storage
Harvest	Draining and manual/ winched netting	Lift nets by hand/winch brail, possible storage	Reduce water level, brail/ drain out, poss. storage	Reduce water level, brail/ drain out, poss. storage

Table 11.4 Water exchange in selected intensive tilapia systems

	Turnover rate (vols/day)		Residence time (h)		
Location	Unit	System	Unit	System	Application
Negev/Aravah (I)	1.5–2	0.05–0.15	12–16	160–480	C (*O. niloticus* × *O. aureus*)
Ginosar (I)	0.2	0.2	120	120	E (*O. niloticus* × *O. aureus*)
Ginosar (I)	8	0.03	3	800	E (*O. niloticus* × *O. aureus*)
Solar aquafarms (U)	1.5	0.01	16	2400	C (*O. niloticus*)
Simaron (U)	4	0.2	6	120	C (red tilapia, blue)
Sde Eliyahu (I)	24	0.36	1	66.6	C (red tilapia)

Source: van Rijn (unpub) Key: I = Israel, U = USA, C = commercial, E = experimental

Production intensity and system carrying capacity

A range of inter-linked descriptors can be used, and it is important to distinguish standing stocks (i.e. biomass at a particular time) from output, i.e. weight produced per time. Maximum values of these factors *per se* are not nearly as critical as the levels at which stock are capable of performing effectively in terms of survival, growth, feeding efficiency, throughout the production period.

Flow rates and operating concentrations; essential parameters – O_2, solids and metabolites such as carbon dioxide and ammonia – must remain within the stocks' tolerance levels (Losordo, 1997). Unless more detailed data are available to relate metabolic effect to fish size, temperature, feed type and/or activity level (Losordo and Westers, 1994), rates can be assumed to vary with feeding rate, using standard constants (Muir, 1982; Colt and Orwicz, 1991). Oxygen is usually the first limitation; at normal pH levels, ammonia is the next, though at low pH, CO_2 may be more critical. The mean outlet concentration can be calculated as:

$QC_i + L = QC_o$, hence $C_o = C_i + L/Q$, where
Q = flowrate through the rearing unit, e.g. $l\ min^{-1}$
L = change in parameter (loading) by fish, e.g. $mg\ min^{-1}$
(positive if added – e.g. ammonia, solids, negative if removed, e.g. oxygen)
$C_{i,o}$ = inlet and outlet concentrations, e.g. $mg\ l^{-1}$ set at desired levels

Where more detailed data are available, particularly for cyclical activity and feeding patterns (maximum levels are typically 1.5–3 times average, some 2–3 h after feeding), flows can be more carefully adjusted, and varied diurnally, as well as being modified as fish grow and feeding rates change. The degree of mixing within tanks or raceways may also affect the pattern of concentration within the holding volume, and the time distribution of waste output under fluctuating conditions.

Density, volumetric stocking density (ρ_v). Defined in terms of fish mass per unit volume (M/V e.g. kg m^{-3}) or per unit area (M/A, kg m^{-2}). The volumetric term is most common for intensive systems, though area-based density is sometimes used to compare with others. It can also be extended to describe output, e.g. in terms of M/V/T, i.e. kg m^{-3} yr^{-1}. Maximum density depends on a number of factors, including fish size, flow rate, behavioural factors, size, configuration and hydrodynamics of the rearing unit. Changes in behaviour and metabolism as stock grow typically mean that density can increase from fry to harvest-size fish. Most intensive tilapia systems can maintain a density of 25–40 kg m^{-3}, although densities of up to 100 kg m^{-3} have been used, and experience with other species would suggest that with sufficiently high O_2 levels, higher densities might be feasible (though possibly not cost-effective).

Loading, or flow-related stocking density (ρ_f). Defined as fish mass per unit water flow (M/Q, e.g. kg l^{-1} min^{-1}), or as the inverse water flow per mass, Q/M, e.g. l min^{-1} kg^{-1} or m^3 h^{-1} t^{-1}. Loading is linked to unit volume, density and flow-rate, defining the relationship between biomass and water for metabolic requirements. A high loading implies that water is used to its maximum in supplying oxygen and picking up wastes. This is critical in determining a system's capability to support stock. As metabolic rate varies inversely with individual size, loadings for juveniles are usually lower than for harvest-size fish. Typical values are 0.2–0.5 kg $l^{-1}min^{-1}$ for fry and juveniles, increasing to 0.8–1.2 kg l^{-1} min^{-1} for larger fish, and 2–3 kg $l^{-1}min^{-1}$or more with additional oxygen. Watanabe *et al.* (1993) showed no growth effect between 15, 25 and 35 tilapia m^{-3} in 23 m^3 tanks, growing from 5.36 to 462 g in 150 d, with a food conversion ratio (FCR) of 1.8 and a final loading of 2.78 kg $l^{-1}min^{-1}$. In raceways, Balarin and Haller (1982) describe outputs from 7.1–50 kg m^{-3} at 1.0–3.3 kg $l^{-1}min^{-1}$. For recycle systems, two loadings can be defined, for the holding unit, and for the whole system. Thus if a system is operated at 1 kg l^{-1} min^{-1} within its tanks, and 90% of water is recycled, with only 10% of water entering as intake, the equivalent (total system) loading would be 1 kg per 0.1 l min^{-1}, or 10 kg l^{-1} min^{-1} of intake water.

Residence time and turnover. The theoretical mean residence time θ is defined as system volume divided by flow rate (V/Q or V/(V/T), and is expressed in e.g. minutes or hours. This value has already been used to characterize production systems in Table 11.4. The turnover rate is the inverse, expressed as the number of times water is exchanged per day or per hour. Residence time can also be calculated directly from stocking density and loading, as:

θ, min = (ρ_v) (ρ_f); where
ρ_v = kg m^{-3}
ρ_f = kg $l^{-1}min^{-1}$

The practical design of holding units will determine whether or not the theoretical θ value is attained. If the holding volume is not fully mixed, actual residence time may be reduced; the difference between theoretical and actual θ may then be used as a measure of the degree of mixing, and holding unit hydraulic regimes may be adjusted.

Feed assimilation capacity. Feeding rate is a function of temperature, fish size and stocking density. As feed is one of the primary production cost elements, good feeding efficiency, as defined by the FCR is essential. Feed input must also be matched to the system capacity. As noted earlier, metabolite production can be related to feeding, and so feed load can become a measure of system capacity, as described by feed assimilation, the daily feed application rate that can be maintained without deterioration of water quality beyond target criteria (Colt and Orwicz, 1991; Watten, 1992). Within appropriate ranges, this is independent of fish size, growth rate and other factors, and defines the ability of the system to handle metabolic load, whether through open water flow, supplemental aeration or oxygenation, or through recycle. This is typically expressed as kg day^{-1}, or on a system area or volume basis, e.g. kg m^{-2} day^{-1}, or kg m^{-3} day^{-1}. If the feed rate for the stock is defined e.g. $\times$ % bodyweight day^{-1}, the term can be converted back to a stocking density.

Cumulative oxygen consumption. Colt and Orwicz (1991) also describe system carrying capacity in terms of the cumulative oxygen consumption (COC), the maximum amount of oxygen that can be consumed across the rearing unit without one of the water quality criteria being exceeded. The term is typically expressed in kg O_2 day^{-1}, or related to system area or volume, as kg m^{-2} day^{-1}, or kg m^{-3} day^{-1}, respectively. The COC depends on pH, whether the system is flow-through or recycle, and/or is supplemented with oxygen or aeration. For open systems, COC will be limited by pH at levels <6, by carbon dioxide at intermediate pH, and by un-ionized ammonia at pH >9.

Production to capacity ratio (P/C). A key parameter describing overall performance (Muir, 1982), this is the ratio of output (e.g. tonnes per unit time – usually annually) to maximum carrying capacity (e.g. tonnes of biomass safely held as defined through COC or feed assimilation) of the rearing unit. For a given system, any factor increasing output, including growth rate, survival and maximizing space use will increase the P/C. This emphasizes the role of selection for growth, of maintaining optimum temperatures, of efficient feeds and feeding practices, and of other factors that minimize stress. For tilapia, P/C ratios of >4.5 are possible (Summerfelt *et al.*, 1993) and ratios >3 may in many cases be necessary for profitable intensive systems. Practical stock management strategies need to be targeted to achieve such ratios, and require regular stock supply and good inventory control.

Recycle system parameters

In these systems, waste streams leave the fish holding units and are returned via various treatment processes to the inlet, to which fresh water may be added continuously or periodically. In some cases, treatment may be carried out within the holding unit. Wastes that build up – suspended solids and metabolites such as carbon dioxide and ammonia – are removed to varying degrees, or re-enter the system, where they must remain within the tolerance levels of the stock. Typical components include biofilters for ammonia removal, settling tanks or screen filters for solids removal, and other systems such as foam fractionaters, denitrifying filters, ozone or UV units (see Muir, 1982; Timmons and Losordo, 1994; Losordo, 1997). Basic parameters of recycle flow can be defined as:

% recycle = $(Q_R/Q_T) \times 100$,
% per day = $Q_{F(daily)}/V$, where V = system volume (m^3) = $Q_F/\theta Q_T$
Recycle residence time = $\theta Q_T/Q_F$; where
Q_R = flow rate running through recycle (water treatment) stages, e.g. $l\ min^{-1}$, $m^3\ h^{-1}$
Q_T = total system flow rate, e.g. $l\ min^{-1}$, $m^3\ h^{-1}$
Q_F = flow rate of fresh (incoming) water, e.g. $l\ min^{-1}$, $m^3\ h^{-1}$

Another useful index is the 'water reuse rate' (B. Bawden, 1997, personal communication), defined as: Q_f/Q_o, where Q_o, is the minimum flow required under open-flow (flow-through) conditions. In intensive systems, this would typically be based on oxygenated systems, and would be the minimum flow required to maintain water quality within target ammonia or solids levels. Q_o, can in turn be derived from the biomass, the feed rate, and the oxygen consumption associated with feed uptake.

As suggested above, oxygen depletion is usually the first parameter to be corrected before water can be used further. At normal pH, ammonia is the next limitation, usually removed through nitrification in biofilters, though at low pH levels, CO_2 may be more critical. Though tilapia themselves are relatively tolerant, suspended solids removal may be important for various treatment processes. The amount and/or type of treatment required can be calculated for a particular reuse level, or conversely, the degree of potential recycling available can be determined for a given type and efficiency of treatment. Flow rates are determined firstly by the metabolite balances in the rearing unit using a similar approach to that for flow-through, modified in long residence time units by the changes occurring within the holding unit itself and secondly by the flow rates required for efficient operation of the treatment units.

Recycle system loading. This parameter links with the concept of mass loading, and relates mass flow per time of metabolites usually ammonia or solids to critical dimensions, such as filter volume or surface area, settling volume or screen area. Typical values are expressed as M/V/T or M/A/T, thus gNH_3–N m^{-3} filter volume day^{-1}. For ammonia removal, nitrification is substrate limited, i.e. concentrations are usually below the half-saturation concentration (K_s) for nitrification, whose rate is generally assumed to be 1st order (i.e. directly related to concentration). However, ammonia oxidation rates per filter surface area can be maximized by increasing flow through filters. Conversion efficiency based on the difference between influent and effluent concentration is reduced, but over a defined range, the mass of ammonia oxidized per time increases with flow rate. Kaiser and Wheaton (1983) found ammonia oxidation rate to be inversely related to hydraulic detention time in biofilters, suggesting that removal efficiency is more closely related to total ammonia load than its concentration.

Practical design elements

Basic production parameters for intensive tilapia systems include the provision of a secure, controllable production environment, effective access to feeds, the elimination of negative behavioural interactions, suitable water exchange, good self-cleaning properties, and simple and effective husbandry and harvest methods. These parameters are similar to those of any other production system, though differences may arise in that many tilapia are very territorial, and their feeding habits may not lend them ideally to complete feeds. A further issue is that high temperature water, particularly if saline, may have very little surplus oxygen to meet metabolic requirements, and so systems in more extreme environments may require special attention to oxygen budgets.

Typical production objectives involve the introduction and growth of fry introduced at a specific age, through to market size; either in single crop system or more commonly, using multiple cropping. Production and growth of tilapia in intensive systems are similar to those in other systems, assuming environment and feeds are satisfactory. In practice, to maintain suitable P/C ratios, a system of increasing tank or raceway sizes is commonly used, and production periods may be divided into part-cycles of similar time-span. Thus with a typical growth period of 240 days for market size stock of 350 g, three part-cycles of 80 days may be used, and production units may be sized as outlined in the example in Table 11.5, for a target production of 200 t per cycle.

Table 13.5 System sizing based on 200 t tilapia production in three cycles

Parameter	0–80 days	80–160 days	160–240 days
Size range	5–50 g	50–200 g	200–350 g
Max biomass	28.5 t	114.3 t	200 t
Max stock density, kg m^{-3}	20	30	40
System volume	1430 m^3	3810 m^3	5000 m^3
Units (approx)	48 × 30 m^3	48 × 80 m^3	48 × 105 m^3

Such a system could have new stocks introduced into each section every 80 days, or e.g. with 48 units at each size, a 20 day cycle of 12 tanks each could run, if fry were available at suitable timing, with 12 tanks cropping 50 t over every 20 day period. With 80-day batches, and a nominal 360 day year, the P/C ratio would be $(200 \times 360/80):(28.5+114.3+200) = 2.63$. This would be the same with a 20 day $\times$ 12 tanks system, but if stocks were moved every 20 days, i.e. with 12 stages, the P/C ratio could be substantially increased. However, this becomes limited by the practicalities of transferring stocks without stress and excess labour cost, and of efficient tank allocation without too many tank sizes. In practice, grading and partial harvesting will also increase efficiency. Increased growth rates will in particular improve utilization; thus reducing the cycle period to 60 days would increase P/C to 3.51. In other intensive systems, such as those based on circulated ponds, stocks may be introduced or separated less frequently, and a single pond may be used for a complete cycle.

Growth rates, mass balances and oxygen budgets. Feed rates and types are the primary driving force in intensive tilapia production, and are also closely tied in with metabolic effects and with the planning of water and oxygen budgets. Specific metabolic effects have not been extensively researched at commercial levels, but simple mass balances can be run. Thus with a biomass of 10 t and a feeding rate of 3% bodyweight day^{-1}, 300 kg of food would be applied. With an oxygen uptake of $0.25 \times$ food input, this would create a demand of 75 kg, an average 3.125 kg h^{-1}, typically ranging from 2–6 kg h^{-1}. Based on an input level of say 8 mg l^{-1} and an allowable minimum concentration of 4 mg l^{-1}, this demand would require an average 780 m^3 h^{-1} or water flow, corresponding to a loading of 1.3 l min^{-1} kg^{-1}, or 0.77 kg l^{-1} min^{-1}. In such conditions, where O_2 demand (in-out) is 4 mg l^{-1}, total ammonia output, at a typical $0.025 \times$ feed, would reach 0.4 mg l^{-1}, and suspended solids, at a typical $0.25–0.3 \times$ feed, perhaps 4–5 mg l^{-1}.

Aeration or oxygenation can be used to supplement oxygen, and it may be quite feasible to reduce water flows to 30% or less, with similar oxygen levels, while tripling or more the ammonia and solids levels. Depending on pH, flows of as little as 10% may be possible before ammonia or solids become limiting. For conventional low-medium residence time systems, water is discharged on a continuous basis, and wastes removed from the system. In high residence time systems, oxygen balances are maintained by aeration or oxygenation, and ammonia and solids may accumulate, break down and add to respiratory demands. Depending on system design and operating conditions, ammonia may be taken up in algal metabolism and/or be oxidized through nitrification to nitrite and nitrate. At least part of the solids may collect within the rearing volume, some of which may be resuspended, the remainder settling. For ammonia, suspended and settled solids remaining within the system, oxygen consumption through respiration may equal or even exceed that of the tilapia, and oxygen budgets need to be developed accordingly (see, e.g. Boyd, 1995).

Water supply. Significant quantities of water may be required for intensive systems, and water supplies, whether derived from surface or borehole, from rivers, lakes or coastal margins, need to be carefully designed, and sized to meet all anticipated needs. Intakes are required to be properly positioned, secure, with cleanable screens, and water flow, whether pumped or gravity induced, needs to be controllable both in the system, and at each main holding unit. Supply pumps, usually electric or diesel centrifugal or axial types, need to be backed up adequately and installed so they can be easily maintained. Given the capital and operating cost of pumping significant quantities of water, systems need to be carefully designed to minimize head loss. Flow at key points is usually monitored and alarmed. A critical design feature is the maximum safe time between water stoppage and oxygen deficit or other water quality limit. This will determine both the nature of backup to be used, and how it might be activated.

Tanks and raceway configuration, water distribution and hydrodynamics. Fig. 11.2 shows typical configurations and flow patterns of commonly used systems. The supply and drainage of water will be determined by the factors outlined earlier, and associated with the hydrodynamics of rearing units, which range from plug

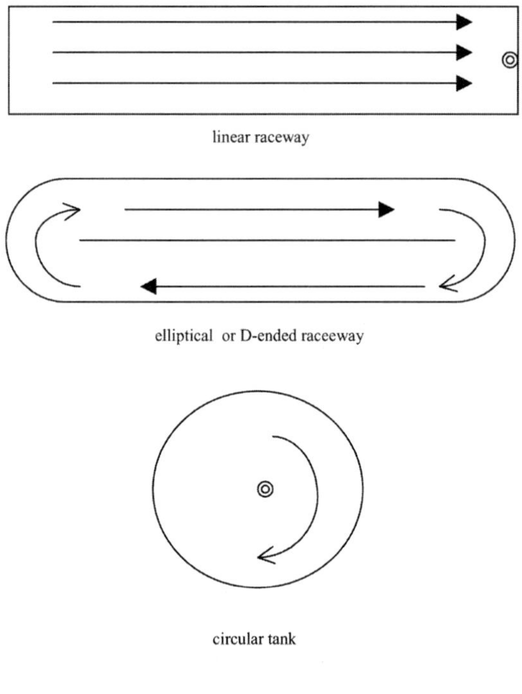

Fig. 11.2 Typical tank or raceway configurations.

flow to completely mixed conditions, most units being intermediate between these extremes. Hydrodynamics are largely determined by the shape and volume/flow relationships of the unit, and that of circular tanks is generally superior to the linear-dominated patterns of rectangular tanks and raceways. However, a cross flow system can improve mixing in rectangular units (Watten and Beck, 1987). An octagonal sided tank configuration has also been proposed to improve mixing and flow characteristics (Losordo *et al.*, 1995), though shapes such as these may be less efficient in space utilization. However, rectangular tanks may be also subdivided as a 'retrofit' to create circular flow cells (Watten and Honeyfield, 1995). As noted earlier, the use of theoretical and actual residence times is a useful indicator of the degree of mixing. The design of water inlets and outlets can be critical in imparting and distributing water velocity, and in creating suitable scouring conditions for the removal of solids, and their configuration can be important for head loss. Usually the aim is to disperse flow as effectively as possible in vertical and horizontal directions, and devices such as spray bars and baffles are common.

Raceway culture systems are operated with water flow rates designed to supply sufficient oxygen for fish metabolic requirements above critical threshold values. If the water in raceways or other systems is supplemented by aeration or pure oxygen, flow must be sufficient to remove metabolites, such as carbon dioxide, ammonia and suspended solids. Balarin and Haller (1982) provided an analysis of relative advantages of different tank and raceway systems for tilapia; those with a return flow, such as circular tanks and elliptical raceways, were more effective in maintaining target environments and velocities, which were also significantly affected by inlet and outlet characteristics. Elliptical tanks designed with effective drainage outlets, capable of removing solids on a continuous basis, offer additional advantages of good use of space. Though basic linear raceways may be simple to build, if operated at lower stocking densities, flows may be insufficient for self cleaning, and if operated at high densities, water quality gradients may make downstream conditions poorer than ideal. However, the potential to make use of sloping ground, and to reuse water to solids or ammonia limits with simple step by step aeration may make them particularly suitable for certain sites and operating objectives

Dual outlets are used in some systems, in which waste solids are selectively removed from the base typically with as little as 5% of the total outflow (Losordo, 1997) – enabling far more efficient solids removal, while the remaining flow is drawn from the upper water column and can be discharged directly or recycled with a minimal need for solids reduction.

System construction. Most systems are constructed in a similar way to those used for other species, as similar constraints of costs, durability and surface smoothness apply. Though GRP (glass fibre) and various forms of proofed steel panel present very high quality surfaces, these are costly, and cast reinforced concrete or concrete block, or even plastered brick, are more common, particularly for larger volumes. Conventional earth ponds may also be feasible, provided water velocities used do not cause scouring and erosion of walls and pond floors. To avoid this in high flow-energy systems, lined earth ponds may be used, and materials such as soil cement have also been considered. Ponds are particularly used for high-retention time photosynthesizing systems. Depending on climatic conditions, more intensive flow-through systems and high intensity recycle systems may be housed in structures of various sorts, from simple shade-netting 'tents' to polythene or glass-walled greenhouses, to purpose-built insulated industrial buildings. In some cases, heat balances have to be carefully assessed, to ensure that temperatures can be maintained efficiently. Owing to heat generation from equipment and stock, intensive recycle systems in particular may have cooling problems.

Water supply and drainage systems are more commonly constructed of plastic or concrete pipe, or pre-cast or concrete block channels. These have to be sized carefully to ensure adequate water flow under intended operating conditions, and where supply is to be controlled gates or valves are commonly required. As large bore plastic valves are very expensive, these are usually limited in number, and control provided by simpler devices where possible. Though single inlets are common, splash pipe, spreader bar or weir inlets – multiple inlets may be used, particularly along the length of a raceway, where additional 'freshening' water supplies may be routed. As an alternative to a conventional single drain, usually with floor or standpipe outlet, multiple outlets may be employed, for example, to remove collected solids. In specially designed systems, the main water stream can be used to move fish to a central collecting point for harvest or redistribution. In some cases, intermittently acting siphon drains can be used, to provide short periods of high flow and scour (Balarin and Haller, 1982). If these or other devices are to be used, care has to be taken that stock are not damaged by excessive velocities, and by trapping dorsal spines, which are typically erect when the fish is alarmed. Water quality may be pre-controlled, primarily through removal of excess silt and in dissolved gas control, e.g. if borehole water is used, requiring oxygenation and/or stripping of nitrogen and other undesired gases.

Ancillaries. Automatic feeders, water quality monitors, alarms, and grading and handling equipment are common, as with other types of intensive aquaculture. Feeders may be particularly important, and need to be designed to carry appropriate quantities of feed for the management period – typically 1–2 days to maintain feed quality (avoiding degradation through temperature change and particle breakdown), disperse feed effectively, and ration it as required to maintain intended growth rates without waste. Belt and vibrator feeders are commonly used for early stages, while vibrator, screw-feed or water/air conveyed systems are common for on-growing. Feed types and applications are reviewed in Chapter 9. Grading and handling systems vary with unit size and configuration, and in-tank systems (e.g. movable screens or grids) or external units (e.g. pumped or brailed systems) may be chosen. In many intensive system locations, land may be expensive and layouts are as compact as possible. However, a well-designed system requires adequate access for water supply, air/oxygen, power and drainage systems and should have suitable points for grading and harvest transfers, whether by pump, trailer or truck. Monitors and alarms are particularly important for oxygen-injected or more complex systems, typically covering oxygen levels, and for recycle units, flow, sub-component levels and operating status, power supply and possibly temperature. Water supply and drainage systems may also be covered by flow and level monitors and, where waste discharge is monitored, output flow and quality may also be monitored.

Recycle system components. The design of recycle systems depends on the degree of water reuse, and on the environmental and operating conditions proposed (Muir, 1982; van Rijn, 1996; Losordo, 1997). Table 11.6 provides an outline of typical system components and their practical design criteria. For conventional systems with a relatively short residence time, individual treatment units can be dimensioned according to the proposed solids, ammonia and CO_2 loading, and arranged in various configurations. Commonly, waste water flows sequentially through solids removal, biofiltration and then additional treatments, before returning to holding units. In more sophisticated systems, waste streams may be split, with the smaller, more concentrated stream of solids moving to solids removal, now most commonly using self-cleaning screens, then rejoining the main flow for biofiltration. Additional, more specialized (and more expensive) stages such as ozonation, fine filtration, etc. can be operated on main or side streams, depending on requirements, and various sumps can be positioned to allow water to be selectively distributed to different treatment areas. Aeration or oxygenation is primarily applied in the holding units and in header tanks, though supplementary inputs may also be provided to biofilters to maintain efficient nitrification. Water is usually circulated using centrifugal pumps, and it is important to minimize the pumping head to reduce energy costs, and if possible, to avoid having to pump at several different points in the system. In some cases, airlift pumps may be used for lower-head applications, with the added advantage of contributing to oxygen supply.

Systems with a long residence time in holding units may be more complex, and treatment design depends primarily on the degree of transformation of metabolites within the holding volumes. Where active bacterial and algal production occurs, ecological management demands can be significant, and key requirements of additional treatment usually involve the separation of excess microbial biomass, possible solids/turbidity reduction and/or oxygen supplementation, and return to the holding units.

Table 11.6 Practical design criteria for recycle tilapia systems

Function	System component	Basic design criteria	Construction
Solids removal	Settling tank/pond	10–100 m^3 day^{-1} for medium to coarse particles, 1–1.5 m deep	Concrete or earth walled pond or linear raceway
	Screen filter	for 40–200 μm particles	GRP/stainless steel
	Drum filter	for 40–200 μm particles	GRP/stainless steel
Ammonia removal	Flooded biofilter	50–100 m^3 m^{-3} day^{-1}, 1–2 m deep	Concrete, GRP, steel container
	Trickling biofilter	20–50 m^3 m^{-3} day^{-1}, 1.5–2.5 m deep	Concrete, timber, GRP
	Fluidized bed filter	100–300 m^3 m^{-3} day^{-1}, 1–3 m deep	Concrete, GRP steel cylinder
	Bead filter	50–100 m^3 m^{-3} day^{-1}, 1–1.5 m deep	Usually GRP
	Biodisc/biodrum	Depends on proprietary system	GRP, plastic, stainless steel
CO_2 removal	Aeration	50–100 m^3 m^{-3} day^{-1}, 1–2 m deep	Diffuser in aeration tank,
	Lime dosing	Depends on water chemistry	or degassing tower
Nitrate removal	Denitrifying filter	20–500 m^3 m^{-3} day^{-1}, 1.5–3 m deep	Concrete, GRP, steel container
Dissolved organics removal	Ozonation	300–600 m^3 m^{-3} day^{-1}, 1–2 m deep, dosage 1–6 g m^{-3}	Proprietary units, GRP, stainless steel
	Foam fractionation	100–400 m^3 m^{-3} day^{-1}, 1–2 m deep	Usually GRP, polycarbonate, ABS
Fine solids removal	Screen or drum filter	150–300 m^3 m^{-3} day^{-1} for 10–20 μm particles	GRP/stainless steel
	Sand or cartridge filter	50–100 m^3 m^{-3} $day^{-1}$for 5–20 μm particles 1–2 m deep, 5–10 m pressure drop	GRP/stainless steel
Sterilization	Ozonation	300–600 m^3 m^{-3} day^{-1}, 1–2 m deep, dose 0.5–2 g m^{-3} residual 0.1–0.4 g m^{-3}	Proprietary units
	Ultraviolet	30–50 W s cm^{-2} for typical organisms; 50–100 m^3 m^{-3} day^{-1}	Proprietary units

Husbandry

While intensive systems offer advantages in ease of access, and the means to concentrate stocks for management action, the requirement for high productivity to provide the necessary return on inputs requires, in turn, a high level of management. However, certain procedures can be simplified by system design, and to varying degrees, the use of mechanized and automated systems.

Stock selection. A broad categorization can be made of pre-harvest features of the tilapia stocks – mainly biologically defined characteristics – and post-harvest features that also incorporate the effects of the husbandry and harvest process. Pre-harvest features include growth rate, environmental tolerance, adaptability to rearing conditions and husbandry techniques, and are primarily related to the genotype; market/post-harvest features include external appearance, body shape and conformation, size distribution, body fat, flavour and texture, and can be controlled or at least manipulated through the production process. In most cases, species and strains for intensive tilapia production have been selected on the basis of their growth rates in the environmental conditions concerned, external appearance, and their related market demand. A secondary consideration, in high temperature and salinity environments, may be the ease by which stocks can reproduce, and the costs of early rearing, though this is not a widespread constraint. In practice therefore, most intensive systems employ similar strains to those used in other systems, though high value market targets may sometimes favour various colour morphs. As noted elsewhere, the ability to supply stocks on a regular basis is an important and possibly dominant factor.

Food supply. This is critical in intensive systems, and is usually a major part of production cost. Key factors include the aim for complete feeds, to optimize costs/growth and to ensure that foodstuffs are stable, palatable, efficient and unwasted. Requirements in different production environments, e.g. metabolic demands in different temperature and activity regimes may be important; and related to this, waste outputs and their implications for water supply, drainage and water reuse (Cowey and Cho, 1991; Hankins *et al.*, 1994). Most tilapia feeds are formulated to meet standard nutritional requirements, but are not specifically 'fine-tuned' for individual system regimes, for example to meet greater energy needs in higher current conditions, or to minimize ammonia loading for biofilters. Feed particle size is also important and needs to link in with stock grading so that correct sizes are given. The use of lower cost feed ingredients with coarser particle sizes may lead to poor palatability and feeding response, reducing performance and increasing wastage, which may be particularly problematic for recycle systems which may be overloaded as a result.

Environment and health management. Though a common objective is to control water quality, and to limit disease introduction, particularly in recycle systems, the practical difficulties of maintaining optimum environments and of limiting entry of pathogens, combined with the close proximity between stocks and the rapidity by which disease might spread, may necessitate close control. Where removal of infected stock and/or improvement of culture conditions (improving water quality and/or reducing stock stress) is insufficient, treatments may be required, either in-feed, or using dips or baths. Key distinctions are that high densities may favour the rapid transfer of pathogens within stock, and that prompt and efficient diagnosis and treatment becomes more critical for overall security. Bacterial and parasite infections may be particularly problematic. For example, heavy losses associated with *Streptococcus* have been recorded in intensive systems in Israel (van Rijn, personal communication), usually associated with poor water quality and heavy organic loading conditions which are relatively common in many intensive or recycle systems. It has also been reported as a major problem in the USA (Plumb, 1997). While the ecological simplicity of intensive tank systems limits potential access for intermediate hosts, and reduces the complication or interference effects of treating stocks, interactions with the biological components of recycle systems may be far more complex. In these cases, biofilters may act as reservoirs for pathogens or hosts, and may be dangerously disrupted if too vigorously treated, risking the deterioration of environmental conditions – a difficult feedback situation (see Timmons and Losordo, 1994). The restoration of treatment function may take days or weeks, depending on temperature and other operating parameters, and would require reduced system loading, and hence lowered feed and growth during that period. The removal of certain resistant parasites such as *Ichthyophthirius* may be particularly problematic.

Cropping and harvesting

Though intensive systems may be relatively easy to harvest compared with others, the requirement for efficiency means that they may need to be operated with closely managed cropping regimes. The aim in most cases is to maximize the production: capacity ratio, maximizing the productive or assimilative capacity of the system and hence minimizing effective production cost per unit output (McNown and Seirig, 1983). A range of options may be considered (see also Fig. 11.3).

(1) *Single batch.* The most basic cropping system. A cohort of fish is stocked as juveniles, cultured to market size and harvested. A new cohort of fish is then stocked and the process repeated. Net production is equal to biomass harvested minus biomass stocked. A major disadvantage is that the carrying capacity of the unit is not reached until just prior to harvest, and much of the capacity of the unit is unused over the culture period. Watten (1992) defines this unused production capacity in terms of the capacity of the system to assimilate feed.

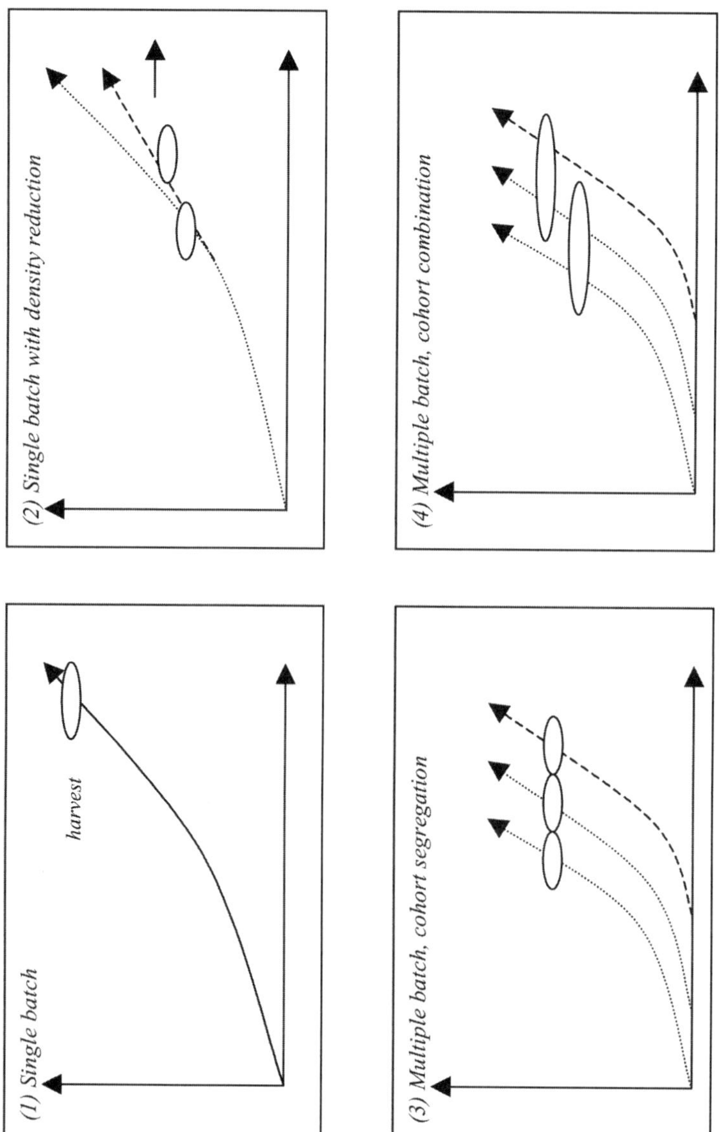

Fig. 11.3 Cropping patterns in intensive tilapia culture.

(2) *Single batch with periodic density reduction.* Many batch systems employ some form of density manipulation to improve capacity use. Fish can be stocked as juveniles at a relatively high density, and as they grow and feeding rate approaches the assimilative capacity of the unit, density is periodically reduced. This results in continued growth, as feed allowance per fish can be maintained. The selection of the appropriate interval and timing of density reduction includes factors of fish size, growth rate, population size distribution, feed allowance, feed assimilation capacity, water exchange rate, volume of culture unit, and labour required for manipulating density. The goal is to maximize growth subject to constraints of water flow available and labour for moving fish.

Paessun and Allison (1984) used a sequential cropping system in four identical recycle units with a common water supply. Tilapia were stocked into one unit, cultured to carrying capacity, divided into two units and grown again to carrying capacity. Van Gorder (1994) describes a system employing three density manipulations, with fifteen identical culture units in geometrically increasing multiples. One unit of juveniles supported two 2nd stage, four 3rd stage and eight 4th stage units. Stock rates (i.e. numbers per volume) in the 1st stage were twice as high as the 2nd stage, four times the 3rd and eight times the 4th stage rates. Every 6 weeks, 8 tanks were harvested, density was reduced by 50% in 7 tanks, and 1 tank stocked with juveniles, providing a regular, continuous output. The P/C ratio was 2.2. Alternately, fish can be moved to units of differing volume or flow. Thus, juveniles can be grown in small, easily managed units at high density, and as carrying capacity is reached, moved to a larger unit to reduce density, or flow increased to reduce loading. An advantage is that inventories, and the health and vigour of stock, can be easily assessed during movement. Grading also improves the homogeneity of stocks, for better targeting of feed types and rates.

(3) *Multiple batch with cohort segregation.* Here, cohorts are stocked at regular intervals into compartments within a culture unit, partitioned to separate individual groups. In some cases, these can be adjusted as fish grow, to increase compartment volume available to each cohort. Individual cohorts are cultured to market size and harvested. The system is operated near the feed load capacity and harvests occur at regular intervals. A model of such a system showed that increase in production increased hyperbolically with the number of cohorts (Watten, 1992). In a hypothetical management situation with four cohorts the P/C was 2.05. The model was sensitive to changes in duration of the culture period and to growth rate, mortality and size at stocking. Van Gorder (1994) describes a variation in which different culture units containing different size groups of fish share a water supply and treatment infrastructure. The cropping system allows the maintenance of feeding rates near the assimilative capacity of the overall system. Harvest of one unit and subsequent stocking with fingerlings does not greatly reduce the unused feed assimilative capacity of the production system.

(4) *Multiple batch with cohort combination.* Cohorts are stocked at regular intervals into a unit, but are not isolated from one another. Grading devices are used to harvest market size fish. A multiple-batch cropping system with two to three cohorts of fish has been implemented as the industry standard in commercial channel catfish culture in earthen ponds (Tucker *et al.*, 1992). Advantages of this system are similar to those described the previous method, although the uniformity of harvested fish size is greater. The P/C ratio of such a system, modelled on stocking every 8 weeks and harvest every 2 weeks can exceed 4.5 (Summerfelt *et al.*, 1993). A principal disadvantage is the stress to fish and the additional labour required for grading, which may be more difficult in round tanks than raceways. Within each unit, a wide range of sizes is present at any time, and the model assumes that interactions between large and small fish do not affect growth rate, which may be questionable for territorial fish such as tilapia. Antagonistic interactions among fish will likely increase the size variability within and between cohorts. Finally, use of this cropping system makes inventory control difficult. It is difficult to identify whether fish removed at harvest are fast growing fish from a recently-stocked cohort or slow growing fish from previously-stocked cohorts.

Rakocy (1997) describes a range of trials with recycle systems at 40 kg m^{-3} and 6 kg l^{-1} m^{-1}, supplementing rearing tanks with 22 W m^{-3} of aeration. Some systems are loaded at very high rates (54–67 kg l^{-1} m^{-1}) and intermediate density (27 kg m^{-3}) with aeration supplemented at 10 W m^{-3} (W. Engler, personal communication). In such highly loaded systems water flow may serve purposes other than removal of metabolites, in the latter case providing heat to the rearing unit. A stock-splitting scheme is used, where density is reduced at ~ 6-week intervals, from an initial density of 100×85 g fish m^{-3}. Sequential stock reductions to 67, 45 and 30 fish m^{-3} are made at each interval. Average final weight was 859 g after 26 weeks, with final density of 21.1 kg m^{-3}, estimated productivity of 82.0 kg m^{-3} yr^{-1} and a P/C ratio of 4.02. In the other study, with small (4.4 m^3) tanks, fish were stocked at 182 m^{-3}, yielding an average weight of >454 g. Final density was 81.4 kg m^{-3}, productivity ~176.5 kg m^{-3} yr^{-1} and the P/C ratio 2.17. Assuming that water quality in the return water does not limit growth, the loading was 1.16 kg l^{-1} min^{-1}.

11.4 PRACTICAL DEVELOPMENT OF INTENSIVE AND RECYCLE SYSTEMS

Initial concepts and current directions

Intensive tilapia production has a relatively long history, from early trials (Uchida and King, 1962) producing *O. mossambicus* as baitfish in simple raceways. The main impetus for development, however arose in the 1970s, as principles of intensive aquaculture were being established, and as the tilapia's rapid growth and apparent robustness and versatility made it an attractive candidate species. Its potential as a high value species capable of supporting the rather higher production costs was critical. Pioneering work in Kenya (Balarin and Haller, 1982) defined many of the basic criteria and showed that tilapia could adapt to commercial stock densities in simple flow-through systems. However, there were problems of feeds and their quality, and of developing a sufficiently continuous fingerling supply from the single site. Intensive tank techniques were also being developed in Israel, particularly for over-wintering late fingerlings and broodstock, and tilapia was also being proposed as an alternative for channel catfish in experimental intensive systems in the USA. Here, interest arose in view of its prospects in the warmer waters of southern states, and the then relatively limited appeal of channel catfish in major markets, resulting in the development of a number of intensive 'fish barn' concepts (Muir, 1975), in some cases using simple aeration and biofiltration for recycling. Though intensive production was shown to be technically feasible, and tilapia entered local markets, rising costs of production after the energy price rises of the early 1970s made intensive tilapia culture generally uncompetitive.

Improved stocks and system designs, better feeds and reducing relative prices of energy, together with the attraction of environmentally managed systems near urban markets, have revitalized interest in intensive units. Recycle systems are being developed in desert, colder water or peri-urban environments, with major output targets. The last decade has seen an acceleration in commercial investment, though not without technical or financial difficulties. Key targets are related to production efficiency, water sparing, and standardized output to meet market specifications, and a range of experimental and commercial units has been designed to respond to perceived opportunities. The following sections describe intensive and recycle production in key regions, particularly in Israel and the USA, where some of the most significant developments have taken place. As many units have alternated between flow-through and recycle operation, or have evolved from one to the other, they are described together.

The Israeli experience

Intensive tilapia culture in Israel was initiated in the early 1980s by testing monoculture of tilapia in small (200 m^2), earthen-bottom ponds. Water addition was limited to account only for evaporation and seepage loss and the mode of operation was therefore similar to that for the static ponds commonly used at the time. It was demonstrated (Sarig and Arieli, 1980) that tilapia could be cultured at significantly higher densities than carp, and yields as high as 15–30 t ha^{-1} were obtained in a 200-day period at stocking rates between 2–8 fish m^{-3}. Higher densities (10 fish m^{-3}) coincided with growth stagnation associated with unfavourable water quality. However, using similar ponds in which organic matter was removed daily using drainage pipes, yields of more than 50 t ha^{-1} were obtained over the same period (Zohar *et al.*, 1984), using 50–60% daily clean water exchange. Based on these results, 50 m^3 concrete octagonal ponds were constructed, with sloping bases and a central outlet allowing more efficient removal of organic matter. A continuous circular motion of water, obtained by paddlewheel aerators, caused organic matter to accumulate at the outlet from where it was discharged once to several times daily. Initially these were operated as flow-through ponds with 4–6 daily changes, at maximum densities of 50 kg m^{-3} at harvest, i.e. 2.5 t per pond. Individual growth rates were similar to those for more extensive culture, though FCRs were significantly higher (Zohar *et al.*, 1985). Follow-up studies revealed that these ponds were characterized by low O_2 (minimum 2 mg l^{-1}) and relatively high ammonia and nitrite concentrations, up to 5 mg and 2 mg N l^{-1}, respectively (van Rijn *et al.*, 1986).

Present day intensive tilapia culture ponds are of a similar design, varying from 50 to 500 m^3, either plastic-lined or concrete, and operated at various residence times. However, very few are operated as flow-through systems, and most commonly, water is recirculated through so called 'satellite' ponds for water purification. As most intensive culture units are constructed in existing fish farms, conventional fish ponds or reservoirs usually serve this function (Mires *et al.*, 1990; Mires and Amit, 1992), and in most of these, fish are cultured to some extent to secure a balanced pond ecosystem. However, the stocking density is low compared with that for conventional grow-out ponds. The ratio between volumes of intensive and satellite ponds varies widely. With low satellite:intensive ratios, periodical elevations in ammonia and nitrite to toxic levels are frequently experienced. Increased inorganic N in the system is often preceded by sudden falls in phytoplankton biomass in the satellite pond (Mires and Amit, 1992). Studies on phytoplankton dynamics in conventional fishponds (Rimon and Shilo, 1982; Shilo and Rimon, 1982) showed that phytoplankton 'crashes' are typical in highly eutrophic conditions in which the natural carrying capacity is exceeded due to nutrient loading. Although no similar studies have been conducted in satellite ponds, excessive nutrient loading is likely to be a key factor underlying periodical oscillations in phytoplankton biomass. Increasing the volume of treatment area (satellite pond) relative to the intensive pond often improves the water-purification performance and stability. Very few design criteria are available and today's systems largely result from empirical trials conducted by pond managers. Based on commercial systems presently in operation, it can be concluded that for tilapia grow-out at densities up to 20 kg m^{-3} at ambient water temperatures between 24 and 29°C, a minimum intensive: satellite pond ratio of 1:10 is required. Minimum water use is around 3 m^3 kg^{-1} produced.

In addition to increasing the area of satellite ponds, other methods for reducing ammonia and nitrite have been examined; for example, levels can be significantly reduced by combining nitrification and bacterial assimilation of inorganic N within the culture ponds. To enhance these processes, ponds were operated at relatively long retention times ($\sim$5 days) to allow a build-up of organic matter. Paddlewheels and airlifts were used to secure complete suspension of organic matter within the water column. As a result of the increased surface area provided by the organic particles and their long retention times, nitrification was significantly higher than in 'cleaner' short retention ponds (Avnimelech *et al.*, 1986, 1992). Since much of the organic material is retained within the pond, a considerable breakdown of organic matter and consequent conversion of organic N to ammonia by ammonifying bacteria takes place. Nitrification alone was found to be insufficient to convert all the liberated ammonia to nitrate. The problem was solved by stimulating growth of heterotrophic bacteria and hence, promoting conversion of inorganic ammonia to bacterial protein (ammonia assimilation). Organic carbon addition in the form of flour or other carbonaceous compounds resulted in a shift in the nutritional status of the water and caused inorganic nitrogen rather than organic carbon to become the limiting factor for bacterial growth. This resulted in the efficient conversion of inorganic N into bacterial biomass and to some extent to protein sparing, as it was it was found by stable carbon tracing, that under these conditions some of the bacterial protein was incorporated by the fish (Avnimelech *et al.*, 1989, 1994).

Recycle systems

Commercial farms with water treatment carried out mainly within the culture ponds themselves are operated in southern Israel, where year-round tilapia culture is carried out in greenhouses, at water temperatures as low as 19°C in winter and as high as 29°C in summer. These facilities comprise 300 m^3 raceways or circular ponds supported by sedimentation basins. Typically, two 100 m^3 ponds, operated in series as primary and secondary sedimentation basins, are used to support between 1800 to 2100 m^3 of intensive culture ponds. Water from each culture pond is recirculated through the sedimentation basins at a rate of 450 – 600 m^3 day^{-1} (1.5 to 2 volume changes day^{-1}). Between 5 and 15% of make-up water is added to the system daily; daily water losses due to evaporation and sludge removal account for up to 2% of total water volume, and overflow water is used for irrigation. One to two paddlewheel aerators are used for oxygen supply and suspension of organic matter within the culture ponds. These are exclusively used for grow-out, with typical initial stocking weights of around 100 g. Maximum stocking densities are 15 kg m^{-3} and maximum productivity is 25 kg m^{-3} yr^{-1}. Daily feeding levels are usually 2% of biomass using 30% protein feed pellets. Minimum water use for the production of 1 kg of tilapia is 2.19 m^3.

To date, very little information is available on the microbiological processes taking place in these systems. Phytoplankton concentrations in these culture systems are usually low, presumably due to a combination of tilapia grazing and use of greenhouse covers, which limit photosynthetic energy penetration. It seems therefore that water quality is mainly mediated by heterotrophic and autotrophic bacterial activity. The mode of operation resembles that of an activated sludge system in which, among other processes, breakdown of organic matter, ammonification, nitrification and denitrification take place within the same compartment. Concentrations of suspended solids in these systems are as high as 150 mg l^{-1}. Values of other water-quality parameters are: up to 5 g NH_4–N l^{-1}, up to 10 mg NO_2-N l^{-1}, up to 40 mg NO_3-N l^{-1} and pH values between 7.0 and 7.4.

Very few attempts have been made to develop and operate tank or raceway based intensively recirculated tilapia systems, though pilot plants are currently being evaluated. The only commercial installation is at kibbutz Sde Eliyahu. This indoor system, operated throughout the year, is used for grow-out of red tilapia (*O. niloticus* × *O. aureus*), and covers a total area of 900 m^2, housing raceways with a total volume of 480 m^3. Ten × 16 m^3 raceways are used to rear tilapia from 40 to 200 g and ten × 32 m^3 raceways are used from 200 g to market size (~450 g). Fish are kept at densities of ~300 fish m^{-3} in the first and ~150 m^{-3} in the second stage. The system is designed so that incoming water passes all raceways before discharge; the system can therefore be viewed as one subdivided culture basin. Additional channels between the raceways allow easy transfer of fish from one to another. Total water volume, including internal water channels and treatment facilities, is 800 m^3. The retention time of water in the raceways is ~1 hour while that for the overall system is 2.8 days. Water temperature ranges from 23°C in winter to 28°C in summer. Water leaving the overall culture basin is mechanically filtered using a 60 μm drum filter, and is then led through a submerged biofilter and finally enriched with liquid oxygen before return to the culture basin. The maximum stocking density of the overall system is around 45 kg m^{-3}. Fish are fed at 2% bodyweight daily with 40% protein diets. The system has been operational since 1993 and based on production parameters thus far it seems feasible to obtain a maximum annual yield of 70 t. Typical water quality parameters are: suspended solids 25 mg l^{-1}; pH ~7.0; ammonia and nitrite <2 mg l^{-1}; oxygen >5–6 mg l^{-1}.

A prototype system has also been developed at Ginosar experimental station, comprising 12 basins of 5 m^3 each from which water is recycled through a mechanical filtration unit followed by aerobic trickling filter and liquid oxygen enrichment. In addition, sludge captured by the mechanical filter is led to a sedimentation pit for sludge digestion. Water from the system is added to the sedimentation pit from which it is withdrawn into a denitrifying fluidized bed reactor for nitrate removal. Details of this system have been presented elsewhere (Arviv and van Rijn, 1994; van Rijn *et al.*, 1995; Aboutboul *et al.*, 1995). Growth performance of tilapia in this system is presently being studied.

Intensive and recycle tilapia production in the United States

It is difficult to generalize about the design and operation of intensive tilapia culture systems in the United States, as designs are idiosyncratic and farms are located in climatic regions ranging from northern, cold temperate/sub-arctic, to warm, arid sub-tropical (American Tilapia Association, 1996). Systems are located in enclosed structures, greenhouses and may optionally include small open ponds for fry and fingerling production. Some have access to heated water from low-grade geothermal sources or from process cooling water used in power generation or industrial production. Hydraulically, systems are managed in recycle or flow-through modes, with many variations within this range. Farm ownership patterns also vary. Most larger farms are corporate financed to varying levels, but there are many smaller (<0.5 million lb yr^{-1} i.e. $\lesssim 250$ t yr^{-1}) owner-operator facilities. Though there is a substantial range of size, the differences in sites, systems and financing may make it difficult to generalize about scale efficiencies.

Solar Aquafarms is currently the largest producer of tilapia in the United States, with an estimated annual production of 1300–1800 t. It is located in S. California, but unlike many other tilapia farms in the region, does not have access to heated water. Production tanks are located in greenhouses and are operated using activated sludge water treatment. Tanks are aerated intensively (>40 hp $acre^{-1}$ or 75 kW ha^{-1}, 7.5 W m^{-2}) but solids are not removed. The farm is severely restricted in water supply (120 gal min^{-1} or 540 l min^{-1}), and tanks have long retention times (>16 h) with abundant algae growth. Control of fish inventory is difficult. The farm has a long history of corporate involvement and is most recently associated with Chiquita Brands.

Several other farms produce $\sim$1.5 million lb or 680 t annually ($\sim$13 t per week). Fish N' Dakota, in Beulah, North Dakota has access to thermal effluent as well as heated air used to inflate a flexible plastic dome covering the facility, operated as a recycle system. Blue Ridge Fisheries in Virginia is a recycle facility enclosed in a steel and concrete building, with 44 rectangular tanks of 114 m^3 (18.3 m $\times$ 4.3 m $\times$ 1.4 m deep). Water passes to a solids removal basin (38 m^3) using tube settling and a 76-m^3 biofilter basin equipped with a wastewater industry RBC (rotating biological contacter). Water is then pumped through a U-tube for O_2 injection before returning to the tanks. Ozone (corona discharge) is added as a side-stream process. Water is recirculated at 3.8 m^3 min^{-1} with a 5-hp (3.8 kW) pump, giving a rearing tank residence time of 30 min. Boeuf River Fisheries in Eudora, Arkansas operates a 6 month cycle during the warmer summer period. Fish are cultured in raceways and water is recycled to a reservoir pond for treatment. Fish are over-wintered in a converted pig barn that also serves as a hatchery.

Myron Kloubec also operates raceways with heated water inputs from a local power supply utility, producing 20–25 000 lb per week ($\sim$ 500 t yr^{-1}). In Pacific Aquafarms, in Niland, California near the Salton Sea, fish are produced in small (1/8 to 1/10 acre – 400–500 m^2) circular ponds, constructed in desert sand using a soil cement technique. Following excavation of pond bottoms, cement was broadcast over the soil, tilled to mix the cement with the sand, moistened, and compacted. Flow-through ponds have a residence time of 24–33 h, aerated intensively (40 hp $acre^{-1}$ or 0.01 hp (7.5 W) m^{-2}). Carrying capacity is $\sim$200 t ha^{-1} or 25 kg m^{-3}. Maximum daily feed rates range from 240 to 300 g m^{-3}.

Tanks of ~200 m^3 are also operated with liquid oxygen and paddlewheel aeration for CO_2 removal. The carrying capacity of these tanks ranges from 53 to 59 kg m^{-3} or maximum feed rates of 636–742 g m^{-3} day^{-1}. In total, Salton Sea farmers accounted for >2700 t (6 million lb) in 1995.

Smaller scale producers have also lined tanks with cages to reduce territoriality, and prevent breeding. Using an expandable framework, cage size can be increased during grow-out to accommodate increased fish size, or the entire cage can be moved if necessary. Effective costs for tanks and water are very small compared with those for feed and labour, thus favouring lower cost/lower protein diets and reducing the amount of grading and restocking. The development of intensive greenhouse production systems in the southern USA, such as Tiltech and others, is also notable (Lutz, 1997), using simple plywood or plastic lined rectangular tanks, typically 150–160 m^3, with water turnover of 60–120% of tank volume $hour^{-1}$, supplementary aeration, solids removal and water filtration, using sand or bead filters. Tanks are commonly partitioned with simple cage frame structures, allowing batches of fish to be moved and graded along the large tanks. Phytoplankton production, though contributing to stock feeding, can be difficult to manage, and can contribute to night-time respiration loads; shaded systems may be more effective (Rakocy *et al.*, 1992). Temperature management is also critical, depending in part on the water source and the extent of recycle adopted.

Other intensive and recycle systems

Elsewhere, these systems have been operated with different approaches to density and loading, usually with high aeration and water exchange, varying degrees of solids removal and nitrification and gas management. Though most systems have been developed in warm water areas, the potential of heated wastewater and recycle systems has been exploited to extend production into cooler zones.

Tropical and subtropical zones

The role of the Bamburi farm in Kenya in developing intensive tilapia concepts has been noted (Balarin and Haller, 1982, op cit). These authors reported densities of up to 50 kg m^{-3} in circular raceways, and loadings of 1.67 kg l^{-1} min^{-1} a residence time of ~33 min, and described other systems operated at intermediate density (25 kg m^{-3}) and very high loading (12–16 kg l^{-1} min^{-1}), usually supplemented with oxygen. Lauenstein (1978) described raceways supplied with geothermal water operated at 16 to 64 kg m^{-3} and a loading of 1.2–3.6 kg $l^{-1}min^{-1}$ a residence time of ~55–75 min. Cheong *et al.* (1987) described the growth of 'Taiwanese red' tilapia from 0.78 to 438 g in 239 days (1.82 g day^{-1}) in recycle systems at up to 49.3–50.2 kg m^{-3}, more moderate growth rates than those for 'Florida red' tilapia. In Zambia, the Tate and Lyle sugar company adopted intensive techniques for tilapia production in its sugar plantation areas (Dickson, 1987) using process waste based feeds. Smaller scale intensive systems were also developed in Malawi (Stewart, 1993) and in South Africa (Caulton, 1979)

A range of intensive systems has been developed in the Mediterranean and Middle East. A pilot unit was developed in Malta in the late 1970s, with small-scale production in agricultural water tanks (C. Agius, personal communication). Though technically successful, production was negligible. In Jordan a highly intensive production system with circular steel tanks, using oxygenated water supplies from the River Jordan was designed in 1985 to produce some 200 t annually with loadings of 8–10 kg $l^{-1}min^{-1}$ or more (Muir, 1986, unpublished). However, due to problems of fry supply and the operation of the intensive tanks, the project did not succeed. In the Kuwait Institute for Scientific Research (KISR), trials on intensive salt or brackish water tilapia culture had been carried out since the early 1980s (Al-Ahmad *et al.*, 1986), initially using simple aerated intensive tanks and return flow systems in greenhouses (Al-Ameeri, 1988), subsequently developing towards more sophisticated recycle systems for integrating with borehole based irrigated agriculture. However, Muir and Hopkins (1988, unpublished) reviewed the financial viability of various intensive scenarios, concluding that even with the favourable investment and operating cost conditions available at that time, considerable development in both technology and market acceptance would be required to meet cost-benefit criteria.

During this period, Muir (1986, 1987, unpublished) also reviewed a range of intensive seawater and borehole-based intensive and recycle production proposals in Saudi Arabia for various commercial developers, concluding that few would meet commercial criteria unless very favourable production incentives were available. However, integrated freshwater systems, with multiple/sequential use of water and better production performance, might offer better prospects, and this approach forms the basis for the present production systems in the country (Al-Hinty and Siddiqui, 1993; F. Bukhari, personal communication). However, technical and market problems remain. Intensive production has also been carried out experimentally in Egypt (Sadek *et al.*, 1992) but its further development has been limited.

Though intensive pond and cage aquaculture in SE Asia has developed strongly (see other chapters), little use has been made of tank-based open flow and recycle systems, other than for intensive hatchery production (D. Little, personal communication), where simple biofilter and aeration systems may be used to provide stable water supplies in rural and larger commercial hatchery centres.

In recent years steady growth in intensive tilapia farming has occurred in South and Central America. Most production is conducted in areas with abundant water resources in earthen or concrete-lined flow-through ponds. Water use typically exceeds 50 m^3 kg^{-1} output. In Colombia, the Colapia company operated from 1991 to 1996, resuming in 1998, producing Florida red tilapia in a total culture area of 30 ha. The culture complex comprises paddlewheel aerated earthen ponds, using river water, with an exchange of 30% of pond volume day^{-1}. Maximum stocking densities are 10 kg m^{-3} and annual output is ~4000 t. Water temperatures fluctuate between 22 and 26°C, and usage is ~52 m^3 kg^{-1} tilapia.

In Costa Rica, Aquacorporacion Internacional (*aka* Rain Forest Aquaculture), originally established in 1992, produces *O. niloticus* in concrete-lined ponds without auxiliary aeration. Ponds are flushed with water diverted from irrigation canals, at 24–27°C, with stocking densities up to 35 kg m^{-3} and exchange rates of up to 300% total pond volume day^{-1}, i.e. a mean residence time of 8 h. Annual production is 4300 t, with 93 m^3 water used kg^{-1} output. Table 11.7 outlines the typical stock pattern. An interesting feature of this scheme is the relatively simple, conventional fingerling production scheme coupled with the hyper-intensive, raceway-type grow-out system (cf. the density figures).

Table 13.7 Stock management scheme for Aqua Corporacion Internacional, Costa Rica

Parameter	Reproduction	Nursery 1	Growth phase Nursery 2	Grow-out 1	Grow-out 2
Avg wt (g)	n/a	0.02–1	1–50	50–300	300–900
Growth (g/d)	n/a	0.03	0.2–0.4	1.5–1.7	3.5–3.9
Cropping per (d)	26–30	30–35	120–130	130–140	120–160
Density (no/m^2)	2–3	500–1000	10–60	150–300	80–125
Survival (%)	99	60–70	60–80	85–95	95–99
Feed (% protein)	30–35	40–45	35–40	30–35	30
Feed conversion	n/a	1.5	1.5	1.7	1.9
Feed rate (% bwd)	1–2	20–40	5–10	2–5	1–2

In Ecuador, several farms have been established, with a total production area of ~300 ha. Florida red tilapia are cultured with shrimp at annual temperatures fluctuating between 24 and 30°C. Ponds operate at an exchange rate of 30% total volume day^{-1}, using estuarine water. Fish are kept at relatively low densities (up to 0.8 kg m^{-3}), and so although at intensive water use levels, yield is at the semi-intensive level. Production is currently estimated at ~1500 t.

In Jamaica, the Florida red tilapia producer Jamaica Aquaculture has operated since 1986 in a 100 ha farm with a total annual production of ~2500 t. River water (24–27°C) is used to replenish earthen ponds at 30% of total water volume daily. Paddle wheels are used for aeration. Average maximum stocking densities are 10 kg m^{-3} and overall water usage is 75 m^3 kg^{-1} produced.

Teichert-Coddington and Green (1997) describe intensive commercial red tilapia culture in earthen ponds in Sula Valley, Honduras, for export, commencing in 1989 (producing 1990); other units commenced in 1991–92, mainly in N. Honduras. Average temperatures range from 28–30°C in lowlands, to 20–22°C in higher areas. Production is exported as fillets, or sold locally as whole gutted fish. However, with 3–32 t ha^{-1} per crop and water exchange up to 20% 50% day^{-1}, this corresponds more to semi-intensive or intensive pond production. Aqua Corporacion de Honduras has been in operation since 1996, with output planned for 4000 t in several stages, each with a grow-out area of 1.2 ha (24×500 m^2 concrete rectangular raceways) and a 2.4 ha nursery area (6×0.4 ha ponds). Selected red tilapia hybrids (mainly Israeli ND59) are cultured at up to 50 kg m^{-3} using a 'stock and split' strategy. Based on first cycle results, projected annual targets are 1000 t per stage. Water is derived from a large mountain lake, collecting mainly rain and river water; temperatures range between 24 and 28°C. From this, the water is diverted in a large canal through two hydroelectric plants, run through the fish farm and returned into the river. Flows are adjusted according to stocking densities and may reach 20 volume exchanges daily. Water utilization for production of 1 kg of fish is estimated at 95 m^3.

Several smaller scale and experimental recycle systems have been developed using tilapia. Watanabe *et al.* (1997) describe a recycle hatchery system in the Bahamas using 6×6.5 l incubators, each carrying up to 12 000 eggs (1846 l^{-1}); fry production is carried out in 8 troughs with recycled water supply, using foam fractionation, sedimentation, biofilter and circulator. They noted that the trough conditions, with top surface area:vol ratio of 4.6:1, were better than those for cylindroconical tanks, at a:v = 2.5:1, with 5.6–8.4 fry produced per litre. Rakocy (1997) describes integration of tilapia production with vegetable hydroponics in recycle systems. Using fish-rearing, solids removal and lettuce production, key criteria were developed, including:

- an optimum ratio of 57 g m^2 day^{-1} feed application to plant growing area;
- the raft hydroponics had excess treatment capacity at 180 g m^{-2} day^{-1}, and hence excellent water quality.

Temperate zones

Intensive tilapia production in cooler areas has essentially been limited by water temperatures, and hence by the need to obtain heated water supplies at suitable cost. However, the relative ease of culture and high growth rate has favoured tilapia as a target species. Via its links with the Bamburi tilapia operation in Kenya, the UK Blue Circle Cement Co established an interest in tilapia as a heated water species, and sponsored a range of initial trials. However, outline assessments suggested that production was unlikely to viable in the UK (Muir, unpublished, 1982). Slightly later, the UK White Fish Authority[1] carried out pilot trials in an established seawater system at Hunterston, Scotland, heated with nuclear power station cooling water, but did not develop the project. A number of distilleries, breweries and steel works have also considered tilapia aquaculture, but have rarely gone beyond initial feasibility stages. However, in 1995, Pisces Aquaculture developed a medium-large scale intensive tilapia facility using waste heat from a textile manufacturing plant. Though the farm developed rapidly, reaching initial targets, and successfully marketed product through one of the leading UK supermarket chains, production cost problems and difficulties with agreeing longer-term site and water access conditions led to the closure of the project (J. Young, personal communication). Probably the most successful tilapia production system has been that at the nuclear power plant at Tihange, Belgium, described earlier, where a system of intensive elliptical tanks was developed in the early 1980s and continues in production.

11.5 PROBLEMS AND CONSTRAINTS

The above accounts suggest that the success of intensive tilapia aquaculture has been at best qualified, and that many issues remain to be resolved. In the United States, perhaps the most promising environment for its establishment, progress has been irregular, marked by some spectacular business failures, many enterprises having a life-span of no more than a few years. A number of constraints can be identified. In some cases, inadequate design or operating procedures have been difficulties, particularly if associated with commercial confidentiality, and a reluctance to exchange information and develop good system and practice criteria. Though this is improving, as stimulated by the recent successful series of technical conferences in Roanoke, Virginia (see e.g. Libey and Timmons, 1996), a broad design and management approach is still to emerge. At present, high investment and operating costs, inefficiencies of systems and components, and the availability and quality of seed and feed remain critical problems. As intensive systems tend to have relatively high overhead costs, factors leading to under-capacity can have critical financial effect. There is clearly a need for continued practical research into intensive tank and raceway systems, into recycle unit processes, control systems, and into improving and fine-tuning the husbandry and operational features of these systems. As with other aquaculture sectors, a 'critical mass' of producers may be needed to encourage secondary investment into targeted research and development.

---

[1] now the Sea Fish Industry Authority, SFIA.

Although the tolerance of tilapia to disease is well known, culture in recycle and other intensive systems presents special problems of stress and disease treatment, and a number of producers have had severe disease problems. *Mycobacterium* has been particularly insidious in some recycle facilities for tilapia and hybrid striped bass, and currently there are no effective therapeutants, although a number of drugs are under investigation. Concern also relates to the lengthy approval procedures for new drugs by government agencies, which may threaten the survival of many smaller producers who do not have the financial reserves to weather devastating disease outbreaks over long periods. In Israel, bacterial diseases are commonly encountered in intensive systems. Most of these are caused by opportunistic bacteria (e.g. *Streptococcus*, *Vibrio* and *Pasteurella*), invading fish already weakened by other disease or environmental processes. The non-specific nature of most of these diseases makes them unpredictable and complicates their diagnosis. *Streptococcus*, in particular, has plagued the industry with estimated morbidities ranging from 30 to 60% of the nation's total tilapia production. In 1989 this was estimated to mean a loss of about $3 million (Bercovier *et al.*, 1991). Presently, no therapeutic agents are available and most research has focused on the development of vaccines. As commonly observed with other infections by opportunistic bacteria, unfavourable environmental conditions leading to stress are closely correlated to the outbreak of *Streptococcus* infections (Bunch and Berjerano, 1997).

In the US, the problem of the lengthy approval processes is not limited to therapeutants. The need to manage the fecundity of tilapia is well established, and many producers have been participating in an investigative new animal drug (INAD) programme through the Food and Drug Administration, to evaluate the efficacy of methyl-testosterone (MT) for sex reversal of tilapia in a practical farm setting. University participants are also conducting efficacy and drug clearance testing programmes. Although the programme has been beset by problems related to feed formulation and violations of use, it appears that MT is likely to be approved within 5 years. Though a coordinated programme to develop populations of all male tilapia, such as the 'GMT YY supermales' (see Chapter 7), which will obviate the need for MT, is yet to be developed, this may become a more practical route.

In the USA, other regulations are also perceived to burden the intensive sector, particularly those related to discharge of waste waters. Effluent discharges from aquaculture are regulated by the US Environmental Protection Agency, whose local permit processes can be at least time consuming and at worst, discourage development altogether. Tilapia is an exotic species to N. America, although there are now self-sustaining populations in California (*O. mossambicus*), Florida (*O. aureus*) and other SE states, and many states have imposed restrictions on their transport and culture. California limits culture to *O. mossambicus*, although *O. niloticus* is known to be present in the Colorado River and at a commercial farm. Louisiana limits tilapia to recirculating systems that must not discharge water to the environment and must be located above the 100-year flood plain. There are many other examples.

Producers also cite marketing aspects as constraints (see Chapter 12). Within the catfish industry, off-flavours associated with effects of certain cyanobacteria have led to strict quality control programmes at processing plants to identify off-flavour fish. However, the US intensive tilapia industry is much smaller, geographically dispersed and focuses on supplying live fish to largely urban markets. A quality assurance programme will be needed to assure the freedom from taint for all fish reaching the market, as poor consumer impressions from relatively small quantities of off-flavoured fish would be sufficient to reduce market demand. Other problems relate to low prices when new farms or recently expanded operations begin marketing at prices that undercut local competition to gain market share. Prices also decline when wholesalers compete for market outlets.

At a more thematic level, sustainability issues may be problematic. Thus, in addition to metabolic and resource access energy supplied by intensive feeds, pumps and aerators represent significant energy inputs into intensive tilapia culture. In addition, heat may be supplied to maintain temperatures, either in flow through or recycle systems. Though cost penalties may not currently be too severe, particularly where energy costs are low, intensive systems are comparatively demanding. In terms of related parameters such as ecological footprint (Roberts and Muir, 1995) intensive systems are also unfavourably positioned, placing greater demands on external resources than other forms of production.

11.6 FUTURE DIRECTIONS

Over the last few decades a steady rise has been observed in world wide tilapia production, and with it, culture methods allowing growth of tilapia at extremely high densities. As in any other sector, the prospect of profit is the main driving force, in turn based on maximizing output and minimizing production cost. The industry has seen rapid development over the last two decades, with new production methods and systems, and a steady increase in interest, as system technology improvements combine with better stock performance and widening market acceptability. However, in many regions, the economic feasibility of intensive tilapia production systems is largely based on high-valued product. Although dependable and standardized output is an asset, the major challenge in most cases is the ability to compete with more widely available products from other, potentially lower cost production systems. At present, financial viability is largely confined to flow-through systems, in areas with ample water supply, though the 'semi-natural' photosynthetically driven recycle systems in Israel also show competitive merit, and may become increasingly important as land and water costs rise and as production techniques are refined.

In an increasingly open market environment, potential will be largely determined by comparative production costs and supply availability, in terms of timing and market size. While market advantages may be found in siting intensive or recycle systems near major conurbations and central distribution points, increasing efficiency in transport and distribution, and in the management of semi-intensive and cage culture systems, may counteract these gains. However, the effect of various advances in production methodology may help maintain or even improve the competitive position and hence their potential. A number of directions can be identified, including:

- Biotechnical advances which may offer advantages in intensive systems – e.g. in growth performance, environmental tolerance, better and cheaper feeds, more effective systems of disease management. However, such advantages may not be uniquely applicable to intensive systems, and may therefore simply reset the baselines for all types of tilapia production.
- Technical developments, such as better tank or raceway design conferring management or production efficiency advantages, better and more reliable systems of water and waste management, and better and more efficient systems of water reuse and disposal may improve reliability and productivity and reduce costs. Discernible gains in performance can already be seen, though costs and risks are still relatively high, and 'system' approaches, in which all the biological, technical and management features are linked together, are still to properly emerge.
- Location opportunities such as seawater environments; a potential exists for species such as *O. mossambicus*, but this has been examined only to a limited extent. Preliminary trials on *Oreochromis* hybrids in seawater (37–41 ppt salinity) at 14–20 kg m^{-3} (Lahav and Ra'anan, 1997) gave good results with daily growth rates as high as 2.2 g when water temperature exceeded 24°C. However, cage culture may be the most immediate beneficiary of these developments.
- Market issues such as the potential acceptability of controlled and 'environmentally managed' supplies, with standard product configurations to meet specialist demands, and better control e.g. of flesh quality, colour, texture, etc, may benefit intensive producers in the longer term, though practical issues such as the absolute control of off-flavours will be essential.

The opportunities for pond or large tank based flow-through and simple water reuse systems are apparent, and indeed many of the present intensive pond systems (Chapter 10) may develop to merge with the systems described here, using relatively simple and low-cost treatment technology (e.g. algal ponds, activated sludge). This is particularly the case if costs/access to water supplies become more critical and/or if environmental regulations impose more stringent standards on effluents. However, technically sophisticated recycle systems are still under question, and at this stage viability has not really been proven. As tilapia is more commonly a lower-priced fish, the high investment and operating expenses of current generation of commercial scale recycle technology are unlikely to be sustained. However, research efforts are steadily growing, technically sound and more reliable systems are available, and if future R&D efforts can meet their aims of reducing construction and operating costs, such systems may become more commercially convincing.

REFERENCES

Aboutboul, Y., Arbiv, R., van Rijn, J. *et al.* (1995) Anaerobic treatment of intensive fish culture effluents: Volatile fatty acid mediated denitrification. *Aquaculture* **133**, 21–32.

Al-Ahmad, T., Hopkins, K.D., Ridjha, M., Al-Ahmed, A. and Hopkins, M. (1986) Tilapia culture in Kuwait: final report. Kuwait Institute for Scientific Research, Kuwait, and International Center for Living Aquatic Resources Management (ICLARM), Philippines, 136 pp.

Al-Ameeri, A.A. (1988) *Protein Intake, Ammonia Excretion and Growth of* Oreochromis spiluris *in Seawater*, PhD thesis, University of Stirling.

Al-Amoudi, M.M. (1987) Acclimation of commercially cultured Oreochromis species to seawater: an experimental study. *Aquaculture* **65**, 333–342.

Al-Hinty, H.M. and Siddiqui, A.Q. (1993) Aquaculture in Saudi Arabia, in *Proceedings of the Aquaculture Symposium: Technology and Investment Opportunities*, 11–14 April, Riyadh, Kingdom of Saudi Arabia.

American Tilapia Association (1996) Newsletter, Summer 1996, 5.

Arviv, R. and van Rijn, J. (1994) Performance of a treatment system for inorganic nitrogen removal from intensive aquaculture systems. *Aquacultural Engineering* **14**, 189–203.

Avnimelech, Y., Weber, B., Hepher, B., Milstein, A. and Zorn, M. (1986) Studies on circulated fish ponds: Organic recycling and nitrogen transformation. *Aquaculture and Fisheries Management* **17**, 231–342.

Avnimelech, A., Mokady, S. and Schroeder, G.L. (1989) Circulated ponds as efficient bioreactors for single-cell protein production. *Bamidgeh* **41**, 58–66.

Avnimelech, Y., Mozes, N. and Weber, B. (1992) Effects of aeration and mixing on nitrogen and organic matter transformations in simulated fish ponds. *Aquacultural Engineering* **11**, 157–169.

Avnimelech, Y., Kochva, M. and Diab, S. (1994) Development of controlled intensive aquaculture systems with a limited water exchange and adjusted carbon to nitrogen ratio. *Bamidgeh* **46**, 119–131.

Balarin, J.D. and R.D. Haller. (1982) The intensive culture of tilapia in tanks, raceways and cages, in *Recent Advances in Aquaculture* (eds J.F. Muir and R.J. Roberts), Croom Helm, London, pp. 265–355.

Balarin, J.D. and Hatton, J.P. (1979) *Tilapia. A Guide to their Biology and Culture in Africa*, University of Stirling, Scotland.

Baird, D.J, Beveridge, M.C.M., Kelly, L.A. and Muir, J.F. (1996) (eds) *Aquaculture and Water Resource Management*, Blackwell Science, Oxford.

Bercovier, H., Eldar, A. and Bejerano, I. (1991) Streptococcal diseases in fish: pathophysiology and etiological agent. The Moshe Shilo Memorial Bat Sheva Symposium on Adaptation of Microorganisms to their Environment, Jerusalem, Israel.

Beveridge, M.C.M. (1996) *Cage Aquaculture*, 2nd Edition, Fishing News Books, Oxford.

Boyd, C.E. (1995) *Bottom Soils, Sediment and Pond Aquaculture*, Chapman and Hall, New York.

Bunch, E.C. and Bejerano, I. (1997) The effect of environmental factors on the susceptibility of hybrid tilapia *Oreochromis niloticus* × *Oreochromis aureus* to streptococccosis. *Bamidgeh* **49**, 67–76.

Caulton, M.S. (1979) The biology and farming of tilapia in Southern Africa, Fish Development Corporation, Gingindlovu, S. Africa.

Chervinski, J. (1982) Environmental physiology of tilapias, in *The Biology and Culture of Tilapias* (eds R.S.V. Pullin and R.H.Lowe-McConnell), ICLARM Conference Proceedings, Manila, Philippines, pp. 119–128.

Cheong, L., Chan, F.K., Wong, J. and Chou, R. (1987) Observations on the culture of red tilapia (*Oreochromis niloticus* hybrid) in seawater under intensive tank condition using a biodrum. *Singapore Journal of Primary Industries* **15**(1), 42–56.

Colt, J. and Orwicz, K. (1991) Modeling production capacity of aquatic culture systems under freshwater conditions. *Aquacultural Engineering* **10**, 1–29.

Costa-Pierce, B.A. (1995) Roles of reservoir fisheries in interactive, land/water ecosystem planning for resettlement, The World Bank, Washington DC.

Cowey, C.B. and Cho, C.Y. (1991) (eds) *Proceedings of the 1st International Symposium on Nutritional Strategies and Aquaculture Waste*, University of Guelph, Guelph, Ontario, Canada.

Dickson, M.W. (1987) The supply of vitamins in feed for intensive tilapia foraming in Zambia. *Aquaculture and Fisheries Management* **18**(2), 221–230.

Hankins, J.A, Summerfelt, S.T. and Durant, M.D. (1995) Impacts of feeding and stock management strategies upon fish production within water recycle systems, in *Aquacultural Engineering and Waste Management* (ed. M.B. Timmons), NE Regional Agricultural Engineering Service, (NRAES) Cooperative Extension, Ithaca, NY, USA, pp. 70–86.

Jauncey, K. and Ross, B. (1982) *A Guide to Tilapia Feeds and Feeding*, Institute of Aquaculture, University of Stirling, Scotland.

Kaiser, G.E. and Wheaton, F.W. (1983) Nitrification filters for aquatic culture systems: state of the art. *Journal of the World Mariculture Society* **14**, 302–324.

Lahav, E. and Ra'anan, Z. (1997) Salinity tolerance of genetically produced tilapia (*Oreochromis*) hybrids. *Bamidgeh* **49**, 160–165.

Lauenstein, P.C. (1978) Intensive culture of tilapia with geothermally heated water, in *Culture of Exotic Fishes Symposium Proceedings* (eds R.O. Smitherman, W.L. Shelton, J.H. Grover), Fish Culture Section, American Fisheries Society, Auburn, AL, pp. 82–85.

Libey, G.S. and M.B. Timmons (1996) Successes and failures in commercial recirculating aquaculture. *Proceedings from the Successes and Failures in Commercial Recirculating Aquaculture Conference, Roanoke, Virginia, July 19–21, 1996*, NE Regional Agricultural Engineering Service (NRAES), Cooperative Extension, Ithaca, NY, USA.

Losordo, T.M. (1997) Tilapia culture in intensive recirculating systems, in *Tilapia Aquaculture in the Americas* (eds B.A. Costa-Pierce and J.E. Rakocy, J.E.), Vol 1, World Aquaculture Society, Baton Rouge, Louisiana, USA, pp. 185–211.

Losordo, T.M. and Westerman, P.W. (1994) An analysis of biological, economic, and engineering factors affecting the cost of fish production in recirculating aquacultural systems, in *Design of High-Density Recirculating Aquaculture Systems*, Louisiana Sea Grant, Baton Rouge, LA, pp. 1–9.

Losordo, T.M. and Westers, H. (1994) System carrying capacity and flow estimation, in *Aquaculture Water Reuse Systems: Engineering, Design and Management* (eds M.B. Timmons and T.M. Losordo), Developments in Aquaculture and Fisheries Sciences 27, Elsevier, Amsterdam, pp. 9–60.

Losordo, T.M., Westerman, P.W., Schei, I., DeLong, D.P., Twaroska, J.G. and Milam, D.L. (1995) An evaluation of the Ecofish tank system technology for use in the intensive production of tilapia utilizing water reuse technology, Technical Report, North Carolina State University, Raleigh, NC, USA.

Lutz, C.G, (1997) Greenhouse tilapia production in Louisiana. *Aquaculture Magazine* Nov/Dec 1997, 28–36.

McNown, W. and Seireg, A. (1983) Computer aided optimum design and control of staged aquaculture systems. *Journal of the World Mariculture Society* **14**, 417–433.

Melard, C. and Philippart, J.C. (1980) Intensive culture of *Sarotherodon niloticus* in Belgium. EIFAC Symposium on New Developments in Utilisation of Heated Effluents and of Recirculation Systems for Intensive Aquaculture. EIFAC/80/Symp-E/11, 28 pp.

Mires, D. and Amit, Y. (1992) Intensive culture of tilapia in quasi-closed water-cycled flow-through ponds – the Dekel aquaculture system. *Bamidgeh* **44**, 82–86.

Mires, D., Amit, Y., Avnimelech, Y., Diab, S. and Cochaba, M. (1990) Water quality in a recycled intensive fish culture system under field conditions. *Bamidgeh* **42**, 110–121.

Muir, J.F. (1975) Waste recycling systems in fish farming. *Fish Farming International* **2**(2), 14–15, 48.

Muir, J.F. (1982) Recirculated water systems in aquaculture, in *Recent Advances in Aquaculture* (eds J.F. Muir and R.J. Roberts), Croom Helm, London, pp. 358–446.

Paessun, M.A. and Allison, R. (1984) Maximizing tilapia production in recirculating systems by sequential rearing. *Aquaculture* **42**, 185–9.

Philippart, J.-Cl. and Ruwet, J.-Cl. (1982) Ecology and distribution of tilapias, in *The Biology and Culture of Tilapias* (eds R.S.V. Pullin and R.H. Lowe-McConnell), International Center for Living Resources Management, Manila, Philippines, pp. 15–59.

Plumb, J.A. (1997) Infectious diseases of tilapia, in *Tilapia Aquaculture in the Americas* (eds B.A. Costa-Pierce and J.E. Rakocy), Vol 1, World Aquaculture Society, Baton Rouge, Louisiana, USA, pp. 212–228.

Popper, D. and Lichatowich, T. (1975) Preliminary success in predator control of *Tilapia mossambica*. *Aquaculture* **5**, 213–214.

Pullin, R.S.V. (ed.) (1988) *Tilapia Genetic Resources for Aquaculture*, International Center Living Aquatic Resources Management, Manila, Philippines.

Rakocy, J.E. (1989) Tank culture of tilapia. Publication No 282, US Department of Agriculture, Southern Regional Aquaculture Center, Stoneville, Mississippi, US.

Rakocy, J.E. (1997) Integrating tilapia culture with vegetable hydroponics in recirculating systems, in *Tilapia Aquaculture in the Americas* (eds B.A. Costa-Pierce and J.E. Rakocy), Vol. 1, World Aquaculture Society, Baton Rouge, LA, pp. 163–184.

Rakocy, J.E., Cole, W.M., Bailey, D.S. and Rangel, A.M. (1992) The effect of phytoplankton on water quality and tilapia production in closed recirculating systems with solids removal and fixed film biofiltration, in *Aquaculture '92: Growing towards the 21st Century*, World Aquaculture Society, Baton Rouge, Louisiana, pp. 189–190.

Ray, L.E. (1978) Production of tilapia in catfish raceways using geothermal water, in *Culture of Exotic Fishes* (eds R.O. Smitherman, W.L. Shelton and J.H. Grower), Symposium Proceedings, Fish Culture Section, American Fisheries Society, Auburn, Alabama, pp. 86–89.

Rimon, A. and Shilo, M. (1982) Factors which affect the intensification of fish breeding in Israel. 1. Physical, chemical and biological characteristics of the intensive fish ponds in Israel. *Bamidgeh* **34**, 87–100.

Roberts, R.J. and Muir, J.F. (1995) 25 years of world aquaculture: sustainability, a global problem, in *Sustainable Fish Farming* (eds H. Reinertsen and H. Haaland), AA Balkema, Rotterdam, pp. 167–181.

Sadek, S., Kallafalah, H. and Adell, F. (1992) Tilapia (*Oreochromis niloticus*) biomass yield in a commercial farm using circular tanks. *Journal of Applied Ichthyology* **8**, 193–202.

Sarig, S. and Arieli, Y. (1980) Growth capacity of tilapia in intensive culture. *Bamidgeh* **32**, 57–65.

Shilo, M. and Rimon, A. (1982) Factors which affect the intensification of fish breeding in Israel. 2. Ammonia transformation in intensive fish ponds. *Bamidgeh* **34**, 101–114.

Stewart, J.A. (1993) An economic analysis of the viability of fish farming in Malawi, Technical Report, University of Stirling, Stirling, UK, 82 pp.

Summerfelt, S.T., Hankins, J.A., Summerfelt, S.R. and Heinen, J.M. (1993) Modeling continuous culture with periodic stocking and selective harvesting to measure the effect on productivity and biomass capacity of fish culture systems, in *Techniques for Modern Aquaculture* (ed. J.-K. Wang), American Society of Agricultural Engineers, St. Joseph, MI, pp. 581–595.

Teichert-Coddington, D.R. and Green, B.W. (1997) Experimental and commercial culture of tilapia in Honduras, in *Tilapia Aquaculture in the Americas* (eds B.A. Costa-Pierce and J.E. Rakocy), Vol 1, World Aquaculture Society, Baton Rouge, Louisiana, USA, pp. 142–162.

Timmons, M.B. and Losordo, T.M. (1994) (eds) *Aquaculture Water Reuse Systems: Engineering, Design and Management*. Developments in Aquaculture and Fisheries Sciences 27, Elsevier, Amsterdam.

Tucker, C.S., Steeby, J.A., Waldrop, J.E. and Garrard, A.B. (1992) Effects of cropping system and stocking density on production of channel catfish in ponds, Bulletin 988. Department of Agricultural Services, Mississippi State University, Mississippi, USA.

Uchida, R.N. and King, J.E. (1962) Tank culture of tilapia. *US Fish and Wildlife Service Fisheries Bulletin* **62**, 21–52.

Van Gorder, S. (1994) Optimizing production by continuous loading of recirculating systems, in *Design of High-Density Recirculating Aquaculture Systems*, Louisiana Sea Grant, Baton Rouge, LA, pp. 10–15.

van Rijn, J. (1996) The potential for integrated biological treatment systems in recirculating fish culture: a review. *Aquaculture* **139**, 181–201.

van Rijn, J., Stutz, R., Diab, S. and Shilo, M. (1986) Chemical, physical and biological parameters of superintensive concrete fish ponds. *Bamidgeh* **38**, 35–43.

van Rijn, J., Fonarev, N. and Berkowitz, B. (1995) Anaerobic treatment of fishculture effluents: Digestion of fish feed and release of volatile fatty acids. *Aquaculture* **133**, 9–20.

Watanabe, W.O., Clark, J.H., Dunham, J.B., Wicklund, R.I. and Olla, B.L. (1990) Culture of Florida red tilapia in marine cages: the effect of stocking density and dietary protein on growth. *Aquaculture* **90**, 123–124.

Watanabe, W.O., Chan, J.R., Smith, S.J., Wicklund, R.I. and Olla, B.L. (1993) Production of Florida red tilapia in flow-through seawater pools at three stocking densities, in *The Third International Symposium on Tilapia in Aquaculture* (eds R.S.V. Pullin, J. Lazard, M. Legendre, J.B. Amon Kothias and D. Pauly), ICLARM Conference Proceedings 41.

Watanabe, W.O, Olla, B.L., Wicklund, R.I. and Head, W.D. (1997) Saltwater culture of the Florida Red Tilapia and other saline tolerant tilapias: a review, in *Tilapia Aquaculture in the Americas* (eds B.A. Costa-Pierce and J.E Rakocy), Vol 1, World Aquaculture Society, Baton Rouge, Louisiana, USA, pp. 55–141.

Watten, B.J. (1992) Modeling the effects of sequential rearing on the potential production of controlled environment fish-culture systems. *Aquacultural Engineering* **11**, 33–46.

Watten, B.J. and Beck, L.T. (1987) Comparative hydraulics of a rectangular cross-flow rearing unit. *Aquacultural Engineering* **6**, 127–140.

Watten, B.J. and Honeyfield, D.C. (1995) Hydraulic characteristics of a rectangular mixed-cell rearing unit, in *Aquacultural Engineering and Waste Management* (ed. M.B. Timmons), NE Regional Agricultural Engineering Service, Cooperative Extension, Ithaca, NY, USA, pp. 112–126.

Zohar, G., Rappaport, U., Avnimelech, Y. and Sarig, S. (1984) Results of the experiments carried out in the Genosar experimental station in 1983. Cultivation of tilapia in high densities and with periodic flushing of the pond water. *Bamidgeh* **36**, 63–69.

Zohar, G., Rappaport, U. and Sarig, S. (1985) Intensive culture of tilapia in concrete tanks. *Bamidgeh* **37**, 103–111.

Chapter twelve

Economics and marketing

J. A. Young and J. F. Muir

12.1 INTRODUCTION

Whatever the reader's background, the subject matter of this chapter, the marketing and economics of tilapia, will almost certainly have played some part in their other connections with the species. As for other species, market and economic issues are pervasive and regardless of context and production system have fundamental implications for the transition from initial R&D through to commercial introduction, growth of output, product diversification, product obsolescence and the search for subsequent replacements. Decisions are commonly determined on the basis of input and output prices in specified tilapia markets and those for competing substitutes. The complexity and interdependence of these phenomena are such that only key issues can be noted within this single chapter.

We have attempted a balance between conceptual issues facing those interested in the marketing and economics of tilapia, and the basic description and analysis of trends in markets and associated products. The diverse aquaculture potential of tilapia has been described earlier, and supports the broad premise that the species are important and may be expected to have significant future prospects. A brief sectoral overview of production and market statistics is given, but emphasis is primarily placed upon the parameters involved rather than unnecessary and quickly obsolete detail. Likewise, comparative cost data are given for illustrative purposes, but present day financial profiles of individual farms are quickly outdated. Market and industry intelligence, via web sites and elsewhere, enables such temporary values to be readily updated. We therefore begin with an overview of the marketing environment within which tilapia must compete and then focus upon the

M.C.M. Beveridge and B.J. McAndrew (eds), Tilapias: Biology and Exploitation, 447–487.

supply characteristics for tilapia. To gain some perspective on the competitive standing of the fish, its production economics are discussed. Thereafter the demand for tilapia products is explored by assessing the nature of consumer satisfaction and market development, focusing on the EU and US. These are finally brought together to outline potential directions and sectoral implications.

12.2 THE TILAPIA MARKETING ENVIRONMENT

The development of the tilapia sector within the world food supply situation has been notable; FAO figures (Table 12.1) illustrate its recent growth and its component elements, demonstrating a compound growth rate of some 11.2% annually over 1986–1995, approximately similar to the growth of the aquaculture sector as a whole (FAO, 1997). More significantly, perhaps, production of *Oreochromis niloticus*, representing more than 75% of 1996 production, had grown by some 18.4% annually over that period.

Table 12.1 Tilapia production from fisheries and aquaculture (t)

Species	1986	1990	1992	1994	1996	1996%	% y^{-1}
O. mossambicus	17 228	42 912	49 327	51 872	50 000	6.24	11.2
O. niloticus	110 974	224 984	321 800	432 274	603 034	75.30	18.4
O. aureus	14 950	18 678	16 990	11 797	11 877	1.48	−2.3
O. spilurus	10	1	0	20	70	0.01	21.5
Other tilapia	109 558	121 387	123 770	123 035	135 842	16.96	2.1
Total	252 722	408 111	511 887	618 998	800 823	100	12.2

Source: Developed from FAO (1998)
Notes: other tilapia = tilapia nei, *O. macrochir*, *O. andersonii*, *T. zillii*

As Table 12.2 indicates, the bulk of current production is in inland environments and systems, and although detailed statistics are not available to validate the estimates given, which are based on known national production characteristics, most of this is likely to be in pond production. The greatest part is likely to be semi-intensive (see Chapter 9), i.e. using moderate amounts of fertilizer to enhance natural productivity. However, cage culture is also increasing in importance and may be expected to contribute further in the future. Given the dominance of the fresh water *O. niloticus* it is unsurprising that much of the growth in tilapia production has been associated with inland environments, but with the introduction of more saline tolerant species and strains, this may gradually change.

Table 12.2 Key tilapia production environments

Production system	1996 production	5-y trend	Notes
Inland	732 756 t	14.9%	based on *O. niloticus, aureus*, other tilapia; usually ponds, increasingly cages; small contribution from intensive systems
Coastal	68 064 t	3.0%	based on *O. mossambicus, spilurus*, some *niloticus*; ponds and cages; some intensive systems

A summary of the tilapia aquaculture producing countries is provided in Table 12.3, which demonstrates its ubiquity and the relative importance of a small group of major centres, particularly China, Indonesia, Philippines, Thailand and Taiwan, although the rapid sectoral growth rate in countries such as Colombia, Malaysia, USA, Costa Rica, Saudi Arabia and Venezuela is also notable. Longer term growth has been very positive, and many countries now producing significantly had barely entered production until the 1980s. It is also noteworthy, as discussed further, that many producer countries primarily serve domestic markets, and are geographically distant from major international markets, but that trade flows, particularly to N America and Europe, are increasing and might be expected to influence future change.

In overall terms, the wild caught supply of tilapia, unlike many other species groups, is recorded to have grown, from 879 645 t in 1990 to 1 318 508 t in 1996, an average increase of 7% y^{-1}. Including aquaculture, total output has risen from 1 287 756 t to 2 119 331 t, (~2% of global fishery supply) over the same period, an average annual growth of 8.7%, the share of aquaculture rising from 31.7% to 37.8%. Although recorded fishery output has grown, given widespread evidence of fishing pressure in inland and coastal waters, and given the advantages of relative stability of supply and quality, future supplies are increasingly likely to derive from various forms of aquaculture.

The recorded growth in aquaculture has also been testament to at least some degree of success in terms of market opportunity, profitability and further investment. As with other aquaculture species, this has also been assisted by technical progress, in genetics, reproduction, nutrition, husbandry and disease management, as well as in system design and management. For the more recent development of tilapia cage culture, and in the relatively small intensive sector, this change may be at least partly documented. However, in pond production, some degree of substitution may have occurred as producers moved towards tilapia production, either as a monoculture or as increasingly observed, in various polycultures. Nonetheless, an overall increase of some 400 000 t over a six year period to 1996, doubling aquaculture output, represents a significant change.

Table 12.3 Key tilapia producer countries and their relative status

Key producer countries	1986	1996	1996%	%/y	Key features, markets
Asia					
China	29 500	394 303	49.24	29.6	Mainly ponds, local markets and regional exports
Indonesia	24 011	78 400	9.79	12.5	Pond, cage culture, inland, coastal; USA, M East markets
Thailand	19 245	76 054	9.50	14.7	Mainly ponds, some coastal; USA, M East markets
Philippines	55 819	76 000	9.49	3.1	Ponds, cages, inland, coastal; USA, M East markets
Taiwan	48 704	44 756	5.59	–0.8	Mainly ponds, inland, some coastal; regional markets
Malaysia	575	11 087	1.38	34.0	Ponds, cages, mining pools; M East and European markets
Sri Lanka	4 000	3 500	0.44	–1.3	Mainly inland ponds, some cages in reservoirs
Laos	0	1800	0.22	6.3	Ponds and some cages, local markets
Japan	4 113	1 479	0.18	–9.7	Ponds and cages, mainly inland; local
	185 967	687 379	85.83		
M East/N Africa					
Egypt	25 000	27 854	3.48	1.1	Mainly inland ponds, some intensive systems
Israel	3 238	6 399	0.80	7.0	Inland ponds, some intensive; European markets
Saudi Arabia	6	2 852	0.36	60.0	Intensive tank systems, groundwater, some cages
Syria	563	1 588	0.20	10.9	Ponds
	28 807	38 693	4.84		

Table 12.3 (cont)

Key producer countries	1986	1996	1996%	%/y	Key features, markets
Sub Saharan Africa					
Zambia	687	4 403	0.54	20.4	Mainly inland ponds
Nigeria	3 274	3 040	0.38	–0.7	Mainly ponds, possibly some cages
	3 961	7 443	0.92		
C and S America					
Mexico	13 110	15 830	1.98	1.9	Ponds, cages, intensive systems; USA markets
Colombia	300	14 026	1.75	46.9	Ponds, intensive systems; USA markets
Cuba	14 942	11 870	1.48	–2.3	Ponds, cages, inland and coastal
Costa Rica	50	4 100	0.51	55.4	Ponds, intensive systems; USA markets
Jamaica	1 442	2 800	0.35	6.9	Cages, ponds, some intensive, coastal; USA markets
Venezuela	0	1 700	0.21	83.2	Ponds, some cages; local, USA markets
	29 844	50 326	6.28		
N America					
USA	0	7 242	0.90	80.3	Ponds, intensive tanks, raceways

Developed from FAO (1998)

An underlying question for such expansion is its potential to occupy greater share of supply, and to become more important in regional and global terms. Although the number of species produced in aquaculture is gradually increasing (FAO, 1998), there is evidence of focus around certain groups, for reasons of technical capability, market attributes and consumer familiarity. Despite the current greater diversity of the aquaculture sector, compared with its agricultural equivalent, future production sectors, already grouped to some extent around freshwater carp, salmon, mussels, catfish and tilapia, may become even more concentrated, and a species group such as tilapia may become a major component of supply.

In a review of aquaculture development potential in Europe, Young and Muir (1995), noted that of various expansion routes, product development around existing species may be one the more promising. In the early 1980s, ICLARM proposed tilapia as the prime candidate for the 'aquatic chicken' because of its desirable product attributes (Maclean, 1984; ICLARM, 1984). These were stated to include simplicity of rearing, hardiness, versatility, undemanding feed requirements, with minimal dependence on fish meal and oil resources, firm flesh texture and neutral flavour. Such characteristics potentially favour widespread acceptance in a range of different product formats in different market segments. While the aquatic chicken is likely to remain more geographically restricted than its terrestrial rival in terms of production sites, increased and relatively inexpensive trade might well promote penetration of a similarly broad range of markets. The scope for such growth in supply may be encouraged by the increasing potential created by simple genetic manipulation (see Chapter 7) and the reduced availability of other traditional species, especially from capture fisheries. This supply push might be further induced through demand pull as noted by Haylor *et al.* (1994), which assessed tilapia, channel catfish and *Clarias* catfish to be favourably placed in respect of product attributes sought by the market.

As the generalized case in Fig. 12.1 outlines, the relationships between the demand for a product (e.g. tilapia in general, or a particular species or tilapia product in a specific market), and the supply will determine the prices and quantities traded, and hence ultimately the value of that particular sector. In this particular case, with demand and supply meeting at P_1Q_1 (with total sectoral value = $P_1 \times Q_1$), improvements in supply efficiency alone could move production to P_2Q_2 and expansion of markets could move the equilibrium to P_3Q_3. The objectives of expansion would normally be to increase sectoral value, and in particular the financial margins of individual enterprises, although other factors, such as the competitive behaviour of firms, may stimulate expansion irrespective of potential profitability or overall sectoral value.

It has been a common observation from the launch of all major aquaculture species that increased availability eventually results in reduced prices, to the extent that producers may face unprofitability, and may merge or leave the market, reducing the propensity to further investment (Young and Muir, 1997).

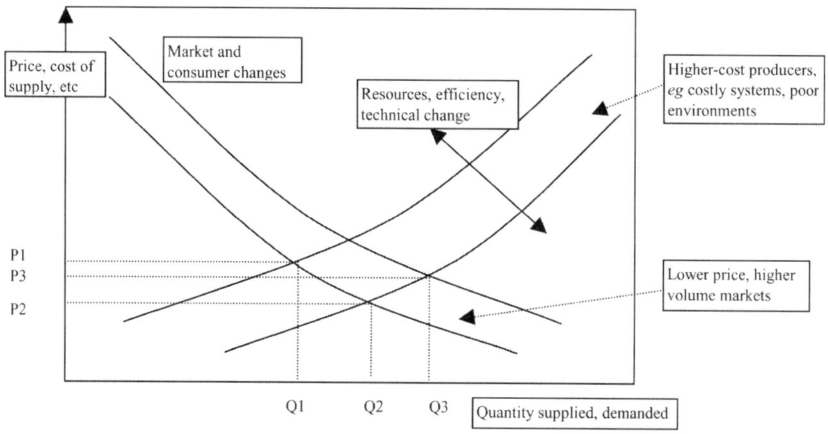

Fig. 12.1 Generalized demand and supply potential for the tilapia sector.

As the figure illustrates, changes in technical efficiency common in many sectors may reduce the cost of supply, and hence increase the opportunity for sales and production, maintaining profitability and investment opportunity, at least in the shorter term. The extent of this increase, and the prices at which supply will be sold, will depend on the shape of the curves, which may vary with product and market, and may change over time. Market and consumer changes are also important, and in the absence of significant technical change or – as may be the case, with increasing regulation or resource scarcity driving costs upwards – these may be the primary means for production and market supply to increase.

It may be a truism to propose that a production sector exists and has potential only to the extent that markets exist for its outputs, and that markets are therefore the primary determinants of change. However, this can easily be overlooked when describing the future of an aquaculture-dominated sector, driven as it may be by interests in biological understanding and technical innovation. Regardless of the future for technical change, people must be prepared to purchase tilapia products in sufficient quantities and at appropriate prices, generating sufficient profits to ensure continued investment and further sectoral development. This will have significance not only for immediate trends for tilapia culture, and the forms in which this may occur, but also in addressing the intriguing question of whether tilapia is capable of taking on the major role noted earlier. There will also be important implications for the structure of the sector, and for international competition. Thus, if markets drive towards unified product form with low producer profit margins, opportunities for smaller scale producers may diminish, particularly if biotechnology gains are preferentially available to agro-industrial producers. Here, the broader economic issues of employment opportunity, social impact and equity may become significant, and expansion may bring unwelcome externalities. Resource use and environmental implications of change may also need to be considered, and scale and location of development, and type and structure of production, will all be important. Although the ubiquity and diversity of tilapia culture suggests potential for development across a range of fronts, from artisanal to agro-industrial, in a broad range of locations and economies, this may not be certain.

12.3 CURRENT SUPPLY FEATURES

Earlier chapters have described the fundamental characteristics of the tilapia, their natural distribution and supply sources, and the range of production systems currently employed. These have identified the environments and habitats in which tilapia can be found or produced, and have considered the potential and constraints of different sources of supply. Though primarily based on inland, freshwater environments, development of tilapia culture in saline waters offers interesting options for greater expansion (Suresh and Lin, 1992; Watanabe *et al.*, 1997) From this basis it is useful to summarize the current supply situation, identify key production sources and systems, and describe the main avenues by which the supply system operates. The production sector is currently very diverse and is primarily rural, with a significant number of smaller operators, although the number of agro-industrial producers is gradually increasing, as tilapia become more interesting as a product, whether for domestic or export markets. This process has also nurtured managerial skills and is likely to accelerate structural diversification. However, with a mix of domestic and export markets, much of the traditional sector continues to be viable and this may support retention of small scale tilapia producers.

The increased availability of all-male genetic technology, such as the ICLARM 'GIFT', the UK 'GMT' and the various commercial red hybrids (see Chapter 3) has had a notable effect on reliability and quality of production. Improved productivity, even in very simple systems, coupled with the prospect for marketing attractive, well-coloured fish, often under locally evolved names, has expanded potential for producers. Further extension of the production base may arise through genetic improvements enabling survival and growth in cool winter climes, as cost restrictions of short production seasons, and the need to overwinter, *inter alia*, can severely limit viability. Likewise, the development of more saline tolerant stocks would significantly extend potential for cage or pond culture in coastal areas. Other developments may relate to the improved ability to use lower cost vegetable dietary substrates, removing any dependence on capture fisheries for oils and meal, although this is already achieved in well managed semi-intensive systems, sometimes using agricultural or other wastes as fertilizers. However, the acceptability of modified food products within different markets is variable, as will be discussed later.

To date, the potential to expand tilapia production has been available to a range of individuals and enterprises, according to their means and circumstances; the sector encompasses both small-scale artisanal producers and large corporations, and occupies a range of ecosystems and production niches. Production of tilapia and other species is considered to be an important element in rural economic development in many areas (Wijkstrom, 1991), provided seed and other resources can be accessed. Whilst the network approaches promoted by GIFT and GMT programmes in the Philippines and elsewhere promise to bring technical improvements into the hands of smaller producers, their risk-prone circumstances, low input levels, and the availability of cheap, locally produced fry of uncertain quality, may lock them into longer-term disadvantage. This suggests that increased tilapia production may move steadily towards the more organized producers, whether corporate or co-operative.

The resource demands and sustainability of aquaculture have increasingly been examined (Folke and Kautsky, 1991; Beveridge *et al.*, 1994; Stewart, 1995) with respect to the implications of current and anticipated production. For tilapia, with a great proportion of production in semi-intensive, low to medium input systems, present and potential ecological impacts could be relatively modest. However, moves to intensify production through more resource demanding feeds, higher operational energy requirements and greater concentrations of inputs and non-marketed outputs (i.e. wastes) could lead to substantially changed impacts. The introduction of tilapia, or non-indigenous strains may also create significant biodiversity impact. With more concentrated ownership and greater substitution of labour input, the balance of social benefit may also be unfavourably shifted. Table 12.4 summarizes the typical range of current impacts and demonstrates the wide variation between systems. Trade-offs may also exist between resource inputs. Thus, at a simple level, extensive systems require substantially more productive land/water area, usually require less water turnover (although water may be turned over naturally through seasonal hydrological processes), involve lower levels of feed, fertilizer and external energy, and have a smaller ecological footprint (most of which is represented by the productive area itself). To increase productivity immediately starts to draw in more inputs, the resulting marginal gains of output depending on the location and the system employed.

Based on these and similar depictions of resource demands, the potential implications of various levels of development for tilapia can be considered and described, and various options for different pathways, e.g. expansion, intensification, system modification may be explored. For illustration, three scenarios are proposed, the outputs of which are shown in Table 12.5.

Table 12.4 Resource impacts of tilapia production

Factor (value)	Typical values for different methods	Implications	Trends
Water use $10^3\ m^3\ t^{-1}$	2–5 recycle 20–50 extensive 10–20 semi-intens 50–300 cages 50–200 intensive	Competitive access; higher yields and integration can improve utilization; timing of use may be critical; rainfed systems may be inefficient; varied impacts of modified water quality;	Water sparing; access constraint increasing; potential of reservoirs, integration, waste recovery, coastal areas; recycling may be expensive; key regions in future may emerge due to resource access
Land area ha t^{-1}	5–50 extensive 0.2–0.5 semi-intens 0.05–0.2 intensive <0.005 high intens	Competition for land of suitable conformation, soils, near water supplies and services; expensive in unsuitable areas; intensify rather than expand	Intensification widespread, in rural and peri-urban areas; urban encroachment; large areas in some regions; with market access these may develop strongly
Labour use t $person^{-1}$	0.2–2 extensive 0.5–2 semi-intens 1–10 intensve 20–50+ high intens	Minimal routine inputs; maintenance and harvest key needs; families may be widely involved, often uncosted; also community, cooperative systems	Agro-industrial systems reduce labour, but may be opportunities to integrate tilapia into artisanal activities; service and downstream sectors may increase
Feed use t t^{-1}	0.2–0.6 semi-intens 0.8–1.5 intensive	Access to key raw materials; particularly fish based; local byproducts, etc.	More diversified raw material sources better feeds, management systems
Waste discharge t t^{-1}	0.3–0.5 susp.solids 0.2–0.5 BOD 0.02–0.05 diss-N 0.005–0.01 diss-P	Options for siting, environmental capacity, potential impacts, enrichment, use in integration; variable outputs; ponds seasonal, vs tanks continuous	Rising nutrient in many systems; greater regulation, demand to control; improved management, recovery methods, integration may be required
Industrial energy gJ t^{-1}	0.4 extensive 2 semi-intensive 25+ intensive	Resource subsidy indicator – mainly linked with feeds, and pumps, aerators, system materials, packing, transport	Efficiency falling with intensification; photosynthetic driven systems with minimal postharvest inputs reduce demands
Ecological footprint	1–100 × productive water area	Indicator of resource input/waste rehabilitation capacity required	As with industrial energy; waste absorption may be critical in some locations

Table 12.5 Outline development projections

Scenario	Year	% y^{-1}	Production	t ha^{-1}	10^3 ha	10^3 m^3 t^{-1}	10^6 m^3 y^{-1}	FCR	Food, 10^3 t	Notes
Current	1996		800 000	2.5	320	30	24 000	0.25	200	Current baseline
A-1	2000	10	1 171 280	3	390	30	35 140	0.25	290	6 × current output,
A-2	2015	5	2 435 010	3.5	696	28	68 180	0.3	730	3 × area, 5 × water,
A-3	2030	5	5 062 210	5	1012	25	126 560	0.35	1 770	9 × food; significant
B-1	2000	8	1 088 390	3	363	30	32 650	0.25	270	4.5 × current level,
B-2	2015	5	2 262 680	4	566	25	56 570	0.3	680	2 × area, 3 × water,
B-3	2030	3	3 525 180	6	588	20	70 500	0.3	1 060	5 × food; lower impact
C-1	2000	15	1 399 210	3	466	35	48 970	0.3	420	> 110 × output, 50 ×
C-2	2015	20	21 557 660	4	5389	30	646 730	0.35	7 550	area, 90 × water, 180 ×
C-3	2030	10	90 051 700	6	15009	25	2 251 290	0.4	36 020	food; huge impact

(A) current average rate of growth stabilizes, average yields steadily increase, water use gradually declines, food conversion declines as systems intensify;

(B) lower rate of development, yields increase more strongly, water use drops more rapidly, food conversion levels off with more efficient use;

(C) massive expansion, tilapia becoming a dominant component for global aquaculture and fishery supplies, yields as with (B), water use and FCR higher.

In this specific example, scenarios A and B, although demonstrating dramatic changes in overall resource inputs, represent no more than simple projections from current trends, including some element of declining growth as resources become more expensive and/or markets become saturated. Scenario C describes what is a challengingly massive change in output, driven for example by huge changes in demand, and/or static supplies of substitutes. However, it serves to illustrate the potential implications of extreme case conditions. Depending on likely directions within specific regions or countries, or arising from technology or market changes, variants on scenarios such as this can be developed with different yield and production system mixes. Although these scenarios are simplistic, they outline implications of various development directions, and can guide investment and development policy, at regional, national or local area level. Key issues could include:

- the existing mix of production systems and resource availability, and their marginal productivity in different systems, and under different options of technology change;
- the source and applications of investment to continue expansion and direct change;
- the role and effectiveness of institutional and commercial research and development in identifying and promoting sectoral improvements;
- the existence of enabling environments in countries and localities concerned; including legislative frameworks, administrative process, market, technical and planning support, environmental protection, definition and security of tenure;
- the need for a clear appreciation of what approaches work well, which areas are most appropriate to develop, and how they should be promoted.

12.4 COMPARATIVE PRODUCTION ECONOMICS

Materials from previous sections have described the basic features of different production systems and the choices of species, water distribution, nutrient and feed input, and management required. From these can be derived performance parameters, and hence comparative production costs. Conventionally, capital and operating costs can be defined to determine typical cost ranges, comparative features of production, and the potential for change in cost subject to different installation or operational variables. Studies have been carried out for a range of tilapia production enterprises, from simple rural aquaculture in low-input ponds in Africa (Stewart, 1993; Engle *et al.*, 1993; Engle, 1997a) to semi-intensive ponds (Fermin, 1985; Green *et al.*, 1990; Head and Watanabe, 1995), integrated production systems (Lovshin and Pretto, 1983; Fermin, 1985; Lovshin *et al.*, 1986; Hatch and Engle, 1987), salt water ponds (Head *et al.*, 1996) cage culture systems (Gonzales, 1984; Oliva, 1985; Brass *et al.*, 1990) to tank systems (Hopkins *et al.*, 1986; Al Ahmad *et al.*, 1996), heated effluent systems (Nelson *et al.*, 1981) and industrial scale systems with sophisticated technologies and expensive operating inputs (Losordo and Westerman, 1994). A range of hatcheries (Yater and Smith, 1985; Escover *et al.*, 1987; Head and Watanabe, 1995) has also been assessed. General rules found for tilapia and other production sectors (Muir, 1995; STAQ, 1996) include:

Capital costs

- these include holding facilities, water supplies, feeding and harvesting, transport and handling facilities, feed production and fish processing equipment, buildings, services and infrastructure; holding facilities are usually the most important (and sometimes the only) element;
- cages and enclosures are usually the cheapest (typically \$500–1000 t^{-1} installed capacity) followed by earth ponds (\$800–2000 t^{-1}), lined ponds (\$1000–3000 t^{-1}), tanks and raceways (\$2000–8000 t^{-1}); recycle systems are generally the most capital intensive (\$5000–15000 t^{-1}), although more sophisticated offshore cage culture systems can be comparable;
- basic infrastructure and ancillaries may add 10–50% to holding system costs; water supplies commonly being the most expensive; for more complex systems, ancillaries and infrastructure costs may be up to 200–300% of holding system costs;
- improved capacity utilization can have significant effects, and better fry supply, multiple cropping systems and better management control can also give better returns;
- scale effects can be noted, related to unit size of holding systems and to reduced unit infrastructure costs and although these may often be subject to physical site limitations; small-scale artisanal production systems can be relatively inexpensive, using simple materials and taking best advantage of local site opportunities;
- hatchery capital costs, expressed, for example, as cost m^{-2} are generally several times as great as ongrowing capital costs, due to more complex and smaller scale holding units and operational facilities; in some cases this will favour specialization and centralization. However, low cost hatcheries, using ponds and hapas (net bags), are widely feasible and may be operated at a very small scale.

Operating costs

- these include feed, fertilizer, seed, labour, energy, capital charges, maintenance, rents, medication, packing, transport, communications, administration and management, specialist services;
- feed and fertilizer is usually the major cost (typically 40–75%), followed by seed (5–25%) and labour (5–15%); for highly intensive systems, capital amortization and energy costs may also be important; for small scale artisanal systems with self-supply of fry, and uncosted family labour inputs, operating costs may be negligible;
- production methods based on fertilizer or supplementary feeds are usually cheaper than those using complete feeds; those entirely based on natural productivity (e.g. cages in reservoirs) may be even cheaper;
- labour productivity effects may not be too significant, and in small-scale artisanal production may be uncosted or negligible, with profits being recovered to owner-operator families
- hatchery production costs, expressed, for example, as cost kg^{-1}, are generally several times as high as those for ongrowing, due to higher cost facilities, more expensive feeds, and greater skilled labour input; however, margins are also generally higher, across a range of production scales, and improvements in efficiency, e.g. through multiple cycles, can bring important benefits.

In considering the relative attractions of different systems it is common to prepare a financial analysis in which future revenue streams are compared with present day investment, either by a commercial enterprise, or in terms of cost-benefit for an institutional investor – including various social, environmental or other factors as relevant. Using appropriate discounts for cost of capital, it is possible to determine the NPV (net present value), or less accurately, an IRR (internal rate of return) i.e. the discount rate at which cost and benefits over the defined period will balance out. Using a more theoretical approach, production functions can be developed, to relate revenue from a range of cases to potential input factors, without any *a priori* concept of their relationships. This can be particularly useful in describing sectoral characteristics, but studies in the aquaculture sector are rather limited and are in any case constrained by the rapid technical and managerial change encountered.

In practice, financial or economic analyses applied to individual projects tend to be very specific to the cases involved, and cannot be extrapolated with any great conviction. It is also difficult to relate these from one location to another, or across national boundaries. Nonetheless, there are useful indicators to be derived from selected cost analyses. As internationalization becomes an increasing factor in input costs and product markets, so may comparative performance be measured, if only in broad terms, using such analyses, for tilapia and for other aquaculture species. The following sections outline some of the earlier and more recent analyses of tilapia production, and give at least a flavour of the financial performance of various tilapia enterprises in the specific conditions defined.

Earlier studies

As with other aquaculture sectors, a range of studies has been carried out for tilapia culture, in a number of different locations and production contexts. Although usually undertaken to determine local feasibility, they have normally had at least some regional significance. Following the promotion of tilapia culture in the late 1960s and 70s (e.g. Semakula and Makoro, 1968; Pagan, 1970) in tropical rural development and for commercial production, a number of studies were carried out, particularly in Asia (see, e.g. Sevilleja 1985; Smith *et al.*, 1985; Escover *et al.*, 1985, 1987; Bimbao and Smith, 1988); and in Latin America (Lovshin and Pretto, 1983; Hatch and Engle, 1987). In most cases these indicated attractively positive returns on investment, and had encouraged further development in rural tilapia production, although as Smith *et al.* (1985) noted, full costing of operators' labour made some projects unviable. In the Philippines, Tidon and Librero (1978) in Pullin *et al.* (1994) showed a simple return on investment of 45–210% in monoculture and 7–610% for polyculture. Gonzales (1984) indicated a 819% return for a freshwater cage hatchery, and Yater and Smith (1985) a return of 14–140% for a freshwater pond hatchery. In more commercial conditions, analyses of tilapia production economics in Israel identified key targets of productivity, the importance of water resources and anticipated in many respects the present day concern for input costs. Muir (1985, 1986, unpublished) carried out a number of commercial feasibility studies for intensive salt water tilapia production in the Middle East, showing that unless investments and feed costs were subsidised through pioneer project status, production would be non-viable. A series of hypothetical studies in Kuwait (Hopkins *et al.*, 1985, 1986) covering various aspects of intensive tank and cage culture came to broadly similar conclusions.

Recent studies

The promotion of tilapia in rural aquaculture lost some momentum by the 1980s due to problems of over-breeding and stunting, and disappointing yields, together with increasing doubts about the effectiveness of aquaculture outside its more traditional areas. In many countries, political instability and the decline of centralized support services also made the promotion of aquaculture more uncertain. However, a number of projects maintained activities, exploring more effective if less ambitious approaches, based on more realistic expectations of the role of tilapia culture, and a number of studies have reviewed its potential. In sub-Saharan Africa, Stewart (1993) in Dickson and Brooks (1997) described the economics of production of tilapia in artisanal and commercial (estate) farm system in Malawi, indicating the importance of local support services, fry and fertilizer, and the development of local markets. In these examples, polycultures with *Clarias* catfish were usually more profitable. Within the same project, Vincke and Schmidt (1991) suggested that financial returns from fishfarming based on local *Oreochromis* species exceeded those for coffee, dairy or poultry, although Johnson (1992) subsequently proposed that returns were far less certain. Her analysis also indicated the wider range of social factors in uptake and management. Broadly similar findings concerning economic viability and the key dependence on social factors and local support and markets were developed within the ICLARM[1] and ALCOM[2] studies in Africa. A study based in Luapula Province, Zambia considered the social factors involved in decisions to participate in and continue with rural aquaculture, noting the importance of otherwise unrecorded non-financial issues such as family context, social status and ceremony (Harrison, 1994; Harrison *et al.*, 1994). In assessing the potential of mixed species productivity gains in temporary wetland ponds in N. Nigeria, Thomas (1994) noted the dichotomy between apparent economic gain, and lack of social acceptance within the 'participating' communities. Engle (1997b) assessed the decision issues involved in low-input tilapia production in Rwanda, and reviews on rural aquaculture (Martinez-Espinosa, 1992) have noted the potential of small scale aquaculture, but also the range of external factors which may be involved in successful uptake and longer term adoption. Although, as in earlier studies, fully costed artisanal labour inputs may not always be supportable, analyses based on marginal returns for underemployed families, and recognizing other non-market issues, suggest a wider range of potential (Stewart, 1993).

In more commercial developments, an assessment of cage tilapia production in Lake Victoria, Uganda, suggested significant potential for developing export markets, provided locally produced feeds, the major cost input, could be produced for $<\$300\ t^{-1}$. In Zimbabwe, the recently established Lake Harvest enterprise, using cage culture in Lake Kariba, aims to exploit similar potential, with a major production target ($\sim 5000\ t\ y^{-1}$) for export to major European markets (M. Dickson, personal communication). Outside sub-Saharan Africa, Stewart (1997) discussed the viability of tilapia production case studies in Egypt, based on monoculture or various polycultures in association with saline groundwater use, concluding that unless values of externalities (reduced

---

1 International Center for Living Aquatic Resource Management
2 Aquaculture for Local Community Development Programme

impacts of salinization) could be included, aquaculture could not give sufficient returns at current market prices. Also in Egypt, Sadek *et al.* (1992) analysed intensive tank based tilapia aquaculture, at annual production levels of ~21–22 kg m^{-2} showing losses of ~50–100% of sales price of \$1.30–1.35 kg^{-1}, and concluding that far higher productivity levels would be required.

Torres and Navera (1985) and Bimbao *et al.* (1993) describe the economics and marketing of tilapia in various parts of the Philippines, demonstrating a variety of practice and profitability, but a positive outlook. In a review of the tilapia sector in Asia, Mohan Dey and Eknath (1997) noted the continued variety of production systems, but an increasing dominance of certain species and systems in specific locations. Thus more than 80% of farmed tilapia production in China arose from pond polyculture systems, with almost 100% use of *O. niloticus*. In the Philippines, production is based on mixed-sex tilapia monoculture, primarily *O. niloticus*, 45% and 35% in freshwater ponds and cages, respectively. In Indonesia, mixed sex *O. mossambicus* accounted for 75% of production, with a more recent development of monosex and red tilapia with the introduction of cages and artificial feeds. In Thailand, some 80% of production arises from pond culture, the remainder from rice paddy and ditches, with the commercial farmers increasingly dominating production. Profitability across the range of practices in Asia was noted to vary. In Thailand, in 1992, a comparison of tilapia with catfish and shrimp culture showed that while returns for tilapia were lower, the rate of return on investment was higher. The cash requirement for the less intensive tilapia culture, at \$9.5 per 100 m^2 was much lower than that for catfish and shrimp, at \$105 and \$126 respectively. In Bangladesh, Ahmed *et al.* (1995) noted profit over variable cost for Nile tilapia of \$12.60 compared with \$17.34 and \$5.94 per 100 m^2 for carp polyculture and silver barb respectively in low input seasonal ponds. Tilapia was more profitable than extensive shrimp culture (Rahman *et al.*, 1994). In the Philippines, Mohan Dey and Eknath (1997) quote 1994 studies in Luzon with net profits of \$231.33 and \$10.25 per 100 m^2 for cage and pond tilapia culture respectively, with equivalent production costs of \$0.45 and \$0.75 kg^{-1}, feed cost being the dominant element. Tilapia culture was more profitable than milkfish (*Chanos chanos*) culture, but less so than shrimp culture. However, input costs were \$50 and \$6 per 100 m^2 for shrimp and tilapia respectively. In Taiwan, in 1992, at \$34 profitability of tilapia per production cycle was marginally higher than milkfish at \$33 per 100 m^2. With intensive levels of 12 t ha^{-1} production costs were \$0.95 kg^{-1} or \$118 per 100 m^2 (Lo and Hwang, 1994). Net income from tilapia culture in China and Vietnam was reported by Mohan Dey and Eknath (1997) at about \$200 ha^{-1} y^{-1} and \$350 ha^{-1} production season^{-1} (6–8 months) respectively. These authors proposed that although net profits were higher in Taiwan than elsewhere, production costs were much higher, and Thailand was likely to have the overall competitive advantage.

In Thailand, Kwei Lin and Yi (1998) assessed a range of options for small-scale pond aquaculture of *O. niloticus* using chicken manure (CM), triple superphosphate (TSP), CM plus urea or urea and TSP, urea and TSP, urea, TSP and feed or feed only. Based on input supply and market prices in Thailand during 1996, unit costs ranged from \$0.12 to \$0.74 kg^{-1}. At an average market price of \$0.48 kg^{-1} and except for the unprofitable totally fed system, returns increased with input levels. Highest returns were obtained with supplemental feeding beyond 100 g; otherwise chicken manure at 75 kg dry matter ha^{-1} $week^{-1}$ plus urea and TSP to 28 kg N and 7 kg P ha^{-1} $week^{-1}$ provided the best net income. Further analysis of pond size, at 200 m^2, 400 m^2 and 800 m^2 showed little influence.

Within a major review of saltwater tilapia production, Watanabe *et al.* (1997) analysed a commercial scale recycle brackish water hatchery for Florida red tilapia in the Bahamas, proposing a breakeven (BE) of \$0.16–0.07 per sex reversed fry at 1.0 and 2.5×10^6 annual production respectively. At 2 m y^{-1} an IRR of 15% was obtained at \$0.11 fry^{-1}. Transport costs were an important marginal factor in profitability in this area, and vertical integration with growout was proposed to be more effective. A proposed 20 ha seawater pond operation in Puerto Rico was also analysed, using pumped seawater at 26–29°C. Based on pilot project results, with \$4.53–5.18 kg^{-1} dressed out product price, 25×0.8 ha earth ponds (60.2% of total capital costs – TCC), stocked at 3.5 m^{-2} of 0.85 g fish would be harvested after 160–220 days at an average 545 g yielding 13.35 t ha^{-1} $crop^{-1}$. Water and drainage, ponds, buildings, aeration represented 27.2, 23.0, 12.6 and 10.6% of TCC, and imported feed, processing and distribution, and sex reversed fry represented 32.4, 16.3 and 14.7% of total operating costs (TOC) respectively. At \$4.55 kg^{-1} the project showed a positive cashflow at year 6, an IRR of 18.2% and a BE price of \$3.86 kg^{-1}, suggesting marginal effectiveness; it was suggested that a stocking density of 4 m^{-2} would be better (Watanabe *et al.*, 1997).

A further assessment was provided of an intensive tank system of 6240 m^3 production volume in the same location in Puerto Rico. This comprised 9×30 m^3 nursery tanks and 52×120 m^3 grow-out systems with supplementary aeration, plus a 930 m^3 sedimentation basin and a 1 ha stabilization pond. With a 155 hp (116 kW) axial pump for seawater, two 35 hp (26 kW) pumps for brackish water from coastal wells and a 15 hp (11 kW) blower, a maximum standing crop of 125.5 t was proposed, with a flow of 62.7 m^3 min^{-1}. Installation costs of \$834 115 (~\$6650 t^{-1} capacity) were 27% higher than those for ponds, tanks representing 41.6%, and the water system 22.5% of TCC respectively. Based on 22 g fingerlings stocked at 75 m^{-3}, raised to 572 g in 182 days with a final density of 39 kg m^{-3}, annual production would be 487.2 t. Feed, processing and distribution, and fry accounted for 34.2%, 17.7% and 12.5% of TOC respectively. The BE was \$3.54 kg^{-1}, and the IRR was 27.8%.

The same authors also describe an assessment of intensive saltwater cage culture in Haiti by Brass *et al.* (1990, 1991), including artisanal and commercial scale operations. With 1 m^3 cages stocked with 10 g fry at 300 and 600 m^{-3}, artisanal production was considered to be unviable based on a cost of \$0.66 kg^{-1} for imported feed and an FCR of 2:1. At the commercial scale, based on $50 \times 7\ m^3$ cages producing 225 g fish in 120 days, with a maximum 70 kg m^{-3}, a BE price of \$1.89 kg^{-1} was proposed, at an FCR 2.0, providing a small profit margin. Feed and fry were 53.9% and 12.9% of TOC. The need for locally produced dicts was apparent, particularly as 50–60% of feed costs related to transport and duty. In a recent study on options for mixed stocking of *O. spiluris* with Red Sea bream (*Sparidentex hasta*) in commercial seacage culture in Kuwait, comparing 300 t annual production of seabream, with 200 t of seabream and 100 t of tilapia, Cruz *et al.* (1998) noted an increase of IRR from 11.6% to 13.1%, though both cases exhibited a negative NPV under the study conditions.

In an assessment of economics of semi-intensive freshwater pond production of *O. niloticus* in Honduras, Teichert-Coddington and Green (1997) examined 41 different feeding and fertilization approaches, based on stocking rates not exceeding 3 m^{-2} and 150 day growing cycles. Incomes exceeded variable costs in all cases except where phosphate was the sole input and in one case where inorganic fertilizers were used. Negative returns to total costs (excluding depreciation on ponds) only arose when using inorganic fertilizer or with weekly chicken manure (CM) application of <1000 kg ha^{-1}, with fish at 1 m^{-2}. Profitability was generally higher when stocking $>2\ m^{-2}$ and fertilizer was more profitable than feeding at low stock rates. Greatest yields (5.3 t^{-1} ha^{-1} 150 $days^{-1}$) were obtained with feeds, but output levels of 3.7 t^{-1} ha^{-1} 150 $days^{-1}$ with CM and nitrogen were at least as profitable. Profitability was generally better when feeds were substituted where possible with CM, although feeds were needed for finishing 300 g fish and high quality feeds were essential for 600–900 g fish (Green *et al.*, 1994).

A number of recent studies have been carried out on intensive aquaculture. A review by van Rijn (1996) considered the potential of recycle systems in Israel, including those for tilapia culture. STAQ (1996, unpublished) revisited the analysis of intensive waste heat based tilapia culture in the UK, concluding that as in previous decades, and in spite of improved technology, production would only be marginally viable, primarily because of potential competition from low cost imports of steadily improving quality. For recycle systems, Losordo and Westerman (1994) proposed a total investment cost of commercial recycle system at \$4.33 kg^{-1} of production (\$4330 t^{-1}) and a production cost of \$2.79 kg^{-1}. Costs were most sensitive to changes in productivity, a 10% increase resulting in a 5% reduction in production costs.

Overall conclusions

The most recent development of larger scale tilapia producers, increasingly linked with major urban and international markets, suggests that economies of scale may become a more important factor in the structure and development of the sector. It is clear that subject to site capacity, individual holding units, as in other forms of aquaculture, will demonstrate declining capital costs per unit volume, and if combined with mechanization may show reduced labour and other operating costs per unit output. Capital intensive, high-input systems are also likely to have infrastructure and management-related economies of scale, as suggested by common target production levels of at least 200–500 t y^{-1} for most current commercial tilapia projects. However, as earlier noted, the diversity of opportunities for tilapia culture and the flexibility of system choices, with a wide range of relative factor costs, provides an unusual potential for heterogeneity.

At this stage advantages would seem to lie with semi-intensive systems, particularly with good quality fry supplies and reliable conditions for growth. The main constraints are the land and water resources involved. Because of its simplicity, flexibility and low cost, cage culture appears particularly attractive, and has been the focus of significant growth in some locations, in both artisanal and commercial sectors (Anon, 1997a). However, it may become self-limiting if expanded too rapidly beyond environmental capacity. Rural aquaculture for tilapia also appears favourable provided resources – particularly seed and fertility inputs – although not required at high levels, are available on a sufficiently reliable basis. Where market conditions appear favourable, more intensive systems may also be competitive and effective, although system design and management efficiency are likely to be critical, and the dependence on feeds increases sensitivity to raw materials costs, an area of some vulnerability (Anon, 1997b). Nonetheless, significant investment in this sector can be noted (CDC, 1998).

Although some indications can be made, it is difficult to pronounce on whether 'high potential' production areas can be identified, whether using strategic approaches such as that originally adopted by Bell and Canterbury (1976), based on geographic, social and economic criteria, more recent studies based on informal structured assessments (Haylor *et al.*, 1994; Muir, 1995), or using formal procedures such as GIS, as in recent reviews of aquaculture potential in Africa (Kapetsky, 1994; Aguilar-Manjarrez and Nath, 1998.) Although trends in international trade and transportation have most recently been towards liberalization, simplification and reduced cost, a range of economic distortions including subsidies, imposts, tariff and non-tariff barriers may constrain wider competition. Continued political, social or financial instability may also restrict potential in some locations. Environmental costs may also become more important, potentially affecting consumer attitudes. These issues apart, large land/water systems in the tropics, with good political and institutional conditions, human resources, communications, infrastructure and services may have particular advantages, and might be expected to contribute more strongly, both to supplying local rural and urban markets and in contributing to strategic demands in developed countries.

12.5 THE MARKET FOR TILAPIA PRODUCTS

The market for tilapia products is inextricably linked to the wider market for both captured and cultured fish species. Furthermore, since individual food consumers will frequently substitute between fish and other protein sources, marketing and economic considerations are soon enmeshed within a much wider and more complex set of food choice variables (Logue, 1986; Marshall, 1995). While these non-fish food substitutes cannot be discussed within the confines of this chapter it is important to recognize their presence and implicit importance in making tilapia marketing decisions.

The price impact of declining availability of traditional species, consequent on continuing high levels of fishing effort, will have important ramifications for the distribution of the demand for fish. Higher real prices following reduced availability of capture supplies will tend to concentrate fish consumption even more within the countries and markets which are willing to pay higher prices. Acceptance of higher fish prices has undoubtedly been influenced by the increasingly favourable consumer perception of fish over recent years. The precise reasons for this vary within individual markets but encompass many attributes, such as healthiness, naturalness and convenience that are considered positive for fish within the mindset of the contemporary food consumer. This perception has also been enforced in a number of products by the incorporation of quality marks at the generic and brand levels (Monfort, 1998). In addition, food safety concerns and developments within some of the more traditional substitutes, notably red meat, have added to the effect.

Fish has been the recipient of a world-wide upsurge in demand, as reflected in per capita consumption levels which have increased by 46% since 1960; in developing countries the rate of growth, at 85%, has been even more dramatic albeit from a lower base. However, it would be incautious to assume that these trends will necessarily continue. The tendency for extreme price fluctuations, especially within the capture fisheries sector, may promote uncertainty and invite longer term consumer resistance. Increased prices are liable to be sustainable only if consumer perceptions of the fish products are comparatively favourable. Within developed country markets and increasingly within segments of developing countries, quality and environmentally friendly attributes are seen as a vital and integral part of the food fish product. Features such as a sustainable production basis are no longer perceived as an isolated mechanism for product differentiation, but are increasingly viewed as a prerequisite for market acceptance (McGovern, 1998).

With the possible exception of channel catfish in the US, the most significant aquaculture contributions to the fish supply imbalance in developed country markets have occurred within High Unit Value species (HUV) where high prices driven by initial scarcity have encouraged technical solutions to production problems. The limited extent of species diversity has been encouraged by the cyclical process whereby initial high market prices have given way to falling real prices brought about through increased supplies. Producers have thus become involved in a series of diversification manoeuvres to both broaden the range of species and to provide a platform for further expansion through added-value product diversification (Young and Muir, 1995). The lag between launch and mature market prices has been expected to provide a 'honeymoon period' in which returns to initial R&D costs can be recouped.

However, doubt may be cast over continuation of this scenario as consumers are faced with a widening array of farmed products at declining relative prices. Since fish consumers seem content to eat only a comparatively limited range of species, their propensity to graze more widely into newly launched highly priced species may reduce as the choice of product expands. Greater price resistance during the initial launch period may thus act as a brake on species diversification and perhaps encourage the search for aquaculture contributions that are not cast in the HUV mould.

12.6 PROSPECTIVE TILAPIA CONSUMERS

The role of tilapia in the international market context is interesting because of its diverse profile and comparative cost advantages, identified earlier, which combined may offer more scope than with many other species. Such is the ubiquity of tilapia that many of the world's population already have some familiarity and a degree of favourable response. Little (1998) has suggested at least three broad market categories each with their attendant supplies. The highest profile market is that of the intensive industrial system primarily supplying developed country markets with a range of tilapia products spanning live to added value processed forms. Significant commercial operations constitute a second category serving the expanding urban markets in S and SE Asia. Less commonly noted is the market comprising the marginal, poorest people who depend upon the highly localized smallholder production systems. Although consumer 'choice' may have a rather different meaning in such groups, the role of the tilapia product may be even more significant in terms of relative importance to its consumers.

Only very few fish species command such a diversity of positions in the global market and it is useful to consider generalized implications for marketing tilapia in such a range of segments, as noted by Muir (1995). The absolute level and composition of expenditure on food will clearly reflect wider disposable incomes and will be influenced by a plethora of societal, cultural and many other values.

The observation of Engel's Law, whereby in all economies the proportion of rising incomes spent on food will tend to decrease, will further complicate the pattern of consumption and the returns available to producers. Conventionally, distinctions can also be made between household (individual or family) and public or institutional (catering/community) food product consumption. At least some of the following generalizations might be expected:

- for zero/negligible disposable income, access to supplies is random; basic sustenance is the aim, opportunity is critical, quality may be of marginal concern, provided there are no extreme health risks; lack of purchasing power limits the commercial response of supply, unless payment is available in kind; institutionally this corresponds only to emergency/disaster relief situations;
- for low disposable income: the primary concern is usually price if basic quality features are satisfied; consumption will vary with availability; except for festive/other reasons, response to price may be to substitute; at the institutional level, this corresponds generally to mass catering requirements and to the criteria for food welfare programmes.
- for moderate disposable incomes, choice becomes rather wider, and regular consumption of preferred products, if at modest price, becomes feasible, while occasional purchases of more desirable, highly priced products become possible, and increasingly common; perceptions of quality become more important, as do factors such as image and association; institutionally, this corresponds to modest private sector catering outlets, better food stalls, etc.
- for high income levels, choice is increasingly unrestricted, and issues of basic consumption become far less critical; quality and image are more important, diversity and variety are also significant, and new product development and presentation becomes more essential in maintaining interest and developing market opportunity; at the public level this corresponds to the higher-quality catering outlets.

Individuals or societies may move between these levels over time, and it is the integration of these episodes that shapes the overall, longer-term response, as indeed it describes overall nutritional welfare. Translating these profiles into patterns of fish consumption, Table 12.6 provides an overview of the types of species that we might anticipate being consumed at different price levels.

The broad indicators below are based upon whole 'round' prices and this provides scope for further positioning of product between bands. Producers, processors and other channel intermediaries may seek to add value by 'finishing' live stocks, grading, filleting, smoking, enrobing, etc. The propensity to consume at different price levels will also be influenced by the availability of alternative protein sources in the market and wider issues pertaining to

Table 12.6 Generalized price/value levels of aquaculture

Price level/ trend	Typical price $/kg	Species	Notes
Very low, subsistence, stable	<0.50	Small carps, undersized tilapia; silver and grass carps when not liked; fish produced in waste ponds where social constraints; various lower grade cockles and clams; seaweeds	Commonly produced/collected as small, secondary crops in ponds, paddy, lake margins; also local coastal community activities; most likely to be consumed locally
Low, variable	0.50–<1.00	Silver and grass carps, small tilapia, small-medium carps, barbs and catfish during peak fishing periods, smaller milkfish, some poorer quality mullet, cockles, clams, some mussels, razor clams, better seaweeds	Usually main crop products, or poorer/lower grades of these; typically extensive or semi-intensive pond, bed or longline systems, may be artisanal or more commercial systems; may enter households/local markets or be sold elsewhere
Low-medium	1.00–<3.00	Main carp species, larger tilapia, channel and clarias catfish, low to mid-grade rainbow trout, poorer quality/small shrimps, prawns, better clams, mussels, oysters, poorer scallops; frogs	Main crop products, most systems from extensive to simple intensive; produced in ponds, cages, lines, baskets; usually for main markets, although local sales increasing in many areas
High, stable or declining	3.00–7.00	Better carp species, select tilapia, eels, mid-better trout, main salmon species, poorer breams, yellowtails; main shrimp species, scallops, better clams and oysters	Main crop products, usually but not always from commercial systems, semi-intensive to intensive, ponds, tanks, raceways, cages; normally to main markets, except gradeouts, etc
Very high, specialized	>7.00	Marine fish; seabass, bream, grouper, *Lates*, jacks/yellowtails, puffer, tuna	Highly valued markets, often small size, may be seasonal; often high-cost, specialized production methods, or at least entry barriers

nutrition. It is beyond the scope of this chapter to discuss the role of fish in nutrition, or the range of regional and national characteristics with respect to its consumption. Nonetheless, having identified something of the composition of the different markets for tilapia, more specific attention is next focused upon the EU and North American markets. Along with Japan these constitute the most important markets for fish internationally as well as being the most important for international trade in tilapia.

12.7 MARKETING TILAPIA IN THE EU

Apart from the number of potential consumers, over 350 million, the composition and contemporary profile of the EU bodes well as a prospective target for tilapia producers (Gentles and Skeldon, 1994; Muir *et al.*, 1996). Average per capita earnings are high, and fish is both affordable and increasingly in demand by a more health conscious consumer. Changed working habits, less-frequent shopping, less formal meal occasions and other sociocultural phenomena have encouraged the increased concentration of food distribution and retailing, stricter quality standards and more intensive competition from substitute foods. Such developments have invoked quite radical innovations in the requirements of fish products and their marketing channels. Once the preserve of northern Europe, these trends are now also growing rapidly in the Mediterranean countries (Paquotte, 1998).

Despite increased internationalization, entrenched differences within the EU remain, with significant implications for tilapia producers. In terms of population size, four countries have more than 55 million inhabitants each while six have under 20% of this. The largest importers of fish for consumption are France, Germany, UK and Italy, while Portugal then Spain have the highest per capita consumption levels. Supply to the EU is heavily reliant upon imports, including farmed fish, and is likely to remain so. In addition to the deficit within more traditional mass market segments, imports also meet the demand from a growing array of niche markets. These consist of various ethnic market segments, demand arising from increased foreign travel and that generated by the variety-seeking fish connoisseur. The prospective opportunities for the tilapia producer may best be set out using the standard structure of retail and catering sectors. While the relative importance of these sectors varies within the individual member states, in all cases each represents a significant component of the overall market.

The structure of the EU market for fish

The structure of the EU catering sector is highly fragmented, with many enterprises and a low concentration of ownership. A fundamental distinction may be made between the 'profit' and the 'cost' sectors, within which a variety of different types of outlet will be found, e.g. upmarket hotels and restaurants within the profit sector through to non-profit institutions such as prisons and hospitals in the cost sector. The importance of different product attributes will also be reflected in the marketing channels used to service these outlets. Within the cost-focused institutions, emphasis will tend to be on frozen product typically serviced by fish processors or food wholesalers, rather than through specific fish wholesale markets (Young and Maddock, 1990). Wholesale fish markets, such as Billingsgate, Rungis and Merca Madrid containing a number of trading firms specializing in a particular product range, will be more important for profit sector buyers focusing on fish, and are likely to be the more important for tilapia.

As elsewhere, the EU food retail market has witnessed dramatic change, most notably with the growth of the supermarket chains and their commensurate buying power. While their importance generally remains less in the southern EU states, a similar underlying trend can be recognized. Supermarkets have latterly moved into the fresh and chilled fish sectors, strengthening their existing presence in frozen and canned fish, and thus have further reduced the share of the specialist fish retailer. Some supermarkets have shown particular enthusiasm for more varied fresh fish products, including tilapia, where displays provide an in-store attraction not matched by many alternative products. Such positioning may also be used to command premium prices. However, logistics support for supermarket chains is now highly sophisticated and the task of satisfying this, combined with strict volume and product specifications, can restrict entry. Although small buyers may be vulnerable given the imbalance in power, supermarkets have become more inclined towards the concept of relationship marketing whereby both parties seek to work to mutual advantage. This may extend to co-operation through the new product development process, and indeed through to vertical integration (Costas, 1997).

While the number of specialist fish retailers has declined dramatically in most EU markets those remaining have tended to be amongst the more innovative, and quality-focused. Supplies are typically purchased through a diversity of sources in conventional wholesale markets to offer a more specialist product range. By emphasizing their position as the source of superior quality, more varied product retailers have sought to occupy important niche sectors. In some instances wholesale market traders have also developed a niche by focusing upon provision of more exotic species supplied to specialists, supermarkets and profit sector caterers. Such strategies have been successful in offsetting the decline in product volume brought about by the increasing role of the supermarket sector.

The tilapia markets of the EU

The market for tilapia within the EU remains comparatively small although dynamic: in 1990 it was estimated to be around 1000 t whereas by the mid-decade it had expanded to around twice this volume (Anon, 1994, 1995a,b). Traditionally, the main EU market has been within ethnic African communities in Paris, London, Amsterdam, Rotterdam, Brussels and other European cities; smaller markets also exist within the Chinese and Asian communities. Within ethnic markets the over-riding preference is for fresh round fish so that traditional quality indicators such as gills, eyes and firm belly can be observed. Extension from these origins has been encouraged by some supermarket chains keen to promote a more diverse product range. At the premium end of the market, chilled product has been imported; elsewhere, frozen fish have been used. The EU market typically requires tilapia grown to a minimum weight of 350 g, which can usually be reached within one season. However, a clear market preference exists for 450–700 g fish, which adds the cost of a second growing season. Beyond this average market target of 600 g there is also a more limited segment for fish up to 2 kg.

The increased popularity of the tilapia product is in no small part due to the confluence of the basic product attributes and those sought by the fish consumer. In particular, its white flesh, firm texture and rather neutral taste are consistent with the traditional preferences of the European consumer, especially those in the north. While there are significant sectors of the EU population who may be more adventurous and tolerant in their fish attribute aspirations, they too are generally willing to consume at least some part of their fish intake in this more mundane format. Moreover, at least for the foreseeable time, the lesser familiarity of tilapia provides some differentiating attractions.

While differentiation in the EU market is accentuated when sold in the whole format, the wider preference for fish to be prepared has created a need for some degree of processing. Yield reductions through mechanical processes are generally less favourable than for many other species and this may be a significant cost disadvantage. The widespread development of machine filleting has yet to emerge, as the number is so far limited, and the hand-filleted yield results in a product with a price level similar to Dover sole, which is very unlikely to be commercially viable. Fillet yields vary according to the type of cut, or more exactly the bone content. At best a 40% fillet yield can be obtained, and where J or V-cuts are used to remove pin bones, unacceptable in some markets, this can fall to 28%. At the extreme, when coupled with deep skinning to remove subcutaneous fat, as little as 22% might be obtained (Popma and Lovshin, 1996). Combined with the higher production costs of the larger fish needed to gain acceptable fillet weights, price may deter the scale of further expansion within the supermarket sector. Although smoked fillets have met with favourable consumer response, yield implications are also financially forbidding. As with many other aquaculture products, the scope for adding value in much other than the whole form may meet with severe price resistance; farm gate prices of around $3/kg are sensitive to competing substitutes.

Some solution to this problem may be found in processing at source to take advantage of lower labour and resultant transport costs. However, processing operations that are capable of satisfying EU health and hygiene legislation, including HACCP demands, require significant investment (Gutting, 1998; FAO/NACA/WHO Study Group, 1999). When coupled with the need for supply chain channel management and the associated logistics support, barriers soon preclude all but the larger scale industrial producers. If combined with advances in strain selection, and optimal growing conditions to increase harvest weight and lower the production cost, target prices for tilapia fillet may be steadily reduced to offer significantly widened opportunities. Although offering filleted products may mean stepping down from premium price positions, there is clear scope to adopt a diversification strategy incorporating both product types. This would be a rational market development, as the positioning expectations of tilapia should not be overstated. Tilapia is extremely unlikely to be perceived within the EU consumer's top tier of fish products. Even in the more exclusive live niches, at best tilapia might rank within the middle market having favourable attributes of taste and competitive price. A longer term approach of differentiating a specialized, high attribute whole fish product, with a quality fillet and value added sector, is likely to be most attractive for producers and markets alike.

12.8 MARKETING TILAPIA IN NORTH AMERICA

The North American market (primarily the US) is amongst the world's largest, wealthiest and most developed concentrations of demand for quality food products. Although there is only very limited scope to increase the volume of food consumed, there remain considerable marketing opportunities through substitute and added-value products. The opportunities for fish products have been particularly favoured since the late 1980s due, *inter alia*, to increased consumer concern towards healthy eating and the movement away from many of the traditional substitute foods. Especially in the last decade, the changing balance of capture and farmed fish supplies has increasingly introduced the consumer to aquaculture product, primarily catfish, salmon, trout, shrimp and tilapia. Imported supplies have grown in importance through greater comparative advantage, not least because of constraints on domestic production.

Seafood consumption in the USA has increased steadily from 1970, and market forecasts suggest that consumption will increase to about 9 kg per capita per annum by the end of the 1990s, compared to only 5.9 kg per capita in 1980. Despite this growth trend there would appear to be scope for yet further expansion, as this is still low compared with US meat and poultry consumption and EU fish consumption. However, at 9 kg per capita this will require an increase of some 2 million tonnes of live weight product.

The structure of the US market for fish

The US market exhibits a rather similar division between fish retailing and the fish catering/service sectors as is found in the northern EU countries. Within the retail market many of the trends are very similar, not surprising given the sociocultural phenomena that have driven the wider changes within the food market. The most notable change has also been the growth and subsequent dominance of the supermarket chains. Within the US supermarket the concept of a specialist fish counter is more deeply established and the US consumer has greater experience of the wide range of in-store fresh product (O'Dierno, 1995). The US consumer's preferences are also broadly similar to those of the EU fish consumer: firm white flesh of fairly neutral odour and taste. The mainstay of consumption is also quite narrowly focused upon a few key species, although US consumers tend to be more diverse in range and open to expand preferences. For example, channel catfish is now the third most preferred 'seafood' product after shrimp and lobster (Anon, 1997c).

Over time, the products preferred by the US consumer have become more polarized. On the one hand there has been the ongoing extension to the added value route of yet further processing, while on the other, the consumer has been increasingly responsive to concerns over 'naturalness' and so has tended to demand product unadulterated in any way whatsoever. Traditionally, US consumers have led European trends towards reduced preparation times and more convenience products, although there are now signs that this is becoming reversed. Thus, the introduction of modified atmosphere packaging, although developed as a product concept in the US, is far more prevalent in northern Europe and is also gaining ground within younger age groups in southern Europe. Overall, however, this diversity within the N. American market is potentially to the advantage of tilapia since it suggests scope for alternative entry routes.

As in Europe, the structure of the catering/food service sector is very diverse and consists of a multitude of individual enterprises. Notwithstanding the large number of independent outlets, restaurant chains are an important feature within the US market and these target a range of market positions. Here, a notable number of restaurant chains focus upon the fish component of their service, whereas within the EU this is relatively rare. Such focus provides a ready market opportunity for new fish products, as the parameters of the meal options are more curtailed, yet there is a strong competitive value in developing customer choice.

The tilapia markets of the US

Tilapia is seen by many within the US fish trade to be the major species market developed in the 1990s (Shaw, 1995; Anon, 1996a, 1997d; Engle, 1997). Consumption of tilapia in the US is currently estimated at around 37 000 t and is now regarded as the third species from aquaculture after shrimp and Atlantic salmon (Lister, 1998). The market for tilapia has developed quite rapidly, based upon a range of hybrids from the three most common species *Oreochromis niloticus, O. aureus* and *O. mossambicus*, which have characteristics of high yield, tolerance to cold water temperatures and (through variants) red skin colour respectively (Anon, 1996b). The greater control over feeds and water quality afforded through aquaculture has been significant in stimulating market demand, which began with red tilapia whose appearance attracted restaurants serving premium fish.

Domestic US production was estimated at 7500 t in 1997, almost one-fifth of the total supply base (USDA, 1997). The live market for tilapia continues to be a driving force in the domestic industry; it is estimated that around 50% of domestic production is sold in this form, a comparative advantage which is not easily challenged. In addition to the food fish market, tilapia are also important in the ornamental sector. The vast majority of US supplies is imported and sourced in two main regions according to product type. SE Asian exporters predominate in the frozen fillet market with Taiwan, Indonesia and Thailand being the main suppliers. In the case of fresh fillets, Costa Rica, Ecuador, Colombia, Jamaica and Honduras are the main producers benefiting from geographical proximity. It can be anticipated that other countries in Central and Latin America will also seek to compete in this market.

As is the case within the EU there are significant ethnic markets for tilapia in the major urban centres of the US; this is especially so amongst Asians, the fastest growing ethnic group in the US, and is thus particularly significant on the west coast and into Canada (Kohler, 1994). Clearly, these markets may also be attractive to imported product, although the premium niches and live trade are likely to be more difficult. As with other fish species, price levels vary considerably according to the product form (live, fresh/chilled, frozen and product type (round, dressed, fillet, etc). In addition, grade size and quality, especially but not only freshness, will further impact upon price at exchange; given the variabilty of market prices, the websites referenced in this chapter and their links may be consulted for current prices. At the time of writing US pond-side prices currently average around \$2.75 kg^{-1}; an attractive level when set against some of the production costs mentioned earlier. Skinless, boneless fillets are currently in the \$6.80–7.70 kg^{-1} range, which would provide no great incentive over whole fish at fillet yields of $\sim$35%. As might be anticipated, live fish tend to command a premium price; at present this runs somewhere between \$3.85–5.50^{-1}, depending on season and delivery point.

The standard size sought by the market is within the 1–1.5 lbs range (450–680 g) although there is a market for larger fish to about 3 lbs (1.4 kg). Tilapia may be sold live, fresh or frozen as whole, gutted, gutted and scaled, headed and gutted, and in fillets: skinless and boneless but also as skin-on. Fillet size is generally quite small at between 2–5 oz (57–142 g). Skin-on boneless fillets can be expected to yield not more than 40% and even less than this from other species strains and in different product forms. In general, frozen fillets sell at prices some 11–14% below the fresh product: \$6.00–6.60 kg^{-1} is a more realistic price range for good quality frozen fillets. Increased competition within this sector is likely as SE Asian producers attempt to secure a greater share of the US market. However domestic producers can retaliate by focusing upon non-price attributes, such as flesh colour and quality (Picchietti, 1996).

In broad terms three grades of tilapia can be identified in the US market (Sipe, 1992). Category A product covers product cultured and harvested in the best quality environment, then purged and processed within a chilled, or cold, chain. The resultant product is aimed at the premium market segments that compete with products such as grouper, snapper and swordfish. Lower production costs and increasing volumes should also enable a greater focus upon the more mainstream traditional quality whitefish products. Category B fish are cultured with lesser emphasis on quality. The resultant inconsistencies in flavour, texture and other attributes command a correspondingly lower price. The focus of this product is more to the price sensitive consumer, but there is also the danger that the presence of inferior product on the market can create confusion. Wild fish constitute category C. Although properly handled, these fish may vary in taste due to the irregular environmental conditions of their habitat. They are primarily marketed in US Asian communities who prefer live or freshly frozen whole tilapia, but would tend to purchase category A product when available. This suggests that the market for category C product, and indeed category B, will be much less significant in future and thus will be important in determining future policy directions.

A major key to market expansion will be the production of tilapia at low cost, while maintaining stringent quality control throughout the marketing chain. Increased volumes of tilapia combined with lower production cost are likely to pull tilapia through from the upmarket restaurant sectors into the wider retail and catering market sectors. With the gradual introduction of A-quality tilapia into food counters at prices similar to those of chicken and turkey, the stage is set for a significant uptake. Indeed there are already some signs that this is happening as the current tilapia market is thought to be split 60:40 in favour of retail over catering. At the same time there is scope to develop further some of the niche markets. The high price paid in the live sector is of obvious attraction. Elsewhere, growth in ethnic catering sectors such as Thai, Indonesian and Japanese sushi bars provide yet further opportunities.

12.9 CHALLENGES FOR TILAPIA MARKETING

Having considered some of the more important trends evident within the EU and US markets and earlier identified the key features of the markets elsewhere, it remains to highlight primary determinants of tilapia's future prospects. In keeping with the earlier caveats of this chapter, our aspirations to global coverage necessitate generalizations that may differ in the more localized context.

Quality issues

Evidence from a variety of markets has indicated the value of the neutral taste of tilapia. However, the feeding habits of the fish can induce off-flavours which may necessitate purging prior to harvest. If tilapia are cultured in intensive or recycle systems, where cyanobacteria or other algae are relatively unimportant, they can still become off-flavour, although causative organisms are less clear. Tilapia are well known to graze periphyton, so communities that develop on tank walls may lead to off-flavours. Exudates from sediments or from biological treatment systems may also be factors. Producers have recognized the need for quality assurance programmes to prevent adverse consumer reactions. However, regulation and enforcement is potentially problematic, especially with live and unpackaged products where consumers tend to make generic species associations rather than to focus on single brands. The difficulty is all the greater where tilapia product may enter the market from a range of sources, not all of which are identifiable through to the consumer level via brands or other standard mechanisms for product differentiation. In such circumstances only relatively small amounts of off-flavoured fish may be sufficient to significantly reduce market demand.

LDC trade issues

The historic role for developing countries has been to act primarily as a seller of raw material, leaving further processing to be undertaken within the importing country, or some other intermediate break of bulk destination. This situation has commonly been perpetuated by poor quality control, lack of corporate resources and limited market information, coupled with trade and other barriers to entry. However, in the case of processed tilapia products there would appear to be comparative advantage in adding value at source rather than at destination. Comparatively cheap skilled labour, with limited mechanized competition, closer proximity to high quality raw material and transport savings through shipping only edible product yield might be expected to continue as important financial determinants. While this may require substantial infrastructure investment, such costs must be set against not only the internalization of profits to the exporting firm but also the potential welfare benefits to the region.

Although the significance of these potential advantages should not be underestimated, the fact that they have so far remained within the remit of the developed nations indicates the difficulty in surmounting traditional power bases. When coupled with legislative change such as HACCP and other product health and safety standards, the constraints to such investments increase. Such barriers are certainly likely to require that producers capable of meeting the necessary investment criteria will need to operate on a large-scale industrial basis. Indeed even for those producers supplying basic unprocessed products, similar responses may be required in order to overcome legislative and other technical barriers, increasingly focused around HACCP measures.

Green issues

Fish consumption levels in the US have perhaps been more affected by green consumer issues than elsewhere, although these concerns are now apparent in many other markets. Media publicity about problems with red tides, pesticides, marine pollution, chemical residues in flesh and other environmental problems have all adversely affected consumption. Both the US and EU markets have also witnessed much more proactive lobbying of various food interest groups, notably concerning beef and poultry but also including fish. Despite these consumer concerns, the scale and wealth of the US and EU markets attract the interests of seafood marketing organizations from many nations which are targeted through a number of promotion channels, including generic, species, regional and brand specific communications.

The implications of the green consumer for tilapia and other species, may be debatable (Anon, 1998). Some resistance to the concept of farmed fish can always be expected for reasons such as animal welfare, protection of native species, concern with chemicals, vaccines, etc. Where the fish has been introduced into the local production systems and has been subject to genetic manipulation, the products may be particularly unacceptable to certain consumer groups. Wider trends within the market for food suggest that such issues are likely to become increasingly important in the future. This is perhaps of greatest significance within the higher priced live market sectors where a more 'natural' product is sought. However, in many other respects tilapia is no worse placed than other farmed species and indeed may be much better placed than most. Its ability to feed low on the food chain in 'natural' systems, and its possible association with employment and welfare gain in poor countries may be considered as positive attributes. The extent to which the finer points of tilapia's greener credentials can be utilized is however more difficult to determine. Issues such as non-fish feeds, tolerance of poorer water conditions and other production characteristics may be adversely interpreted by some consumer groups which otherwise remain happy to buy tilapia.

Marketing orientation issues

Mention should be made of the tendency for many, although not all, within the tilapia sector to have focused less on the marketing function and more on output. As in many other fisheries, tilapia has tended to be driven by production aims and technical progress rather than by proactivity to the needs of the market. With notable exceptions, simple but fundamental market issues such as reaction to shape and skin colour, the presence of bones and other attributes, have tended to take secondary importance to technical solutions and production price driven strategies. This is not altogether surprising since the imbalance of market power, the size of producers relative to markets, distance, and the intricacies of the diverse market segments make it problematic to gather market data and understand diverse buyer behaviour. However, if tilapia is to compete effectively within the market for fish – and the wider market for food – it will be vital to promote a stronger marketing orientation.

12.10 POTENTIAL DIRECTIONS AND CONCLUSIONS

The production and cost characteristics of tilapia, its consumer acceptance and compatibility with various market segments suggest that a wide-ranging scenario for development can be considered (Muir and Young, 1997). Faced with a widely observed deficit in future decades (Ruckes, 1994; Westlund, 1995), tilapia seemingly has the capacity to expand and contribute to global fish supplies on a number of different fronts. The prospects for further expansion also seem to have strengths in a number of different, and pivotal, characteristics. In terms of technical change, genetic manipulation seems set to bring gains in respect of the productivity, performance and environmental tolerance. As noted earlier, *O. niloticus* and its hybrids are now the most commonly cultured tilapia, and this focus may increase, perhaps with the development of other strains capable of extending good performance into colder areas and into coastal or marine waters. Notwithstanding the probable need to reconcile green consumer considerations within some market segments, these developments are likely to further enhance competitive cost advantage while retaining opportunities for 'green appeal' within the market.

In addition to these supply-side characteristics tilapia also has a number of possibilities resulting from its more overt appeal to a range of different fish consumers. Underpinning its attractions is a set of attributes pertaining to white flesh, neutral taste and firm texture available at a competitive price and, somewhat uniquely, in a variety of compositions. Tilapia products have been shown to have the capacity to reach from the most marginal consumer markets within developing countries through to premium segments of international trade, in both live and added value formats. When combined with the emergent production options an interesting matrix of product opportunities can be developed as shown in Table 12.7 below (adapted from STAQ, 1996).

Detailed evaluation of the various outcomes suggested above is futile as individual situations will tend to favour different pathways, or some combination thereof. Regardless of how the scenario evolves in different locations, the heterogeneity of options seems common throughout. Unlike many other species where diversification might be forecast within relatively narrow parameters, tilapia seems to harbour the capacity to adapt to a much wider variety of situations.

There may be a far more difficult challenge of proving sustainable delivery at a price affordable by some of the world's poorest consumers. Tilapias have the potential to play a more important role, not least because of the shortage of other significant candidate species in many areas. Development may come about

Table 12.7 Scenarios for future development

Species diversity; low genetic-linked gain Wide range of species primarily occupying HUV markets, some being forced to operate at very low margins; increasing competition from imports, but moderate opportunity for replacing appreciated capture fishery stocks with 'natural' products. Possible opportunities for small scale producers; low to moderate growth potential, potentially good conservationist image associated with unmodified stocks, although careless strain management could be negative.	*Species diversity; high genetic-linked gain* Wide range of species, but limited degree of product development; flexible opportunity to occupy HUV markets or compete with lower cost, e.g. capture fisheries and other food product markets. Moderate to high growth, but constrained by acceptance of product form. Biotechnology skills/costs may favour larger producers, but licensing may allow small-scale producers to participate. Neutral-poor conservationist image, may be improved with biodiversity support programmes.
Product diversity; low genetic-linked gain Limited degree of further development of other species – only to the point of niche markets, and developing high quality specialized profiles for producers; otherwise widening product base for 2–3 main species, with steadily levelling prices. Moderate growth, with dominance by primary species, and steadily increasing penetration of traditional markets, possibly disadvantaging traditional suppliers.	*Product diversity; high genetic-linked gain* Some degree of development of other species, but substantial development of product forms based on steadily reduced base prices of key species. Potentially very high growth, with major food supply consequences, and primary role of large scale producers and specialist biotechnology suppliers. Very limited opportunity for specialist and small scale producers, reduced pressure on wild fisheries due to low prices may improve conservation impact, reducing fishery employment.

through a combination of planned programmes and increasingly, those focused upon indigenous micro-enterprises. Scaling up from this, options may rise to service growing urban markets, in SE Asia, Latin America and Africa with the advantage of closer proximity, abundant demand and lower logistics barriers than would be the case for international markets.

Barriers notwithstanding, tilapia has also clearly demonstrated competitive advantages in international markets, reflected in its remarkable growth within some of the world's richest food markets. Within live market segments the product is likely to continue to be supplied more from the domestic supply base, but elsewhere the value added sector is likely to favour the raw material production sites. Expansion by means of such a strategy has already been proven viable elsewhere within the wider seafood industry and there is little reason to expect any significant divergence in approach for tilapia. To return to an earlier motif: with so many roads to follow there may be no need for the aquatic chicken to cross any.

REFERENCES

Aguilar-Manjarrez, J. and Nath, S.S. (1998) A strategic reassessment of fish farming potential in Africa. *CIFA Technical Paper* No 32, FAO, Rome, 170 pp.

Al-Ahmad, T., Hopkins, K.D., Ridha, M., Al-Ahmed, A. and Hopkins, M. (1986) Tilapia culture in Kuwait: final report. Kuwait Institute for Scientific Research, Kuwait, and International Center for Living Aquatic Resources Management (ICLARM), Philippines, 136 pp.

Ahmed, M., Abdur Rab, M. and Bimbao, M.P. (1995) Aquaculture technology adoption in Kapasia Thana, Bangladesh: some preliminary results from farm record-keeping data. *ICLARM Technical Report* 44, 34 pp.

Anon (1994) Tilapia update. *Seafood Leader* **14**(2), 163–162.

Anon (1995a) Finfish Tilapia. *Seafood Leader* **15**(2), 110–114.

Anon (1995b) Tilapia. *Seafood Leader* **15**(5), 118.

Anon (1996a) Annual tilapia situation and outlook. *Aquaculture Magazine* **22**(4), 11–16.

Anon (1996b) Tilapia update. *Seafood Leader* **16**(2), 130–131.

Anon (1997a) Tilapia fillets from Indonesia to the world. *Fish Farmer* May/June, 30–31.

Anon (1997b) Pisces reasons behind its failure. *Fish Farmer* May/June, 29–30.

Anon (1997c) Catfish production rides high in USA. *Fish Farmer* **25**, 27.

Anon (1997d) Tilapia situation and outlook report. *Aquaculture Magazine* Sept/Oct, 7–12.

Anon (1998) Transgenic salmon fail to gain consumer confidence. *Seafood International* **13**(2), 8.

Bell, F.W. and Canterbery, E.R. (1976) *Aquaculture for the Developing Countries: A Feasibility Study*, Ballinger, Cambridge, USA.

Beveridge, M.C.M., Ross. L.G. and Kelly, L.A. (1994) Aquaculture and biodiversity. *Ambio* **23**(8), 497–502.

Bimbao, M.P. and Smith, I.R. (1988) Philippine tilapia industry growth and potential, in *The Second International Symposium on Tilapia in Aquaculture* (eds R.S.V. Pullin, T. Bhukaswan, K. Tonguthai and J.L. Maclean), ICLARM Conference Proceedings 15, pp. 539–551.

Brass, J.L., Rust, M.B., Olla, L. and Wicklund, R.I. (1990) Preliminary investigations into the socioeconomic feasibility of saltwater cage culture of Florida red tilapia in Haiti. *Journal of the World Aquaculture Society* **21**(3), 192–200.

Brass, J.L., Olla, B.L. and Wicklund, R.I. (1991) Social, economic and cultural considerations for saltwater cage culture of Florida red tilapia in northeastern Haiti. *Gulf and Caribbean Fisheries Institute* **40**, 398–404.

Brummett, R.E. and Noble, R. (1995) Aquaculture for African smallholders. *ICLARM Technical Report 46*, ICLARM, Manila, 69pp.

CDC (Commonwealth Development Corporation) (1998) Press release. Major investment in Costa Rican Aquaculture, 4 Feb 1998.

Costas, R. (1997) Retailers driving change. *Seafood International* **12**(12), 20–22.

Cruz, E.M., Al-Ameeri, A.A. and Kraynick, R. (1998) Economic feasibility analysis of tilapia (*Oreochromis spilurus* Gunther) mariculture in a seabream (*Sparidentex hasta* Valenciennes) farm in Kuwait. *Asian Fisheries Science* **11**, 11–18.

Dickson, M.C. and Brooks, A.C. (1997) Fish farming in Malawi: a case study of the Central and Northern Regions Fish Farming Project. Govt of Malawi and European Commission,70 pp.

Engle, C.R. (1997a) Marketing tilapias, in *Tilapia Aquaculture in the Americas* (eds B.A. Costa-Pierce and J.E. Rakocy), Vol. 1, World Aquaculture Society, Louisiana, pp. 244–258.

Engle, C.R. (1997b) Optimal resource allocation by fish farmers in Ruanda. *Journal of Applied Aquaculture* **7**(1), 1–17.

Engle, C.R., Brewster, M. and Hitayezu, F. (1993) An economic analysis of fish production in a subsistence rural economy: the case of Rwanda. *Journal of Aquaculture in the Tropics* **8**, 151–165.

Escover, E.M., Salon, O.T. and Lim, C.P. (1985) Tilapia marketing in Bicol, Philippines, in *Philippine Tilapia Economics* (eds I.R. Smith, E.B. Torres and E.O. Tan), ICLARM Conference Proceedings, Manila, Philippines, 15, pp. 192–205.

Escover, E.M., Salon, O.T. and Smith, I.R. (1987) The economics of tilapia fingerling production and marketing in the Philippines. *Aquaculture and Fisheries Management* **18**, 1–13.

FAO (1997) Recent Trends in Aquaculture Production, 1984–1995, FAO, Rome.

FAO (1998) Aquaculture Production Statistics. FAO Fisheries Circular, No. 815, Rev. 10, Rome.

FAO/NACA/WHO. (1999) Food Safety Issues Associated with Products from Aquaculture, Report of a Joint FAO/NACA/WHO Study Group, WHO Technical Report Series 883.

Fermin, F. (1985) The introduction of integrated backyard fishponds in lowland Cavite, Philippines, in *Philippine Tilapia Economics* (eds I.R. Smith, E.B. Torres and E.O. Tan), ICLARM Conference Proceedings, Manila, Philippines, 15, pp. 151–164.

Folke, C. and Kautsky, N. (1991) Ecological economic principles for aquaculture, in *Nutritional Strategies and Aquaculture Waste* (eds C.B. Cowey and C.Y. Cho), University of Guelph, Canada, pp. 207–222.

Gentles, P. and Skeldon, S. (1994) Consumer trends in European fish consumption. Paper presented to the Fisch 94 International aquaculture and marketing seminar Bremen, June.

Gonzales, E. (1984) Small-scale tilapia cage technology adopted in fishing villages in Laguna Lake, Philippines. *Aquaculture* **41**, 161–169.

Green, B.W., Teichert-Coddington, D.R. and Phelps, R.R. (1990) Response of tilapia yield and economics to varying rates of organic fertilization and season in two Central American countries. *Aquaculture* **90**, 279–290.

Green, B.W., Teichert-Coddington, D.R and Phelps, R.R. (1994) Development of semi-intensive aquaculture technologies in Honduras. Research and Development Series No 39, Auburn University, Alabama, 47 pp.

Gutting, R. (1998) The HACCP era. *Seafood Business* **17**(2), 8–10.

Harrison, E. (1994) Aquaculture in Africa: socio-economic dimensions, in *Recent Advances in Aquaculture* (eds J.F. Muir and R.J. Roberts), Vol. 4, Blackwell Science, Oxford, pp. 240-299.

Harrison, E., Stewart, J.A., Stirrat, R.L. and Muir, J.F. (1994) *Fish Farming in Africa – What's the Catch?* Summary report of ODA-supported Research Project Aquaculture Development in Sub-Saharan Africa. ODA (DFID) London.

Hatch, U. and Engle, C. (1987) Economic analysis of aquaculture as a component of integrated agro-aquaculture systems: some evidence from Panama. *Journal of Aquaculture in the Tropics* **2**, 93–105.

Haylor, G.S., Young, J.A., Muir, J.F. and Scott, D.C.B. (1994) Commercial aquaculture in Africa. Commonwealth Development Corporation, London.

Head, W.D. and Watanabe, W.O. (1995) Economic analysis of a commercial scale recirculating brackish water hatchery for Florida red tilapia. *Journal of Applied Aquaculture* **5**, 1–23.

Head, W.D., Zerbi, A. and Watanabe, W.O. (1996) Economic evaluation of commercial scale, saltwater pond production of Florida red tilapia in Puerto Rico. *Journal of the World Aquaculture Society* **27**, 275–289.

Hopkins, K.D., Adel-Halim, M.M., Hopkins, M.C., Dan, N. and Maheshwari, G. (1985) Tilapia culture in Kuwait. A feasibility study. Kuwait Institute for Scientific Research, Report No KISR1637, Kuwait.

Hopkins, K.D., Hopkins, M., Leclerq, D. and Al-Ameeri, A.A. (1986) Tilapia culture in Kuwait: a preliminary economic analysis of production systems. *Kuwait Bulletin of Marine Science* **7**, 45–64.

Hopkins, M.L. and Hopkins, K.D. (1986) Tilapia marketing tests in Kuwait, in *The First Asian Fisheries Forum* (eds J.L. Maclean, L.B. Dizon and L.V. Hosillos), Asian Fisheries Society, Manila, Philippines, pp. 433–436.

ICLARM. (1984) Introducing the tilapias. *ICLARM Newsletter* **7**(1), 3.

Johnson, F.C. (1992) People, fish and ponds: a study of aquaculture in Malawi. Report to the Central and Northern Regions Fish Farming Project, Govt of Malawi and European Commission.

Kapetsky, J.M (1994) A strategic assessment of warm-water fish farming potential in Africa. CIFA Technical Paper, No 27, Rome, FAO, 67 pp.

Kohler, S.T. (1994) Using census data and geographical information systems to identify target markets for aquaculture products in the USA. *World Aquaculture* **27**(4), 23–35.

Kwei Lin, C. and Yi, Y. (1998) Economics of tilapia production in Thailand. Paper presented at the Asian Fisheries Society Conference, Chiang Mai, Thailand, December 1998.

Lazaga, J.F. and Roa, L.L. (1983) Financial and economic analysis of grow-out tilapia cage-farming in Laguna de Bay. Paper presented at Tilapia Economics Workshop, Laguna Philippines. PCARRD and ICLARM, 17 pp.

Lee, C.S. (1983) Production and price changes in tilapia industry: the case of Taiwan, 1965–1982. *Journal of Agricultural Economics* **34**, 191–209.

Lister, P. (1998) Tilapia takes off. *Seafood Business* **17**(1), 12–16.

Little, D.C. (1998) Options in the development of the aquatic chicken. *Fish Farmer* July/August, 35–37.

Lo, M.C. and Hwang, T.C. (1994) The internal structure and development of aquaculture in Taiwan, in *Socioeconomics of Aquaculture* (eds Y.C. Shang, P.S. Leung,

C.S. Lee, M.S. Su and I.C. Liao), Tunkang Marine Laboratory Proceedings, Vol. 4, Tungkang, Taiwan, pp. 49–67.

Logue, A.W. (1986) *The Psychology of Eating and Drinking*, WH Freeman, New York.

Losordo, T.M. and Westerman, P.W. (1994) An analysis of biological, economic, and engineering factors affecting the cost of fish production in recirculating aquacultural systems, in *Design of High-Density Recirculating Aquaculture Systems*, Louisiana Sea Grant, Baton Rouge, LA, pp. 1–9.

Lovshin, L.L. and Pretto, R. (1983) A strategy for the use of tilapias in rural Latin America: the Panamanian integrated approach, in *International Symposium on Tilapia in Aquaculture* (eds L.I. Fischelson and Z. Yaron, Z), Nazareth, Israel, pp. 494–505.

Lovshin, L.L., Schwartz, N.B., de Castillo, V.G., Engle, C.R. and Hatch, U.L. (1986) Cooperatively managed rural Panamanian fish ponds: the integrated approach. Research and Development Series No 33, Auburn University, Alabama, 47 pp.

Maclean, J.L. (1984) Tilapia the aquatic chicken. *ICLARM Newsletter* **7**(1), 17.

Marshall, D. (1995) *Food Choice and the Consumer*, Blackie, Glasgow.

Martinez-Espinosa, M. (1992) Rural aquaculture: from myth to reality. *FAO Aquaculture Newsletter* No 2, 13–15 Nov 1992, FAO Rome.

McGovern, D. (1998) Counter attack. *Seafood Business* **17**(3), 1 and 22–28.

Mohan Dey, M. and Eknath, A.E. (1996) Current trends in the Asian tilapia industry and the significance of genetically improved breeds, in *Sustainable Aquaculture* (eds K.P.P. Nambiar and T. Singh), Proceedings of the Infofish-Aquatech 96 conference, Kuala Lumpur, Malaysia, 25–27 Sept 1996, Infofish. Kuala Lumpur, pp. 59–78.

Montfort, C. (1998) Making consumers aware of quality. *Seafood International* **13**(3), 39–40.

Muir, J.F. (1995) Perspectives on Aquaculture: Aquaculture and food security, 224 pp, Mimeo. Study document commissioned for FAO, Rome.

Muir, J.F. and Young, J.A. (1997) Aquaculture and Marine Fisheries: will capture fisheries remain competitive? Paper presented to NAFO 19th Annual Meeting: Symposium. What future for capture fisheries? Memorial University, St Johns, Newfoundland, Canada, 10–12 September.

Muir, J.F., Young, J.A. and Smith, A.P. (1996) Aquaculture, economics and development: a European perspective, in *Proceedings of the VIIth Annual Conference of the European Association of Fisheries Economists*, CEMARE, Portsmouth, UK 10–12 April 1995, pp. 195–210.

Nelson, R.G., Behrends, L.L., Waddell, E.L. and Burch, D.W. (1981) Tilapia culture in heated effluents: economic aspects, in *Third Waste Heat Management and Utilisation Conference* (eds S. Sengupta and S.S. Lee), Hemisphere, New York, pp. 397–404.

O'Dierno, L.J. (1995) Understanding the retail market for farm raised fish and seafood products *Aquaculture Magazine 21*(4), 7–13.

Oliva, L.P. (1985) The economics of tilapia cage culture in Mindanao, Philippines, in *Philippine Tilapia Economics* (eds I.R. Smith, E.B. Torres and E.O. Tan), ICLARM Conference Proceedings, Manila, Philippines, 15, pp. 83–106.

Pagan, F.A. (1970) Cage culture of tilapia. *FAO Fisheries Bulletin* **3**, 6, FAO Rome.

Paquotte, P. (1998) New species in Mediterranean aquaculture: is it an answer to the market demand for differentiated products? International Symposium on new species for Mediterranean aquaculture, Alghero, Italy 22–24 April.

Picchietti, M. (1996) Tilapia marketing maturity *Aquaculture Magazine* **22**(5), 19–26.

Popma, T.J. and Lovshin, L.L. (1996) Worldwide prospects for production of commercial aquaculture, R&D Series No 44, Auburn University, 23 pp.

Pullin, R.S.V., Bimbao, M.A.P. and Bimbao, G.B. (1994) World outlook on tilapia farming. First International Symposium on Aquaculture, June 1994, Boca del Rio, Mexico.

Rahman, A., Islam, I, Roy, L. and Islam, K.S. (1994) Shrimp culture and the environment in the coastal region. Bangladesh Institute of Development Studies, Dhaka, Bangladesh, 114 pp.

Ruckes, E. (1994) Can aquaculture fill the market gap? *FAO Aquaculture Newsletter* **6**, 17–18, FAO Rome.

Sadek, S., Kallafalah, H and Adell, F. (1992) Tilapia (*Oreochromis niloticus*) biomass yield in a commercial farm using circular tanks. *Journal of Applied Ichthyology* **8**, 193–202.

Sarig, S. (1988) Is an increase in yields in intensive commercial fish farming always accompanied by an increase in profitability? *Bamidgeh* **40**(4), 122–124.

Semakula, S.N. and Makoro, J.T. (1968) The culture of *Tilapia* species in Uganda. *FAO Fish Rep.* **44**, 161–164.

Sevilleja, R.C. (1985) Tilapia production in freshwater fishponds of Central Luzon, pp 83–106, in *Philippine Tilapia Economics* (eds I.R. Smith, E.B. Torres and E.O. Tan), ICLARM Conference Proceedings, Manila, Philippines, 15, pp. 83–106.

Shaw, D. (1995) Tilapias time has come. *Seafood Leader* **15**(1), 54–57.

Sipe, M. (1992) Tilapia marketing in the USA. *Infofish* **3**/92, 23–25.

Smith, I.R., Torres, E.B. and Tan, E.O (1985) (eds) *Philippine Tilapia Economics*, ICLARM Conference Proceedings, 15, Manila, Philippines.

STAQ. (Stirling Aquaculture) (1996) The European Aquaculture Industry; key trends in science and technology. Report to the Science and Technology Office of Assessment, European Parliament.

Stewart, J.A. (1993) An economic analysis of the viability of fishfarming in Malawi, University of Stirling, Stirling, UK, 82 pp.

Stewart, J.A. (1995) *Assessing Sustainability of Aquaculture Development*, Unpublished PhD Thesis, University of Stirling, Stirling.

Stewart, J.A. (1997) Aquaculture potential in saline groundwater. *Aquaculture News* **23**, 36–37, Institute of Aquaculture, University of Stirling.

Suresh, A.V. and Lin, C.K. (1992) Tilapia culture in saline waters: a review. *Aquaculture* **106**, 201–226.

Teichert-Coddington, D.R. and Green, B.W. (1997) Experimental and commercial culture of tilapia in Honduras, in *Tilapia Culture in the Americas* (eds B.A. Costa-Pierce and J.E. Rakocy), Vol. 1, World Aquaculture Society, Baton Rouge, Louisiana, USA, pp. 142–162.

Thomas, D.H.L. (1994) Socio-economic and cultural factors in aquaculture development: a case study from Nigeria. *Aquaculture* **119**, 329–343.

Tidon, A and Librero, A. (1978) A socio-economic study of tilapia farming in the Philippines. SEAFDEC/PCARR Research Paper Series, 10, Los Banos, Laguna, Philippines.

Torres, E.B. and Navera, E.R. (1985) Tilapia marketing in Central Luzon and Metro Manila, Philippines, pp 181–191, in *Philippine Tilapia Economics* (eds I.R. Smith, E.B. Torres and E.O. Tan), ICLARM Conference Proceedings, Manila, Philippines, 15, pp. 181–191.

USDA (1997) Aquaculture Outlook. Washington, DC 20005-4789. Economic Research Service, US Department of Agriculture.

van Rijn, J. (1996) The potential for integrated biological treatment systems in recirculating fish culture: a review. *Aquaculture* **139**, 181–201.

Vincke, M.M.J. and Schmidt, U.W. (1991) Report on the mid-term evaluation of the Central and Northern Regions fish farming project. 17/3–28/4/91 EEC/Govt of Malawi, 67 pp.

Watanabe, W.O. (1991) Saltwater culture of tilapia in the Caribbean. *World Aquaculture* **22**, 49–54.

Watanabe, W.O, Olla, B.L., Wicklund, R.I. and Head, W.D. (1997) Saltwater culture of the Florida Red Tilapia and other saline tolerant tilapias: a review, in *Tilapia Culture in the Americas* (eds B.A. Costa-Pierce and J.E. Rakocy), Vol. 1, World Aquaculture Society, Baton Rouge, Louisiana, USA, pp. 55–141.

Westland, L. (1995) Apparent historical consumption and future demand for fish and fishery products – exploratory calculations. Paper to International Conference on Sustainable Contribution of Fisheries to Food Security, Kyoto, Japan, 4–9 December, FAO.

Wijkstrom, U. (1991) How fish culture can stimulate growth: Conclusions from fish farmer surveys in Zambia. ALCOM Report No. 9, GCP/INT/436/SWE/REP/9 FAO, Rome.

Yater, L.R. and Smith, I.R. (1985) Economics of private tilapia hatcheries in Laguna and Rizal Provinces, Philippines, in *Philippine Tilapia Economics* (eds I.R. Smith, E.B. Torres and E.O. Tan), ICLARM Conference Proceedings 15, pp. 15–32.

Young, J.A. and Maddock, S. (1990) The UK catering market for fishery products Globefish Research Programme No 8, 41 pp.

Young, J.A. and Muir, J.F. (1995) Diversity in adversity? The case of the UK aquaculture sector, in *Proceedings of the 7th Biennial Conference of the International Institute of Fisheries Economics and Trade*, Taiwan, July 1994, Vol. 11, Taipei, Taiwan ROC, pp. 82–95.

Young, J.A. and Muir, J.F. (1997) The emergent role of aquaculture products within the wider supply base for fish. *Proceedings of Agra Europe Conferenc. Successful Strategies for Marketing Fish and Seafood Products*, London, 24–25 September 1997.

Websites

http://www.ag.arizona.edu/azaqua/ista/ata: American Tilapia Association
http://www.ag.arizona.edu/azaqua/ista/tilapia$.txt: Tilapia aquaculture wholesale market prices
http://www.aquaculture-online.org/text/market/mark.pric.html: Market price data
http://www.businessauctions.com/cgi-bin/productListing.cgi/directsales/small/Tilapia: Tilapia auction
http://www.fao.org/waicent/faoinfo/fishery/fishery.htm: Fishery statistics
http://www.fishfarming.com/tilapia.html: Tilapia farming
http://www.fishfarming.com/imports.html: US Tilapia import trends
http://www.fishmart.com/world/: US NMFS Statistics and Economics Division
http://home.istar.ca/-gadus/links.html: Useful index of over 2000 fisheries sites
http://www.remora.ssp.nmfs.gov/trade/index.html: US NMFS Foreign Trade Statistics

Species index

Index

Fish and Fisheries Series

1. R.J. Wootton: *Ecology of Teleost Fishes.* ISBN 0-412-31730-3
2. M.A. Keenleyside: *Cichlid Fishes Behaviour.* Ecology and Evolution. ISBN 0-412-32200-5
3. I.J. Winfield and J.S. Nelson (eds.): *Cyprinid Fishes Systematics.* Biology and Exploitation. ISBN 0-412-34920-5
4. E. Kamler: *Early Life History of Fish.* An Energetics Approach. ISBN 0-412-33710-X
5. D.N. MacLennan and E.J. Simmonds (eds.): *Fisheries Acoustics.* ISBN 0-412-33060-1
6. T.J. Hara: *Fish Chemoreception.* ISBN 0-412-35140-4
7. T.J. Pitcher: *Behaviour of Teleost Fishes.* Second Edition. PB; ISBN 0-412-42940; HB; ISBN 0-412-42930-3
8. C.R. Purdom: *Genetics and Fish Breeding.* ISBN 0-412-33040-7
9. J.C. Rankin and F.B. Jensen (eds.): *Fish Ecophysiology.* ISBN 0-412-45920-5
10. J.J. Videler: *Fish Swimming.* ISBN 0-412-40860-0
11. R.J.H. Beverton and S.J. Holt (eds.): *On the Dynamics of Exploited Fish Populations.* ISBN 0-412-54960-3
12. G.D. Pickett and M.G. Pawson (eds.): *Sea Bass.* ISBN 0-412-40090-1
13. M. Jobling: *Fish Bioenergetics.* ISBN 0-412-58090-X
14. D. Pauly: *On the Sex of Fish and the Gender of Scientists.* A Collection of essays in fisheries sciences. ISBN 0-412-59540-0
15. J. Alheit and T. Pitcher (eds.): *Hake.* Fisheries, Products and Markets. ISBN 0-412-57350-4
16. M. Jobling: *Environmental Biology of Fishes.* ISBN 0-412-58080-2
17. P. Moller: *Electric Fishes.* History and Behavior. ISBN 0-412-37380-7
18. T.J. Pitcher and P.J.B. Hart (eds.): *The Impact of Species Changes in African Lakes.* ISBN 0-412-55050-4
19. J.F. Craig: *Pike.* Biology and Exploitation. ISBN 0-412-42960-8
20. N.V.C. Polunin and C.M. Roberts (eds.): *Reef Fisheries.* ISBN 0-412-60110-9
21. R.C. Chambers and E.A. Trippel (eds.): *Early Life History and Recruitment in Fish Populations.* ISBN 0-412-64190-9
22. S.J.M. Blaber: *Fish and Fisheries in Tropical Estuaries.* ISBN 0-412-78500-5
23. T.J. Pitcher, P.J.B. Hart and D. Pauly (eds.): *Reinventing Fisheries Management.* 1999 ISBN 0-7923-5777-9

Fish and Fisheries Series

24. R.J. Wootton: *Ecology of Teleost Fishes.* Second Edition. 1999
PB; ISBN 0-412-64200-X; HB; ISBN 0-412-84590-3
25. M.C.M. Beveridge and B.J. McAndrew (eds.): *Tilapias: Biology and Exploitation.* 2000
PB; ISBN 0-7923-6391-4 HB; ISBN 0-412-80090-X

KLUWER ACADEMIC PUBLISHERS – BOSTON / DORDRECHT / LONDON